Pure Mathematics 2

*Also by J. K. Backhouse and S. P. T. Houldsworth,
revised by P. J. F. Horril*

Pure Mathematics 1

(See page 457 of Book 2 for a list of contents.)

Pure Mathematics
Book 2
Third Edition

J. K. Backhouse, M.A.
Tutor, Department of Educational Studies, University of Oxford
Formerly Head of the Mathematics Department, Hampton Grammar School

S. P. T. Houldsworth, M.A.
Lately Headmaster, Sydney Grammar School
Formerly Assistant Master at Harrow School

B. E. D. Cooper, M.A.
Headmaster of St Bartholomew's School, Newbury

This edition revised by
P. J. F. Horril, M.A.
Head of the Mathematics Department, Nottingham High School

LONGMAN GROUP LIMITED
*Longman House, Burnt Mill, Harlow, Essex CM20 2JE, England
and Associated Companies throughout the World*

First published 1963
SI edition 1971
Third edition 1985
ISBN 0 582 35387 4

*Set in Times Mathematics 569
by H Charlesworth & Co Ltd, Huddersfield*

*Produced by Longman Group (F.E.) Ltd.
Printed in Hong Kong*

Contents

Chapter 14 Differential equations

Chapter 15 Second order linear differential equations with constant coefficients

Chapter 16 Approximations — further expansions in series

Chapter 21 Further vector methods 390

Answers

Contents of Book 1

Index

Preface

This book is the sequel to *Pure Mathematics 1* by J. K. Backhouse and S. P. T. Houldsworth, revised by P. J. F. Horril. The two books bring the work up to the standard required for examinations in Mathematics and Further Mathematics at Advanced level; some carefully selected S-level topics have also been included.

The revised edition takes into account the changes in syllabuses which have taken place since the original version was produced in 1963. However the authors hope that the artificial and damaging distinction between so-called 'modern' and 'traditional' mathematics will soon disappear and that school mathematics will once again be seen as a unified subject.

Readers familiar with the earlier editions will note that less emphasis has been placed on heavy manipulation of algebraic and trigonometrical identities and that certain topics have been omitted because they have gone out of fashion. On the other hand, there are several entirely new chapters, namely

    Chapter 8,   Further matrices and determinants
    Chapter 15,   Second order differential equations
    Chapter 21,   Further vector methods.

Chapters 8 and 21 follow up the work done on these topics in Book 1, and Chapter 15 continues the work started in the revised and expanded Chapter 14 of Book 2; there is also a section on differential equations in Chapter 17, Numerical methods. The introduction to complex numbers now appears in Book 1, and Book 2 contains a sequel to it (Chapter 20). This edition contains an index, and there is also a table of the contents of Book 1.

The exercises have been extensively revised and include a wide selection of questions from recent GCE papers.

The individual reader has been kept in mind, and he or she is advised to work through the questions in the text marked **Qu.**; the class teacher will find that many of these are suitable for oral work. In the exercises, certain questions have been marked with an asterisk *; this indicates that they contain a useful result or method for which room has not been found in the text.

Once again, I would like to thank the authors of the original books for entrusting me with the task of revision; their help with preparing the drafts and reading the proofs has been invaluable. I would also like to thank everyone from the Longman Group who has helped with the production of these new books.

My thanks are also due to the Master and Fellows of Selwyn College,

Cambridge, and the President and Fellows of St John's College, Oxford, because much of the work involved was done while I was resident at their respective colleges. I should also like to acknowledge the valuable help of my colleagues and pupils who read the new material. Lastly, I must thank my wife and family, without whose support and patience the project would never have been completed.

Nottingham   P. J. F. Horril
June 1984

Note on degree of accuracy of answers

In order to avoid tedious repetition in the wording of questions the following conventions are observed throughout the book, unless there are specific instructions to the contrary.

(a) When possible an exact answer is given. To this end it is normally appropriate to retain surds and π in the answers where they occur. (The word *exact* is used here in the rather limited sense of being derived from the data without any intervening approximation.)

(b) When an answer is not exact, it is given correct to three significant figures, or, if it is an angle measured in degrees, to the nearest tenth of a degree.

Acknowledgements

We are grateful to the following examining bodies for permission to reproduce questions from past examination papers:

University of Cambridge Local Examinations Syndicate (C); Joint Matriculation Board (JMB); University of London, School Examinations Department (L); Oxford and Cambridge Schools Examination Board (O & C) and University of Oxford Delegacy of Local Examinations (O).

Questions from the above bodies are indicated by the letters shown in brackets.

Mathematical notation

The following notation is used in this book. It follows the conventions employed by most GCE Examining Boards.

1. Set notation

$\in$	is an element of.
$\notin$	is not an element of.
$\{a, b, c, \ldots\}$	the set with elements $a, b, c \ldots$
$\{x: \ldots\}$	the set of elements x, such that $\ldots$
$n(A)$	the number of elements in set A.
$\varnothing$	the empty set.
$\mathscr{E}$	the universal set.
A'	the complement of set A.
$\mathbb{N}$	the set of natural numbers (including zero) $0, 1, 2, 3 \ldots$
$\mathbb{Z}$	the set of integers $0, \pm 1, \pm 2, \pm 3, \ldots$
$\mathbb{Z}^+$	the set of positive integers $+1, +2, +3 \ldots$
$\mathbb{Q}$	the set of rational numbers.
$\mathbb{R}$	the set of real numbers.
$\mathbb{C}$	the set of complex numbers.
$\subseteq$	is a subset of.
$\subset$	is a proper subset of.
$\cup$	union.
$\cap$	intersection.
$[a, b]$	the closed interval $\{x \in \mathbb{R}: a \leqslant x \leqslant b\}$.
(a, b)	the open interval $\{x \in \mathbb{R}: a < x < b\}$.

2. Miscellaneous symbols

$=$	is equal to.
$\neq$	is not equal to.
$>, <$	is greater than, is less than.
$\geqslant, \leqslant$	is greater than or equal to, is less than or equal to.
$\approx$	is approximately equal to.

3. Operations

$a + b$	a plus b.
$a - b$	a minus b.
$a \times b$, ab, $a.b$	a multiplied by b.
$a \div b$, $\dfrac{a}{b}$, a/b	a divided by b.
$\displaystyle\sum_{i=1}^{i=n} a_i$	$a_1 + a_2 + a_3 + \ldots + a_n$.

4. Functions

$f(x)$	the value of the function f at x.
$f: A \to B$	f is a function which maps each element of set A onto a member of set B.
$f: x \mapsto y$	f maps the element x onto an element y.
f^{-1}	the inverse of the function f.
$g \circ f$ or gf	the composite function $g(f(x))$.
$\lim\limits_{x \to a} f(x)$	the limit of $f(x)$ as x tends to a.
δx	an increment of x.
$\dfrac{dy}{dx}$	the derivative of y with respect to x.
$\dfrac{d^n y}{dx^n}$	the nth derivative of y with respect to x.
$f'(x), f''(x), \ldots f^{(n)}(x)$	the first, second, ... nth derivatives of $f(x)$.
$\int y\,dx$	the indefinite integral of y with respect to x.
$\displaystyle\int_a^b y\,dx$	the definite integral, with limits a and b.
$[F(x)]_a^b$	$F(b) - F(a)$.

5. Exponential and logarithmic functions

e^x or $\exp x$	the exponential function.
$\log_a x$	logarithm of x in base a logarithms.
$\ln x$	$\log_e x$.
$\lg x$	$\log_{10} x$.

6. Circular and hyperbolic functions

$\sin x$, $\cos x$, $\tan x$	the circular functions sine, cosine, tangent.
$\operatorname{cosec} x$, $\sec x$, $\cot x$	the reciprocals of the above functions.
$\sin^{-1} x$ or $\arcsin x$	the inverse of the function $\sin x$ (with similar abbreviations for the inverses of the other circular functions).
$\sinh x$ etc.	the hyperbolic functions.

7. Other functions

$\sqrt{a}$	the positive square root of a.

$|a|$	the modulus of a.
$n!$	n factorial; $n! = n \times (n-1) \times (n-2) \times \ldots \times 3 \times 2 \times 1$. $(0! = 1)$
$\binom{n}{r}$	$\dfrac{n!}{r!(n-r)!}$ when $n, r \in \mathbb{N}$ and $0 \leqslant r \leqslant n$,
$\binom{n}{r}$	$\dfrac{n(n-1)\ldots(n-r+1)}{r!}$ when $n \in \mathbb{Q}$ and $r \in \mathbb{N}$.

8. Complex numbers

i	the square root of -1.
z or w	a typical complex number, e.g. $x + iy$, where $x, y \in \mathbb{R}$.
$\mathrm{Re}(z)$	the real part of z; $\mathrm{Re}(x + iy) = x$.
$\mathrm{Im}(z)$	the imaginary part of z; $\mathrm{Im}(x + iy) = y$.
$|z|$	the modulus of z; $|x + iy| = \sqrt{(x^2 + y^2)}$.
$\arg(z)$	the argument of z.
z^*	the complex conjugate of z.

9. Matrices

$\mathbf{M}$	a typical matrix $\mathbf{M}$.
$\mathbf{M}^{-1}$	the inverse of a matrix $\mathbf{M}$ (provided it exists).
$\mathbf{M}^{\mathrm{T}}$	the transpose of matrix $\mathbf{M}$.
$\det(\mathbf{M})$	the determinant of a square matrix $\mathbf{M}$.
$\mathrm{adj}(\mathbf{M})$	the adjoint of a square matrix $\mathbf{M}$
$\mathbf{I}$	the identity matrix.

10. Vectors

$\mathbf{a}$	the vector $\mathbf{a}$.
$|\mathbf{a}|$ or a	the magnitude of vector $\mathbf{a}$.
$\hat{\mathbf{a}}$	the unit vector with the same direction as $\mathbf{a}$.
$\mathbf{i}, \mathbf{j}, \mathbf{k}$	unit vectors parallel to the Cartesian coordinate axes.
$\overrightarrow{AB}$	the vector represented by the line segment AB.
$|\overrightarrow{AB}|$ or AB	the length of the vector $\overrightarrow{AB}$.
$\mathbf{a} \cdot \mathbf{b}$	the scalar product of $\mathbf{a}$ and $\mathbf{b}$.
$\mathbf{a} \wedge \mathbf{b}$	the vector product of $\mathbf{a}$ and $\mathbf{b}$.

Chapter 1

Integration

Introduction

1.1   In Book 1† we dealt with the differentiation of powers of x, polynomials, products and quotients, composite functions, trigonometrical functions, and we also discussed implicit functions and parameters.

Now that we come to extend the scope of integration we find that it is not, unfortunately, merely a matter of putting into reverse the techniques for differentiation; we have learned a technique for differentiating $(3x^2 + 2)^4$ as it stands, but can we integrate this function without first expanding it? Even consider the simple function x^n; we can differentiate this whenever $n \in \mathbb{Q}$, but we must bear in mind the gap which remains to be filled later in this book when we discover how to deal with $\int x^{-1}\,dx$.

Integration is, in fact, less susceptible than differentiation to concise systematic treatment. It presents a broad front, and the reader's experience of it will gradually expand, so that by quick recognition of an increasing number of forms of *integrand* (i.e. the function to be integrated) there is developed the power to discriminate between the many possible lines of attack.

Recognising the presence of a function and its derivative

1.2   The very first thing to search for in any but the simplest integrands is the presence of a function and its derivative; with this, we may often guess the integral to be a certain composite function, check by differentiation, and adjust the numerical factor. Two examples follow to illustrate this method.

Example 1   *Find* $\int x(3x^2 + 2)^4\,dx.$

[We note that the x outside the bracket is a constant $\times$ the derivative of the expression inside the bracket. We deduce that the integral is a function of $(3x^2 + 2)$.]

---

† *Pure Mathematics 1*, J. K. Backhouse, S. P. T. Houldsworth and P. J. F. Horril, published in 1985 by Longman and hereafter referred to as Book 1.

$$\frac{d}{dx}\{(3x^2 + 2)^5\} = 5(3x^2 + 2)^4 \times 6x = 30x(3x^2 + 2)^4$$

$$\therefore \frac{d}{dx}\left\{\frac{1}{30}(3x^2 + 2)^5\right\} = x(3x^2 + 2)^4$$

Hence

$$\int x(3x^2 + 2)^4 \, dx = \frac{1}{30}(3x^2 + 2)^5 + c$$

Example 2   *Find* $\int \sin^2 4x \cos 4x \, dx$.

[We note that $\cos 4x$ is a constant $\times$ the derivative of $\sin 4x$, and we deduce that the integral is a function of $\sin 4x$.]

$$\frac{d}{dx}\{\sin^3 4x\} = 3(\sin 4x)^2 \times \cos 4x \times 4 = 12 \sin^2 4x \cos 4x$$

Hence

$$\int \sin^2 4x \cos 4x \, dx = \frac{1}{12} \sin^3 4x + c$$

Qu. 1   Differentiate:

(a) $(2x^2 + 3)^4$,   (b) $\sqrt{(x^2 - 2x + 1)}$,   (c) $\dfrac{1}{(2x - 1)^2}$,

(d) $\sin(4x - 7)$,   (e) $\tan^3 x$,   (f) $\cos^2 3x$.

Qu. 2   Find the following integrals, and check by differentiation:
(a) $\int x(x^2 + 1)^2 \, dx$,   (b) $\int (2x + 1)^4 \, dx$,   (c) $\int (x^2 + 1)^3 \, dx$,
(d) $\int \frac{1}{2} \sin 3x \, dx$,   (e) $\int x^2 \sqrt{(x^3 + 1)} \, dx$,   (f) $\int \sec^2 x \tan x \, dx$.

Pythagoras' theorem. Odd powers of sin x, cos x, etc.

1.3   Pythagoras' theorem in the forms

$$\cos^2 x + \sin^2 x = 1, \quad \cot^2 x + 1 = \operatorname{cosec}^2 x, \quad \text{and} \quad 1 + \tan^2 x = \sec^2 x$$

(see Book 1, §16.6), may be used to change some integrands to a form susceptible to the method of §1.2. In particular, it enables us to integrate odd powers of $\sin x$ and $\cos x$.

Example 3   *Find* $\int \sin^5 x \, dx$.

$$\int \sin^5 x \, dx = \int \sin^4 x \sin x \, dx$$
$$= \int (1 - \cos^2 x)^2 \sin x \, dx$$
$$= \int (\sin x - 2 \cos^2 x \sin x + \cos^4 x \sin x) \, dx$$
$$\therefore \int \sin^5 x \, dx = -\cos x + \tfrac{2}{3} \cos^3 x - \tfrac{1}{5} \cos^5 x + c$$

Qu. 3   Find: (a) $\int \sin^3 x \, dx$,   (b) $\int \cos^5 x \, dx$.

Qu. 4   Find: (a) $\int \cos^3 x \sin^2 x \, dx$,   [Write $\cos^3 x$ as $\cos x(1 - \sin^2 x)$.]
   (b) $\int \sin^3 x \cos^2 x \, dx$.

Qu. 5   Find $\int \sec x \tan^3 x \, dx$. $\left[\text{Remember } \dfrac{d}{dx} (\sec x) = \sec x \tan x. \right]$

Even powers of sin x, cos x†

1.4   Two very important formulae derived from the double-angle formulae are $\cos^2 x = \frac{1}{2}(1 + \cos 2x)$ and $\sin^2 x = \frac{1}{2}(1 - \cos 2x)$. (See Book 1, §17.3).

Their use in integrating even powers of sin x and cos x is illustrated in the latter part of Exercise 1a, which also gives practice in the use of other formulae, including the factor formulae. (See Book 1, §17.8.)

Exercise 1a

1 Differentiate:

(a) $(5x^2 - 1)^3$,   (b) $\dfrac{1}{(2x^2 - x + 3)^2}$,   (c) $\sqrt[3]{(x^2 + 4)}$,

(d) $\cot 5x$,   (e) $\cos(5x - 1)$,   (f) $\sin^2 \dfrac{x}{3}$,

(g) $\tan \sqrt{x}$,   (h) $\sec^2 2x$,   (i) $\sqrt{\operatorname{cosec} x}$.

Find the following integrals in Nos. 2–4:

2 (a) $\int x(x^2 - 3)^5 \, dx$,   (b) $\int (3x - 1)^5 \, dx$,   (c) $\int x(x + 2)^2 \, dx$,

(d) $\int \dfrac{x}{(x^2 + 1)^2} \, dx$,   (e) $\int \dfrac{x + 1}{(x^2 + 2x - 5)^3} \, dx$,

(f) $\int (2x - 3)(x^2 - 3x + 7)^2 \, dx$,   (g) $\int \dfrac{2x}{(4x^2 - 7)^2} \, dx$,

(h) $\int 2x \sqrt{(3x^2 - 5)} \, dx$,   (i) $\int (x^3 + 1)^2 \, dx$,

(j) $\int \dfrac{x^2 - 1}{\sqrt{(x^3 - 3x)}} \, dx$,   (k) $\int \dfrac{x - 1}{(2x^2 - 4x + 1)^{3/2}} \, dx$,

(l) $\int (2x^2 - 1)^3 \, dx$.

3 (a) $\int 3 \cos 3x \, dx$,   (b) $\int \sin(2x + 3) \, dx$,
(c) $\int \cos x \sin x \, dx$,   (d) $\int \frac{1}{3} \cos 2x \, dx$,
(e) $\int \sin 3x \cos^2 3x \, dx$,   (f) $\int \sec^2 x \tan^2 x \, dx$,
(g) $\int \sec^5 x \tan x \, dx$,   (h) $\int \cos x \sqrt{\sin x} \, dx$,

---

†This section and the latter part of Exercise 1a may with advantage be delayed and done in conjunction with later parts of the chapter.

(i) $\int x \csc^2 x^2 \, dx,$ (j) $\int \dfrac{\cos \sqrt{x}}{\sqrt{x}} \, dx,$

(k) $\int \csc^3 x \cot x \, dx.$

4 (a) $\int \cos^3 x \, dx,$ (b) $\int \cos^5 \dfrac{x}{2} \, dx,$

(c) $\int \sin^3 2x \, dx,$ (d) $\int \cos^3 (2x + 1) \, dx,$
(e) $\int \sin^5 x \cos^2 x \, dx,$ (f) $\int \cos^3 x \sin^3 x \, dx,$
(g) $\int \sec^4 x \, dx,$ (h) $\int \csc x \cot^3 x \, dx,$
(i) $\int \tan^5 x \sec x \, dx.$

5 Find $\int \tan x \sec^4 x \, dx,$ (a) as a function of sec x, (b) as a function of tan x, and show that they are the same.

6 Show that the integral given in No. 3(c) may be obtained in three different forms.

Nos. 7 onwards may be delayed (see footnote to §1.4).

7 Express (a) $\sin^2 \dfrac{x}{2}$ in terms of cos x,   (b) $\cos^2 3x$ in terms of cos $6x$.

8 Find (a) $\int \cos^2 x \, dx,$ (b) $\int \sin^2 \dfrac{x}{2} \, dx,$ (c) $\int \cos^2 3x \, dx.$

9 Express $\sin^4 x$ in terms of cos $2x$, and $\cos^2 2x$ in terms of cos $4x$. Show that $\int \sin^4 x \, dx = \frac{3}{8}x - \frac{1}{4} \sin 2x + \frac{1}{32} \sin 4x + c.$
10 Find $\int \cos^4 x \, dx.$
11 Find the following integrals:

(a) $\int \sin^2 x \, dx,$ (b) $\int \cos^2 \dfrac{x}{3} \, dx,$ (c) $\int \sin^4 2x \, dx,$ (d) $\int \cos^4 \dfrac{x}{2} \, dx.$

12 Write down a formula for cos x in terms of $\cos \dfrac{x}{2}$, and show that

$$\int \frac{1}{1 + \cos x} \, dx = \tan \frac{x}{2} + c$$

13 Find the following integrals:

(a) $\int \sqrt{(1 + \cos x)} \, dx,$ (b) $\int \dfrac{\cot x}{\sqrt{(1 - \cos 2x)}} \, dx,$

(c) $\int \sin 2x \sin^2 x \, dx,$ (d) $\int 2 \sin x \cos \dfrac{x}{2} \, dx.$

14 (a) Factorise sin $3x$ + sin x. (See Book 1, §17.8.)
   (b) Express 2 sin $3x$ cos $2x$ as the sum of two terms.
   (c) Find $\int \sin 3x \cos 2x \, dx.$
15 Find the following integrals:

(a) $\int \sin x \cos 3x \, dx,$ (b) $\int 2 \cos \dfrac{3x}{2} \cos \dfrac{x}{2} \, dx,$ (c) $\int \sin 4x \sin x \, dx.$

Changing the variable

1.5 In Example 1 we found that

$$\int x(3x^2 + 2)^4 \, dx = \frac{1}{30}(3x^2 + 2)^5 + c$$

The integral is a function of $(3x^2 + 2)$. If we write $3x^2 + 2$ as u, then the integral is a function of u; this suggests that we might make the substitution $u = 3x^2 + 2$ in the integrand, and *integrate with respect to u*. Let us see how this can be done.
Let

$$y = \int x(3x^2 + 2)^4 \, dx$$

then

$$\frac{dy}{dx} = x(3x^2 + 2)^4$$

If $u = 3x^2 + 2$, x may be expressed as a function of u. Then, by the chain rule,

$$\frac{dy}{du} = \frac{dy}{dx} \times \frac{dx}{du}$$

$$\therefore \frac{dy}{du} = x(3x^2 + 2)^4 \frac{dx}{du}$$

Integrating with respect to u,

$$y = \int x(3x^2 + 2)^4 \frac{dx}{du} \, du$$

But $u = 3x^2 + 2$, $\therefore \dfrac{du}{dx} = 6x$ and $\dfrac{dx}{du} = \dfrac{1}{6x}$.

$$\therefore \int x(3x^2 + 2)^4 \, dx = \int x(3x^2 + 2)^4 \frac{1}{6x} \, du$$

$$= \int \tfrac{1}{6} u^4 \, du$$

$$= \tfrac{1}{30} u^5 + c$$

$$= \tfrac{1}{30}(3x^2 + 2)^5 + c$$

Qu. 6 Find $\int \sin^2 4x \cos 4x \, dx$; put $u = \sin 4x$.
Qu. 7 Find $\int \sin^5 x \, dx$; put $u = \cos x$.

Comparing the foregoing text and questions with the solutions of Examples 1, 2 and 3, it might appear that we have merely introduced a more cumbersome technique; however, the power of changing the variable lies in its application to a wide class of integrals not susceptible to the method of §§1.2, 1.3.
In general, let $f(x)$ be a function of x, and let

$$y = \int f(x) \, dx$$

Then

$$\frac{dy}{dx} = f(x)$$

If u is a function of x, then by the chain rule

$$\frac{dy}{du} = \frac{dy}{dx} \times \frac{dx}{du}$$

$$\therefore \frac{dy}{du} = f(x)\frac{dx}{du}$$

$$\therefore y = \int f(x)\frac{dx}{du}\,du$$

$$\therefore \int f(x)\,dx = \int f(x)\frac{dx}{du}\,du$$

Thus an integral with respect to x may be transformed into an integral with respect to a related variable u, by using the above result, and substituting for $f(x)$ and $\dfrac{dx}{du}$ in terms of u.

Example 4   *Find* $\int x\sqrt{(3x-1)}\,dx$.

Sidework:

$$\int x\sqrt{(3x-1)}\frac{dx}{du}\,du = \int \frac{1}{3}(u^2+1)u\,\frac{2u}{3}\,du$$

Let $\sqrt{(3x-1)} = u$.

$$= \int (\tfrac{2}{9}u^4 + \tfrac{2}{9}u^2)\,du$$

$$x = \tfrac{1}{3}(u^2+1).$$

$$= \tfrac{2}{45}u^5 + \tfrac{2}{27}u^3 + c$$

$$\frac{dx}{du} = \frac{2u}{3}.$$

$$= \tfrac{2}{135}u^3(3u^2 + 5) + c$$

$$\therefore \int x\sqrt{(3x-1)}\,dx = \tfrac{2}{135}(3x-1)^{3/2}(9x+2) + c$$

Qu. 8   Find the following integrals, using the given change of variable:
(a) $\int x\sqrt{(2x+1)}\,dx$, $\sqrt{(2x+1)} = u$,
(b) $\int x\sqrt{(2x+1)}\,dx$, $2x+1 = u$,
(c) $\int x(3x-2)^6\,dx$, $3x-2 = u$.

Exercise 1b

1 Find the following integrals, using the given change of variable:
   (a) $\int 3x\sqrt{(4x-1)}\,dx$, $\sqrt{(4x-1)} = u$,
   (b) $\int x\sqrt{(5x+2)}\,dx$, $5x+2 = u$,

(c) $\int x(2x - 1)^6 \, dx,$ $\qquad\qquad$ $2x - 1 = u,$

(d) $\int \dfrac{x}{\sqrt{(x - 2)}} \, dx,$ $\qquad\qquad$ $\sqrt{(x - 2)} = u,$

(e) $\int (x + 2)(x - 1)^4 \, dx,$ $\qquad\qquad$ $x - 1 = u,$

(f) $\int (x - 2)^5(x + 3)^2 \, dx,$ $\qquad\qquad$ $x - 2 = u,$

(g) $\int \dfrac{x(x - 4)}{(x - 2)^2} \, dx,$ $\qquad\qquad$ $x - 2 = u,$

(h) $\int \dfrac{x - 1}{\sqrt{(2x + 3)}} \, dx,$ $\qquad\qquad$ $\sqrt{(2x + 3)} = u.$

2 Repeat Nos. 1(a) and 1(d) using a different change of variable in each case.

3 Of each of the following pairs of integrals, one should be found by a suitable change of variable, the other written down at once as a composite function of x, as in Example 1 on p. 1:

(a) $\int x\sqrt{(3x - 4)} \, dx$ $\qquad$ and $\qquad$ $\int x\sqrt{(3x^2 - 4)} \, dx.$

(b) $\int x(x^2 + 5)^6 \, dx$ $\qquad$ and $\qquad$ $\int x(x + 5)^6 \, dx,$

(c) $\int \dfrac{x}{\sqrt{(x - 1)}} \, dx$ $\qquad$ and $\qquad$ $\int \dfrac{x}{\sqrt{(x^2 - 1)}} \, dx.$

4 Find the following integrals, using a suitable change of variable only when necessary:

(a) $\int x\sqrt{(2x^2 + 1)} \, dx,$ $\qquad\qquad$ (b) $\int \dfrac{3x^2 - 1}{(x^3 - x + 4)^3} \, dx,$

(c) $\int 2x\sqrt{(2x - 1)} \, dx,$ $\qquad\qquad$ (d) $\int \cos^3 2x \, dx,$

(e) $\int \sin x\sqrt{\cos x} \, dx,$ $\qquad\qquad$ (f) $\int \cot^2 x \, \mathrm{cosec}^2 x \, dx,$

(g) $\int 2x(4x^2 - 1)^3 \, dx,$ $\qquad\qquad$ (h) $\int \dfrac{x}{\sqrt{(2x^2 - 5)}} \, dx,$

(i) $\int \dfrac{3x}{\sqrt{(4 - x)}} \, dx,$ $\qquad\qquad$ (j) $\int \dfrac{\sin \sqrt{x}}{\sqrt{x}} \, dx.$

Definite integrals and changing the limits

1.6 The method of changing the variable is also applicable to definite integrals. It is usually more convenient to change the limits to those of the new variable at the same time.

As a reminder that one must be ever watchful for the presence of a function and its derivative in an integrand, two examples of this type are also given here.

Example 5   *Evaluate* $\displaystyle\int_{1/2}^{3} x\sqrt{(2x+3)}\,\mathrm{d}x.$

$$\dagger\int_{x=1/2}^{x=3} x\sqrt{(2x+3)}\,\frac{\mathrm{d}x}{\mathrm{d}u}\,\mathbf{d}u = \int_{4}^{9} \tfrac{1}{2}(u-3)u^{1/2}\,\tfrac{1}{2}\,\mathrm{d}u \qquad \text{Let } 2x+3 = u.$$

$$x = \tfrac{1}{2}(u-3).$$

$$= \int_{4}^{9} (\tfrac{1}{4}u^{3/2} - \tfrac{3}{4}u^{1/2})\,\mathrm{d}u \qquad \frac{\mathrm{d}x}{\mathrm{d}u} = \tfrac{1}{2}.$$

$$= \left[\tfrac{1}{10}u^{5/2} - \tfrac{1}{2}u^{3/2}\right]_{4}^{9}$$

x	u
3	9
$\frac{1}{2}$	4

$$= (24.3 - 13.5) - (3.2 - 4)$$

$$= 11.6$$

Example 6   *Evaluate* (a) $\displaystyle\int_{2}^{3} \frac{x}{\sqrt{(x^2-3)}}\,\mathrm{d}x,$ (b) $\displaystyle\int_{0}^{\pi/4} \cos^3 x \sin x \,\mathrm{d}x.$

(a) $\displaystyle\int_{2}^{3} \frac{x}{\sqrt{(x^2-3)}}\,\mathrm{d}x = \left[(x^2-3)^{1/2}\right]_{2}^{3}$

$$= (9-3)^{1/2} - (4-3)^{1/2}$$

$$= \sqrt{6} - 1$$

(b) $\displaystyle\int_{0}^{\pi/4} \cos^3 x \sin x \,\mathrm{d}x = \left[-\tfrac{1}{4}\cos^4 x\right]_{0}^{\pi/4}$

$$= (-\tfrac{1}{4} \times \tfrac{1}{4}) - (-\tfrac{1}{4})$$

$$= \tfrac{3}{16}$$

Exercise 1c

1 Evaluate the following definite integrals by changing the variable and the limits:

(a) $\displaystyle\int_{2}^{3} x\sqrt{(x-2)}\,\mathrm{d}x,$ (b) $\displaystyle\int_{0}^{1} x(x-1)^4\,\mathrm{d}x,$

(c) $\displaystyle\int_{1}^{2} \frac{x}{\sqrt{(2x-1)}}\,\mathrm{d}x,$ (d) $\displaystyle\int_{1}^{2} (2x-1)(x-2)^3\,\mathrm{d}x,$

(e) $\displaystyle\int_{-3/8}^{0} \frac{x+3}{\sqrt{(2x+1)}}\,\mathrm{d}x.$

†The reader should note that, in practice, this integral will of course first be written down as given (i.e. as an integral with respect to x). When it is decided to change the variable, dx is changed to $\frac{\mathrm{d}x}{\mathrm{d}u}\,\mathrm{d}u$; it is then necessary to specify that the limits are still those of x.

2 Evaluate the following definite integrals either by writing down the integral as a function of x, or by using the given change of variable:

(a) $\displaystyle\int_0^{\pi/6} \sec^4 x \tan x \, dx$ $\quad$ (sec $x = u$),

(b) $\displaystyle\int_0^{\pi/2} \sin^5 x \, dx$ $\quad\quad$ (cos $x = u$),

(c) $\displaystyle\int_{\pi/6}^{\pi/2} \frac{\cot x}{\sqrt{\csc^3 x}} \, dx$ $\quad$ (cosec $x = u$).

3 Evaluate:

(a) $\displaystyle\int_0^{1/2} \frac{x}{\sqrt{(1-x^2)}} \, dx$, $\quad\quad$ (b) $\displaystyle\int_0^4 2x\sqrt{(4-x)} \, dx$,

(c) $\displaystyle\int_{-1}^0 x(x^2-1)^4 \, dx$, $\quad\quad$ (d) $\displaystyle\int_0^{\pi/4} \sec^4 x \, dx$,

(e) $\displaystyle\int_{1/2}^1 \frac{x-2}{(x+2)^3(x-6)^3} \, dx$, $\quad$ (f) $\displaystyle\int_{-1}^2 (x+1)(2-x)^4 \, dx$,

(g) $\displaystyle\int_{-\pi/2}^{\pi/2} \cos^3 x \, dx$, $\quad\quad$ (h) $\displaystyle\int_{5/3}^{8/3} \frac{x+2}{\sqrt{(3x-4)}} \, dx$,

(i) $\displaystyle\int_0^{\pi/2} \sin x\sqrt{\cos x} \, dx$.

4 Calculate the area enclosed by the curve $y = x/\sqrt{(x^2-1)}$, the x-axis, $x = 2$ and $x = 3$.

5 Calculate the area under $y = \sin^3 x$ from $x = 0$ to $x = 2\pi/3$.

6 Calculate the volume of the solid generated when the area under $y = \cos x$, from $x = 0$ to $x = \pi/2$ is rotated through four right angles about the x-axis. (See §1.4.)

7 The area of a uniform lamina is that enclosed by the curve $y = \sin x$, the x-axis, and the line $x = \pi/2$. Find the distance from the x-axis of the centre of gravity of the lamina. (See §1.4.)

Integration using the inverse trigonometrical functions

1.7 The inverse trigonometrical functions were introduced in Book 1, §18.7. Some readers may need to revise this topic before proceeding further; Qu. 9–12 are included for this purpose.

Qu. 9 The following angles lie between 0 and 90° inclusive. Express them in degrees, and in radians in terms of π:

(a) $\tan^{-1} 1$, $\quad\quad$ (b) $\sin^{-1}\frac{1}{2}$, $\quad\quad$ (c) $\frac{1}{2}\sin^{-1} 1$,

(d) $\cos^{-1}\frac{1}{2}$, $\quad\quad$ (e) $\frac{1}{2}\cos^{-1}\frac{1}{2}$, $\quad\quad$ (f) $\cos^{-1} 1$,

(g) $2 \cos^{-1} \dfrac{\sqrt{3}}{2}$,   (h) $\frac{1}{3} \cos^{-1} 0$,   (i) $\frac{2}{3} \cot^{-1} 1$,

(j) $\sec^{-1} 2$,   (k) $2 \operatorname{cosec}^{-1} \sqrt{2}$.

Qu. 10   Express the following angles in radians, leaving π in the answers:
(a) $20°$,   (b) $70°$,   (c) $150°$,   (d) $300°$,   (e) $405°$.

Qu. 11   Express the following angles in degrees:
(a) 1 radian,   (b) 0.03 radian,   (c) 1.25 radians,
(d) 0.715 radian,   (e) $\pi/5$ radian.

Qu. 12   Express the following (acute) angles in radians:
(a) $2 \sin^{-1} 0.6$,   (b) $\tan^{-1} 1.333$,   (c) $\frac{2}{3} \cos^{-1} 0.3846$.

The inverse sine function may be written as arcsin x, or as $\sin^{-1} x$. Both forms are in current use and both will be used in this book to familiarise the reader with them.

The expression $\sqrt{(1 - x^2)}$ may be reduced to a rational form by changing the variable to u, where $x = \sin u$; thus

$$\sqrt{(1 - x^2)} = \sqrt{(1 - \sin^2 u)} = \sqrt{\cos^2 u} = \cos u$$

This is used in the following example.

Example 7   *Find* $\displaystyle\int \dfrac{1}{\sqrt{(1 - x^2)}}\, dx.$

$$\int \frac{1}{\sqrt{(1 - x^2)}} \frac{dx}{du}\, du = \int \frac{1}{\sqrt{(1 - \sin^2 u)}} \cos u\, du \qquad \text{Let } x = \sin u.$$

$$= \int \frac{1}{\cos u} \cos u\, du \qquad\qquad \frac{dx}{du} = \cos u.$$

$$= u + c$$

$$= \arcsin x + c$$

$$\int \frac{1}{\sqrt{(a^2 - b^2 x^2)}}\, dx$$

1.8   The reader should check that the integral found in §1.7 is not susceptible to the change of variable $\sqrt{(1 - x^2)} = u$; $\displaystyle\int \dfrac{1}{\sqrt{(1 - x^2)}}\, dx$ merely becomes $\displaystyle\int \dfrac{-1}{\sqrt{(1 - u^2)}}\, du$. However, changes of variable involving a trigonometrical substitution, such as was successfully applied in this case, open the way to finding a very important group of integrals. Here are two examples of the type of substitution we shall be using.

If $x = 5 \sin u$,

$$\sqrt{(25 - x^2)} = \sqrt{(25 - 25 \sin^2 u)} = \sqrt{\{25(1 - \sin^2 u)\}} = 5 \cos u$$

If $x = \dfrac{\sqrt{3}}{2}\sin u$,

$$\sqrt{(3 - 4x^2)} = \sqrt{(3 - 4 \times \tfrac{3}{4}\sin^2 u)} = \sqrt{\{3(1 - \sin^2 u)\}} = \sqrt{3}\cos u$$

Qu. 13   Reduce each of the following to the form $k\cos u$, and give u in terms of x in each case:

(a) $\sqrt{(9 - x^2)}$,   (b) $\sqrt{(1 - 25x^2)}$,   (c) $\sqrt{(4 - 9x^2)}$,
(d) $\sqrt{(7 - x^2)}$,   (e) $\sqrt{(1 - 3x^2)}$,   (f) $\sqrt{(3 - 2x^2)}$.

We see that to deal with $\sqrt{(a^2 - b^2x^2)}$ we write

$$a^2 - b^2 x^2 \quad \text{as} \quad a^2 - a^2\sin^2 u$$

thus $b^2 x^2 = a^2\sin^2 u$, and $x = (a/b)\sin u$. Note that $u = \arcsin(bx/a)$ and, for the substitution to be valid, and of use, u must be real and not $\pi/2$, so $|bx| < |a|$; this condition is implicit in $\sqrt{(a^2 - b^2x^2)}$ being real and not zero.

Example 8   *Find* $\displaystyle\int \frac{1}{\sqrt{(9 - 4x^2)}}\,dx$.

$$\int \frac{1}{\sqrt{(9 - 4x^2)}}\,\frac{dx}{du}\,du = \int \frac{1}{\sqrt{(9 - 9\sin^2 u)}} \times \frac{3}{2}\cos u\,du \qquad 9 - 4x^2.$$
$$9 - 9\sin^2 u.$$

$$= \int \frac{1}{3\cos u} \times \frac{3}{2}\cos u\,du \qquad\qquad \text{Let } x = \tfrac{3}{2}\sin u.$$

$$= \int \tfrac{1}{2}\,du \qquad\qquad\qquad\qquad \frac{dx}{du} = \frac{3}{2}\cos u.$$

$$= \tfrac{1}{2}u + c$$

$$= \frac{1}{2}\arcsin\left(\frac{2x}{3}\right) + c$$

(This answer could also be written $\dfrac{1}{2}\sin^{-1}\left(\dfrac{2x}{3}\right) + c$.)

Qu. 14   Find the following integrals:

(a) $\displaystyle\int \frac{1}{\sqrt{(4 - x^2)}}\,dx$,   (b) $\displaystyle\int \frac{1}{\sqrt{(1 - 3x^2)}}\,dx$,   (c) $\displaystyle\int \frac{1}{\sqrt{(16 - 9x^2)}}\,dx$.

Qu. 15   Prove that $\displaystyle\int \frac{1}{\sqrt{(a^2 - b^2x^2)}}\,dx = \frac{1}{b}\arcsin\left(\frac{bx}{a}\right) + c$.

$$\int \frac{1}{a^2 + b^2 x^2}\,dx$$

1.9   In §1.8 we made use of Pythagoras' theorem in the form $\cos^2 u + \sin^2 u = 1$; we shall now find that an alternative form, $1 + \tan^2 u = \sec^2 u$, helps to effect other useful changes of variable.

Qu. 16 Find $\displaystyle\int \frac{1}{1+x^2}\,dx$ by taking x as tan u.

Qu. 17 Reduce each of the following to the form $k \sec^2 u$, and give u in terms of x in each case:
(a) $9+x^2$,   (b) $1+4x^2$,   (c) $25+9x^2$,
(d) $3+x^2$,   (e) $1+5x^2$,   (f) $7+3x^2$.

Qu. 18 Find the following integrals:

(a) $\displaystyle\int \frac{1}{4+x^2}\,dx$,   (b) $\displaystyle\int \frac{1}{1+16x^2}\,dx$,   (c) $\displaystyle\int \frac{1}{3+4x^2}\,dx$.

Example 9 *Evaluate* $\displaystyle\int_{\sqrt{3}/2}^{3/2} \frac{1}{3+4x^2}\,dx.$

$$\int_{x=\sqrt{3}/2}^{x=3/2} \frac{1}{3+4x^2}\frac{dx}{du}\,du$$

$3+4x^2.$

$3+3\tan^2 u.$

$$=\int_{\pi/4}^{\pi/3} \frac{1}{3(1+\tan^2 u)}\frac{\sqrt{3}}{2}\sec^2 u\,du$$

Let $x=\dfrac{\sqrt{3}}{2}\tan u.$

$$=\int_{\pi/4}^{\pi/3} \frac{\sqrt{3}}{6}\,du$$

$\dfrac{dx}{du}=\dfrac{\sqrt{3}}{2}\sec^2 u.$

$$=\left[\frac{\sqrt{3}\,u}{6}\right]_{\pi/4}^{\pi/3}$$

$$=\frac{\sqrt{3}}{6}\left(\frac{\pi}{3}-\frac{\pi}{4}\right)$$

$$=\frac{\sqrt{3}\pi}{72}$$

x	$\tan u$	u
$\dfrac{3}{2}$	$\sqrt{3}$	$\dfrac{\pi}{3}$
$\dfrac{\sqrt{3}}{2}$	1	$\dfrac{\pi}{4}$

Exercise 1d

(Nos. 1–4 are revision exercises on Book 1, §18.7.)

1 The following angles lie between 0 and 90° inclusive. Express them in degrees, and in radians in terms of π:

(a) $\arccos \dfrac{1}{\sqrt{2}}$,   (b) $\operatorname{arccot} 1$,   (c) $\dfrac{1}{\sqrt{3}}\operatorname{arccot} \sqrt{3}$,

(d) $\arcsin \dfrac{\sqrt{3}}{2}$,   (e) $\sqrt{3}\arcsin \tfrac{1}{2}$,   (f) $\tfrac{1}{3}\operatorname{arcsec} \sqrt{2}$,

(g) $\tfrac{3}{2}\arctan 1$,   (h) $\tfrac{1}{2}\operatorname{arccosec} 2$.

2 Express the following angles in radians:
(a) 32°,   (b) $60^\circ\ 21'$,   (c) $5^\circ\ 41'$,   (d) $235^\circ\ 16'$.

3 Express the following angles in degrees:
(a) 2 radians,   (b) 0.08 radian,   (c) 1.362 radians,
(d) $\pi/6$ radian.

4 Express the following (acute) angles in radians:

(a) $\sin^{-1} 0.8$,   (b) $\frac{1}{2} \cos^{-1}\left(\dfrac{5}{13}\right)$,   (c) $2 \tan^{-1} 0.625$.

(In this question, the alternative notation, $\sin^{-1} x$, etc., is employed to give the reader some practice in using it.)

5 Express the following in the form $k \cos u$, and give u in terms of x in each case:
(a) $\sqrt{(16 - x^2)}$,   (b) $\sqrt{(1 - 9x^2)}$,   (c) $\sqrt{(9 - 4x^2)}$,
(d) $\sqrt{(10 - x^2)}$,   (e) $\sqrt{(1 - 6x^2)}$,   (f) $\sqrt{(5 - 3x^2)}$.

6 Find the following integrals:

(a) $\displaystyle\int \frac{1}{\sqrt{(25 - x^2)}} \, dx,$ (b) $\displaystyle\int \frac{1}{\sqrt{(1 - 4x^2)}} \, dx,$ (c) $\displaystyle\int \frac{1}{\sqrt{(4 - 9x^2)}} \, dx,$

(d) $\displaystyle\int \frac{1}{\sqrt{(3 - x^2)}} \, dx,$ (e) $\displaystyle\int \frac{1}{\sqrt{(1 - 7x^2)}} \, dx,$ (f) $\displaystyle\int \frac{1}{\sqrt{(2 - 3x^2)}} \, dx.$

7 Express the following in the form $k \sec^2 u$, and give u in terms of x in each case:
(a) $16 + x^2$,   (b) $1 + 9x^2$,   (c) $4 + 3x^2$,
(d) $2 + x^2$,   (e) $1 + 3x^2$,   (f) $5 + 2x^2$.

8 Find the following integrals:

(a) $\displaystyle\int \frac{1}{25 + x^2} \, dx,$ (b) $\displaystyle\int \frac{1}{1 + 36x^2} \, dx,$ (c) $\displaystyle\int \frac{1}{16 + 3x^2} \, dx,$

(d) $\displaystyle\int \frac{1}{5 + x^2} \, dx,$ (e) $\displaystyle\int \frac{1}{1 + 6x^2} \, dx,$ (f) $\displaystyle\int \frac{1}{3 + 10x^2} \, dx.$

9 Find the following integrals:

(a) $\displaystyle\int \frac{1}{9 + 2x^2} \, dx,$ (b) $\displaystyle\int \frac{3}{\sqrt{(4 - 5x^2)}} \, dx,$

(c) $\displaystyle\int \frac{1}{\sqrt{(3 - 2x^2)}} \, dx,$ (d) $\displaystyle\int \frac{2}{3 + 5x^2} \, dx.$

10 Evaluate the following integrals, leaving π in your answers:

(a) $\displaystyle\int_1^{\sqrt{3}} \frac{2}{1 + x^2} \, dx,$ (b) $\displaystyle\int_0^{\sqrt{2}} \frac{1}{\sqrt{(4 - x^2)}} \, dx,$ (c) $\displaystyle\int_{1/2}^1 \frac{3}{\sqrt{(1 - x^2)}} \, dx,$

(d) $\displaystyle\int_0^3 \frac{1}{9 + x^2} \, dx,$ (e) $\displaystyle\int_1^{1/6} \frac{1}{\sqrt{(1 - 9x^2)}} \, dx,$ (f) $\displaystyle\int_{-2}^{\sqrt{3}} \frac{1}{5\sqrt{(4 - x^2)}} \, dx.$

11 (a) Find $\int \dfrac{1}{\sqrt{(9-x^2)}}\,dx$ using (i) $x = 3 \sin u$, (ii) $x = 3 \cos u$.

(b) Evaluate $\displaystyle\int_{3/2}^{3} \dfrac{1}{\sqrt{(9-x^2)}}\,dx$ using (i) $x = 3 \sin u$, (ii) $x = 3 \cos u$.

12 Find the following integrals, using the given change of variable:

(a) $\displaystyle\int \dfrac{1}{\sqrt{\{4-(x+1)^2\}}}\,dx, \quad x + 1 = 2 \sin u,$

(b) $\displaystyle\int \dfrac{1}{9+(x-3)^2}\,dx, \quad x - 3 = 3 \tan u.$

13 Find the following integrals:

(a) $\displaystyle\int \dfrac{1}{(x+3)^2+25}\,dx,$ (b) $\displaystyle\int \dfrac{1}{\sqrt{\{4-(x-1)^2\}}}\,dx,$

(c) $\displaystyle\int \dfrac{1}{3(x-2)^2+5}\,dx,$ (d) $\displaystyle\int \dfrac{1}{\sqrt{\{9-3(x+1)^2\}}}\,dx.$

*14 (a) $2x^2 - 12x + 21$ may be written $2(x^2 - 6x + 9) + 21 - 18 = 2(x-3)^2 + 3$.
   Write the following expressions in the form $a(x+b)^2 + c$ (see Book 1,
   §10.3):
   (i) $x^2 - 6x + 16$,   (ii) $3x^2 - 12x + 14$,   (iii) $2x^2 - 4x + 5$.
(b) Find the following integrals:

(i) $\displaystyle\int \dfrac{1}{x^2-2x+5}\,dx,$ (ii) $\displaystyle\int \dfrac{1}{2x^2+4x+11}\,dx,$

(iii) $\displaystyle\int \dfrac{1}{x^2-4x+13}\,dx,$ (iv) $\displaystyle\int \dfrac{1}{4x^2-8x+7}\,dx.$

15 (a) $1 + 6x - 3x^2$ may be written $4 - 3(x^2 - 2x + 1) + 3 = 4 - 3(x-1)^2$.
   Write the following expressions in the form $a - b(x+c)^2$:
   (i) $3 - 2x - x^2$,   (ii) $5 + 4x - x^2$,   (iii) $7 + 2x - 2x^2$.
(b) Find the following integrals:

(i) $\displaystyle\int \dfrac{1}{\sqrt{(3-2x-x^2)}}\,dx,$ (ii) $\displaystyle\int \dfrac{1}{\sqrt{(1+8x-4x^2)}}\,dx,$

(iii) $\displaystyle\int \dfrac{1}{\sqrt{(12+4x-x^2)}}\,dx,$ (iv) $\displaystyle\int \dfrac{1}{\sqrt{(-2x^2+12x-9)}}\,dx.$

16 Evaluate:

(a) $\displaystyle\int_{2}^{3} \dfrac{1}{x^2-4x+5}\,dx,$ (b) $\displaystyle\int_{-1}^{1} \dfrac{1}{\sqrt{(3-2x-x^2)}}\,dx.$

*17 Find the following integrals by writing each integrand as two fractions:

(a) $\displaystyle\int \dfrac{3-x}{\sqrt{(1-x^2)}}\,dx,$ (b) $\displaystyle\int \dfrac{2x+3}{\sqrt{(4-x^2)}}\,dx.$

18 Show that $\int \sqrt{(1-x^2)}\,dx = \frac{1}{2}\sin^{-1}x + \frac{1}{2}x\sqrt{(1-x^2)} + c$.

Find the following integrals:

(a) $\displaystyle\int \frac{x^2}{\sqrt{(1-x^2)}}\,dx$,   (b) $\displaystyle\int \frac{1}{(x^2+9)^2}\,dx$,   (c) $\displaystyle\int \frac{x}{\sqrt{(4-x^4)}}\,dx$.

19 Show that $\displaystyle\int \frac{1}{(1-x^2)^{3/2}}\,dx = x(1-x^2)^{-1/2} + c$.

Find the following integrals:

(a) $\displaystyle\int \frac{1}{(1-9x^2)\sqrt{(1-9x^2)}}\,dx$,   (b) $\displaystyle\int \frac{1}{x^2\sqrt{(1-x^2)}}\,dx$,

(c) $\displaystyle\int \frac{1}{x\sqrt{(x^2-1)}}\,dx$.

Trigonometrical functions of numbers

1.10   At this stage it is advisable to discuss some of the implications of the definite integrals evaluated in Exercise 1d.

Let us find the area under the curve $y = \dfrac{1}{\sqrt{(9-4x^2)}}$ from $x=0$ to $x=1$ (see

Fig. 1.1). The element of area $= y\,\delta x = \dfrac{1}{\sqrt{(9-4x^2)}}\,\delta x$, and the required area is

$$\int_{x=0}^{x=1} \frac{1}{\sqrt{(9-4x^2)}}\frac{dx}{du}\,du$$

$$9 - 4x^2.$$
$$9 - 9\sin^2 u.$$

$$\approx \int_0^{0.7297} \frac{1}{\sqrt{\{9(1-\sin^2 u)\}}}\frac{3}{2}\cos u\,du$$

Let $x = \frac{3}{2}\sin u$.

$$= \left[\frac{1}{2}u\right]_0^{0.7297}$$

$$\frac{dx}{du} = \frac{3}{2}\cos u.$$

x	$\sin u$	u
1	2/3	0.7297 rad.
0	0	0

$= 0.365$,   correct to three
significant figures

Now if we retain x as our variable (see Example 8), the integral is evaluated as

$$\left[\frac{1}{2}\sin^{-1}\frac{2x}{3}\right]_0^1 = \frac{1}{2}\sin^{-1}\frac{2}{3}$$

It may at first sight seem surprising that the number 0.365 can measure, at one and the same time, units of area and the angle $\frac{1}{2}\sin^{-1}\frac{2}{3}$ in radians. These two aspects must be reconciled, and it is apparent that we must extend our definitions of the trigonometrical ratios to include the functions of *numbers*, as well as of *angles*.

Thus $\sin u$ may be considered as a function of the number u which, as u moves

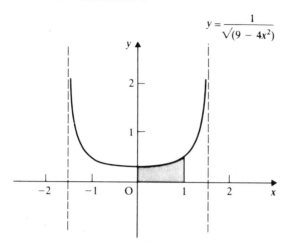

$$y = \frac{1}{\sqrt{(9 - 4x^2)}}$$

Figure 1.1

from $-\infty$ to $+\infty$, oscillates between -1 and $+1$ with period 2π. (See Book 1, §16.2.) The sine of a given *number* is the same as the sine of the *angle* given by that number of *radians*; thus

$$\sin 2 = \sin (2 \text{ radians}) \approx \sin 114° \, 35' = \sin 65° \, 25' \approx 0.9093$$

[$\sin 2 \approx 0.9093$ can easily be obtained from a calculator, but make sure the calculator is in radian mode.]

Qu. 19   Find the values of
(a) $\cos (\pi/6)$,   (b) $\sin \frac{1}{2}$,   (c) $\tan 1.2$,   (d) $\cos 3$,   (e) $\sec 6$.
Qu. 20   Find the following numbers between $-\pi/2$ and $\pi/2$ inclusive:
(a) $\sin^{-1} 1$,   (b) $\tan^{-1} (-1)$,   (c) $\sin^{-1} (\frac{1}{2})$,   (d) $\tan^{-1} 2$.

Exercise 1e

1 Find the value of
   (a) $\cos (\pi/12)$,   (b) $\sin 1.5$,   (c) $\tan 0.0806$,   (d) $\operatorname{cosec} \frac{5}{4}$.
2 Prove that $\arcsin x + \arccos x = \pi/2$.
3 Find the following numbers, in terms of π, or to three significant figures:

   (a) $\tan^{-1} 3$,   (b) $\sqrt{3} \tan^{-1} (\sqrt{3})$,   (c) $\frac{2}{3} \sin^{-1} \frac{\sqrt{2}}{2}$,

   (d) $\cos^{-1} (-0.375)$,   (e) $\sec^{-1} \pi$.

4 Find x in terms of y if

   (a) $\tan^{-1} x = \tan^{-1} y + \frac{\pi}{4}$,   (b) $\cos^{-1} x = \cos^{-1} y - \frac{\pi}{6}$.

5 Sketch the graph of $y = 1/\sqrt{(16 - x^2)}$, and calculate the area under the curve
   (a) from $x = 0$ to $x = 2$, (b) from $x = 2$ to $x = 3$.

6 Sketch the graph of $y = 1/(1 + x^2)$ from $x = -3$ to $x = +3$, and calculate the area under the curve (a) from $x = 0$ to $x = 1$, (b) from $x = 1$ to $x = 2$.

7 A particle moves along a straight line so that t s after starting it is x m from a point O on the straight line, where $x = 10 \cos t$.
(a) How far from O is it after $0, \pi/2, \pi, 3\pi/2, 2\pi$ s?
(b) When is it first at a distance 5 m from O?
(c) When is it first 5 m on the negative side of O?

8 A particle moves along a straight line so that t s after the start it is x m from a point O on the line, where $x = 5 \sin \frac{1}{2}t$.
(a) How far from O is it after $0, \pi, 2\pi, 3\pi, 4\pi$ s?
(b) How long does it take to travel the first 3 m from O?

9 Evaluate the following, correct to three significant figures:

(a) $\int_1^3 \frac{1}{1 + x^2} \, dx,$
(b) $\int_0^{0.8} \frac{1}{\sqrt{(1 - x^2)}} \, dx,$

(c) $\int_1^2 \frac{1}{4 + 25x^2} \, dx,$
(d) $\int_1^{4/3} \frac{1}{\sqrt{(25 - 9x^2)}} \, dx,$

(e) $\int_{1/2}^{\sqrt{3}} \frac{1}{3 + 4x^2} \, dx,$
(f) $\int_{-1}^{-1/2} \frac{1}{\sqrt{(4 - 2x^2)}} \, dx.$

Exercise 1f (Miscellaneous)

Find the integrals Nos. 1–20, which are arranged by types in the order in which they occur in the chapter:

1 $\int x^2 \sqrt{(x^3 - 1)} \, dx.$

2 $\int \frac{x}{(x^2 - 1)^3} \, dx.$

3 $\int \sin 2x \cos^2 2x \, dx.$

4 $\int \sec^2 x \sqrt{\cot x} \, dx.$

5 $\int \cos^3 4x \, dx.$

6 $\int \sin^2 \frac{x}{3} \, dx.$

7 $\int \cos^4 2x \, dx.$

8 $\int \sqrt{(1 - \cos x)} \, dx.$

9 $\int \sin \frac{x}{3} \cos^2 \frac{x}{6} \, dx.$

10 $\int \cos 3x \cos 2x \, dx.$

11 $\int x(3x - 7)^4 \, dx.$

12 $\int \frac{x}{\sqrt{(5 + x)}} \, dx.$

13 $\int \frac{1}{\sqrt{(6 - 5x^2)}} \, dx.$

14 $\int \frac{1}{1 + 8x^2} \, dx.$

15 $\int \frac{1}{\sqrt{(5 - 4x - x^2)}} \, dx.$

16 $\int \frac{1}{3x^2 + 6x + 5} \, dx.$

17 $\int \frac{x + 1}{\sqrt{(5 - x^2)}} \, dx.$

18 $\int \sqrt{(9 - x^2)} \, dx.$

19 $\int \frac{1}{(4 - x^2)^{3/2}} \, dx.$

20 $\int \frac{1}{x^2 \sqrt{(16 - x^2)}} \, dx.$

21 Find (a) $\displaystyle\int \frac{x}{\sqrt{(1-x^2)}}\,dx$, (b) $\displaystyle\int \frac{2}{1+x^2}\,dx$, (c) $\displaystyle\int x\sqrt{(1-x^2)}\,dx$,

(d) $\displaystyle\int \frac{2}{\sqrt{(1-x^2)}}\,dx$, (e) $\displaystyle\int \frac{2-x}{\sqrt{(1-x^2)}}\,dx$.

Find the integrals Nos. 22–45:

22 $\displaystyle\int \frac{x+2}{\sqrt{(3x-1)}}\,dx$.   **23** $\displaystyle\int \tfrac{1}{2}x\sqrt{(x^2+2)}\,dx$.   **24** $\displaystyle\int \frac{1}{x^2\sqrt{(1-x^2)}}\,dx$.

25 $\displaystyle\int \cos^2 \frac{x}{2}\sin^3 \frac{x}{2}\,dx$.   **26** $\displaystyle\int \frac{x}{3\sqrt{(4-x^2)}}\,dx$.   **27** $\displaystyle\int \frac{3}{\sqrt{(36-x^2)}}\,dx$.

28 $\displaystyle\int (x-1)^2(x+3)^5\,dx$.   **29** $\displaystyle\int \frac{x^2}{\sqrt{(4-x^2)}}\,dx$.   **30** $\displaystyle\int \frac{2+x}{\sqrt{(9-x^2)}}\,dx$.

31 $\displaystyle\int \frac{1}{3x^2-12x+16}\,dx$.   **32** $\displaystyle\int \sin^2 \frac{x}{5}\,dx$.   **33** $\displaystyle\int \sqrt{(\cos 3x+1)}\,dx$.

34 $\displaystyle\int \sqrt{(4-x^2)}\,dx$.   **35** $\displaystyle\int \sec^5 x \tan x\,dx$.   **36** $\displaystyle\int \frac{x^2}{\sqrt{(4-9x^6)}}\,dx$.

37 $\displaystyle\int \sin^5 2x\,dx$.   **38** $\displaystyle\int \frac{1}{(1+x^2)^2}\,dx$.   **39** $\displaystyle\int \sin \frac{3x}{2}\cos \frac{5x}{2}\,dx$.

40 $\displaystyle\int \frac{\tan x}{\sqrt{(\cos 2x+1)}}\,dx$.   **41** $\displaystyle\int \frac{x}{1+x^4}\,dx$.   **42** $\displaystyle\int \cos x\sqrt{\cos 2x}\,dx$.

43 $\displaystyle\int \frac{1}{\cos^2 x+4\sin^2 x}\,dx$ (put $\tan x = u$).

44 $\displaystyle\int \frac{1}{x\sqrt{(x^2-9)}}\,dx$.

45 $\displaystyle\int \frac{1}{(1-x)\sqrt{(1-x^2)}}\,dx$ (put $x = \cos u$; show integral $= \cot \dfrac{u}{2}+c$).

Evaluate Nos. 46–50:

46 $\displaystyle\int_0^3 2x\sqrt{(5x+1)}\,dx$.   **47** $\displaystyle\int_{\pi/4}^{\pi/2} \cos x\,\mathrm{cosec}^3 x\,dx$.

48 $\displaystyle\int_0^\pi \sin \frac{x}{2}\cos \frac{x}{2}\,dx$.   **49** $\displaystyle\int_{-2\pi}^0 \cos^4 \frac{x}{4}\,dx$.

50 $\displaystyle\int_0^{\sqrt{3}} \frac{1}{(4-x^2)^{3/2}}\,dx$.

Chapter 2

Exponential and logarithmic functions

Exponential functions

2.1 The word *exponent* is often used instead of *index*, and functions in which the variable is in the index (such as 2^x, $10^{\sin x}$) are called **exponential functions.**†

The graph of $y = a^x$

2.2 Let us first consider the function 2^x. A table of values follows, and a sketch of $y = 2^x$ is given in Fig. 2.1.

Table of values, $y = 2^x$

x	-3	-2	-1	0	1	2
2^x	$\frac{1}{8}$	$\frac{1}{4}$	$\frac{1}{2}$	1	2	4

As $x \to -\infty$, $2^x \to 0$, and so the curve approaches the x-axis but does not meet it.

Qu. 1 Copy and extend the above table to include values of 1.5^x (from $x = -3$ to $x = +3$), and of 2.5^x, 3^x (both from $x = -2$ to $x = +2$). Sketch, with the same axes, the graphs of $y = 1^x$, $y = 1.5^x$, $y = 2^x$, $y = 2.5^x$, $y = 3^x$. What do you notice about the gradient of $y = a^x$ at $(0, 1)$ as a takes different values greater than 1?

Qu. 2 How would you deduce the shape of the graph of $y = (\frac{1}{2})^x$ from Fig. 2.1?

---

†The graph of $y = 10^x$ is usually encountered during the elementary introduction to logarithms; in Book 1, §9.5 equations such as $2^x = 3$ are solved. It is probably only in these contexts that the reader has previously met exponential functions.

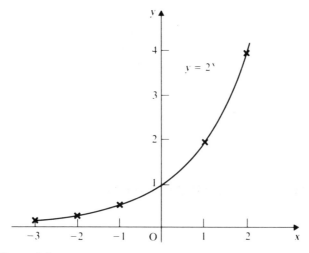

Figure 2.1

The gradient of $y = a^x$ at (0, 1); a limit

2.3   We shall confine our attention for the time being to exponential functions of the form a^x, where a is taken to be a constant real number greater than 1. Since $a^0 = 1$, the graph of $y = a^x$ (Fig. 2.2) passes through the point A(0, 1), and we let the gradient of the curve at this point be **m**.

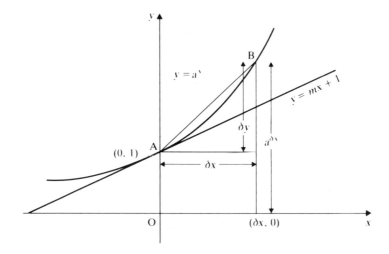

Figure 2.2

With the usual notation, if B is the point $(\delta x, a^{\delta x})$, then the gradient of AB is

$$\frac{\delta y}{\delta x} = \frac{a^{\delta x} - 1}{\delta x}$$

Now as $\delta x \to 0$, the gradient of AB $\to m$.

It follows that the **limit**, as $\delta x \to 0$, of $\dfrac{a^{\delta x} - 1}{\delta x}$ is m, the gradient of $y = a^x$ at (0, 1).

The form of $\dfrac{\mathrm{d}}{\mathrm{d}x}(a^x)$

2.4 The limit just established enables us to investigate the gradient of $y = a^x$ at any point $P(x, y)$ on the curve. With the usual notation, if Q is the point $(x + \delta x, y + \delta y)$,

$$y + \delta y = a^{x + \delta x}$$
$$\therefore \; \delta y = a^{x + \delta x} - a^x = a^x(a^{\delta x} - 1)$$

$\therefore$ the gradient of PQ, $\dfrac{\delta y}{\delta x} = a^x\left(\dfrac{a^{\delta x} - 1}{\delta x}\right)$ \hfill (1)

Now as $\delta x \to 0$, the gradient of PQ $\to$ the gradient of the tangent at P; also, since we have shown that

$$\left(\frac{a^{\delta x} - 1}{\delta x}\right) \to m$$

then the R.H.S. of (1) $\to ma^x$.

$$\therefore \; \frac{\mathrm{d}y}{\mathrm{d}x} = ma^x$$

Thus $\dfrac{\mathrm{d}}{\mathrm{d}x}(a^x) = ma^x$, **where m is the gradient of $y = a^x$ at the point (0, 1).**

We have already noted (see Qu. 1) that as a increases, the gradients of the curves $y = a^x$ at (0, 1) increase; for every value of a there is an appropriate value of m, and it is reasonable to suppose that we should be able to express m in terms of a. However for the time being we must be satisfied with some numerical approximations for m which we will now proceed to find.

Approximate derivatives of 2^x and 3^x

2.5 The following table was used to draw the graphs of $y = 2^x$ and $y = 3^x$ in Fig. 2.3; $y = m_2 x + 1$ and $y = m_3 x + 1$ are the respective tangents at (0, 1).

Table of values for $y = 2^x$ and $y = 3^x$

x	-2	$-\frac{3}{2}$	-1	$-\frac{1}{2}$	$-\frac{1}{4}$	0	$\frac{1}{4}$	$\frac{1}{2}$	1	$\frac{3}{2}$	2
2^x	0.25	0.35	0.5	0.71	0.84	1	1.19	1.41	2	2.83	4
3^x	0.11	0.19	0.33	0.58	0.76	1	1.32	1.73	3	5.20	9

Qu. 3   (a) Measure the gradients of the two tangents in Fig. 2.3, and deduce approximate expressions for $\dfrac{d}{dx}(2^x)$ and $\dfrac{d}{dx}(3^x)$.

(b) Now calculate the gradient of $y = 2^x$ where $x = 1$, and the gradient of $y = 3^x$ where $x = \frac{1}{2}$. Check from the graph.

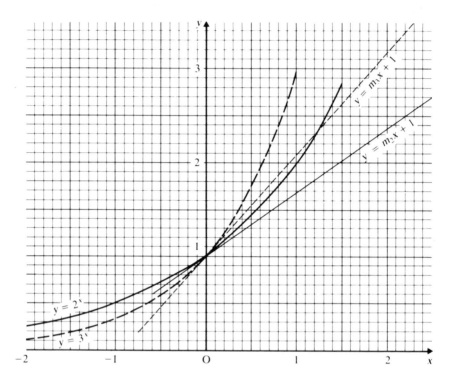

Figure 2.3

Qu. 4   Tangents were drawn to a graph of $y = 2^x$, and their gradients were measured and entered in the following table:

x	-3	-2	-1	0	1	$\frac{3}{2}$	2	$\frac{5}{2}$
$y = 2^x$	0.125	0.25	0.5	1	2	2.83	4	5.66
$\dfrac{dy}{dx}$	0.08	0.18	0.37	0.62	1.33	1.90	2.82	3.68

Confirm graphically that these results indicate that $\dfrac{dy}{dx} \propto y$, and deduce an approximate expression for $\dfrac{d}{dx}(2^x)$.

The exponential function e^x

2.6 It has been established in the previous section that

$$\frac{d}{dx}(2^x) \approx 0.7 \times 2^x$$

and

$$\frac{d}{dx}(3^x) \approx 1.1 \times 3^x$$

Since, in general, $\dfrac{d}{dx}(a^x) = ka^x$, these results suggest that for simplicity we should find a value of a between 2 and 3 for which $k = 1$; this number is called **e**, its value is approximately 2.71828, and it will be found to play a vital part in the further development of mathematics from this point.

Let us now summarise what we know about e.

Definition

e *is the number such that the gradient of* $y = e^x$ *at* (0, 1) *is* 1. e^x *is called the* **exponential function**.

Thus $\dfrac{d}{dx}(e^x) = e^x$, or if $y = e^x$, $\dfrac{dy}{dx} = y$.

Also $\int e^x\,dx = e^x + c$.

Since x may be any real number, the *domain* of the exponential function is $\mathbb{R}$. However e^x is always positive, so the *range* of the exponential function is $\mathbb{R}^+$. (See Book 1, §2.8.)

Qu. 5 Letting 1 cm represent 0.1 on each axis, plot the graph of $y = e^x$, taking values of x at intervals of 0.05 from -0.5 to $+0.5$, and making use of tables or a calculator for e^x and e^{-x}. Obtain the gradient of the tangent to the curve at the point given by $x = 0.08$ (a) by drawing and measurement, (b) by measuring the ordinate, (c) by differentiation.

Example 1 *Find* $\dfrac{dy}{dx}$ *when* $y = e^{3x^2}$.

[Here we have a composite function of x. This example is written out in full as a reminder of the technique involved, but the reader should be able to differentiate in one step.]

$$y = e^{3x^2}$$

Let $u = 3x^2$, then $y = e^u$.

$$\therefore \frac{du}{dx} = 6x \quad \text{and} \quad \frac{dy}{du} = e^u$$

Now, by the chain rule, $\dfrac{dy}{dx} = \dfrac{dy}{du} \times \dfrac{du}{dx} = e^u \times 6x$.

$$\therefore \dfrac{dy}{dx} = 6x\, e^{3x^2}$$

Example 2   *Find* (a) $\int e^{x/2}\, dx$, (b) $\int x^2\, e^{x^3}\, dx$, (c) $\dfrac{d}{dx}(e^{3y})$.

(a) Since $\dfrac{d}{dx}(e^{x/2}) = \tfrac{1}{2}e^{x/2}$, then $\int e^{x/2}\, dx = 2e^{x/2} + c$.

(b) Since $\dfrac{d}{dx}(e^{x^3}) = 3x^2\, e^{x^3}$, then $\int x^2\, e^{x^3}\, dx = \tfrac{1}{3}e^{x^3} + c$.

(c) $\dfrac{d}{dx}(e^{3y}) = \dfrac{d}{dy}(e^{3y}) \times \dfrac{dy}{dx} = 3e^{3y} \times \dfrac{dy}{dx}$.

Qu. 6   Differentiate with respect to x:
(a) $(2x^3 + 1)^5$,   (b) $\sin (2x^3)$,   (c) e^{2x^3},   (d) e^{y^2},
(e) e^{-x^2},   (f) $e^{\tan x}$,   (g) e^{vx},   (h) $e^{\sin y}$.

Qu. 7   Find the following integrals, and check by differentiation:

(a) $\displaystyle\int \dfrac{x}{(x^2 + 1)^2}\, dx$,   (b) $\int x \sin (x^2)\, dx$,   (c) $\int x\, e^{x^2}\, dx$,

(d) $\int \sin x\, e^{\cos x}\, dx$,   (e) $\int 2e^{x/3}\, dx$,   (f) $\int 3e^{2x}\, dx$,

(g) $\int \tfrac{1}{2}x\, e^{3x^2}\, dx$,   (h) $\int \mathrm{cosec}^2\, 2x\, e^{\cot 2x}\, dx$.

Exercise 2a

1 Make rough sketches of the graphs of the following functions:
   (a) e^{2x},   (b) 2^{-x},   (c) $2^{1/x}$,   (d) 2^{x^2},   (e) $3e^x$,   (f) e^{x+1}.
2 With the same axes, sketch the graphs of $y = e^{\sin x}$ and $y = e^{\cos x}$; are these functions periodic?
   (a) Is the function $e^{\sin x}$ odd, even or neither?
   (b) Is the function $e^{\cos x}$ odd, even or neither?
3 Is the function e^{-x^2} odd, even or neither? If its domain is $\mathbb{R}$, what is its range? Sketch the graph of $y = e^{-x^2}$.
4 Use the following table of values to draw the graph of $y = 3^x$, taking 2 cm to represent 1 unit on each axis:

x	-2	$-\tfrac{7}{4}$	$-\tfrac{3}{2}$	$-\tfrac{5}{4}$	-1	$-\tfrac{3}{4}$	$-\tfrac{1}{2}$	$-\tfrac{1}{4}$	0
$y = 3^x$	0.11	0.15	0.19	0.25	0.33	0.44	0.58	0.76	1

x	$\tfrac{1}{4}$	$\tfrac{1}{2}$	$\tfrac{3}{4}$	1	$\tfrac{5}{4}$	$\tfrac{3}{2}$	$\tfrac{7}{4}$	2
$y = 3^x$	1.32	1.73	2.28	3	3.95	5.20	6.84	9

Draw the tangents to the curve at the points given by $x = -\frac{3}{2}, -1, -\frac{1}{2}, 0,$ $\frac{1}{2}, 1, \frac{3}{2}$, measure their gradients, and confirm graphically that $\dfrac{dy}{dx} = ky.$

Deduce an approximate expression for $\dfrac{d}{dx}(3^x).$

Differentiate with respect to x in Nos. 5–8.

5 (a) $4e^x$,   (b) e^{3x},   (c) e^{2x+1},   (d) e^{2x^2},
   (e) e^{-2x},   (f) e^{3y},   (g) e^{x^2+3},   (h) e^{x-2},
   (i) $e^{5/x}$,   (j) $e^{\sqrt[3]{x}}$,   (k) e^{ax^2+b},   (l) $e^{\sqrt{t}}$.

6 (a) $e^{\cos x}$,   (b) $e^{\sec x}$,   (c) $e^{3\tan y}$,   (d) $e^{\sin 2x}$,
   (e) $e^{-\cot x}$,   (f) $e^{\csc^2 x}$,   (g) $e^{\sqrt{\cos x}}$,   (h) $e^{a\sin bx}$,
   (i) $e^{\sin 3t}$,   (j) $e^{\tan x^2}$.

7 (a) $e^{\sqrt{(x^2+1)}}$,   (b) $e^{(1-x^2)^{-1}}$,   (c) $e^{\sin^2 4x}$,   (d) $e^{\tan(x^2+1)}$,

   (e) $e^{\sec^2 3x}$,   (f) $\dfrac{1}{e^{\csc x}}$,   (g) $\dfrac{1}{e^{x-2}}$,   (h) $e^{x\sin x}$,

   (i) e^{xy},   (j) e^{e^x}.

8 (a) $x^2 e^x$,   (b) $\dfrac{e^x}{x}$,   (c) $\dfrac{x}{2}e^{\sin x}$,   (d) $e^{x^2}\csc x$,

   (e) $\dfrac{e^x}{\sin x}$,   (f) $\dfrac{\cos x}{x\,e^x}$,   (g) e^{xe^x},   (h) $e^{ax}\sec bx$,

   (i) $\dfrac{e^{ax}}{\sin bx}$,   (j) $\tan^n e^x$,   (k) $e^x(\cos x + \sin x)$.

9 Find the following integrals:

   (a) $\int 3e^{x/2}\,dx$,   (b) $\int e^{-x}\,dx$,   (c) $\int e^{x/3}\,dx$,

   (d) $\int 2e^{3x-1}\,dx$,   (e) $\int \dfrac{x}{2}e^{x^2}\,dx$,   (f) $\int x^2 e^{-x^3}\,dx$.

   (g) $\int \sin x\, e^{\cos x}\,dx$,   (h) $\int (1+\tan^2 x)\,e^{\tan x}\,dx$,   (i) $\int \dfrac{e^{\cot x}}{\sin^2 x}\,dx$,

   (j) $\int x^{-2} e^{1/x}\,dx$.

10 Find the equation of the tangent to the curve $y = e^x$ at the point given by $x = a$. Deduce the equation of the tangent to the curve which passes through the point $(1, 0)$.

11 Find the volume of the solid generated by rotation about the x-axis of the area enclosed by $y = e^x$, the axes, $x = 1$.

12 Find $\dfrac{d}{dx}(x\,e^x)$, and deduce $\int x\,e^x\,dx$.

13 Investigate any maximum or minimum values of the function $x\,e^x$, and then sketch the graph of $y = x\,e^x$. Find the equation of the tangent to this curve at the point where $x = -2$.

In Nos. 14–17, A and B are constants; in each case show that the differential equation (see Book 1, §6.1) is satisfied by the given solution.

14 $\dfrac{d^2y}{dx^2} = 4y;$ $y = A\,e^{2x} + B\,e^{-2x}.$

15 $\dfrac{d^2s}{dt^2} + 4\dfrac{ds}{dt} = 0;$ $s = A + B\,e^{-4t}.$

16 $\dfrac{d^2y}{dx^2} + 4\dfrac{dy}{dx} + 3y = 0;$ $y = A\,e^{-x} + B\,e^{-3x}.$

17 $\dfrac{d^2y}{dt^2} - 6\dfrac{dy}{dt} + 9y = 0;$ $y = e^{3t}(A + Bt).$

18 If $f(x) = e^{4x}\cos 3x$, show that $f'(x) = 5e^{4x}\cos(3x + \alpha)$, where $\tan\alpha = \tfrac{3}{4}$. Deduce expressions of a similar form for $f''(x)$ and $f'''(x)$.

19 If $f(x) = e^{5x}\sin 12x$, show that $f'(x) = 13e^{5x}\sin(12x + \beta)$, where $\tan\beta = 12/5$. Write down an expression for $f''(x)$.

20 Show that the 9th derivative of $e^x\sin x$ is $16\sqrt{2}\,e^x\sin(x + \pi/4)$.

Further theory of logarithms

2.7   Since a logarithm is an index (or exponent), the discussion of exponential functions leads naturally to further consideration of logarithms. It is advisable at this stage to restate some of the ideas covered in Book 1, §§9.4, 9.5.

Definition

The logarithm of b to the base a, written $\log_a b$, *is the power to which the base must be raised to equal b.*

Thus, since $10^2 = 100,$ $2 = \log_{10} 100,$
and if $a^x = b,$ $x = \log_a b.$

The reader should already be familiar with the following basic rules:

$\log_c (ab) = \log_c a + \log_c b$
$\log_c (a/b) = \log_c a - \log_c b$
$\log_c (a^n) = n \log_c a$

Remember also that if $y = \log_a x$ then $x = a^y$, and that if we eliminate y from these two equations, we obtain

$x = a^{\log_a x}$

On the other hand, eliminating x gives

$y = \log_a (a^y)$

We shall now show that $\log_a b = \dfrac{\log_c b}{\log_c a}$.

Let $x = \log_a b$,

$\therefore a^x = b$

Taking logarithms to the base c of each side,

$\log_c (a^x) = \log_c b$

$\therefore x \log_c a = \log_c b$

$\therefore x = \dfrac{\log_c b}{\log_c a}$

i.e. $\log_a b = \dfrac{\log_c b}{\log_c a}$ †

Qu. 8 Express as a single logarithm:
(a) $2 \log_{10} a - \frac{1}{3} \log_{10} b + 2$,
(b) $\log_c (1 + x) - \log_c (1 - x) + A$, where $A = \log_c B$.

Qu. 9 Express in terms of $\log_c a$:

(a) $\log_c (2a)$,   (b) $\log_c a^2$,   (c) $\log_c \dfrac{1}{a}$,   (d) $\log_c \dfrac{2}{a}$,

(e) $\log_c \sqrt{a}$,   (f) $\log_c \dfrac{a}{2}$,   (g) $\log_c \dfrac{1}{a^2}$,   (h) $\log_c (2a)^{-1}$.

Qu. 10 Solve the equations:
(a) $3^{2x} = 27$,   (b) $1.2^x = 3$.

Qu. 11 (a) Prove that $\log_2 10 = \dfrac{1}{\log_{10} 2}$.

(b) Evaluate $\log_e 100$ correct to 3 significant figures, taking e as 2.718.

Natural logarithms

2.8 Logarithms to the base e are called *natural logarithms*, or *Napierian logarithms*, in honour of John Napier, a Scotsman, who published the first table of logarithmic sines in 1614. It is not surprising that the idea of logarithms was discovered independently at about the same time by Joost Bürgi, a Swiss; there was a pressing need to reduce labour involved in computation, especially in astronomy and navigation.

Napier's first publication on this topic fired the imagination of Henry Briggs, who visited him to discuss the practical application of the discovery; the fruit of this meeting was the eventual introduction of logarithms to the base 10, or

†The identity $\dfrac{b}{a} = \dfrac{b/c}{a/c}$ provides a mnemonic.

common logarithms, for computation. However, it should be remembered that neither Napier nor Bürgi put forward the concept of a *base*, with the logarithm as an index; this idea does not appear to have been fully developed until about the middle of the eighteenth century. The choice of e as the base, though less convenient than 10 for computation, provides a new function $\log_e x$ of fundamental importance.

Qu. 12 Evaluate $\log_e 2$ correct to three significant figures, taking e as 2.718, and using logarithms to the base 10. Check your answer with a table of natural logarithms, or a calculator.

Exercise 2b

1 Express as a single logarithm:
  (a) $2 \log_{10} a - 2 + \log_{10} 2a$,
  (b) $3 \log_e x + 3 - \log_e 3x$,
  (c) $4 \log_e (x - 3) - 3 \log_e (x - 2)$,
  (d) $\frac{1}{2} \log_e (1 + y) + \frac{1}{2} \log_e (1 - y) + \log_e k$.

2 Express in terms of $\log_e a$:
  (a) $\log_e 3a$,   (b) $\log_e a^3$,   (c) $\log_e (a/3)$,
  (d) $\log_e (1/a^3)$,   (e) $\log_e (3/a)$,   (f) $\log_e (\frac{1}{3} a^{-1})$,
  (g) $\log_e (\sqrt[3]{a})$.

3 Express as the sum or difference of logarithms:
  (a) $\log_e \cot x$,   (b) $\log_e \tan^2 x$,   (c) $\log_e (x^2 - 4)$,

  (d) $\log_e \sqrt{\left(\dfrac{x + 1}{x - 1} \right)}$,   (e) $\log_e (3 \sin^2 x)$.

4 Solve the equations:
  (a) $\frac{3}{2} \log_{10} a^3 - \log_{10} \sqrt{a} - 2 \log_{10} a = 4$,   (b) $\log_{10} y - 4 \log_y 10 = 0$.

5 Solve the equations:
  (a) $2^{2/x} = 32$,   (b) $3^{x+1} = 12$.

6 Evaluate $\log_e 3$ correct to three significant figures, taking e as 2.718, and using logarithms to the base 10.

Solve the equations in Nos. 7–10.

7 $3^{2(1+x)} - 28 \times 3^x + 3 = 0$.

8 $\log_{10} a + \log_a 100 = 3$.

9 $\log_{10} (19x^2 + 4) - 2 \log_{10} x - 2 = 0$.

10 $\log_{10} x + \log_{10} y = 1$, $x + y = 11$.

Notation

The notation $\log_e x$ *has the great virtue that it emphasises that the base of the logarithms is* e. *However, in recent years* $\ln x$ *has been universally adopted as the standard abbreviation (the* n *signifying that these are Napierian, or perhaps natural logarithms). From here onwards we shall use this notation exclusively.*

Qu. 13   Express as a single term:

(a) $\ln x + \ln y - 1$,   (b) $\ln\left(\dfrac{e^3}{x}\right) + \ln\left(\dfrac{x}{e}\right)$.

In the function $\ln x$, the independent variable x *must* be a positive real number (it must not be zero, nor must it be negative), in other words the *domain* of the logarithmic function is $\mathbb{R}^+$ (or some subset of $\mathbb{R}^+$). When $x > 1$, $\ln x$ is positive, it is zero when $x = 1$ and it is negative when $0 < x < 1$; its *range*, in other words, is $\mathbb{R}$. In this context, if the domain is not explicitly stated, it should always be assumed that it has been chosen so that only logarithms of positive numbers are required. For example, in the function $\ln(1 + x)$, it should be assumed that $x > -1$, or again, in the function $\ln(2 - x)$, it should be assumed that $x < 2$ is intended.

The reader is strongly advised to commit the following important identities to memory:

$$\ln(e^x) = x \quad \text{and} \quad e^{\ln x} = x$$

Qu. 14   Simplify:
(a) $e^{2\ln x}$,   (b) $e^{-\ln x}$,   (c) $e^{(1/2)\ln x}$,
(d) $\ln(e^{\sin x})$,   (e) $\tfrac{1}{2}\ln(e^{x^2})$,   (f) $\ln(\sqrt{e^x})$.

The derivative of ln *x*

2.9   Figure 2.4 shows the graph of $y = \ln x$ (or $x = e^y$); this curve is the reflection in the line $y = x$ of the graph of $y = e^x$ (which is shown as a dotted curve). This is to be expected because $\ln x$ and e^x are inverse functions (see Book 1, §2.16).

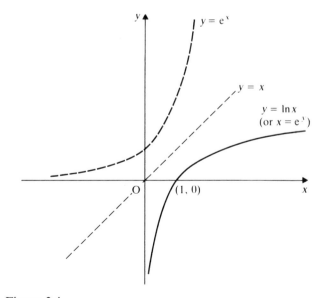

Figure 2.4

To find $\dfrac{d}{dx}$ (ln x), we write $y = \ln x$ as

$$x = e^y$$

Differentiating each side with respect to x,†

$$\frac{d}{dx}(x) = \frac{d}{dy}(e^y) \times \frac{dy}{dx}$$

$$\therefore 1 = e^y \frac{dy}{dx}$$

$$\therefore \frac{dy}{dx} = \frac{1}{e^y} = \frac{1}{x}$$

$$\therefore \frac{d}{dx}(\ln x) = \frac{1}{x}$$

Example 3   *Find $\dfrac{dy}{dx}$ if*

(a) $y = \ln 2x$,   (b) $y = \ln x^2$,   (c) $y = \ln (x^2 + 2)$,   (d) $\ln \dfrac{x}{\sqrt{(x^2 + 1)}}$.

(a) $y = \ln 2x = \ln 2 + \ln x$

$$\therefore \frac{dy}{dx} = \frac{1}{x}$$

(b) $y = \ln x^2 = 2 \ln x$

$$\therefore \frac{dy}{dx} = \frac{2}{x}$$

(c) $y = \ln (x^2 + 2)$

Let $u = x^2 + 2$, then $y = \ln u$.

$$\frac{du}{dx} = 2x \quad \text{and} \quad \frac{dy}{du} = \frac{1}{u}$$

$$\therefore \frac{dy}{dx} = \frac{dy}{du} \times \frac{du}{dx} = \frac{1}{u} \times 2x$$

$$\therefore \frac{dy}{dx} = \frac{2x}{x^2 + 2}$$

†Or we may differentiate each side with respect to y.

$$\frac{dx}{dy} = e^y = x, \qquad \therefore \frac{dy}{dx} = 1 \bigg/ \frac{dx}{dy} = \frac{1}{x}.$$

(d) $y = \ln \dfrac{x}{\sqrt{(x^2 + 1)}} = \ln x - \dfrac{1}{2} \ln (x^2 + 1)$

$\therefore \dfrac{dy}{dx} = \dfrac{1}{x} - \dfrac{1}{2} \times \dfrac{2x}{x^2 + 1} = \dfrac{x^2 + 1 - x^2}{x(x^2 + 1)}$

$\therefore \dfrac{dy}{dx} = \dfrac{1}{x(x^2 + 1)}$

Qu. 15 Differentiate with respect to x:
(a) $(x^2 - 2)^5$,   (b) $\operatorname{cosec} x^2$,   (c) e^{x^2},
(d) $\ln (x^2 - 2)$,   (e) $\ln \sin^2 x$,   (f) $\ln \sin x^2$.

Qu. 16 If $y = \ln \{x \sqrt{(x + 1)}\}$ find $\dfrac{dy}{dx}$ (a) by differentiating the logarithm of a product as it stands, (b) by first writing y as the sum of two logarithms.

Qu. 17 Differentiate with respect to x:
(a) $\ln (3x)$,   (b) $\ln (4x)$,   (c) $\ln (3x + 1)$,
(d) $\ln y$,   (e) $\ln (2x^3)$,   (f) $\ln (x^3 - 2)$,
(g) $\ln (x - 1)^3$,   (h) $\ln (4t)$,   (i) $\ln (3 \sin x)$,
(j) $\ln \cos 3x$,   (k) $\ln (2 \cos^3 x)$,   (l) $\ln (4 \sin^2 3x)$,

(m) $\ln \sqrt{(x^2 - 1)}$,   (n) $\ln \dfrac{x}{(x - 1)^2}$.

$\dfrac{d}{dx}(a^x)$ and $\displaystyle\int a^x \, dx$

2.10 In §2.4 we found that $\dfrac{d}{dx}(a^x) = ma^x$, m being the gradient of $y = a^x$ at $(0, 1)$.

When $a = c$, $m = 1$; we can now find m for other values of a.
   Let $y = a^x$, then

$\ln y = \ln a^x = x \ln a$

Differentiating with respect to x:

$\dfrac{d}{dy}(\ln y) \times \dfrac{dy}{dx} = \dfrac{d}{dx}(x \ln a)$

$\therefore \dfrac{1}{y} \times \dfrac{dy}{dx} = \ln a$

$\therefore \dfrac{dy}{dx} = y \ln a$

$\therefore \dfrac{d}{dx}(a^x) = a^x \ln a$ (1)

Qu. 18 Find $\dfrac{d}{dx}(4^x)$ reproducing the above method in full. Find the gradient of the curve $y = 4^x$ at $(2, 16)$.

Qu. 19 Find the gradient at $(0, 1)$ of the following curves, to four decimal places: (a) $y = 2^x$, (b) $y = 3^x$.

Qu. 20 Differentiate with respect to x: (a) 10^x, (b) 2^{3x+1}.

It follows from (1) on the previous page that

$$\int a^x \ln a \, dx = a^x + k$$

$$\therefore \int a^x \, dx = \frac{a^x}{\ln a} + c$$

Qu. 21 Find $\dfrac{d}{dx}(5^x)$ and deduce $\int 5^x \, dx$.

Qu. 22 Find $\dfrac{d}{dx}(2^{x^2})$ and deduce $\int x \, 2^{x^2} \, dx$.

Qu. 23 Find the following integrals:

(a) $\int 3^{2x} \, dx$,   (b) $\int x^2 \, e^{x^3} \, dx$,   (c) $\int 2^{\tan x} \sec^2 x \, dx$.

Exercise 2c

1 Differentiate with respect to x:

(a) $\ln (4x)$,   (b) $4 \ln x$,   (c) $\ln (2x - 3)$,

(d) $\ln (\tfrac{1}{3}y)$,   (e) $\ln \dfrac{x-1}{2}$,   (f) $\ln x^4$,

(g) $\ln (x^2 - 1)$,   (h) $\ln 3x^2$,   (i) $3 \ln x^2$,

(j) $\ln (x + 1)^2$,   (k) $\ln (2t^3)$,   (l) $\ln \dfrac{1}{x}$,

(m) $\ln (\tfrac{1}{2}x)$,   (n) $\ln \sqrt{x}$,   (o) $\ln \dfrac{1}{2x}$,

(p) $\ln \dfrac{2}{x}$,   (q) $\ln x^{-2}$,   (r) $\log_{10} x$,

(s) $\ln \dfrac{1}{t^3}$,   (t) $\ln \sqrt[3]{x}$.

2 Differentiate with respect to x:

(a) $\ln \cos x$,   (b) $\ln \sin^2 x$,   (c) $\ln \tan 3x$,

(d) $\ln \cos^3 2x$,   (e) $\ln (2 \cot^2 x)$,   (f) $\ln (3 \cos^2 2x)$,

(g) $\ln \tan \dfrac{x}{2}$,   (h) $\ln \sec x$,   (i) $\ln (\sec x + \tan x)$,

(j) $\ln \operatorname{cosec} x^2$,   (k) $\ln \dfrac{\sin x + \cos x}{\sin x - \cos x}$.

3 Find:

(a) $\dfrac{d}{dx} \ln \sqrt{\dfrac{1-x}{1+x}}$,   (b) $\dfrac{d}{dx} \ln \{x\sqrt{(x^2-1)}\}$,

(c) $\dfrac{d}{dx} \ln \dfrac{(x+1)^2}{\sqrt{(x-1)}}$,   (d) $\dfrac{d}{dx} \ln \{x + \sqrt{(x^2-1)}\}$.

4 Differentiate with respect to x:

(a) $\ln t$,   (b) $x \ln x$,   (c) $x^2 \ln x$,

(d) $\dfrac{\ln x}{x}$,   (e) $x \ln y$,   (f) $y \ln x$,

(g) $\dfrac{\ln x}{x^2}$,   (h) $\dfrac{x}{\ln x}$,   (i) $(\ln x)^2$,

(j) $\ln (\ln x^k)$,   (k) $\ln e^{\sin x}$.

5 Differentiate with respect to x:

(a) 5^x,   (b) 2^{x^2},   (c) 3^{2x-1},   (d) $e^{\ln x}$.

6 (a) Find $\dfrac{d}{dx}(3^x)$ and deduce $\int 3^x \, dx$.

(b) Find $\dfrac{d}{dx}(2^{x^2})$ and deduce $\int x \, 2^{x^2} \, dx$.

7 Find the following integrals:

(a) $\int 10^x \, dx$,   (b) $\int 2^{3x} \, dx$,   (c) $\int x \, 3^{x^2} \, dx$,   (d) $\int 2^{\cos x} \sin x \, dx$.

8 Find $\dfrac{d}{dx}(x \ln x)$ and deduce $\int \ln x \, dx$.

9 Find $\dfrac{d}{dx}(x \, 2^x)$ and deduce $\int x \, 2^x \, dx$.

10 Find (a) $\dfrac{d}{dx} \ln (x-2)$,   (b) $\dfrac{d}{dx} \ln (2-x)$.

Sketch on the same axes the graphs of $\ln (x-2)$, $\ln (2-x)$, $y = \dfrac{1}{x-2}$.

11 Sketch on the same axes the following curves:

(a) $y = \ln x$, $y = \ln (-x)$, $y = \dfrac{1}{x}$,

(b) $y = \ln \dfrac{1}{x}$, $y = \ln \left(-\dfrac{1}{x} \right)$, $y = -\dfrac{1}{x}$,

(c) $y = \ln(x - 3)$, $y = \ln(3 - x)$, $y = \dfrac{1}{x - 3}$,

(d) $y = \ln\left(\dfrac{1}{x - 3}\right)$, $y = \ln\left(\dfrac{1}{3 - x}\right)$, $y = \dfrac{1}{3 - x}$.

12 Given that $x \in \mathbb{R}^+$, write down the range of the following functions:
   (a) $y = \ln(1 + x^2)$,   (b) $y = \ln(1/x)$.
   Sketch the graph of each of these functions.

$$\int \frac{f'(x)}{f(x)}\,dx$$

2.11   The result $\displaystyle\int x^n\,dx = \frac{x^{n+1}}{n+1} + c$ holds for all rational values of n, except

$n = -1$; this hitherto puzzling gap may now be filled. Since we have established

that $\dfrac{d}{dx}(\ln x) = \dfrac{1}{x}$, it follows that

$$\int \frac{1}{x}\,dx = \ln x + c$$

or $\displaystyle\int \frac{1}{x}\,dx = \ln(kx)$ where $c = \ln k$

Example 4   *Find the following integrals:*

(a) $\displaystyle\int \frac{1}{2x}\,dx$,   (b) $\displaystyle\int \frac{1}{2x - 1}\,dx$.

(a) $\displaystyle\int \frac{1}{2x}\,dx = \frac{1}{2}\int \frac{1}{x}\,dx$

$\qquad\qquad = \tfrac{1}{2}\ln x + c$

$\qquad\qquad = \ln(k\sqrt{x})$ where $c = \ln k$

(b) $\displaystyle\int \frac{1}{2x - 1}\,dx$

This is best tackled in reverse by guessing the form of the integral.

$$\frac{d}{dx}\{\ln(2x - 1)\} = \frac{2}{2x - 1}$$

$$\therefore \int \frac{1}{2x - 1}\,dx = \tfrac{1}{2}\ln(2x - 1) + c$$

Qu. 24   Find the following integrals:

(a) $\displaystyle\int \frac{2}{x}\,dx$,   (b) $\displaystyle\int \frac{1}{3x}\,dx$,   (c) $\displaystyle\int \frac{1}{3x - 2}\,dx$,   (d) $\displaystyle\int \frac{1}{3x - 6}\,dx$.

Qu. 25   Find the following integrals:

(a) $\displaystyle\int \frac{1}{2x + 3}\,dx,$ using the substitution $u = 2x + 3,$

(b) $\displaystyle\int \frac{1}{1 - x}\,dx,$ using the substitution $u = 1 - x.$

Qu. 26   Evaluate $\displaystyle\int_1^2 \frac{3}{x}\,dx.$

Qu. 27   (a) Show that the answer to Example 4(b) may be written $\ln\{A(2x - 1)^{1/2}\}$, and express c in terms of A.
    (b) If $\frac{1}{2}\ln(x - \frac{1}{2}) + c$ may be written as $\ln\{k\sqrt{(2x - 1)}\}$, express c in terms of k.

An integral of the form $\displaystyle\int \frac{f'(x)}{f(x)}\,dx$ may be reduced to the form $\displaystyle\int \frac{1}{u}\,du$, by the substitution $u = f(x)$.

$$\int \frac{f'(x)}{f(x)}\,dx = \int \frac{f'(x)}{f(x)} \times \frac{dx}{du}\,du$$

$$= \int \frac{f'(x)}{u} \times \frac{1}{f'(x)}\,du$$

$$= \int \frac{1}{u}\,du$$

$$= \ln u + c$$

$$= \ln f(x) + c$$

Hence

$$\int \frac{f'(x)}{f(x)}\,dx = \ln\{k\,f(x)\}$$

From now onwards we must be prepared to recognise, in yet another form, the integrand involving a function of x and its derivative. As before, such an integral may be found by substitution, or usually it may be written down at once.

Example 5   *Find* $\displaystyle\int \frac{x}{x^2 + 1}\,dx.$

Since $\displaystyle\frac{d}{dx}\ln(x^2 + 1) = \frac{2x}{x^2 + 1},$

$$\int \frac{x}{x^2 + 1}\,dx = \tfrac{1}{2}\ln(x^2 + 1) + c$$

$$= \ln\{k\sqrt{(x^2 + 1)}\}   \text{where } K = \ln c$$

Qu. 28   Find the following integrals:

(a) $\displaystyle\int \frac{x^2}{(x^3 - 2)^2}\,dx,$ (b) $\int x^2 \cos x^3\,dx,$ (c) $\int x^2\, e^{x^3}\,dx,$

(d) $\displaystyle\int \frac{x^2}{x^3 - 2}\,dx,$ (e) $\displaystyle\int \frac{x - 1}{x^2 - 2x}\,dx,$ (f) $\displaystyle\int \frac{2x}{3 - x^2}\,dx,$

(g) $\int \cot x\,dx.$

Qu. 29   Find $\displaystyle\int \frac{x}{x - 1}\,dx$

(a) using the substitution $u = x - 1$,
(b) by first dividing the numerator by the denominator.

$$\int_a^b \frac{1}{x}\,dx \quad \textbf{when } \boldsymbol{a, b} \textbf{ are negative}$$

†**2.12**   An important point must be cleared up. Reference to Fig. 2.4 reminds us that as the value of x goes from 0 to $+\infty$, the value of $\ln x$ goes from $-\infty$ to $+\infty$; $\ln x$ *is not defined for negative values of* x. This presents us with an apparent paradox which may be demonstrated in graphical terms as follows.

Referring to the graph of $y = 1/x$ in Fig. 2.5 it is apparent that the two shaded areas are equal in magnitude and of opposite sign. However, we soon get into trouble if we seek to evaluate the appropriate integral with the negative limits; thus, is it true to say that

$$\int_{-2}^{-1} \frac{1}{x}\,dx = \left[\ln x\right]_{-2}^{-1}$$

$$= \text{`}\ln(-1) - \ln(-2)\text{'}?$$

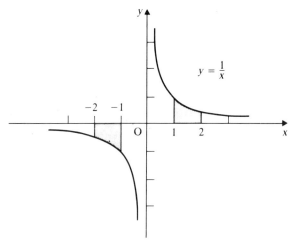

Figure 2.5

†§2.12 should be delayed until after the reader has answered No. 1 of Exercise 2d.

We could now write this as $\ln\left(\dfrac{-1}{-2}\right) = \ln\frac{1}{2} = \ln 2^{-1} = -\ln 2$, and thus obtain a correct figure for the area, but the working is not valid, since the expression '$\ln(-1) - \ln(-2)$' is meaningless.

We surmount this difficulty as soon as we realise that *for negative values of x,* although $\ln x$ is not defined, $\ln(-x)$ *does exist,* and

$$\frac{d}{dx}\ln(-x) = \frac{-1}{-x} = \frac{1}{x} \qquad \text{(See Fig. 2.6)}$$

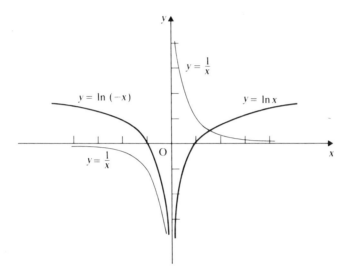

Figure 2.6

Thus, if a and b are negative, $\displaystyle\int_a^b \frac{1}{x}\,dx = \left[\ln(-x)\right]_a^b.$

(Using the modulus sign, we could write this as $\displaystyle\int_a^b \frac{1}{x}\,dx = \left[\ln|x|\right]_a^b.$ This form of the result could be used for a and b both positive *and* for a and b both negative; notice however that a and b must not have opposite signs.)

Hence the left-hand shaded area in Fig. 2.5 $= \displaystyle\int_{-2}^{-1} \frac{1}{x}\,dx$

$$= \left[\ln(-x)\right]_{-2}^{-1}$$

$$= \ln 1 - \ln 2$$

$$= -\ln 2$$

Qu. 30   Evaluate (a) $\displaystyle\int_{-4}^{-3} \frac{1}{x}\,dx$,   (b) $\displaystyle\int_{-1}^{-1/2} \frac{1}{x}\,dx$.

Qu. 31   Evaluate $\displaystyle\int_{-4}^{-2} \frac{1}{x}\, dx$, using the change of variable $x = -u$.

Qu. 32   Can any meaning be assigned to $\displaystyle\int_{-2}^{+2} \frac{1}{x}\, dx$?

Example 6   *Find the area enclosed by the curve* $y = \dfrac{1}{x-2}$ *and*

(a) *the lines* $x = 4$, $x = 5$, *and the x-axis*,
(b) *the line* $x = 1$, *and the axes.*   (Fig. 2.7.)

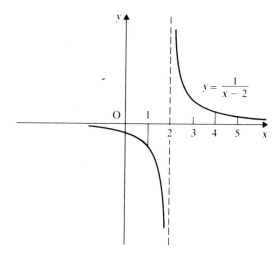

Figure 2.7

(a) The required area $= \displaystyle\int_{4}^{5} \frac{1}{x-2}\, dx$

$$= \left[\ln(x-2)\right]_{4}^{5}$$

$$= \ln 3 - \ln 2$$

$$= \ln \tfrac{3}{2}$$

(b) $\Big($ If we proceed as in (a) but with the new limits, we obtain the meaningless

'$\left[\ln(x-2)\right]_{0}^{1} = \ln(-1) - \ln(-2)$.' We must note that when $x > 2$,

$$\int \frac{1}{x-2}\, dx = \ln(x-2) + c$$

but when $x < 2$,

$$\int \frac{1}{x-2} \, dx = \ln(2-x) + c$$

i.e. when $x > 2$ or $x < 2$, $\left(\int \frac{1}{x-2} \, dx = \ln|x-2| + c \right)$

The required area $= \int_0^1 \frac{1}{x-2} \, dx$

$$= \left[\ln|x-2| \right]_0^1$$

$$= \ln 1 - \ln 2$$

$$= -\ln 2$$

Qu. 33   Find $\dfrac{d}{dx}\{\ln(x-3)\}$ and $\dfrac{d}{dx}\{\ln(3-x)\}$.

Qu. 34   Sketch the curve $y = \dfrac{1}{x-3}$, and evaluate:

(a) $\displaystyle\int_5^6 \frac{1}{x-3} \, dx,$ (b) $\displaystyle\int_{-2}^2 \frac{1}{x-3} \, dx.$

Qu. 35   Sketch the curve $y = \dfrac{1}{2-x}$ and evaluate $\displaystyle\int_3^5 \frac{1}{2-x} \, dx.$

Exercise 2d

1 Find the following integrals:

(a) $\displaystyle\int \frac{1}{4x} \, dx,$

(b) $\displaystyle\int \frac{5}{x} \, dx,$

(c) $\displaystyle\int \frac{1}{2x-3} \, dx,$

(d) $\displaystyle\int \frac{1}{2x+8} \, dx,$

(e) $\displaystyle\int \frac{1}{3-2x} \, dx,$

(f) $\displaystyle\int \frac{x}{1-x^2} \, dx,$

(g) $\displaystyle\int \frac{3x}{x^2-1} \, dx,$

(h) $\displaystyle\int \frac{2x+1}{x^2+x-2} \, dx,$

(i) $\displaystyle\int \frac{2x-3}{3x^2-9x+4} \, dx,$

(j) $\displaystyle\int \frac{x}{x+2} \, dx,$

(k) $\displaystyle\int \frac{3x}{2x+3} \, dx,$

(l) $\displaystyle\int \frac{2x}{3-x} \, dx,$

(m) $\displaystyle\int \frac{x-1}{2-x} \, dx,$

(n) $\displaystyle\int \frac{3-2x}{x-4} \, dx,$

(o) $\displaystyle\int \tan x \, dx,$

(p) $\displaystyle\int \cot \frac{x}{2} \, dx,$

(q) $\displaystyle\int \cot(2x+1) \, dx,$

(r) $\displaystyle\int -\tan \frac{x}{3} \, dx,$

(s) $\displaystyle\int \frac{1-\sin 2x}{x-\sin^2 x}\,dx,$ (t) $\displaystyle\int \frac{1-\tan x}{1+\tan x}\,dx,$ (u) $\displaystyle\int \frac{2+\tan^2 x}{x+\tan x}\,dx.$

2 (a) Sketch the curves $y = \ln(2x-1)$, and $y = \ln(1-2x)$.

(b) Find $\dfrac{d}{dx}\ln(2x-1)$ and $\dfrac{d}{dx}\ln(1-2x)$.

(c) Evaluate $\displaystyle\int_1^2 \frac{1}{2x-1}\,dx$ and $\displaystyle\int_{-2}^0 \frac{1}{2x-1}\,dx.$

3 Sketch the curve $y = \dfrac{1}{x-4}$ and evaluate:

(a) $\displaystyle\int_1^2 \frac{1}{x-4}\,dx,$ (b) $\displaystyle\int_5^6 \frac{1}{x-4}\,dx.$

4 (a) Find $\dfrac{d}{dx}\ln\left(\dfrac{1}{3-x}\right)$ and $\dfrac{d}{dx}\ln\left(\dfrac{1}{x-3}\right).$

(b) Sketch on the same axes the graphs of $y = -\ln(3-x)$, $y = -\ln(x-3)$, $y = 1/(3-x)$, and find the area enclosed by the latter, the lines $x = 5$, $x = 6$, and the x-axis.

(c) Find the area under $y = 1/(3-x)$ from $x = 0$ to $x = 1$.

5 Evaluate the following:

(a) $\displaystyle\int_2^8 \frac{1}{2x}\,dx,$ (b) $\displaystyle\int_1^{4/3} \frac{1}{3x-2}\,dx,$ (c) $\displaystyle\int_1^3 \frac{1}{x-5}\,dx,$

(d) $\displaystyle\int_3^5 \frac{1}{1-2x}\,dx,$ (e) $\displaystyle\int_{-0.25}^{0.25} \frac{1}{2x+1}\,dx,$ (f) $\displaystyle\int_{-\sqrt{2}}^0 \frac{x}{x^2+2}\,dx,$

(g) $\displaystyle\int_0^3 \frac{2x-1}{x^2-x+1}\,dx,$ (h) $\displaystyle\int_4^6 \frac{x}{x-2}\,dx,$ (i) $\displaystyle\int_{-7}^{-5} \frac{x+1}{x+3}\,dx,$

(j) $\displaystyle\int_{-0.5}^0 \frac{2-x}{x-1}\,dx,$ (k) $\displaystyle\int_{\pi/3}^{\pi/2} \cot\theta\,d\theta,$ (l) $\displaystyle\int_0^{\pi/6} \tan 2x\,dx,$

(m) $\displaystyle\int_{\pi/6}^{\pi/4} \frac{\sec^2\theta}{\tan\theta}\,d\theta.$

Exercise 2e (Miscellaneous)

(For integration see Exercises 13e and 13f.)

1 Prove that $\log_a b = 1/(\log_b a)$, where $a,b \in \mathbb{R}^+$. Solve the equation

$$\log_2 x + \log_x 2 = 2.5$$

2 Solve the equation $x^{1.23} = 0.12.$ (L)

3 Solve the equation $\log_a(x^2+3) - \log_a x = 2\log_a 2.$ (O & C)

4 Solve the equations
(a) $6^{x+2} = 2(3^{2x})$,   (b) $\log_x 0.028 = 3$,   (c) $\log_{10} x = \log_5 (2x)$.   (L)

5 If $\log_{10} y = 2 - \log_{10} (x^{2/3})$, express y as a function of x not involving logarithms; hence show that, if $x = 8$, then $y = 25$.   (O & C)

6 Prove that $\log_a x = \dfrac{\log_b x}{\log_b a}$.

Given that $\log_{10} 3 = 0.4771$ and that $\log_{10} e = 0.4343$, calculate the value of ln 0.3, giving your answer correct to three significant figures.

7 Given that $2 \log_y x + 2 \log_x y = 5$, show that $\log_y x$ is either $\frac{1}{2}$ or 2. Hence find all pairs of values of x and y which satisfy simultaneously the equation above and the equation $xy = 27$.   (JMB)

8 (a) The functions f_1 and f_2, each with domain $D = \{x: x \in \mathbb{R}, x > -1\}$, are defined by

$$f_1(x) = \ln (x + 1) \qquad f_2(x) = x^2 + 1$$

For each function state the range. Show that an inverse function f_1^{-1} exists and, using the same axes, sketch the graphs of $y = f_1(x)$ and $y = f_1^{-1}(x)$. Show that an inverse function f_2^{-1} does not exist and suggest an interval such that f_2, restricted to this interval, will have an inverse function.

(b) Functions g_1 and g_2, each with domain $\mathbb{R}$, are defined by

$$g_1(x) = \ln (1 + x^2) \qquad g_2(x) = 1 + x$$

Given that $g_1 g_2$ and $g_2 g_1$ are the composite functions defined on $\mathbb{R}$, find expressions for $g_1 g_2(x)$ and $g_2 g_1(x)$ and state whether each of these functions is odd, even or neither.   (L)

9 Using the same scale and axes, draw the graphs of $y = 2^x$ and $y = 2^{-x}$, between $x = -3$ and $x = +3$. Use your graphs to estimate:
(a) the root of the equation $2^x - 2^{-x} = 3$,
(b) the value of $\sqrt[4]{32} - \sqrt[4]{\frac{1}{32}}$.
Give both results correct to two significant figures.   (L)

10 (a) Two quantities x and y are connected by the equation $y = a\,e^{bx}$, where a and b are constants. If $y = 1$ when $x = 1$, and $y = 3$ when $x = 4$, find the values of a and b, correct to 3 decimal places.
(b) Solve the equation $2^{2x} - 2^{x+2} - 5 = 0$.   (L)

11 A function y is of the form $y = ax^n + bx$ where a, b, and n are constants. When x is equal successively to 1, 3, and 9, the corresponding values of y are 4, 6, and 15. Find two relations between a and b not involving n, and hence find the numerical values of a, b, and n (the last to three places of decimals).   (O & C)

12 Prove that $\log_2 e - \log_4 e + \log_8 e - \log_{16} e + ... = 1$, where e is the base of natural logarithms. (See p. 28.)   (JMB)

13 If $y = A\,e^{-x} \cos (x + \alpha)$, where A and α are constants, prove that

(a) $\dfrac{d^2 y}{dx^2} + 2\dfrac{dy}{dx} + 2y = 0$,   (b) $\dfrac{d^4 y}{dx^4} + 4y = 0$.   (O & C)

14 A particle is moving in a straight line. The displacement x, from an origin O on the line, is given at time t by the equation

$$x = e^{-(3/4)t}(a \sin t + b \cos t)$$

Initially $t = 0$, $x = 4$, $\dfrac{dx}{dt} = 0$. Find the constants a and b. Determine also

(a) the time elapsing from the start before the particle first reaches O,

(b) the time taken from O to attain the greatest displacement on the negative side of the origin.   (JMB)

15 Find the maximum and minimum values of the function $(1 + 2x^2)\,e^{-x^2}$.

(O & C)

16 If $y = e^{-x} \cos x$, determine the three values of x between 0 and 3π for which $\dfrac{dy}{dx} = 0$. Show that the corresponding values of y form a geometric progression with common ratio $-e^{-\pi}$.   (JMB)

17 (a) Find $\dfrac{dy}{dx}$ if $y = \ln\left(\dfrac{3 + 4 \cos x}{4 + 3 \cos x}\right)$.

(b) If $y = e^{4x} \cos 3x$, prove that $\dfrac{d^2 y}{dx^2} - 8\dfrac{dy}{dx} + 25y = 0$.   (O & C)

18 Functions f and g are defined as follows:

$$\text{f}: x \mapsto e^{-x} \quad (x \in \mathbb{R}^+) \qquad \text{g}: x \mapsto \frac{1}{1 - x} \quad (x \in \mathbb{R}, x < 1)$$

Give the ranges of f, g and g∘f.

Give definitions of the inverse functions f^{-1}, g^{-1} and $(\text{g}\circ\text{f})^{-1}$ in a form similar to the above definitions.   (C)

19 Find the real value of x satisfying the equation

$$e^x - e^{-x} = 4$$

Show that for this value of x, $e^x + e^{-x} = 2\sqrt{5}$.   (L)

20 Find the maximum and minimum values of $x^2\,e^{-x}$, and sketch the graph of this function. Find the equation of the tangent to the graph at the point at which $x = 1$.   (L)

21 Show, by means of a sketch graph, or otherwise, that the equation

$$e^{2x} + 4x - 5 = 0$$

has only one real root, and that this root lies between 0 and 1.

Starting with the value 0.5 as a first approximation to this root, use the Newton-Raphson† method to evaluate successive approximations, showing the stages of your work and ending when two successive approximations give answers which, when rounded to two decimal places, agree.   (C)

†Readers who need to revise the Newton-Raphson method will find it in Book 1, Chapter 24.

22 The function f with domain $\mathbb{R}$ is defined by $f(x) = e^{\cos x}$.
   (a) Prove that f is periodic and state the period.
   (b) Determine whether f is an odd function, an even function or a function which is neither odd nor even.
   (c) Sketch the graph of $y = e^{\cos x}$ for $-2\pi \leqslant x \leqslant 2\pi$.
   (d) State, with reason, whether the inverse function f^{-1} exists.
   (e) If f^{-1} does not exist, determine a subset D of $\mathbb{R}$ which is the domain of a function g, with $g(x) = f(x)$, $x \in D$, and g^{-1} exists.   (L)

Chapter 3

Partial fractions

Introduction

3.1   Early training in algebra teaches us how to 'simplify' an expression such as $\dfrac{1}{x-1} - \dfrac{1}{x+1}$ by reducing it to $\dfrac{2}{x^2-1}$ (see Book 1, Appendix).

We have now reached the stage when the reverse process is of value. Given a fraction such as $\dfrac{5}{x^2+x-6}$ whose denominator factorises, we may split it up into its component fractions, writing it as $\dfrac{1}{x-2} - \dfrac{1}{x+3}$; it is now said to be in **partial fractions**. Just one example of the several applications of this must suffice for the present. No change of variable yet discussed would enable us to find $\displaystyle\int \frac{5}{(x-2)(x+3)}\,dx$ as it stands, but using partial fractions,

$$\int \frac{5}{(x-2)(x+3)}\,dx = \int \left\{ \frac{1}{x-2} - \frac{1}{x+3} \right\}\,dx$$

$$= \ln(x-2) - \ln(x+3) + c$$

$$= \ln \left\{ \frac{k(x-2)}{x+3} \right\}$$

Qu. 1   Express each of the following as a single fraction:

(a) $\dfrac{1}{1-x} + \dfrac{2}{1+x}$,   (b) $\dfrac{2x-1}{x^2+1} - \dfrac{1}{x+1}$,   (c) $\dfrac{3}{(x-1)^2} + \dfrac{1}{x-1} + \dfrac{2}{x+1}$.

Qu. 2   Express in partial fractions:

(a) $\dfrac{4}{(x-2)(x+2)}$,   (b) $\dfrac{1}{1-x^2}$,   (c) $\dfrac{1}{2\times 3}$,   (d) $\dfrac{1}{n(n+1)}$.

Unfortunately most partial fractions cannot be obtained by trial and error quite as easily as those in Qu. 2. The reader need only consider attempting Qu. 1 in reverse, to be convinced that we need some technique to find partial fractions;

44

we shall find that this involves us in handling algebraic identities, so we must discuss these briefly.

Identities

3.2 Let us first distinguish clearly between an *equation* and an *identity*. $x^2 = 4$ is an *equation*, which is satisfied only by the two values $x = \pm 2$. But

$$x^2 - 4 \equiv (x + 2)(x - 2)$$

and

$$x^2 + 2x - 2 \equiv (x + 1)(x - 1) + 2(x + 1) - 3$$

are both **identities**, and for them the L.H.S. = R.H.S. *for any value of x*; moreover, if the R.H.S. is multiplied out, the coefficients of x^2, x and the constant term will be identical on each side.

Example 1 *Find the values of the constants A, B, C such that*

$$5x + 3 \equiv Ax(x + 3) + Bx(x - 1) + C(x - 1)(x + 3)$$

First method

Collecting like terms on the R.H.S.,

$$5x + 3 \equiv (A + B + C)x^2 + (3A - B + 2C)x - 3C$$

Equating coefficients of x^2,

$$0 = A + B + C \tag{1}$$

Equating coefficients of x,

$$5 = 3A - B + 2C \tag{2}$$

Equating constant terms,

$$3 = -3C \tag{3}$$

From (3), $C = -1$, and substituting this value into (1) and (2), and solving these equations simultaneously, we obtain $A = 2$, and $B = -1$.

Second method

$$5x + 3 \equiv Ax(x + 3) + Bx(x - 1)\ \ + C(x - 1)(x + 3)$$

Putting $x = 0$,

$$3 = \quad 0 \quad + \quad 0 \quad - \quad 3C\dagger$$
$$\therefore C = -1$$

Putting $x = -3$,

$$-15 + 3 = \quad 0 \quad + B(-3)(-4) + \quad 0$$
$$\therefore -12 = \quad 12B$$
$$\therefore B = -1$$

---

†This should be compared with equation (3) above.

Putting $x = 1$,

$$5 + 3 = A \times 1 \times 4 + \quad 0 \quad + \quad 0$$
$$\therefore A = 2$$

It should be noted that the identity holds for *any* value of x, but we have chosen those particular values which make all but one term on the R.H.S. vanish each time.

Qu. 3 $2x^2 + 9x - 10 \equiv A(x - 3)(x + 4) + B(x + 2)(x + 4) + C(x + 2)(x - 3)$.
(a) Obtain three equations in A, B, C by substituting $x = -1$, 0, 1 in this identity.
(b) Find the values of A, B, C by substituting more convenient values of x.

Qu. 4   Find the values of the constants A, B, C in the following identities:
(a) $22 - 4x - 2x^2 \equiv A(x - 1)^2 + B(x - 1)(x + 3) + C(x + 3)$,   using   the   first method in Example 1,
(b) $5x + 31 \equiv A(x + 2)(x - 1) + B(x - 1)(x - 5) + C(x - 5)(x + 2)$,   using   the second method in Example 1,
(c) $13x - 11 \equiv A(3x - 2) + B(2x + 1)$.

Qu. 5   Put $x = 1$ to find the value of A in the identity

$$x^2 + x + 7 = A(x^2 + 2) + (Bx + C)(x - 1)$$

Now substitute any other values of x to find B and C.

The substitution method is fast, but often it may be combined with the method of equating coefficients for greater speed and simplicity (e.g. having found A in Qu. 5, equate coefficients of x^2 to find B). The latter method also gives us a deeper insight into the nature of identities. Let us consider the statement

'$x^2 - 5x + 8 \equiv A(x + 3) + B(x - 1)^2$'

Applying the method of substitution we obtain $A = 1$, $B = 2$; however, when A and B are given these values we do *not* have an identity!

This apparently alarming breakdown is readily explained when we apply the method of equating coefficients. This shows that for only *two* unknowns we have the *three* equations, $B = 1$, $A - 2B = -5$, and $3A + B = 8$, which are not consistent. Thus we cannot find values for A and B to form the 'identity' given above.

Since we shall soon be concerned with forming identities, the method of equating coefficients will be a valuable check that the number of unknown constants introduced corresponds to the number of equations to be satisfied.

Qu. 6   Can values of A, B, C be found which make the following pairs of expressions identical?
(a) $2x + 3$ and $A(x + 1)(x - 2) + B(x + 1)^2 + C$,
(b) $x^2 - 8x + 30$ and $A(x - 3)^2 + B(x + 2)$.

Exercise 3a

1 Express each of the following as a single fraction:

(a) $\dfrac{3}{x+3} - \dfrac{2}{x-2}$,   (b) $\dfrac{1}{(x+2)^2} - \dfrac{2}{x+2} + \dfrac{1}{3x-1}$,

(c) $\dfrac{4}{2+3x^2} - \dfrac{1}{1-x}$,   (d) $\dfrac{3}{x^2+1} - \dfrac{1}{x-1} + \dfrac{2}{(x-1)^2}$.

2 Express in partial fractions:

(a) $\dfrac{2x}{(3+x)(3-x)}$,   (b) $\dfrac{a}{a^2-b^2}$,   (c) $\dfrac{1}{5\times 6}$,   (d) $\dfrac{1}{p(1-p)}$.

3 Use the first method of Example 1 to find the values of the constants A, B, C in the following identities:

(a) $31x - 8 \equiv A(x-5) + B(4x+1)$,

(b) $8 - x \equiv A(x-2)^2 + B(x-2)(x+1) + C(x+1)$,

(c) $71 + 9x - 2x^2 \equiv A(x+5)(x+2) + B(x+2)(x-3) + C(x-3)(x+5)$,

(d) $2x^3 - 15x^2 - 10$
$$\equiv A(x-2)(x+1) + B(x+1)(2x^2+1) + C(2x^2+1)(x-2).$$

4 Use the second method of Example 1 to find the values of the constants A, B, C in the following identities:

(a) $2x \quad 4 \equiv A(3+x) + B(7-x)$,

(b) $8x + 1 \equiv A(3x-1) + B(2x+3)$,

(c) $4x^2 + 4x - 26 \equiv A(x+2)(x-4) + B(x-4)(x-1) + C(x-1)(x+2)$,

(d) $17x^2 - 13x - 16$
$$\equiv A(3x+1)(x-1) + B(x-1)(2x-3) + C(2x-3)(3x+1).$$

5 Can values of A, B, C, D be found which make the following pairs of expressions identical?

(a) $2x^2 - 22x + 53$
$$\text{and}\quad A(x-5)(x-3) + B(x-3)(x+2) + C(x+2)(x-5),$$

(b) $x + 7$ and $A(x-2) + B(x+1)^2$,

(c) $3x^2 + 7x + 11$ and $(Ax+B)(x+2) + C(x^2+5)$,

(d) $x + 1$ and $A(x-2) + B(x^2+1)$,

(e) $x^3 + 2x^2 - 4x - 2$ and $(Ax+B)(x-2)(x+1) + C(x+1) + D(x-2)$.

6 Find the values of A, B, C if $x^3 - 1$ is expressed in the form

$$(x-1)(Ax^2 + Bx + C)$$

Factorise: (a) $x^3 + 1$,   (b) $x^3 - 8$,   (c) $x^3 + 27$,
(d) $8x^3 - 27$,   (e) $27x^3 + 125$.

7 Express $x^3 + 1$ in the form $x(x-1)(x-2) + Ax(x-1) + Bx + C$.

8 Find the values of a and b if $x^4 + 12x^3 + 46x^2 + ax + b$ is the square of a quadratic expression.

9 Write down the quadratic equation whose roots are α, β. If the same equation may also be written $ax^2 + bx + c = 0$, express $\alpha + \beta$ and $\alpha\beta$ in terms of a, b, c.

10 If α, β, γ are the roots of the equation $px^3 + qx^2 + rx + s = 0$, deduce expressions for $\alpha + \beta + \gamma$, $\beta\gamma + \gamma\alpha + \alpha\beta$, $\alpha\beta\gamma$ in terms of p, q, r, s.

Type I — denominator with only linear factors

3.3 We shall find that in the more straightforward cases which we have to deal with at this stage, partial fractions fall into three main types; each will be illustrated by a worked example, and the reader is strongly advised to work through the questions following each of these, before going on to consider the next type.

In practice, a question of considerable length and complexity may depend upon the correct determination of partial fractions in the early stages; to avoid fruitless labour at a later date, the habit of *checking* partial fractions should be firmly established from the start, and they should be thrown back into one fraction mentally, the numerator obtained being checked with the original.

First we deal with a fraction whose denominator consists of only linear factors.

Example 2 *Express* $\dfrac{11x + 12}{(2x + 3)(x + 2)(x - 3)}$ *in partial fractions.*

Let $\dfrac{11x + 12}{(2x + 3)(x + 2)(x - 3)} \equiv \dfrac{A}{2x + 3} + \dfrac{B}{x + 2} + \dfrac{C}{x - 3}$, where A, B, C are constants to be found. It follows that

$$\frac{11x + 12}{(2x + 3)(x + 2)(x - 3)} \equiv \frac{A(x + 2)(x - 3) + B(x - 3)(2x + 3) + C(2x + 3)(x + 2)}{(2x + 3)(x + 2)(x - 3)}$$

$$\therefore\ 11x + 12 \equiv A(x + 2)(x - 3) + B(x - 3)(2x + 3) + C(2x + 3)(x + 2)$$

Putting $x = 3$,

$$33 + 12 = \quad 0 \quad + \quad 0 \quad + C \times 9 \times 5$$
$$\therefore\ C = 1$$

Putting $x = -2$,

$$-22 + 12 = \quad 0 \quad + B \times (-5) \times (-1) + \quad 0$$
$$\therefore\ -10 = 5B$$
$$\therefore\ B = -2$$

Putting $x = -\frac{3}{2}$,

$$-\tfrac{33}{2} + 12 = A \times \tfrac{1}{2} \times \tfrac{-9}{2} \quad + \quad 0 \quad + \quad 0$$
$$\therefore\ \tfrac{9}{2} = -\tfrac{9}{4} A$$
$$\therefore\ A = 2$$

$$\therefore\ \frac{11x + 12}{(2x + 3)(x + 2)(x - 3)} \equiv \frac{2}{2x + 3} - \frac{2}{x + 2} + \frac{1}{x - 3}$$

$$\left[\text{Since the R.H.S.} \equiv \frac{2(x+2)(x-3) - 2(x-3)(2x+3) + (2x+3)(x+2)}{(2x+3)(x+2)(x-3)}, \text{ we}\right.$$

check the coefficients in the numerator.$\Big]$

Check: Coefficient of $x^2 = 2 - 4 + 2 = 0$.
   Coefficient of $x = -2 + 6 + 7 = 11$.
   Constant term $= -12 + 18 + 6 = 12$.

Qu. 7 Express in partial fractions:

(a) $\dfrac{6}{(x+3)(x-3)}$,   (b) $\dfrac{x}{(2+x)(2-x)}$,   (c) $\dfrac{x-1}{3x^2 - 11x + 10}$,

(d) $\dfrac{3x+1}{(x+2)(x+1)(x-3)}$,   (e) $\dfrac{3-4x}{2+3x-2x^2}$.

Type II — denominator with a quadratic factor

3.4 Fractions which can be split solely into partial fractions are necessarily *proper*, by which is meant that *the degree of the numerator is less than the degree of the denominator*.† Moreover, the partial fractions themselves are always proper.

Bearing this in mind we can now discover how to deal with a fraction having in the denominator a quadratic factor which does not factorise.

Let $\dfrac{3x+1}{(x-1)(x^2+1)} \equiv \dfrac{A}{x-1} + \dfrac{\text{`numerator'}}{x^2+1}$

Then $3x + 1 \equiv A(x^2 + 1) + \text{`numerator'} \times (x-1)$

From our previous work on identities, we see, by equating coefficients that there are *three* equations to be satisfied. It follows that there are *three* constants to determine,‡ and therefore the *numerator* must contain two of them; thus the only way to write the second partial fraction, so that it is proper, is in the form $\dfrac{Bx+C}{x^2+1}$.

---

†With an improper fraction, we divide first, and we obtain a quotient and partial fraction, thus

$$\frac{x^2 + x + 1}{(x-1)(x+2)} \equiv 1 + \frac{3}{(x-1)(x+2)} \equiv 1 + \frac{1}{x-1} - \frac{1}{x+2},$$

and

$$\frac{x^3 + 2x^2 - 7x - 18}{x^2 - 9} \equiv x + 2 + \frac{2x}{(x+3)(x-3)} \equiv x + 2 + \frac{1}{x+3} + \frac{1}{x-3}.$$

‡In general, the number of constants to be found is the same as the degree of the denominator of the original fraction.

Example 3   *Express* $\dfrac{3x+1}{(x-1)(x^2+1)}$ *in partial fractions.*

Let $\dfrac{3x+1}{(x-1)(x^2+1)} \equiv \dfrac{A}{x-1} + \dfrac{Bx+C}{x^2+1}$

$$\therefore \ 3x+1 \equiv A(x^2+1) + (Bx+C)(x-1).$$

Putting $x = 1$, $4 = 2A + 0$, $\therefore A = 2$.

Putting $x = 0$, $1 = A - C$, $\therefore 1 = 2 - C$, $\therefore C = 1$.

Equating coefficients of x^2, $0 = A + B$, $\therefore B = -2$.

$$\therefore \ \frac{3x+1}{(x-1)(x^2+1)} \equiv \frac{2}{x-1} + \frac{1-2x}{x^2+1} \equiv \frac{2}{x-1} - \frac{2x-1}{x^2+1}$$

Check: Coefficient of $x^2 = 2 - 2 = 0$.
Coefficient of $x = -(-2-1) = +3$.
Constant term $= 2 - 1 = +1$.

Qu. 8   Express in partial fractions:

(a) $\dfrac{6-x}{(1-x)(4+x^2)}$,   (b) $\dfrac{4}{(x+1)(2x^2+x+3)}$,

(c) $\dfrac{5x+2}{(x+1)(x^2-4)}$,   (d) $\dfrac{3+2x}{(2-x)(3+x^2)}$.

Type III — denominator with a repeated factor

3.5   Here we take as an example $\dfrac{1}{(x+2)(x-1)^2}$. Written as $\dfrac{1}{(x+2)(x^2-2x+1)}$ this suggests Type II and the partial fractions $\dfrac{A}{x+2} + \dfrac{Bx+K}{x^2-2x+1}$. Certainly we have the correct number of constants to be found to identify this expression with the original fraction; however, the denominator of the second partial fraction factorises, and so we have not gone far enough.

$$\frac{Bx+K}{(x-1)^2} \equiv \frac{B(x-1) + B + K}{(x-1)^2}$$

$$\equiv \frac{B}{x-1} + \frac{B+K}{(x-1)^2}$$

Writing C for $B + K$, we obtain

$$\frac{Bx+K}{(x-1)^2} \equiv \frac{B}{x-1} + \frac{C}{(x-1)^2}$$

This indicates the appropriate form when we have a repeated factor. (See also Qu. 10.)

Example 4   *Express* $\dfrac{1}{(x+2)(x-1)^2}$ *in partial fractions.*

Let $\dfrac{1}{(x+2)(x-1)^2} \equiv \dfrac{A}{x+2} + \dfrac{B}{x-1} + \dfrac{C}{(x-1)^2}$

$\therefore \dfrac{1}{(x+2)(x-1)^2} \equiv \dfrac{A(x-1)^2 + B(x-1)(x+2) + C(x+2)}{(x+2)(x-1)^2}$

$\therefore 1 \equiv A(x-1)^2 + B(x-1)(x+2) + C(x+2)$

Putting $x = -2$, $1 = 9A$, $\therefore A = \frac{1}{9}$.

Putting $x = 1$, $1 = 3C$, $\therefore C = \frac{1}{3}$.

Equating coefficients of x^2, $0 = A + B$, $\therefore B = -\frac{1}{9}$.

$\therefore \dfrac{1}{(x+2)(x-1)^2} \equiv \dfrac{1}{9(x+2)} - \dfrac{1}{9(x-1)} + \dfrac{1}{3(x-1)^2}$

Check: Expressing the R.H.S. as a single fraction with denominator $(x+2)(x-1)^2$, the numerator is $\frac{1}{9}\{(x-1)^2 - (x+2)(x-1) + 3(x+2)\}$.

Coefficient of $x^2 = \frac{1}{9}(1-1) = 0$.
Coefficient of $x = \frac{1}{9}(-2-1+3) = 0$.
Constant term $= \frac{1}{9}(1+2+6) = 1$.

Qu. 9   Express in partial fractions:

(a) $\dfrac{x+1}{(x+3)^2}$,   (b) $\dfrac{2x^2 - 5x + 7}{(x-2)(x-1)^2}$.

Qu. 10   Find the values of A, B, C, D, if

$$\dfrac{x^3 - 10x^2 + 26x + 3}{(x+3)(x-1)^3} \equiv \dfrac{A}{x+3} + \dfrac{B}{x-1} + \dfrac{C}{(x-1)^2} + \dfrac{D}{(x-1)^3}$$

Improper fractions

3.6   As already implied, an *improper* fraction is one whose *numerator is of degree equal to, or greater than, that of the denominator*. To deal with this we first divide the numerator to obtain a quotient and a proper fraction, and then split the latter into partial fractions. Thus

$$\dfrac{x^4 - 2x^3 - x^2 - 4x + 4}{(x-3)(x^2+1)} \equiv x + 1 + \dfrac{x^2 - 2x + 7}{(x-3)(x^2+1)}, \quad \text{etc.}$$

Often, instead of doing long division, it is quicker to proceed as follows:

$$\dfrac{2x^2 + 1}{(x-1)(x+2)} \equiv \dfrac{2(x^2 + x - 2) - 2x + 5}{x^2 + x - 2} \equiv 2 + \dfrac{5 - 2x}{(x-1)(x+2)}, \quad \text{etc.}$$

Qu. 11   Express the following in the form of a quotient and a proper fraction:

(a) $\dfrac{x^3 + 2x^2 - 2x + 2}{(x - 1)(x + 3)}$ (by long division),

(b) $\dfrac{3x^2 - 2x - 7}{(x - 2)(x + 1)}$ (by the short method suggested above).

Qu. 12   Express in partial fractions:

(a) $\dfrac{x^2 - 7}{(x - 2)(x + 1)}$,   (b) $\dfrac{x^3 - x^2 - 4x + 1}{x^2 - 4}$.

Exercise 3b

Express in partial fractions:

1 (a) $\dfrac{x - 11}{(x + 3)(x - 4)}$,

(b) $\dfrac{x}{25 - x^2}$,

(c) $\dfrac{3x^2 - 21x + 24}{(x + 1)(x - 2)(x - 3)}$,

(d) $\dfrac{4x^2 + x + 1}{x(x^2 - 1)}$,

(e) $\dfrac{8x^2 + 13x + 6}{(x + 2)(2x + 1)(3x + 2)}$,

(f) $\dfrac{2x^3 + x^2 - 15x - 5}{(x + 3)(x - 2)}$.

2 (a) $\dfrac{5x^2 - 10x + 11}{(x - 3)(x^2 + 4)}$,

(b) $\dfrac{2x^2 - x + 3}{(x + 1)(x^2 + 2)}$,

(c) $\dfrac{3x^2 - 2x + 5}{(x - 1)(x^2 + 5)}$,

(d) $\dfrac{11x}{(2x - 3)(2x^2 + 1)}$,

(e) $\dfrac{20x + 84}{(x + 5)(x^2 - 9)}$,

(f) $\dfrac{2x^3 - x - 1}{(x - 3)(x^2 + 1)}$.

3 (a) $\dfrac{x - 5}{(x - 2)^2}$,

(b) $\dfrac{5x + 4}{(x - 1)(x + 2)^2}$,

(c) $\dfrac{5x^2 + 2}{(3x + 1)(x + 1)^2}$,

(d) $\dfrac{x^4 + 3x - 1}{(x + 2)(x - 1)^2}$.

4 (a) $\dfrac{3x^3 + x + 1}{(x - 2)(x + 1)^3}$ (see Qu. 10),

(b) $\dfrac{3x^2 + 2x - 9}{(x^2 - 1)^2}$.

5 (a) $\dfrac{x^3 + 2x^2 - 10x - 9}{x^2 - 9}$,

(b) $\dfrac{3(x^2 - 3)}{(x - 1)(x + 2)}$,

(c) $\dfrac{2x^4 - 4x^3 - 42}{(x - 2)(x^2 + 3)}$,

(d) $\dfrac{x^4 - 6x^2 + 3}{x(x + 1)^2}$.

6 $\dfrac{3x + 7}{x(x + 2)(x - 1)}$.

7 $\dfrac{3}{x^2(x + 2)}$.

8 $\dfrac{2x^4 - 17x - 1}{(x - 2)(x^2 + 5)}$.

9 $\dfrac{68 + 11x}{(3 + x)(16 - x^2)}.$

10 $\dfrac{2x + 1}{x^3 - 1}.$

11 $\dfrac{2x^2 + 39x + 12}{(2x + 1)^2(x - 3)}.$

12 $\dfrac{x + 4}{6x^2 - x - 35}.$

13 $\dfrac{x - 2}{x^2(x - 1)^2}.$

14 $\dfrac{7x + 2}{125x^3 - 8}.$

15 $\dfrac{x^2 + 2x + 18}{x(x^2 + 3)^2}.$

16 $\dfrac{1}{x^4 + 5x^2 + 6}.$

17 $\dfrac{1}{x^4 - 9}.$

Summation of series

3.7 An introduction to the summation of series was given in Book 1, Chapter 13. There are some series which may be summed by the use of partial fractions; the method of application is illustrated in the following example.

Example 5 (a) *Express* $\dfrac{2}{n(n + 1)(n + 2)}$ *in partial fractions, and* (b) *deduce that*

$$\frac{1}{1 \times 2 \times 3} + \frac{1}{2 \times 3 \times 4} + \dots + \frac{1}{n(n + 1)(n + 2)} = \frac{1}{4} - \frac{1}{2(n + 1)(n + 2)}$$

(a) Let $\dfrac{2}{n(n + 1)(n + 2)} \equiv \dfrac{A}{n} + \dfrac{B}{n + 1} + \dfrac{C}{n + 2}$

$$\therefore 2 \equiv A(n + 1)(n + 2) + B(n + 2)n + Cn(n + 1)$$

Putting $n = 0, \quad 2 = 2A, \therefore A = 1.$

Putting $n = -1, \quad 2 = -B, \therefore B = -2.$

Putting $n = -2, \quad 2 = 2C, \therefore C = 1.$

$$\therefore \frac{2}{n(n + 1)(n + 2)} \equiv \frac{1}{n} - \frac{2}{n + 1} + \frac{1}{n + 2}$$

Check: Coefficient of $n^2 = 1 - 2 + 1 = 0.$
Coefficient of $n = 3 - 4 + 1 = 0.$
Constant term $= 2.$

(b) If $S \equiv \dfrac{1}{1 \times 2 \times 3} + \dfrac{1}{2 \times 3 \times 4} + \dots + \dfrac{1}{n(n + 1)(n + 2)}$

$$2S \equiv \frac{2}{1 \times 2 \times 3} + \frac{2}{2 \times 3 \times 4} + \dots + \frac{2}{n(n + 1)(n + 2)}$$

From Part (a) it follows that

$$2S \equiv \left(\frac{1}{1} - \frac{2}{2} + \frac{1}{3}\right) + \left(\frac{1}{2} - \frac{2}{3} + \frac{1}{4}\right) + \left(\frac{1}{3} - \frac{2}{4} + \frac{1}{5}\right) + \dots + \left(\frac{1}{n} - \frac{2}{n + 1} + \frac{1}{n + 2}\right)$$

We see that the majority of terms when grouped three together in a different way, such as $\frac{1}{3} - \frac{2}{3} + \frac{1}{3}$, have zero sum. We then have to pick out those terms which remain at the beginning and at the end, and this is most easily done if we set out the working in columns.]

From Part (a)

$$\frac{2}{1 \times 2 \times 3} = \frac{1}{1} - \frac{2}{2} + \frac{1}{3}$$

$$\frac{2}{2 \times 3 \times 4} = \frac{1}{2} - \frac{2}{3} + \frac{1}{4}$$

$$\frac{2}{3 \times 4 \times 5} = \frac{1}{3} - \frac{2}{4} + \frac{1}{5}$$

. . .   .

$$\frac{2}{(n-2)(n-1)n} = \frac{1}{n-2} - \frac{2}{n-1} + \frac{1}{n}$$

$$\frac{2}{(n-1)n(n+1)} = \frac{1}{n-1} - \frac{2}{n} + \frac{1}{n+1}$$

$$\frac{2}{n(n+1)(n+2)} = \frac{1}{n} - \frac{2}{n+1} + \frac{1}{n+2}$$

Adding,

$$2S = \frac{1}{2} - \frac{1}{n+1} + \frac{1}{n+2}$$

$$= \frac{1}{2} - \frac{1}{(n+1)(n+2)}$$

$$\therefore S = \frac{1}{4} - \frac{1}{2(n+1)(n+2)}$$

The reader should also note that as $n \to \infty$, $\dfrac{1}{2(n+1)(n+2)} \to 0$; thus the infinite series $\dfrac{1}{1 \times 2 \times 3} + \dfrac{1}{2 \times 3 \times 4} + \dfrac{1}{3 \times 4 \times 5} + \dots$ is *convergent*, and its *sum to infinity* is $\frac{1}{4}$ (see Book 1, §14.4).

Qu. 13   Show that $\dfrac{1}{1 \times 2} + \dfrac{1}{2 \times 3} + \dfrac{1}{3 \times 4} + \dots + \dfrac{1}{n(n+1)} = 1 - \dfrac{1}{n+1}$.

Integration

3.8   We have already shown in §3.1 how partial fractions may be applied to integration. Two more examples follow.

Example 6   *Find* $\displaystyle\int \frac{2x-1}{(x+1)^2}\,dx.$

Let $\displaystyle\frac{2x-1}{(x+1)^2} \equiv \frac{A}{x+1} + \frac{B}{(x+1)^2}$; we find that $A = 2$, $B = -3$.

$$\therefore \int \frac{2x-1}{(x+1)^2}\,dx = \int \left\{ \frac{2}{x+1} - \frac{3}{(x+1)^2} \right\} dx$$

$$= 2 \ln (x+1) + 3(x+1)^{-1} + c$$

Qu. 14   Find (a) $\displaystyle\int \frac{1}{x^2-9}\,dx,$ (b) $\displaystyle\int \frac{2x+2}{(2x-3)^2}\,dx.$

Qu. 15   (a) Find $\displaystyle\int \frac{x}{4-x^2}\,dx$ without using partial fractions.

(b) Find this integral using partial fractions.

Example 7   *Evaluate* $\displaystyle\int_2^3 \frac{5+x}{(1-x)(5+x^2)}\,dx$ *correct to three significant figures.*

Let $\displaystyle\frac{5+x}{(1-x)(5+x^2)} \equiv \frac{A}{1-x} + \frac{Bx+C}{5+x^2}$; we find that $A = 1$, $B = 1$, $C = 0$.

$$\therefore \int_2^3 \frac{5+x}{(1-x)(5+x^2)}\,dx = \int_2^3 \left\{ \frac{1}{1-x} + \frac{x}{5+x^2} \right\} dx$$

$$= \left[-\ln (x-1) + \tfrac{1}{2} \ln (5+x^2) \right]_2^3$$

$$= (-\ln 2 + \tfrac{1}{2} \ln 14) - (-\ln 1 + \tfrac{1}{2} \ln 9)$$
$$= \tfrac{1}{2} \ln 14 - \ln 2 - \ln 3$$
$$(= \tfrac{1}{2} (\ln 10 + \ln 1.4) - \ln 6)\dagger$$
$$= -0.472 \quad \text{(correct to three significant figures)}$$

Example 7 also revises an important point.   If $x < 1$,

$$\int \frac{1}{1-x}\,dx = -\ln (1-x) + c$$

However, if $x > 1$, as the limits show to be the case here,

$$\int \frac{1}{1-x}\,dx = -\ln (x-1) + c \qquad \text{(See §2.12.)}$$

Qu. 16   Can $\displaystyle\int_0^2 \frac{5+x}{(1-x)(5+x^2)}\,dx$ be evaluated?

---

†This step is only necessary when using four-figure tables.

Qu. 17   Evaluate (a) $\displaystyle\int_2^3 \frac{x-4}{(x+2)(x-1)}\,dx$,   (b) $\displaystyle\int_1^2 \frac{3x^2+2x+2}{(x+1)(x^2+2)}\,dx$.

Exercise 3c

1 Express $\dfrac{2}{n(n+2)}$ in partial fractions, and deduce that

$$\frac{1}{1\times3}+\frac{1}{2\times4}+\frac{1}{3\times5}+\dots+\frac{1}{n(n+2)}=\frac{3}{4}-\frac{2n+3}{2(n+1)(n+2)}$$

2 Express $\dfrac{n+3}{(n-1)n(n+1)}$ in partial fractions, and deduce that

$$\frac{5}{1\times2\times3}+\frac{6}{2\times3\times4}+\frac{7}{3\times4\times5}+\dots+\frac{n+3}{(n-1)n(n+1)}=1\tfrac{1}{2}-\frac{n+2}{n(n+1)}$$

3 For the series given in No. 2 write down (a) the nth term, (b) the sum of the first n terms, (c) the limit of this sum as $n\to\infty$.

4 Prove that the series $\dfrac{2}{1\times2}+\dfrac{2}{2\times3}+\dfrac{2}{3\times4}+\dots$ is convergent, and find its sum to infinity.

5 Find the sum of the first n terms of the following series:

(a) $\dfrac{1}{1\times4}+\dfrac{1}{2\times5}+\dfrac{1}{3\times6}+\dots,$

(b) $\dfrac{1}{2\times4}+\dfrac{1}{4\times6}+\dfrac{1}{6\times8}+\dots,$

(c) $\dfrac{1}{3\times6}+\dfrac{1}{6\times9}+\dfrac{1}{9\times12}+\dots,$

(d) $\dfrac{1}{2\times6}+\dfrac{1}{4\times8}+\dfrac{1}{6\times10}+\dots,$

(e) $\dfrac{1}{1\times3\times5}+\dfrac{1}{2\times4\times6}+\dfrac{1}{3\times5\times7}+\dots,$

(f) $\dfrac{1}{3\times4\times5}+\dfrac{2}{4\times5\times6}+\dfrac{3}{5\times6\times7}+\dots.$

6 Find the sum of the first n terms of the following series, remembering that $2n-1$, $2n+1$, etc. are odd for all integral values of n:

(a) $\dfrac{2}{1\times3}+\dfrac{2}{3\times5}+\dfrac{2}{5\times7}+\dots,$

(b) $\dfrac{1}{1\times3\times5}+\dfrac{1}{3\times5\times7}+\dfrac{1}{5\times7\times9}+\dots,$

(c) $\dfrac{2}{1 \times 3 \times 5} + \dfrac{3}{3 \times 5 \times 7} + \dfrac{4}{5 \times 7 \times 9} + \ldots$

7 Find the following integrals:

(a) $\displaystyle\int \frac{1}{x(x-2)}\,dx,$

(b) $\displaystyle\int \frac{1}{(x+3)(5x-2)}\,dx,$

(c) $\displaystyle\int \frac{7x+2}{3x^3+x^2}\,dx,$

(d) $\displaystyle\int \frac{x}{16-x^2}\,dx,$

(e) $\displaystyle\int \frac{1}{x^2-4x-5}\,dx,$

(f) $\displaystyle\int \frac{x-2}{x^2-4x-5}\,dx,$

(g) $\displaystyle\int \frac{2x^2+2x+3}{(x+2)(x^2+3)}\,dx,$

(h) $\displaystyle\int \frac{22-16x}{(3+x)(2-x)(4-x)}\,dx,$

(i) $\displaystyle\int \frac{4x-33}{(2x+1)(x^2-9)}\,dx,$

(j) $\displaystyle\int \frac{5x+2}{(x-2)^2(x+1)}\,dx,$

(k) $\displaystyle\int \frac{x^2-8x+5}{(2x+1)(x^2+9)}\,dx,$

(l) $\displaystyle\int \frac{6-9x}{27x^3+8}\,dx,$

(m) $\displaystyle\int \frac{x^3-18x-21}{(x+2)(x-5)}\,dx,$

(n) $\displaystyle\int \frac{37}{4(x-3)(1+4x^2)}\,dx.$

8 (It is intended that all the parts of this question should be answered at one sitting, in order to bring out the comparison between the forms.) Find the following integrals:

(a) $\displaystyle\int \frac{1}{1+x^2}\,dx,$

(b) $\displaystyle\int \frac{x}{1+x^2}\,dx,$

(c) $\displaystyle\int \frac{1+x}{1+x^2}\,dx,$

(d) $\displaystyle\int \frac{1}{1-x^2}\,dx,$

(e) $\displaystyle\int \frac{x}{1-x^2}\,dx,$

(f) $\displaystyle\int \frac{x}{\sqrt{(1-x^2)}}\,dx,$

(g) $\displaystyle\int \frac{1}{\sqrt{(1-x^2)}}\,dx,$

(h) $\displaystyle\int \frac{1+x}{\sqrt{(1-x^2)}}\,dx,$

(i) $\displaystyle\int \frac{1}{1-x}\,dx,\quad (x<1),$

(j) $\displaystyle\int \frac{1}{1-x}\,dx,\quad (x>1),$

(k) $\displaystyle\int \frac{x}{1+x}\,dx,$

(l) $\displaystyle\int \frac{1}{(1-x)^2}\,dx,$

(m) $\displaystyle\int \frac{x}{(1-x)^2}\,dx.$

9 Evaluate the following, correct to three significant figures:

(a) $\displaystyle\int_3^5 \frac{2}{x^2-1}\,dx,$

(b) $\displaystyle\int_{-1}^0 \frac{2}{(1-x)(1+x^2)}\,dx,$

(c) $\displaystyle\int_2^3 \frac{x-9}{x(x-1)(x+3)}\,dx,$

(d) $\displaystyle\int_0^3 \frac{13x+7}{(x-4)(3x^2+2x+3)}\,dx.$

10 Find the volume of the solid generated when the area under $y = \dfrac{1}{x-2}$ from $x = 3$ to $x = 4$ is rotated through four right angles about the x-axis. If the solid is made of material of uniform density, where is its centre of gravity?

Chapter 4

The binomial theorem

The expansion of $(1 + x)^n$ when n is not a positive integer

4.1 The binomial theorem that

$$(1 + x)^n = 1 + nx + \frac{n(n-1)}{2!}x^2 + \frac{n(n-1)(n-2)}{3!}x^3 + \dots$$

provided $-1 < x < +1$, was used in Chapter 14 of Book 1, but the general term in the expansion was not discussed for values of n other than positive integers. The term in x^r is found to be

$$\frac{n(n-1)\dots(n-r+1)}{r!}x^r$$

The proof of this, for values of n other than positive integers, is outside the scope of this book. In the case when n is a positive integer, the term in x^r has been shown (Book 1, §14.3) to be

$$^nC_r x^r = \frac{n!}{(n-r)!r!}x^r$$

Dividing numerator and denominator by $(n-r)!$, we obtain

$$\frac{n(n-1)\dots(n-r+1)}{r!}x^r$$

Note that $n(n-1)\dots(n-r+1)$ contains r factors. If the reader can remember that this expression is $n!/(n-r)!$ (when n is a positive integer), it may help him or her to remember that the last factor, $n-r+1$, is 1 greater than $n-r$.

Using the notation

$$\binom{n}{r} = \frac{n(n-1)(n-2)\dots(n-r+1)}{r!}$$

the expansion can be written

$$(1 + x)^n = 1 + \binom{n}{1}x + \binom{n}{2}x^2 + \binom{n}{3}x^3 + \dots$$

Example 1   *Find the general terms in the expansions in ascending powers of x of*
(a) $(1 + x)^{-1}$, (b) $(1 - 2x)^{-3}$.

(a) $(1 + x)^{-1}$. The general term is

$$(-1)(-2) \ldots (-r)x^r/r!$$
$$= (-1)^r 1 \times 2 \ldots rx^r/r!$$
$$= (-1)^r x^r$$

(b) $(1 - 2x)^{-3}$. The general term is

$$(-3)(-4) \ldots (-2 - r) \times (-2x)^r/r!$$
$$= (-1)^r 3 \times 4 \ldots (r + 2) \times (-1)^r 2^r x^r/r!$$
$$= \frac{r!(r + 1)(r + 2)}{2} \times \frac{2^r x^r}{r!}$$
$$= (r + 1)(r + 2)2^{r-1} x^r$$

The expansion obtained in the first part of the last example,

$$(1 + x)^{-1} = 1 - x + x^2 - \ldots + (-1)^r x^r + \ldots$$

is frequently required and should be memorised. Note that the right-hand side is
an infinite geometrical progression with first term 1 and common ratio $-x$,
therefore its sum to infinity is $1/(1 + x)$, provided $-1 < x < +1$ (see Book 1,
§13.9).

The approximation for $(1 - x)^{-1}$ obtained by taking the first three terms of
the binomial expansion, i.e.

$$(1 - x)^{-1} \approx 1 + x + x^2$$

is quite good provided x is small. For instance, when $x = 0.2$,

$$\text{L.H.S.} = 0.8^{-1} = 1.25$$

and $\text{R.H.S.} = 1 + 0.2 + 0.04 = 1.24$

Fig. 4.1 shows the graph of $y = (1 - x)^{-1}$ (continuous curve) and the graph of
$y = 1 + x + x^2$ (broken curve).

We can see from this diagram that the graphs are close together (showing
that the approximation is quite good) for $|x| < 0.5$; they begin to diverge when
$0.5 < |x| < 1$, but when $|x| > 1$, the curves are totally unrelated. The approximation between -1 and $+1$ could be improved by taking an extra term of the
binomial expansion, i.e.

$$(1 - x)^{-1} \approx 1 + x + x^2 + x^3$$

With $x = 0.2$, the R.H.S. is equal to 1.248, which is clearly nearer to the exact
value than is the value we obtained from the previous approximation. The
reader is advised to plot the graphs of $y = (1 - x)^{-1}$ and $y = 1 + x + x^2 + x^3$.

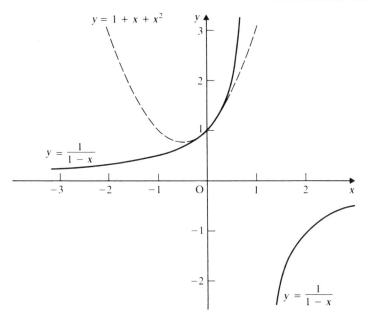

Figure 4.1

These graphs should be quite close together for $|x| < 1$. (Readers who are fortunate enough to have access to a microcomputer with High Resolution Graphics should also try plotting some of the graphs obtained by including further terms of the binomial expansion). However it should be noted that, although the approximation for $|x| < 1$ can be improved by taking more terms of the binomial expansion, outside this interval the binomial expansion is totally useless.

Qu. 1 Write down and simplify the general terms in the expansions of
(a) $(1 + x)^{-2}$,   (b) $(1 - 3x)^{-1}$,   (c) $(1 - \frac{1}{2}x)^{-3}$,   (d) $(1 + x)^{-4}$.

It is worth noting that the coefficients in the expansions of $(1 - x)^{-1}$, $(1 - x)^{-2}, (1 - x)^{-3}, \ldots$ are contained in Pascal's triangle (Book 1, §14.1).

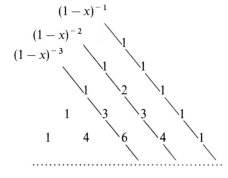

Example 2   *Find the first three terms and the general term in the expansion in ascending powers of x of*

$$\frac{x+5}{(1+3x)(2-x)}$$

Expressed in partial fractions,

$$\frac{x+5}{(1+3x)(2-x)} = \frac{2}{1+3x} + \frac{1}{2-x}$$

(The reader should verify that this is so.)

$$2(1+3x)^{-1} = 2\{1-3x+9x^2 - \ldots +(-1)^r(3x)^r + \ldots\}$$
$$= 2-6x+18x^2 - \ldots +(-1)^r \times 2(3x)^r + \ldots$$
$$(2-x)^{-1} = 2^{-1}(1-\tfrac{1}{2}x)^{-1}$$
$$= \tfrac{1}{2}\{1+\tfrac{1}{2}x+\tfrac{1}{4}x^2 + \ldots +(\tfrac{1}{2}x)^r + \ldots\}$$
$$= \tfrac{1}{2}+\tfrac{1}{4}x+\tfrac{1}{8}x^2 + \ldots +(\tfrac{1}{2})^{r+1}x^r + \ldots$$

Therefore the sum of the two expansions is

$$2\tfrac{1}{2} - 5\tfrac{3}{4}x + 18\tfrac{1}{8}x^2 + \ldots + \{(-1)^r \times 2 \times 3^r + (\tfrac{1}{2})^{r+1}\}x^r + \ldots$$

For the expansion to be valid,

$$-1 < 3x < +1 \quad \text{and} \quad -1 < -\tfrac{1}{2}x < 1$$

Multiplying the pairs of inequalities by $\tfrac{1}{3}$ and -2 respectively,

$$-\tfrac{1}{3} < x < +\tfrac{1}{3} \quad \text{and} \quad 2 > x > -2\dagger$$

Therefore the expansion is valid when $-\tfrac{1}{3} < x < +\tfrac{1}{3}$.

Example 3   *Find the first three terms and the term in x^r in the expansion in ascending powers of x of $(x+2)(1+x)^{12}$.*

$$(1+x)^{12} = 1 + 12x + \frac{12 \times 11}{2!}x^2 + \ldots + {}^{12}C_{r-1}x^{r-1} + {}^{12}C_r x^r + \ldots + x^{12}$$

$$\therefore 2(1+x)^{12} = 2 + 24x + 132x^2 + \ldots + 2x^{12}$$

and

$$x(1+x)^{12} = \qquad x + 12x^2 + 66x^3 + \ldots + x^{13}$$

Adding

$$(x+2)(1+x)^{12} = 2 + 25x + 144x^2 + \ldots + x^{13}$$

To find the term in x^r, we must multiply

the term in x^r in the expansion of $(1+x)^{12}$ by 2 and
the term in x^{r-1} in the expansion of $(1+x)^{12}$ by x.

---

†When an inequality is multiplied by a negative number, the direction of the inequality sign is reversed. (See §6.1.)

Thus the term in x^r is

$$2 \times {}^{12}C_r x^r + x \times {}^{12}C_{r-1} x^{r-1}$$

$$= \left\{ \frac{2 \times 12!}{(12-r)!r!} + \frac{12!}{(13-r)!(r-1)!} \right\} x^r$$

$$= \frac{12!}{(12-r)!(r-1)!} \left(\frac{2}{r} + \frac{1}{13-r} \right) x^r$$

$$= \frac{12!(26-r)}{(13-r)!r!} x^r$$

Qu. 2 Find the terms in x^r in the expansions in ascending powers of x of
(a) $(1-x)(1+x)^{20}$, (b) $(2x+3)(1-x)^{10}$.
Qu. 3 Find the general term in the expansion of $(2x-1)/(1+x)^2$ (a) by the method of Example 3, (b) by expressing the function in partial fractions.

Example 4 *Sum to infinity the series*

(a) $1 - \dfrac{3}{2}x + \dfrac{3 \times 9}{2 \times 4}x^2 - \dfrac{3 \times 9 \times 15}{2 \times 4 \times 6}x^3 + \ldots,$

(b) $\dfrac{1}{4} - \dfrac{1 \times 2}{4 \times 8} + \dfrac{1 \times 2 \times 5}{4 \times 8 \times 12} - \dfrac{1 \times 2 \times 5 \times 8}{4 \times 8 \times 12 \times 16} + \ldots.$

(a) Let $S_1 = 1 - \dfrac{3}{2}x + \dfrac{3 \times 9}{2 \times 4}x^2 - \dfrac{3 \times 9 \times 15}{2 \times 4 \times 6}x^3 + \ldots$

Note that there is a factor of 3^r in the numerator and a factor of 2^r in the denominator of the term in x^r.

$$\therefore S_1 = 1 - \frac{1}{2}(3x) + \frac{1 \times 3}{2^2} \frac{(3x)^2}{2!} - \frac{1 \times 3 \times 5}{2^3} \frac{(3x)^3}{3!} + \ldots$$

$$= 1 + \left(-\frac{1}{2} \right)(3x) + \left(-\frac{1}{2} \right)\left(-\frac{3}{2} \right)\frac{(3x)^2}{2!} + \left(-\frac{1}{2} \right)\left(-\frac{3}{2} \right)\left(-\frac{5}{2} \right)\frac{(3x)^3}{3!} + \ldots$$

$$\therefore S_1 = (1+3x)^{-1/2}, \quad \text{provided } |x| < \tfrac{1}{3}$$

(b) Let $S_2 = \dfrac{1}{4} - \dfrac{1 \times 2}{4 \times 8} + \dfrac{1 \times 2 \times 5}{4 \times 8 \times 12} - \dfrac{1 \times 2 \times 5 \times 8}{4 \times 8 \times 12 \times 16} + \ldots$

Here the denominators are $4^r \times r!$

$$\therefore S_2 = \frac{1}{4} - \frac{1 \times 2}{4^2} \times \frac{1}{2!} + \frac{1 \times 2 \times 5}{4^3} \times \frac{1}{3!} - \frac{1 \times 2 \times 5 \times 8}{4^4} \times \frac{1}{4!} + \ldots$$

By altering some signs, we can arrange that the factors in the numerators become terms of an arithmetical progression.

$$\therefore S_2 = \frac{1}{4} + \frac{1(-2)}{4^2} \times \frac{1}{2!} + \frac{1(-2)(-5)}{4^3} \times \frac{1}{3!} + \frac{1(-2)(-5)(-8)}{4^4} \times \frac{1}{4!} + \dots$$

$$\therefore S_2 = \frac{1}{3} \times \frac{3}{4} + \frac{\frac{1}{3}(-\frac{2}{3})}{2!} \times \frac{3^2}{4^2} + \frac{\frac{1}{3}(-\frac{2}{3})(-\frac{5}{3})}{3!} \times \frac{3^3}{4^3} + \frac{\frac{1}{3}(-\frac{2}{3})(-\frac{5}{3})(-\frac{8}{3})}{4!} \times \frac{3^4}{4^4} + \dots$$

The series has a sum to infinity since $-1 < \frac{3}{4} < +1$.

$$\therefore 1 + S_2 = (1 + \tfrac{3}{4})^{1/3}$$
$$\therefore S_2 = \sqrt[3]{\tfrac{7}{4}} - 1$$

The expansion of $(1 + x)^n$ when $|x| > 1$

4.2 It will be recalled that the expansion of $(1 + x)^n$ in ascending powers of x is only valid for $|x| < 1$. If, however, $|x| > 1$, the function can be expanded in ascending powers of $1/x$.

$$(1 + x)^n = \{x(1 + x^{-1})\}^n = x^n(1 + x^{-1})^n$$

When $|x| > 1$, it follows that $|x^{-1}| < 1$, so that we may write

$$(1 + x)^n = x^n \left\{ 1 + n\left(\frac{1}{x}\right) + \frac{n(n-1)}{2!}\left(\frac{1}{x}\right)^2 + \dots \right.$$

$$\left. + \frac{n(n-1)\dots(n-r+1)}{r!}\left(\frac{1}{x}\right)^r + \dots \right\}$$

Qu. 4 Expand the following in ascending powers of $1/x$, giving the ranges of values of x for which the expansions are valid:

(a) $(1 + x)^{-1}$,   (b) $(2 + x)^{-2}$,   (c) $(1 + 3x)^{-2}$,

(d) $\dfrac{3}{(x - 1)(x - 2)}$,   (e) $\dfrac{x}{x^2 + 1}$.

Exercise 4a

1 Find the first three terms and the general terms in the expansions of the following functions in ascending powers of x. State the ranges of values of x for which the expansions are valid.

(a) $(1 + 3x)^{-1}$,   (b) $(1 - 2x)^{-1}$,   (c) $(1 + x)^{-2}$,

(d) $(1 - \tfrac{1}{2}x)^{-2}$,   (e) $(1 + x)^{-3}$,   (f) $(2 + x)^{-1}$,

(g) $\dfrac{1}{(3 - x)^2}$,   (h) $\dfrac{1}{(2 - 3x)^3}$,   (i) $\sqrt{(1 + x)}$.

2 Express the following functions in partial fractions and find the first three terms and the general terms in their expansions in ascending powers of x. For what values of x are the expansions valid?

(a) $\dfrac{3}{(1-x)(1+2x)}$,   (b) $\dfrac{1}{(1+x)(x+2)}$,   (c) $\dfrac{x-1}{x^2+2x+1}$,

(d) $\dfrac{5}{1-x-6x^2}$,   (e) $\dfrac{x+3}{(x-2)^2}$,   (f) $\dfrac{x+2}{x^2-1}$.

3 Expand the following functions in ascending powers of x, giving the first three terms and the general term, and state the necessary restrictions on the values of x:

(a) $\dfrac{1}{1+x^2}$,   (b) $\dfrac{x}{1-x^2}$,   (c) $\dfrac{1-x}{1+x}$,

(d) $(1+x)(1-x)^{10}$,   (e) $\dfrac{4}{(x+3)(1+x)}$,   (f) $\dfrac{x+5}{(3-2x)(x-1)}$,

(g) $\dfrac{x+7}{(x+1)^2(x-2)}$.

4 Expand the following functions in ascending powers of $1/x$, giving the first three terms and the general terms. State the necessary restrictions on the values of x.

(a) $(2+x)^{-1}$,   (b) $(3-x)^{-3}$,   (c) $(1-2x)^{-2}$,

(d) $\dfrac{x+2}{x+1}$,   (e) $\dfrac{x-1}{(x+2)^2}$,   (f) $\dfrac{1}{x^2-5x+6}$,

(g) $\dfrac{2x+4}{(x-1)(x+3)}$,   (h) $\dfrac{2x}{1-x^2}$,   (i) $\dfrac{1}{1-x+x^2-x^3}$.

5 Expand $(x-2)^{1/2}$ as a series of descending powers of x as far as the third term. By substituting $x=100$, evaluate $\sqrt{2}$ to five significant figures. [Hint: $\sqrt{98}=7\sqrt{2}$.]

6 Obtain $\sqrt[3]{2}$ to five places of decimals by substituting $x=1000$ in the expansion of $(x+24)^{1/3}$ in descending powers of x.

In Nos. 7–10, use the binomial expansion to find the values of

7 $(16.32)^{1/4}$ to five places of decimals.

8 $\sqrt{9.09}$ to six places of decimals.

9 $\dfrac{1}{(10.04)^2}$ to four significant figures.

10 $\dfrac{1}{\sqrt{17}}$ to four places of decimals.

11 Expand the function $(1+2x)^{1/2}(1-3x)^{-1/3}$ in a series of ascending powers of x as far as the term in x^2. [Hint: multiply the first three terms of the expansion of $(1+2x)^{1/2}$ by those of $(1-3x)^{-1/3}$, ignoring terms in x^3 and higher powers of x.]

In Nos. 12–17 expand the functions in series of ascending powers of x as far as the terms indicated.

12 $\dfrac{1}{1 + x + 2x^2}$, (x^3).   [Write $y = x + 2x^2$.]

13 $\dfrac{1}{(1 + 2x + 3x^2)^2}$, (x^3).

14 $\sqrt[3]{(1 + 3x)}\sqrt{(1 + 2x)}$, (x^2).

15 $\dfrac{\sqrt[3]{(1 - x)^2}}{1 + x}$, (x^3).

16 $\dfrac{x}{1 - \sqrt{(1 + 2x)}}$, (x^2).   [Expand the denominator; then divide numerator and denominator by x.]

17 $\dfrac{x^2}{1 - \sqrt[3]{(1 - 3x)}}$, (x^3).

18 The field H on the axis of a bar magnet of moment M at a distance d from its centre is approximately $2M/d^3$. Suppose that in calculating the value of H, values of M and d differ by $\pm 2\%$ and $\pm 1\%$ respectively. What is the greatest possible percentage error in calculating the value of H?

19 If a clock with a seconds pendulum registers x s too few per day, what is the time of one beat of the pendulum?

   One beat of a seconds pendulum takes $\pi(l/g)^{1/2}$ s, where l is the length of the pendulum and g is a constant. If the length of the pendulum increases by 0.04% owing to expansion, calculate the number of seconds it will have failed to register in a day.

20 If a pendulum beats seconds (see No. 19) at a place where $g = 981$ cm/s^2 and is then removed to a place where g is 0.05% less, how many seconds will it have failed to register in a day?

21 The heat H produced by an electric current flowing through a resistance R with potential difference V for a time t is given by $H = JV^2t/R$, where J is a constant. If V, t, R are given percentage increases x, y, z which are so small that the squares and products of x, y, z may be neglected, find the percentage increase in the value of H.

22 The period of oscillation T of a vibration magnetometer is given by the formula

$$T = 2\pi\sqrt{\left(\frac{I}{MH}\right)}$$

If the quantities I, M, H are estimated with errors of p, q, r per cent, respectively, find the corresponding percentage error in T if the squares and products of p, q, r may be neglected.

Sum to infinity the series in Nos. 23–30 stating the necessary restrictions on the value of x.

23 $1 - \frac{1}{2} + \frac{1}{4} - \frac{1}{8} + ...$

24 $1 - x + \dfrac{1 \times 3}{1 \times 2}x^2 - \dfrac{1 \times 3 \times 5}{1 \times 2 \times 3}x^3 + \dfrac{1 \times 3 \times 5 \times 7}{1 \times 2 \times 3 \times 4}x^4 - ....$

25 $1 + 4x + 12x^2 + ... + (n + 1)2^n x^n + ...$

26 $1 - \dfrac{2}{3} + \dfrac{2 \times 3}{3 \times 6} - \dfrac{2 \times 3 \times 4}{3 \times 6 \times 9} + ...$

27 $1 + \dfrac{1}{6} - \dfrac{1 \times 2}{6 \times 12} + \dfrac{1 \times 2 \times 5}{6 \times 12 \times 18} - \dfrac{1 \times 2 \times 5 \times 8}{6 \times 12 \times 18 \times 24} + ...$

28 $1 - 6x + 24x^2 - ... + (-1)^n(n + 1)(n + 2)2^{n-1}x^n + ...$

29 $1 - x - \dfrac{x^2}{2!} - 1 \times 3\dfrac{x^3}{3!} - 1 \times 3 \times 5\dfrac{x^4}{4!} - ....$

30 $1 + \dfrac{1}{4} + \dfrac{1 \times 4}{4 \times 8} + \dfrac{1 \times 4 \times 7}{4 \times 8 \times 12} + ...$

Relations between binomial coefficients

4.3 We first show how the greatest coefficient in a binomial expansion may be found. A similar method may be applied to find the greatest term. In this section n and r represent positive integers.

Example 5 *Find the greatest coefficient in the expansion of* $(2x + 3)^{12}$.

The coefficient of x^r is given by

$$u_r = \frac{12!}{(12 - r)!r!}2^r 3^{12-r}$$

and the coefficient of x^{r+1} is given by

$$u_{r+1} = \frac{12!}{(11 - r)!(r + 1)!}2^{r+1}3^{11-r}$$

The ratio of these coefficients is

$$\frac{u_r}{u_{r+1}} = \frac{r + 1}{12 - r} \times \frac{3}{2}$$

Therefore $u_r < u_{r+1}$ if

$$\frac{r+1}{12-r} \times \frac{3}{2} < 1$$

$$\therefore 3r + 3 < 24 - 2r\dagger$$

$$\therefore 5r < 21$$

$$\therefore r < \tfrac{21}{5} = 4\tfrac{1}{5}$$

That is, $u_1 < u_2$, $u_2 < u_3$, $u_3 < u_4$, $u_4 < u_5$, but $u_5 \not< u_6$. Therefore the coefficient of x^5 is the largest. Its value is $792 \times 2^5 \times 3^7$.

Qu. 5   Find the greatest *term* in the above expansion when $x = 2$.

Now some series involving the binomial coefficients will be considered. For brevity we shall write

$$(1 + x)^n = c_0 + c_1 x + c_2 x^2 + \ldots + c_n x^n$$

So far we have not assigned any meaning to $^nC_0$. Since, in general, $^nC_r = c_r$, it is most convenient to *define* $^nC_0 = 1$; and, if we *define* $0! = 1$, we can write

$$c_r = {}^nC_r = \frac{n!}{(n-r)!r!} \quad \text{for all values of } r \text{ from 0 to } n.$$

It should be noted that c_r is only used for ^nC_r. Other coefficients such as $^{n-1}C_r$ and $^{2n}C_r$ will not be abbreviated in this way.

Example 6   *Find the values of*
(a) $c_0 + c_1 + \ldots + c_n$,
(b) $c_0 - 2c_1 + 3c_2 - \ldots + (-1)^n(n+1)c_n$,
(c) $\tfrac{1}{2}c_0 + \tfrac{1}{3}c_1 + \tfrac{1}{4}c_2 + \ldots + c_n/(n+2)$.

(a) $(1 + x)^n = c_0 + c_1 x + \ldots + c_n x^n$.
Substituting $x = 1$,

$$c_0 + c_1 + \ldots + c_n = 2^n$$

(b) Remember that $\dfrac{d(x^n)}{dx} = nx^{n-1}$.

$$x(1 + x)^n = c_0 x + c_1 x^2 + c_2 x^3 + \ldots + c_n x^{n+1}$$

Differentiating with respect to x,

$$(1 + x)^n \times 1 + x \times n(1 + x)^{n-1} = c_0 + 2c_1 x + 3c_2 x^2 + \ldots + (n+1)c_n x^n$$

Substituting $x = -1$,

$$c_0 - 2c_1 + 3c_2 - \ldots + (-1)^n(n+1)c_n = 0$$

$\dagger 12 - r$ is positive so the inequality sign is unchanged.

(c) Remember that $\int x^n \, dx = x^{n+1}/(n+1) + k$.

$$(1+x)^n = c_0 + c_1 x + c_2 x^2 + \ldots + c_n x^n$$
$$\therefore x(1+x)^n = c_0 x + c_1 x^2 + c_2 x^3 + \ldots + c_n x^{n+1}$$

Integrating the R.H.S. with respect to x between 0 and 1, we obtain

$$\tfrac{1}{2}c_0 + \tfrac{1}{3}c_1 + \tfrac{1}{4}c_2 + \ldots + c_n/(n+2)$$

For the L.H.S., we write

$$x(1+x)^n = \{(1+x) - 1\}(1+x)^n$$

$$\int_0^1 x(1+x)^n \, dx = \int_0^1 \{(1+x)^{n+1} - (1+x)^n\} \, dx$$

$$= \left[\frac{(1+x)^{n+2}}{n+2} - \frac{(1+x)^{n+1}}{n+1} \right]_0^1$$

$$= \frac{2^{n+2} - 1}{n+2} - \frac{2^{n+1} - 1}{n+1}$$

which is the sum of the series.

Alternatively, the integral could have been evaluated by the substitution $u = 1 + x$.

Certain relations between the binomial coefficients may be obtained by equating coefficients (see p. 45). For example, the identity

$$(1+x)^{n+2} \equiv (1+x)^2 (1+x)^n$$

may be expanded in two different ways:
(a) $1 + {}^{n+2}C_1 x + \ldots + {}^{n+2}C_r x^r + \ldots + {}^{n+2}C_{n+2} x^{n+2}$,
(b) $(1 + 2x + x^2)(c_0 + c_1 x + \ldots + c_r x^r + \ldots + c_n x^n)$.

The term in x^r in (b) is obtained by multiplying $c_r x^r$ by 1, $c_{r-1} x^{r-1}$ by $2x$, $c_{r-2} x^{r-2}$ by x^2. Equating coefficients of x^r in the expansions (a) and (b),

$${}^{n+2}C_r = c_r + 2c_{r-1} + c_{r-2} \qquad (2 \leqslant r \leqslant n)$$

Qu. 6 What relations are obtained by equating coefficients of
(a) x^{r+1},   (b) x^{r+2}?

Example 7 *Prove that* $c_0^2 + c_1^2 + \ldots + c_n^2 = {}^{2n}C_n$.

[The expression ${}^{2n}C_n$ suggests the use of $(1+x)^{2n}$, and the terms c_r^2 suggest the square of $(1+x)^n$.]

$$(1+x)^n (1+x)^n \equiv (1+x)^{2n}$$
$$\therefore (c_0 + c_1 x + \ldots + c_n x^n)(c_0 + c_1 x + \ldots + c_n x^n) \equiv (1+x)^{2n}$$

[${}^{2n}C_n$ is the coefficient of x^n on the R.H.S.]

Equating coefficients of x^n,

$$c_0 c_n + c_1 c_{n-1} + \ldots + c_{n-1} c_1 + c_n c_0 = {}^{2n}C_n$$

But $c_r = c_{n-r}$ (see Book 1, §12.4),

$$\therefore c_0^2 + c_1^2 + \ldots + c_{n-1}^2 + c_n^2 = {}^{2n}C_n$$

Exercise 4b

1 Find the greatest coefficients in the binomial expansions of the following:
  (a) $(x + 2)^{10}$,   (b) $(3x + 1)^8$,   (c) $(4x + 3)^{12}$,
  (d) $(2x + 5)^{20}$,   (e) $(x + \frac{2}{3})^{11}$,   (f) $(3x - 2)^9$,
  (g) $(12 - 11x)^{-2}$,   (h) $(7 - 5x)^{-3}$.

2 Find the greatest terms in the binomial expansions of
  (a) $(2x + 3y)^{12}$,   when $x = 1$, $y = 3$;
  (b) $(x + 2y)^{10}$,   when $x = \frac{1}{2}$, $y = \frac{1}{3}$;
  (c) $(4x + 5y)^8$,   when $x = \frac{1}{3}$, $y = \frac{1}{2}$;
  (d) $(3x - 5)^{-2}$,   when $x = 1\frac{1}{2}$.

Prove that

3 $c_0 - c_1 + c_2 - \ldots + (-1)^n c_n = 0$.
4 $c_1 + 2c_2 + 3c_3 + \ldots + nc_n = n \times 2^{n-1}$.
5 $c_0 + 2c_1 + 3c_2 + \ldots + (n + 1)c_n = 2^{n-1}(n + 2)$.
6 $c_1 - 2c_2 + 3c_3 - \ldots + (-1)^{n+1} nc_n = 0$.
7 $2 \times 1c_2 + 3 \times 2c_3 + \ldots + n(n - 1)c_n = n(n - 1)2^{n-2}$.
8 $2c_1 - 6c_2 + \ldots + (-1)^{n+1} n(n + 1)c_n = 0$.
9 $1^2 c_1 + 2^2 c_2 + \ldots + n^2 c_n = n(n + 1)2^{n-2}$.
10 $1^2 c_0 + 2^2 c_1 + \ldots + (n + 1)^2 c_n = (n + 1)(n + 4)2^{n-2}$.
11 $\frac{1}{2}c_0 - \frac{1}{3}c_1 + \ldots + (-1)^n c_n/(n + 2) = 1/\{(n + 1)(n + 2)\}$.
12 $c_0 + \frac{1}{2}c_1 + \frac{1}{3}c_2 + \ldots + c_n/(n + 1) = (2^{n+1} - 1)/(n + 1)$.
13 $\frac{1}{2}c_0 - \frac{1}{6}c_1 + \ldots + (-1)^n c_n/(n + 1)(n + 2) = 1/(n + 2)$.

Exercise 4c (Miscellaneous)

1 Express the function $\dfrac{1 + 2x + 3x^2}{(1 - x)(1 + x^2)}$ in partial fractions.

  If x is so small that powers higher than the third may be neglected, expand
the function in the form $A + Bx + Cx^2 + Dx^3$.   (JMB)

2 Find numbers A, B, and C such that the fraction

$$\frac{2x}{(1 - x)(1 + x^2)} \quad \text{is equal to} \quad \frac{A}{1 - x} + \frac{B + Cx}{1 + x^2}$$

  Hence obtain the expansion of the fraction in ascending powers of x as far
as x^5. Between what values must x lie in order that this expansion may be
valid?   (JMB)

3 (a) Without using tables, find the value of $\dfrac{(\sqrt{5} + 2)^6 - (\sqrt{5} - 2)^6}{8\sqrt{5}}$.

  (b) Expand $(1 - 3x)^{1/3}$ in ascending powers of x as far as the term in x^3. By
taking $x = \frac{1}{8}$, evaluate $\sqrt[3]{5}$ correct to two decimal places.   (C)

4 Express in partial fractions $\dfrac{x^2 + 2x + 8}{x^3(x + 2)}$.

Hence express in partial fractions $\dfrac{y^2 + 7}{(y - 1)^3(y + 1)}$, and expand this expression in ascending powers of y as far as y^2, stating for what values of y this expansion is valid. (C)

5 (a) Express $\dfrac{7x + 3}{(3x - 1)(x + 1)^2}$ in partial fractions, and hence find the coefficient of x^n when this expression is expanded in ascending powers of x.

(b) Write down the first three terms of the binomial expansion of $(1 - \frac{1}{1000})^{1/3}$. Hence evaluate $(37)^{1/3}$ to six decimal places. (O & C)

6 Write down and simplify the first three terms in the binomial expansions of $(1 + x)^{1/2}$ and $(1 + x)^{-1/2}$.

AB is a chord, of length $2ka$, of a circle of radius a. The tangents to the circle at A and B meet in C. Show that, if k is so small compared with unity that k^7 is negligible, the area of the triangle ABC is $a^2k^3 + \frac{1}{2}a^2k^5$. (L)

7 Use the binomial theorem to evaluate $\dfrac{1}{9.84} - \dfrac{1}{9.85}$ to four significant figures.

8 Prove that, if x is so small that its cube and higher powers can be neglected,

$$\sqrt{\dfrac{1 + x}{1 - x}} = 1 + x + \dfrac{1}{2}x^2$$

By taking $x = \frac{1}{9}$, prove that $\sqrt{5}$ is approximately equal to $\frac{181}{81}$. (C)

9 (a) Find the percentage increase in the value of $x^2 y^4 / z$ when the percentage increases in x, y, z are p, q, r, respectively, if the squares and higher powers of p, q, r can be neglected.

(b) Obtain the first four terms in the expansion of $\dfrac{1 + x}{(1 - x)^3}$, (i) in a series of ascending powers of x, (ii) in a series of descending powers of x. (L)

10 (a) By means of the binomial theorem evaluate $(10.02)^{10}$ to the nearest thousand.

(b) Write down the expansion of $(1 - x)^{-2}$ in ascending powers of x and deduce that

$$\sum_{n=0}^{\infty} (a + bn)x^n = \dfrac{a + (b - a)x}{(1 - x)^2}, \quad \text{when } |x| < 1$$

Prove that $\displaystyle\sum_{n=0}^{\infty} \dfrac{8n + 1}{3^{2n}} = \dfrac{9}{4}$. (JMB)

11 Use the binomial series to write down the first four terms of the expansion of $(1 + y)^{-1/2}$ in a series of ascending powers of y.

Hence find, in terms of $\cos\theta$, the coefficients c_1, c_2, c_3 in the expansion of $(1 - 2x\cos\theta + x^2)^{-1/2}$ in the form $1 + c_1 x + c_2 x^2 + c_3 x^3 + \dots$.

Prove that, when $\theta = 0$, every coefficient in the series is equal to $+1$.

[You may assume throughout that the expansions are valid.] (JMB)

12 (a) If x is small compared with unity and powers of x higher than x^6 are neglected, show that

$$\frac{1}{1+x^2} = 1 - x^2 + x^4 - x^6$$

Prove that the error in this approximation is less than x^8.
(b) Expand $\sqrt{(1+x)}$ as far as the term in x^3.
(c) Find the coefficients of x^{2n} and x^{2n+1} in the expansion in ascending powers of $(1-x)^2/(1+x)^2$.   (L)

13 Write down the series for $\sqrt{(1+x)}$ in ascending powers of x as far as the term in x^4.

Show also that the error in taking $\frac{1}{4}(6+x) - \dfrac{1}{2+x}$ as an approximation to $\sqrt{(1+x)}$ when x is small is approximately $x^4/128$.   (C)

14 Prove (do not merely verify) that, if E denotes the function

$$\frac{x^2}{2 - x + 2\sqrt{(1-x)}}$$

then $E = 2 - x - 2\sqrt{(1-x)}$.
Deduce that, if x is small, E is approximately equal to $\frac{1}{4}x^2$.   (O & C)

15 Express $\dfrac{1}{(x+2)^2(2x+1)}$ in partial fractions, and hence expand the expression as a series in ascending powers of x, giving the first four terms and the coefficient of x^n.

Show that, for values of x so small that x^4 may be neglected, the given expression can be represented by $\dfrac{1}{(3x+2)^2} + kx^3$ for some number k independent of x, and find k.   (O & C)

16 Write down, without proof, the binomial expansion for $\sqrt{(1-2x)}$ in ascending powers of x, giving the first three terms and the general term.
Prove that the sum of the first two terms exceeds $\sqrt{(1-2x)}$ by exactly

$$\frac{x^2}{1 - x + \sqrt{(1-2x)}}$$

By putting $x = 0.005$, obtain from the first two terms of the expansion an approximation for $\sqrt{11}$, and determine to how many places of decimals your approximation is correct.   (C)

17 Show that, if x is so small in comparison with unity that x^3 and higher powers can be neglected,

$$\frac{(1-4x)^{1/2}(1+3x)^{1/3}}{(1+x)^{1/2}} = 1 - \tfrac{3}{2}x - \tfrac{33}{8}x^2 \tag{L}$$

18 (a) Find and simplify the term independent of x in the binomial expansion of $\left(x^2 - \dfrac{1}{2x}\right)^9$.

(b) Write down and simplify the first four terms in the expansion in ascending powers of x of $(1 + 3x)^{1/3}$. Hence evaluate $\sqrt[3]{1.03}$ correct to five places of decimals. (JMB)

19 Express $\dfrac{10 - 17x + 14x^2}{(2 + x)(1 - 2x)^2}$ in partial fractions of the form

$$\frac{A}{2 + x} + \frac{B}{1 - 2x} + \frac{C}{(1 - 2x)^2}$$

Hence, or otherwise, obtain the expansion of $\dfrac{10 - 17x + 14x^2}{(2 + x)(1 - 2x)^2}$ in ascending powers of x, up to and including the term in x^3. State the restrictions which must be imposed on x for the expansion in ascending powers of x to be valid. (C)

20 Show that for any positive integral value of n there are two values of a such that the coefficients of the powers of x in the three middle terms of the expansion of $(1 + ax)^{2n}$ are in arithmetical progression. Show also that the product of these values of a is independent of n. (L)

21 If c_r is the coefficient of x^r in the binomial expansion of $(1 + x)^n$, where n is a positive integer, prove that

(a) $c_0 + c_2 + c_4 + \ldots = c_1 + c_3 + c_5 + \ldots = 2^{n-1}$,

(b) $c_0{}^2 + c_1{}^2 + c_2{}^2 + \ldots + c_n{}^2 = (2n)!/(n!)^2$,

(c) $c_0 + 2c_1 + 3c_2 + \ldots + (n + 1)c_n = (n + 2)2^{n-1}$. (L)

22 (a) Find which is the greatest term in the expansion of $(3 + 2x)^{14}$ in ascending powers of x when $x = 5/2$.

(b) Prove that three consecutive terms in the expansion of $(1 + x)^n$, where n is a positive integer, can never be in geometrical progression. (L)

23 (a) If the expansion of $(1 + x)^n$ in ascending powers of x is denoted by

$$(1 + x)^n = c_0 + c_1 x + c_2 x^2 + c_3 x^3 + \ldots + c_r x^r + \ldots$$

write down the values of c_0, c_1, c_2, c_3 in terms of n, and the value of c_r in terms of n and r. State the range of values of x for which the expansion is valid whatever the value of n.

If $n > -1$, determine the range of values of r for which

$$\left| \frac{c_r}{c_{r-1}} \right| < 1.$$

(b) Use the binomial expansion to calculate the value of $(1 + \frac{1}{1000})^{200}$ correct to four places of decimals. (JMB)

24 Show that the first three terms in the expansion in ascending powers of x of $(1 + 8x)^{1/4}$ are the same as the first three terms in the expansion of $(1 + 5x)/(1 + 3x)$.

Use the corresponding approximation

$$(1 + 8x)^{1/4} \approx \frac{1 + 5x}{1 + 3x}$$

to obtain an approximation to $(1.16)^{1/4}$ as a rational fraction in its lowest terms.   (JMB)

25 (a) Show that, when x is small, the expansion in powers of x of the function

$$(1 + x)^p + (1 - x)^p - 2(1 + x^2)^q$$

is of the form

$$a_2 x^2 + a_4 x^4 + a_6 x^6 + \ldots$$

If $a_2 = 0$, find q in terms of p. If, in addition, $a_4 = 0$ and p is not equal to 0, 1, or 2, find the values of p, q, a_6.

(b) Show that, when x is large and positive,

$$(x^2 + x)^{1/2} + (x^2 + 3x)^{1/2} = 2x + 2, \quad \text{approximately.} \qquad \text{(JMB)}$$

Chapter 5

Three-dimensional trigonometry

Introduction

5.1 In Book 1, Chapter 15, we considered a number of problems concerning lines and planes in three dimensions, to which we applied the notation and techniques of vector geometry. In this chapter we shall examine similar problems, but we shall solve them by using pure trigonometry. A mature mathematician must learn to be flexible and to select the best approach to a given problem.

Drawing a clear figure

5.2 To begin with, it needs to be emphasised that some of the questions will be very difficult without a clear figure. In general, the four following basic rules should be adopted:

(a) Parallel lines are drawn parallel.
(b) Vertical lines are drawn parallel to the sides of the paper.
(c) East–West lines are generally drawn parallel to the bottom of the paper, and North–South lines are drawn at an acute angle to East–West lines.
(d) All unseen lines should be dotted in.

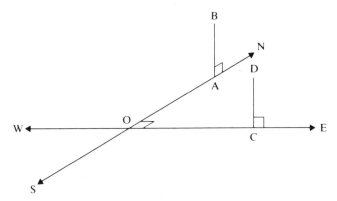

Figure 5.1

In Fig. 5.1, AB and CD are vertical posts. Notice that the angle NOE is marked a right angle, because this is what it represents.

Qu. 1   Copy Fig. 5.1 and draw AC. Mark in all the right angles at A and C.

The angle between a line and a plane

5.3   When calculating the angle between a line and a plane, we are concerned with calculating the angle between two lines: the given line, and another line lying in the given plane. An infinite number of lines can be drawn, lying in the plane and passing through the point in which the given line meets the plane, and each line will yield a different angle. Which line should we take?

In Fig. 5.2, the line QR meets the plane π in O. In order to find the angle between QR and π, take any point P on QR and drop a perpendicular PN to the plane; now join N to O and θ is the angle required. ON is the **projection** of OP onto the plane π.

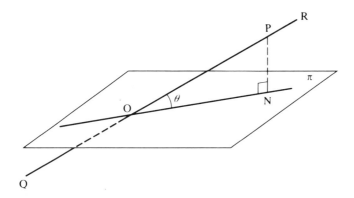

Figure 5.2

Qu. 2   Copy Fig. 5.2 and draw any line lying in π and passing through O. Let M be the foot of the perpendicular from P to this line. Show that angle POM is greater than angle PON.

The result of Qu. 2 shows that θ is the *least* angle between QR and any line which can be drawn in π and passing through O.

Example 1   Fig. 5.3 *represents a rectangular box* 9 cm × 6 cm × 6 cm *with its lid open at an angle of* 30°. *Calculate the angle between* BD' *and the plane* CDD'C'.

BD' meets the plane CDD'C' in D'. Take any other point on BD': B is an obvious point. Drop the perpendicular from B to the plane: BC. The angle we want is BD'C. Select triangle BD'C and mark in lengths (see Fig. 5.4). We must calculate CD' or BD' first. CD' is easier, so draw triangle CC'D' and again mark in lengths.

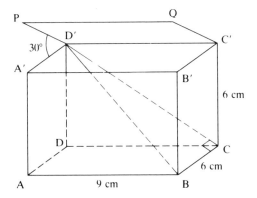

Figure 5.3

In triangle CC'D',

$$CD'^2 = 6^2 + 9^2 = 117$$
$$\therefore CD' = \sqrt{117}†$$

Now mark the length of CD' in triangle BCD'.

$$\cot \theta = \frac{\sqrt{117}}{6}$$

$$\therefore \theta = 29.0°$$

Therefore the angle between BD' and the plane CDD'C' is 29.0°.

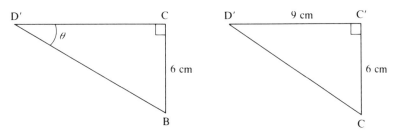

Figure 5.4

Qu. 3   Calculate the angle between
(a) BD' and the plane BCC'B', (b) AC' and BD'.
Qu. 4   Calculate the angle between BP and the plane ABCD.

The angle between two planes

5.4   When calculating the angle between two planes we are again concerned with calculating the angle between two lines, one in each plane.

†If a calculator is being used, intermediate calculations (such as $\sqrt{117}$) should be delayed, unless explicitly required by the question. Writing down intermediate steps can lead to a build-up of rounding errors.

Referring to Fig. 5.5, in order to find the angle between two planes π and π′, we select a point C on their common line AB and draw lines PC and CQ in π and π′, respectively, and at right angles to AB. PCQ is the angle we want. This angle is called the **dihedral** angle of the two planes.

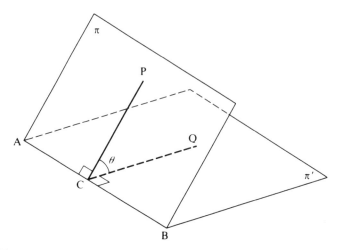

Figure 5.5

Qu. 5 Copy Fig. 5.5 and draw a line PR parallel to AB; join CR. Let P′, R′ be the feet of the perpendiculars from P, R, respectively, to π′. Show that angle RCR′ < angle PCQ.

Example 2 VABCD *is a right pyramid on a square base* ABCD *of side* 10 cm. *Each sloping edge is* 12 cm *long. Calculate the angle between the faces* VAB *and* VBC.

To obtain a good figure first represent the base ABCD as a rhombus, then dot in diagonals to meet at N. Put up the vertical NV and choose V so that AV does not coincide with DV (see Fig. 5.6).

VB is the common line of the two planes. Draw AX perpendicular to VB, then, since the figure is symmetrical about VDB, CX is also perpendicular to VB. The angle we want is AXC, so we must work in triangle AXC.

First we want to find AX and AC.

In triangle VAM (see Fig. 5.7), $VM^2 = 12^2 - 5^2 = 119$.

$$\therefore VM = \sqrt{119} \text{ cm}$$

Area of triangle $VAB = 5\sqrt{119} = \frac{1}{2} \times AX \times 12$.

$$\therefore AX = \frac{5\sqrt{119}}{6} \text{ cm}$$

In triangle ABC (see Fig. 5.7), $AC^2 = 10^2 + 10^2 = 200$.

$$\therefore AC = \sqrt{200} = 10\sqrt{2}$$

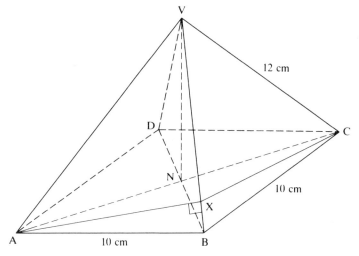

Figure 5.6

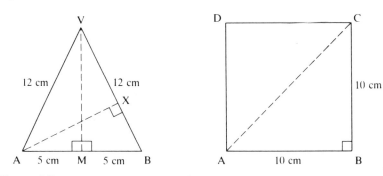

Figure 5.7

In triangle AXN (see Fig. 5.8), $AN = \frac{1}{2}AC = 5\sqrt{2}$.

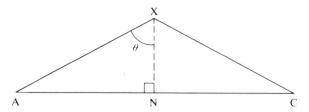

Figure 5.8

$$\sin \theta = \frac{AN}{AX} = \frac{5\sqrt{2}}{\frac{5}{6}\sqrt{119}} = 6\sqrt{\left(\frac{2}{119}\right)}$$

$$\therefore \theta = \arcsin 6\sqrt{\left(\frac{2}{119}\right)}$$

$$= 51.06°$$

$\therefore \angle AXC = 102.1°$

Therefore the angle between the faces VAB and VBC is 102.1°.

Qu. 6 If, in Example 2, X is any point on VB, prove that triangle ABX ≡ triangle BCX. Hence prove that if AX is perpendicular to VB, then CX is also perpendicular to VB.

Qu. 7 If, in Example 2, Y is the mid-point of BC, show that VYN is the angle between VBC and ABCD, and calculate it.

Exercise 5a

1 A cuboid is formed by joining the vertices AA′, BB′, CC′, DD′ of two rectangles ABCD and A′B′C′D′. AB = 6 cm, BC = 6 cm, CC′ = 8 cm. X and Y are the mid-points of AD and CD, respectively. Calculate:
   (a) the angle between XB′ and the base ABCD,
   (b) the angle between the plane XYB′ and the base ABCD,
   (c) the angle between the plane BB′X and the plane BB′Y.

2 A right pyramid VABCD stands on a rectangular base ABCD. AB = 6 cm, BC = 8 cm, and the height of the pyramid is 12 cm. Calculate:
   (a) the angle which a slant edge makes with the base,
   (b) the angle which the slant face VAB makes with the base,
   (c) the angle between the two opposite slant faces VBC and VAD.

3 A hanging lamp is supported by three chains of equal length, fixed to points A, B, C in the ceiling which form an equilateral triangle of side 16 cm, and the lower ends are connected at a point 20 cm below the ceiling. Calculate
   (a) the length of each chain,
   (b) the angle which each chain makes with the ceiling.

4 Two equal rectangles 3 m by 4 m are placed so that the longer sides XY coincide. The angle between their planes is 50°. Find the angle between the diagonals which pass through X.

5 A right pyramid stands on a square base of side 8 cm. The height of the pyramid is 10 cm. Calculate the angle between two adjacent faces.

6 Three mutually perpendicular lines meet at O and equal lengths OA, OB, OC are cut off. Find the inclination of ABC to ABO.

7 O is the middle point of the edge AD of a cube, of side 6 cm, whose faces ABCD, A′B′C′D′ are similarly situated. Calculate (a) the sine of the angle between the plane OCD′ and the face CDD′C′ of the cube, (b) the sine of the angle between the edge DD′ and the plane OC′D′.

8 Calculate the vertical height and the slope of the slant edges and faces of a regular tetrahedron of side 8 cm, which stands on a horizontal base.

9 The roofs of an L-shaped house slope at 45°. What is the inclination to the horizontal of the line in which the two roofs meet?

10 In Fig. 5.3, taking the same dimensions, calculate the angle between CP and the plane C′D′PQ.

11 A solid is formed by placing a pyramid with square base of side 15 cm and height 20 cm on top of a cuboid with the same dimensions of base and the

same height. Calculate the angle which a line drawn from the vertex of the pyramid to a bottom corner of the solid makes with the base.

12 In a regular tetrahedron ABCD, P is the mid-point of AB. Calculate the cosine of the angle between the planes PCD and BCD.

13 In a regular tetrahedron ABCD, Q is the middle point of AD. Find the angle between the line BQ and the plane DBC.

14 In a tetrahedron ABCD, $AC = 13$ cm, $AB = 12$ cm, $BC = 5$ cm, $CD = \sqrt{41}$ cm, $BD = 4$ cm, and $AD = 12$ cm. Calculate the cosine of the angle between the planes ABC and BDC.

15 Three adjacent edges of a rectangular box are $AB = a$ cm, $AD = b$ cm, and $AF = c$ cm. Find the angle between the planes BDF and BAD.

16 In a tetrahedron PQRS, P is vertically above Q, one corner of the horizontal base QRS. $PS = PR = a$, $PQ = 2b$, and $QR = QS = RS = b$. A is the mid-point of PQ. Calculate the sine of the angle between the planes PRS and ARS.

17 A pyramid on a rectangular base has equal slant edges. Prove that a slant edge makes an angle $\cot^{-1}\sqrt{(\cot^2 \alpha + \cot^2 \beta)}$ with the base where α and β are the angles which the slant faces make with the base.

18 In a tetrahedron ABCD, the base ABC is an equilateral triangle of side a cm and the edges DA, DB, DC are all b cm long. X is the centroid of the face ABD. Prove that the angle CX makes with the base is $\arctan \frac{1}{4}\sqrt{(3b^2/a^2 - 1)}$.

19 OA, OB, OC are unequal mutually perpendicular lines. Prove that $\cos \angle BAC = \cos \angle OAB \times \cos \angle OAC$.

20 The base of a tetrahedron is an equilateral triangle. The slant edges are of equal lengths a and make angles θ with each other. Prove that the height of the tetrahedron is $\frac{1}{3}a\sqrt{(3 + 6 \cos \theta)}$.

Algebraic problems in trigonometry

5.5   Apart from the later questions in Exercise 5a we have so far dealt only with numerical examples. We have obtained solutions, but we shall now want to generalise these solutions in order to understand more clearly how the results depend upon what is given. We shall find, also, that the result often suggests to us the best method of proof.

Two useful hints for solving problems are
(a) draw a clear figure marking in the given facts distinctly, and right angles in particular,
(b) the method is often suggested by the result to be proved.

Example 3   *A man notices two towers, one due North and one in a direction N θ E. If the angle of elevation β of both towers is the same but the height of one is twice the height of the other, prove that*

$$\theta = \arccos \frac{5 \cot^2 \beta - \cot^2 \alpha}{4 \cot^2 \beta}$$

where α is the angle of elevation of the top of one tower from the top of the other.

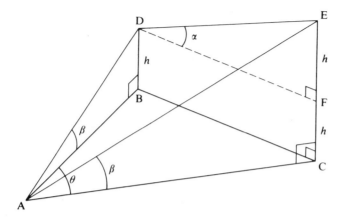

Figure 5.9

The towers are BD, height h, and CE, height $2h$. The horizontal plane through D cuts CE at F (see Fig. 5.9).

∴ CF = FE = h

[The result shows that we need $\cos \theta$, and this suggests that the cosine formula should be applied to triangle ABC. See Book 1, §18.3.]

From triangles ABD, DFE, ACE,

$$\text{AB} = h \cot \beta \qquad \text{BC} = \text{DF} = h \cot \alpha \qquad \text{AC} = 2h \cot \beta$$

By the cosine formula in triangle ABC,

$$\cos \theta = \frac{h^2 \cot^2 \beta + 4h^2 \cot^2 \beta - h^2 \cot^2 \alpha}{2 \times h \cot \beta \times 2h \cot \beta}$$

$$= \frac{5h^2 \cot^2 \beta - h^2 \cot^2 \alpha}{4h^2 \cot^2 \beta}$$

$$\therefore \theta = \arccos \frac{5 \cot^2 \beta - \cot^2 \alpha}{4 \cot^2 \beta}$$

* Qu. 8   Draw a triangle ABC and its circumcircle, marking the centre, O. Show that $\angle \text{BOC} = 2A$, where $\angle \text{BAC} = A$. By considering the isosceles triangle OBC, show that

$$\frac{a}{\sin A} = 2R$$

where R is the radius of the circumcircle.

[Similarly it can be shown that $\dfrac{b}{\sin B}$ and $\dfrac{c}{\sin C}$ equal $2R$. This enables us to prove a more general form of the sine rule (see Book 1, §18.2), namely

$$\frac{a}{\sin A} = \frac{b}{\sin B} = \frac{c}{\sin C} = 2R$$

This form of the sine rule is used in the next example.]

Example 4   *Two vertical walls of equal height cast shadows whose widths are b m and c m when the altitude of the sun is θ. If the angle between the walls is α, prove that their height is*

$$\sqrt{\left\{ \frac{b^2 + c^2 + 2bc \cos \alpha}{\cot^2 \theta \sin^2 \alpha} \right\}} \quad m$$

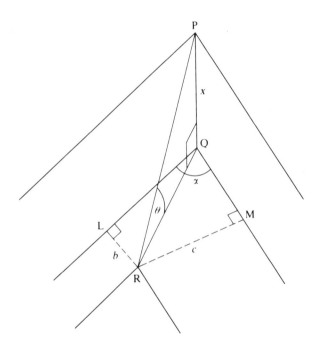

Figure 5.10

PQ is the line in which the two walls meet, and R is the point shadow of P. L and M are the feet of the perpendiculars from R to the two walls. Let x m be the height of each wall (see Fig. 5.10).

Since $\angle QLR = \angle QMR = 90°$, RLQM is cyclic, and QR is the diameter of the circumcircle. $\angle LQM = \alpha$ therefore $\angle LRM = 180° - \alpha$.

By the cosine formula in triangle LMR,

$$LM^2 = b^2 + c^2 - 2bc \cos (180° - \alpha)$$
$$= b^2 + c^2 + 2bc \cos \alpha$$

By the sine formula in triangle QLM,

$$\frac{LM}{\sin \alpha} = 2R$$

where R is the circumradius of triangle LMR and therefore of RLQM.
But from triangle PQR, the diameter of the circumcircle $QR = x \cot \theta$,

$\therefore LM = x \cot \theta \sin \alpha$
$\therefore x^2 \cot^2 \theta \sin^2 \alpha = b^2 + c^2 + 2bc \cos \alpha$

$$\therefore \text{height of wall} = \sqrt{\left\{ \frac{b^2 + c^2 + 2bc \cos \alpha}{\cot^2 \theta \sin^2 \alpha} \right\}} \quad m$$

Exercise 5b

1 The angles of elevation of points A, B from a point P are α and β respectively.
The bearings of A and B from P are S 20° W and S 40° E, and their distances
from P measured on the map are 3 km and 1 km respectively. A is higher
than B. Prove that the elevation of A from B is $\tan^{-1} \dfrac{3 \tan \alpha - \tan \beta}{\sqrt{7}}$.

2 A pole is set up at C, a point due East of A and due North of B. M is the mid-
point of AB. The angles of elevation of the top of the pole from A, M, B are α,
γ, β, respectively.
   Prove that $\cot^2 \alpha + \cot^2 \beta = 4 \cot^2 \gamma$. [Hint: a semi-circle can be drawn on
AB as diameter, with C as a point on the circumference.]

3 A plane slopes down towards the South at an angle α to the horizontal. A
road is made up in the plane in the direction ϕ E of N. Prove that the
inclination of the road to the horizontal is $\tan^{-1} (\tan \alpha \cos \phi)$.

4 From a point on the ground, two points on the top of a horizontal wall, a m
apart, are observed at angles of elevation α and β. The line joining them
subtends an angle θ at the point. Prove that the height of the wall is

$$\frac{a \sin \alpha \sin \beta}{\sqrt{\{\sin^2 \alpha + \sin^2 \beta - 2 \sin \alpha \sin \beta \cos \theta\}}} \quad m$$

5 A mast is erected at a point P. At a point B due West, its angle of elevation
is α, and at a point C due South, its angle of elevation is β. Prove that
its angle of elevation at a point due South of B and due West of C is
$\cot^{-1} \sqrt{(\cot^2 \alpha + \cot^2 \beta)}$.

6 A vertical flagstaff of height y m stands on the top A of a tower. The elevation
of A from a point due South of it is α, and from a point due East is β. The
direct distance between these two points, which are in the horizontal plane
through the foot of the tower, is x m and the elevation of the top of the
flagstaff from the second point is γ. Prove that

$$y^2 = \frac{x^2 (\cos \beta - \cot \gamma)^2}{\cot^2 \gamma (\cot^2 \alpha + \cot^2 \beta)}$$

7 There are two lights, each l m above level ground, and a m apart. A man, whose height is h m, stands anywhere on the ground. Prove that the line joining the ends of his two shadows cast by the lights is of length $ah/(l-h)$ m.

8 The angles of elevation of the top of a tower measured from three points A, B, C are α, β, γ, respectively. A, B, C are in a straight line such that $AB = BC = a$, but the line AC does not pass through the base of the tower. Prove that the height of the tower is

$$\frac{a\sqrt{2}}{\{\cot^2\alpha + \cot^2\gamma - 2\cot^2\beta\}^{1/2}}$$

9 A man observes a flagpole due North of him. He walks in a direction α N of W for a distance of x m and finds the angle of elevation β is the same. Prove that the angle of elevation when he has walked a further distance x m in the same direction is

$$\arctan\frac{\tan\beta}{\sqrt{(1 + 8\sin^2\alpha)}}$$

10 From a point A, a lighted window due North of A has an elevation α. From a point B, due West of A, the angle of elevation is β. Prove that the angle of elevation from the mid-point of AB is

$$\arctan\frac{2}{\sqrt{(3\cot^2\alpha + \cot^2\beta)}}$$

11 A factory is built within a rectangular plot ABCD. The elevations of the tallest chimney on the building from the three corners A, B, C are α, β, γ respectively. Prove that its elevation from the fourth corner D is

$$\text{arccot}\sqrt{(\cot^2\alpha + \cot^2\gamma - \cot^2\beta)}$$

12 A vertical rectangular target faces due South on a horizontal plane. The area of the shadow is $1\frac{1}{4}$ times the area of the target when the sun's altitude is α. Find the bearing of the sun.

13 A vertical tower stands on horizontal ground. From a point P on the ground due South of the tower the angle of elevation of the top of the tower is α. From a point Q on the ground, South-East of the tower, the angle of elevation is β. Prove that the bearing of Q from P is

$$\arctan\frac{\cot\beta}{\sqrt{2}\cot\alpha - \cot\beta}\ \text{E of N}$$

14 A triangle ABC is drawn on a plane sloping at θ to the horizontal. A, B are on the same level and C is below them. CA, CB makes angles α, β with the horizontal plane through AB.
    Prove that, if $\angle ACB = C$,

$$\sin\theta = \frac{\sqrt{\{\sin^2\alpha + \sin^2\beta - 2\sin\alpha\sin\beta\cos C\}}}{\sin C}$$

15 A building at B is x m higher than a building at C, and there is a third building at A, taller than either. The angles of elevation of the tops of the buildings at A, B from the top of that at C are α, β, respectively, and the angle between the vertical planes through CB and CA is θ. The angle between the vertical planes through BA and BC is ϕ. Prove that the building at A is

$$\frac{x \tan \alpha \sin \phi}{\tan \beta \sin (\theta + \phi)} \text{ m higher than the building at C.}$$

16 A vertical post of height h m rises from a plane which slopes down towards the South at an angle α to the horizontal. Prove that the length of its shadow when the sun is S θ W at an elevation β is

$$\frac{h\sqrt{(1 + \tan^2 \alpha \cos^2 \theta)}}{\tan \beta + \tan \alpha \cos \theta} \text{ m}$$

17 A right pyramid of height h stands on a square base of side a. Prove that the angle between adjacent sloping faces is $\arccos \left(\dfrac{-a^2}{a^2 + 4h^2} \right)$.

18 A right pyramid stands on a regular hexagonal base and its slant faces are equal isosceles triangles of base angle α. Prove that
(a) the angle between a slant face and the base is $\arccos (\sqrt{3} \cot \alpha)$,

(b) the angle between adjacent faces is $\arccos \dfrac{2 \sin^2 \alpha - 3}{2 \sin^2 \alpha}$.

19 A vertical flagpole of height one unit stands on top of a vertical tower of height h units. At points on level ground distant x units and y units from the foot of the tower, where $x \neq y$, the flagpole subtends equal angles of magnitude θ. Prove that
(a) $x + y = \cot \theta$,
(b) xy is independent of θ.   (C)

20 One face of a cube has vertices A, B, C, D and the four edges of the cube perpendicular to this face are AA', BB', CC', DD'. Each edge is of length a. The cube rests with the edge AB on a horizontal table and the edge AD inclined at an angle α to the horizontal. Denoting the angles of inclination to the horizontal of the diagonals AC' and DB' by θ and ϕ respectively, prove that

$$3 \cos^2 \theta = 2 - \sin 2\alpha, \quad \text{and that} \quad 3 \cos^2 \phi = 2 + \sin 2\alpha \qquad \text{(C)}$$

Chapter 6

Some inequalities and graphs

Some inequalities

6.1 Anyone who has studied mathematics up to this level will be thoroughly used to manipulating equations, but many readers will not be familiar with inequalities. An inequality is a statement that one number is less than (or greater than) another. Thus the statement, 'The sum of the squares of two numbers is greater than or equal to twice their product', may be written in the form

$$a^2 + b^2 \geqslant 2ab$$

To prove this,

$$\text{L.H.S.} - \text{R.H.S.} = a^2 + b^2 - 2ab = (a - b)^2$$

But $(a - b)^2$ is a square and so is greater than or equal to zero, and the inequality is proved.

Note that the equality occurs only if $a = b$, therefore we may write

$$a^2 + b^2 > 2ab \qquad (a \neq b)$$

(Throughout this chapter we shall be concerned with *real* numbers and the reader should assume that any letters which are used represent real numbers.)

Inequalities may be manipulated in much the same way as equations but with certain important reservations (see Qu. 1–4). The rules that will be used here are

(a) we may add any number (positive or negative) to each side of an inequality,
(b) we may multiply each side of an inequality by any *positive* number,
(c) if each side of an inequality is multiplied by a *negative* number, the inequality is reversed.

We illustrate these three rules by applying them to the inequality $5 < 9$:

(a) $5 + 2 < 9 + 2$ and $5 - 13 < 9 - 13$,
(b) $5 \times 3 < 9 \times 3$, .
(c) $5 \times (-4) > 9 \times (-4)$.

Note particularly that these statements are about algebraic (or directed) numbers. If two such numbers are represented by points on an axis going from left to right, the greater is on the right. Thus $-8 < -4$ and $-20 > -36$.

87

Qu. 1   If $a = x$, then $a^2 = x^2$. Consider the inequality $a < x$ if (a) $a = 3$, $x = 5$,
(b) $a = -7$, $x = 4$. Is $a^2 < x^2$?

Qu. 2   If $b = y$, then $1/b = 1/y$. Consider the inequality $b < y$ if (a) $b = 2$, $y = 4$,
(b) $b = -3$, $y = 1$. Is $1/b < 1/y$?

Qu. 3   If $a = x$, $b = y$, then $a - b = x - y$. Consider the inequalities $a < x$, $b < y$
if $a = 5$, $x = 6$, $b = 4$, $y = 7$. Is $a - b < x - y$?

Qu. 4   If $a < x$, $b < y$, is $ab < xy$? Try $a = -3$, $x = 2$, $b = -4$, $y = 5$.

Example 1   *Find the values of x for which* $\dfrac{x+3}{x-1} > 2$.

(Inequalities such as this are usually easier to handle if one of the sides is zero,
so the first step is to subtract 2 from both sides.)

$$\frac{x+3}{x-1} - 2 > 0$$

Putting the L.H.S. over a common denominator of $(x-1)$, we obtain

$$\frac{x+3-2(x-1)}{x-1} > 0$$

$$\therefore \quad \frac{5-x}{x-1} > 0$$

The numerator (or top line) of this fraction is zero when $x = 5$, and the
denominator is zero when $x = 1$. To obtain the inequality we require, these terms
must have the same signs, so consider the table below:

	$x < 1$	$1 < x < 5$	$x > 5$
$5 - x$	$+$	$+$	$-$
$x - 1$	$-$	$+$	$+$
$\dfrac{5-x}{x-1}$	$-$	$+$	$-$

From this table we can see that the interval required is

$$1 < x < 5$$

[It should be emphasised that x represents a real number and so the statement
$1 < x < 5$ means that x can be *any* number (not just *whole* numbers) between 1
and 5.]

Qu. 5   Sketch the graphs of $y = \dfrac{x+3}{x-1}$ and $y = 2$, and use your diagram to
illustrate the result of Example 1.

Example 2   *For what values of x is the function* $2x^2 + 5x - 3$
(a) *negative*, (b) *positive*?

Let $f(x) = 2x^2 + 5x - 3 = (2x - 1)(x + 3)$. $f(x)$ is zero when $x = \frac{1}{2}$ or $x = -3$, so consider the signs of the factors in the following intervals:

	$x < -3$	$-3 < x < \frac{1}{2}$	$\frac{1}{2} < x$
$x + 3$	$-$	$+$	$+$
$2x - 1$	$-$	$-$	$+$
$f(x)$	$+$	$-$	$+$

Alternatively, sketch the graph of the function $f(x)$. As $x \to \pm \infty$, $f(x) \to +\infty$; $f(\frac{1}{2}) = f(-3) = 0$. The curve is sketched in Fig. 6.1.

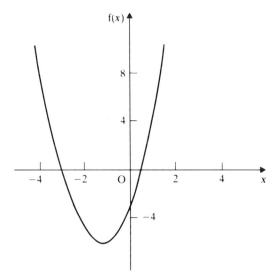

Figure 6.1

By either method, the function is negative if $-3 < x < \frac{1}{2}$ and positive if $x < -3$ or $x > \frac{1}{2}$.

Example 3   *Show that $3x^2 + 10x + 9$ cannot be negative and find its least value.*

Completing the square (see Book 1, §10.3),

$$3x^2 + 10x + 9 = 3(x^2 + \tfrac{10}{3}x + \tfrac{25}{9}) + 9 - \tfrac{25}{3}$$
$$= 3(x + \tfrac{5}{3})^2 + \tfrac{2}{3}$$

Since $(x + \tfrac{5}{3})^2$ is a square, the least value it can take is zero, so that $3x^2 + 10x + 9$ cannot be zero and its least value is $\frac{2}{3}$.

A similar method may be applied to functions of more than one variable.

Example 4   *Show that $a^2 + b^2 + c^2 - bc - ca - ab$ cannot be negative. Under what circumstances is it zero?*

$$a^2 + b^2 + c^2 - bc - ca - ab$$
$$= \tfrac{1}{2}(b^2 + c^2 - 2bc + c^2 + a^2 - 2ca + a^2 + b^2 - 2ab)$$
$$= \tfrac{1}{2}\{(b - c)^2 + (c - a)^2 + (a - b)^2\} \geqslant 0$$

The equality occurs only when each square is zero, that is when $a = b = c$.

Example 5   *If $(x^2 - x + 1)y = 2x$, within what interval does y lie?*

Writing $(x^2 - x + 1)y = 2x$ as a quadratic equation in x,

$$x^2 y - x(y + 2) + y = 0$$

The roots of the equation $ax^2 + bx + c = 0$ are real when $b^2 - 4ac \geqslant 0$ (see Book 1, §10.2). Therefore, for real values of x,

$$\{-(y + 2)\}^2 - 4y^2 \geqslant 0$$
$$\therefore\ -3y^2 + 4y + 4 \geqslant 0$$
$$\therefore\ (2 + 3y)(2 - y) \geqslant 0$$

Hence, if x is real, y lies in the interval $-\tfrac{2}{3} \leqslant y \leqslant 2$ (see Fig. 6.2).

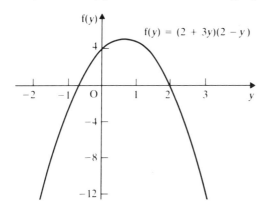

Figure 6.2

Exercise 6a

For what intervals do the following inequalities hold?

1 (a) $\dfrac{x + 1}{2 - x} < 1,$ (b) $\dfrac{x + 1}{2 - x} > 1.$ 2 (a) $\dfrac{4 - x}{x + 2} < 3,$ (b) $\dfrac{4 - x}{x + 2} > 3.$

3 $(3 - x)(x + 2) > 0.$
5 $2x^2 + x - 15 < 0.$
7 $(4x - 3)(x + 1) > 2.$

4 $(2x - 5)(3x + 7) > 0.$
6 $10 + x - 2x^2 < 0.$
8 $(5x - 7)(x - 3) < 16x.$

9 $\dfrac{2x^2 - 7x - 4}{3x^2 - 14x + 11} > 2.$

10 $\dfrac{(x - 1)(x - 3)}{(x + 1)(x - 2)} > 0.$

Prove the following inequalities and find the extreme values of the functions concerned.

11 $x^2 - 5x + 7 > 0.$

12 $4x - x^2 - 5 < 0.$

13 $2x^2 + 3x + 2 > 0.$

14 $5x - 3x^2 - 3 < 0.$

Find the intervals of x and y for which there are no real points on the following loci:

15 $y^2 = x(1 - x).$

16 $3x^2 + 4y^2 = 12.$

17 $y^2 = x(x^2 - 1).$

18 $(x - 2)(x - 3)y = 2x - 5.$

19 $(x^2 + 1)y = 3x + 4.$

For what intervals are the following equations satisfied by real values of θ?

20 $\sin \theta = \dfrac{x - 1}{x + 1}.$

21 $\cos \theta = \dfrac{x + 3}{3 - x}.$

22 $\sin \theta + \cos \theta = x.$

23 Show that $(a + b)^2 \geqslant 4ab.$

24 Verify the identity

$$a^3 + b^3 + c^3 - 3abc = (a + b + c)(a^2 + b^2 + c^2 - bc - ca - ab)$$

and deduce that the arithmetic mean of three unequal positive numbers x, y, z $[\frac{1}{3}(x + y + z)]$ is greater than their geometric mean $[(xyz)^{1/3}]$.

25 Express $5x^2 - 12xy + 9y^2 - 4x + 4$ as the sum of two squares and show that the expression is positive except for one pair of values of x and y.

Rational functions of two quadratics

6.2   In this section we shall be concerned with rational functions of two quadratics, that is, functions of the form

$$\frac{ax^2 + bx + c}{Ax^2 + Bx + C}$$

where a, b, c, A, B, C are constants. The method of the last section will be used.

Example 6   *Sketch the curve* $y = \dfrac{(x - 1)(x + 2)}{(x + 1)(x - 3)}.$

First note the following:

(a) when $y = 0$, $x = 1$ or $x = -2$;

(b) when $x = 0$, $y = \frac{2}{3}$;

(c) when $x = -1$ or $x = 3$, the denominator of the fraction is zero so that there is no corresponding value of y; the function is *discontinuous* at these points. If x differs from 1 or 3 by a small amount, the denominator is small and so y is large. Therefore $y \to \infty$ as $x \to -1$ and $x \to 3$;

(d) given any value of x, other than 1 or 3, there exists one and only one value of y. Note that since the equation is a quadratic in x, there are in general two values of x corresponding to each value of y;

(e) the sign of y may be determined by inspecting the signs of the factors $x + 2$, $x + 1$, $x - 1$, $x - 3$;

	$x < -2$	$-2 < x < -1$	$-1 < x < 1$	$1 < x < 3$	$3 < x$
$x + 2$	$-$	$+$	$+$	$+$	$+$
$x + 1$	$-$	$-$	$+$	$+$	$+$
$x - 1$	$-$	$-$	$-$	$+$	$+$
$x - 3$	$-$	$-$	$-$	$-$	$+$
y	$+$	$-$	$+$	$-$	$+$

(f) $y = \dfrac{x^2 + x - 2}{x^2 - 2x - 3}$.

If x is large, the terms in x and the constants are small compared with x^2 so that $y \approx x^2/x^2 = 1$. If we substitute $y = 1$ in the equation,

$$x^2 - 2x - 3 = x^2 + x - 2$$
$$\therefore x = -\tfrac{1}{3}$$

Therefore the graph crosses $y = 1$ at $(-\tfrac{1}{3}, 1)$.

Our findings are shown in Fig. 6.3; the shading denotes areas where the curve cannot lie (see stage (e)).

The graph is then sketched as in Fig. 6.4.

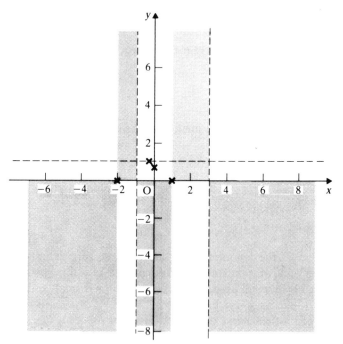

Figure 6.3

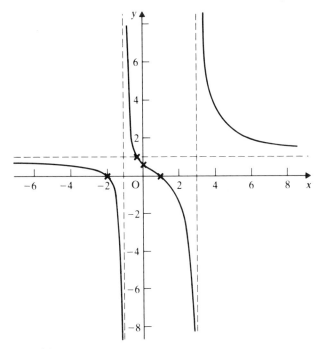

Figure 6.4

The lines $x = -1$, $x = 3$, $y = 1$, which are represented by broken lines, are called **asymptotes**. Note that the curve approaches them ever more closely, without meeting them, as it recedes from the origin. It is possible, however, for the curve to cut an asymptote, as at $(-\frac{1}{3}, 1)$.

Care should be taken to find from which sides the graph approaches the asymptotes. For $x = -1$ and $x = 3$ this was ensured by examining the sign of y. For $y = 1$ the point of intersection with the graph was found. Another method for the latter is to take a second approximation for y, namely

$$\frac{x^2 + x}{x^2 - 2x}$$

If $x > 0$, the numerator is greater than the denominator, so that the graph approaches $y = 1$ from above. On the other hand, when x is large and negative, $y < 1$.

Example 7   *Prove that $(3x - 9)/(x^2 - x - 2)$ cannot lie between two certain values. Illustrate graphically.*

Let $y = \dfrac{3x - 9}{x^2 - x - 2}$.

Regard this equation as a quadratic which gives x in terms of y, then

$$(x^2 - x - 2)y = 3x - 9$$
$$\therefore yx^2 - x(y + 3) + 9 - 2y = 0 \tag{1}$$

When x is *not* real,

$$\{-(y+3)\}^2 - 4y(9-2y) < 0$$
$$\therefore 9y^2 - 30y + 9 < 0$$
$$\therefore 3(3y-1)(y-3) < 0$$
$$\therefore \tfrac{1}{3} < y < 3$$

Therefore there are no real values of y between $\tfrac{1}{3}$ and 3.

Now $y = \dfrac{3(x-3)}{(x+1)(x-2)}$, and we may proceed as in Example 6.

(a) If $y = 0$, $x = 3$.
(b) If $x = 0$, $y = 4\tfrac{1}{2}$.
(c) The lines $x = -1$ and $x = 2$ are asymptotes; the function is discontinuous when $x = -1$ and $x = 2$.
(d) There is only one value of y for each value of x.
(e) The sign of y is obtained:

	$x < -1$	$-1 < x < 2$	$2 < x < 3$	$3 < x$
$x+1$	$-$	$+$	$+$	$+$
$x-2$	$-$	$-$	$+$	$+$
$x-3$	$-$	$-$	$-$	$+$
y	$-$	$+$	$-$	$+$

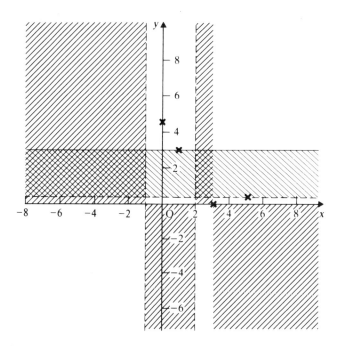

Figure 6.5

(f) As $x \to \infty$, $y \to 0$.

(g) The values of x corresponding to $y = \frac{1}{3}$ and $y = 3$ are found from equation (1).

[Note that $y = \frac{1}{3}$ and $y = 3$ make the discriminant '$b^2 - 4ac$' $= 0$, so that equation (1) on page 93 has equal roots. The sum of the roots is $(y + 3)/y$, therefore $x = \frac{1}{2}(y + 3)/y$.] When $y = \frac{1}{3}$, $x = 5$; when $y = 3$, $x = 1$.

Our findings are shown in Fig. 6.5 and the curve has been sketched in Fig. 6.6.

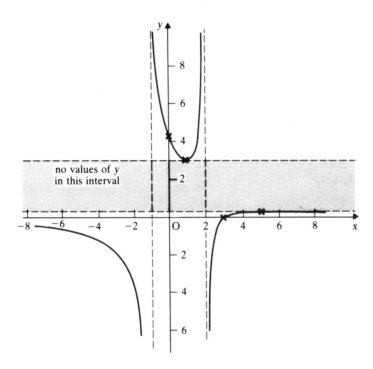

Figure 6.6

Example 8 *Sketch the curve* $y = 2x/(x^2 + 1)$.

(a) The curve cuts the axes only at $(0, 0)$.

(b) As a quadratic in x, the equation is $x^2y - 2x + y = 0$.

For real values of x,

$$(-2)^2 - 4y^2 \geq 0$$
$$\therefore\ 4(1 - y)(1 + y) \geq 0$$
$$\therefore\ -1 \leq y \leq +1$$

(c) When $y = -1$, $x = -1$, and when $y = +1$, $x = +1$. Therefore $(-1, -1)$ is a minimum and $(1, 1)$ a maximum.

(d) As $x \to \infty$, $y \to 0$.

(e) Since $x^2 + 1$ is positive, x and y have the same sign; $x^2 + 1$ is never zero, so the function is continuous.

The curve has been sketched in Fig. 6.7.

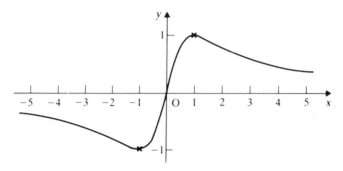

Figure 6.7

Exercise 6b

Sketch the following curves.

1 $y = \dfrac{x-2}{x+3}$.

2 $y = \dfrac{x}{x-1}$.

3 $y = \dfrac{x}{x^2-1}$.

4 $y = \dfrac{x^2}{x^2-1}$.

5 $y = \dfrac{2x-4}{(x-1)(x-3)}$.

6 $y = \dfrac{(x-1)(x+3)}{(x-2)(x+2)}$.

7 $y = \dfrac{x^2-4x+1}{x^2-4x+4}$.

8 $y = \dfrac{(x-1)^2}{x(x-2)}$.

9 $y = \dfrac{(x-2)^2}{x(x+2)}$.

For each of the following curves, find the intervals within which y *cannot* lie. Illustrate graphically.

10 $y = \dfrac{4}{(x-1)(x-3)}$.

11 $y = \dfrac{3x-6}{x(x+6)}$.

12 $y = \dfrac{1}{x^2+1}$.

13 $y = \dfrac{4x^2-3x}{x^2+1}$.

14 $y = \dfrac{(x-3)(x-1)}{(x-2)^2}$.

15 $y = \dfrac{x^2+1}{x^2-x-2}$.

Find the turning points of the following and sketch the curves.

16 $y = \dfrac{x^2-4x}{x^2-4x+3}$.

17 $y = \dfrac{x^2-x-2}{x^2-2x-8}$.

18 $y = \dfrac{x^2-3x}{x^2+5x+4}$.

19 $y = \dfrac{2x^2-9x+4}{x^2-2x+1}$.

Some tests for symmetry

6.3 The remainder of this chapter is devoted to further aids to curve sketching, and the most useful of these is symmetry.

First consider the graph of $y = x^2$ (Fig. 6.8 (i)), which is symmetrical about the y-axis. If the point (h, k) lies on the curve, we have $k = h^2$, and so the point $(-h, k)$ also lies on the curve. In general, if an equation is unaltered by replacing x by $-x$, the curve is symmetrical about the y-axis. The graphs of all *even* functions are symmetrical about the y-axis.

Similarly, if the equation of a curve is unaltered by replacing y by $-y$, there is symmetry about the x-axis (Fig. 6.8 (ii)).

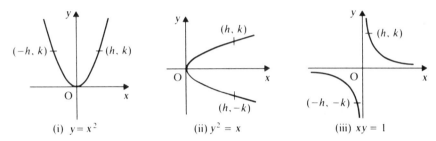

(i) $y = x^2$ (ii) $y^2 = x$ (iii) $xy = 1$

Figure 6.8

Fig. 6.8 (iii) represents the curve $xy = 1$, which has *rotational* symmetry about a *point*, the origin. If (h, k) lies on the locus, so does $(-h, -k)$. In general, if an equation is unaltered when x and y are replaced by $-x$ and $-y$ respectively, the curve is said to have rotational symmetry about the origin. The graphs of all *odd* functions have rotational symmetry about the origin.

Qu. 6 Which of the following show symmetry about (i) the y-axis, (ii) the x-axis, (iii) the origin?
(a) $4x^2 + y^2 = 1$,   (b) $y^2 = x(x + 1)$,
(c) $x^5 + y^5 = 5xy^2$,   (d) $x^2 - 3xy + y^2 = 1$,
(e) $y^2 = x^2(x + 1)(x - 1)$,   (f) $x^2y - x + y^3 = 0$,
(g) $y^2 = \cos x$,   (h) $\tan y = \sin x$.
Qu. 7 Some equations are unaltered by the following substitutions:
(a) $x = y$, $y = x$;   (b) $x = -y$, $y = -x$.
About what lines are the corresponding curves symmetrical?
Qu. 8 Show that a curve which is symmetrical about the x- and y-axes has rotational symmetry about the origin.

Example 9 *Sketch the curve* $x^2 - y^2 = 1$.

(a) The equation shows symmetry about both axes and the origin.
(b) Since $y^2 = x^2 - 1$, y is not real when x is numerically less than 1.
(c) When $y = 0$, $x = \pm 1$.
(d) As x increases in magnitude, so does y.

(e) On differentiation,

$$2y\frac{dy}{dx} = 2x$$

$$\therefore \frac{dy}{dx} = \frac{x}{\sqrt{(x^2-1)}}$$

$$\therefore \text{ as } x \to \pm 1, \quad \frac{dy}{dx} \to \infty$$

and

$$\text{as } x \to \pm \infty, \quad \frac{dy}{dx} \to \pm 1$$

(f) Since $y^2 = x^2 - 1$, when x, y are large, y^2 is nearly equal to x^2. Thus the curve approaches the lines $y = \pm x$.

The curve has been sketched in Fig. 6.9.

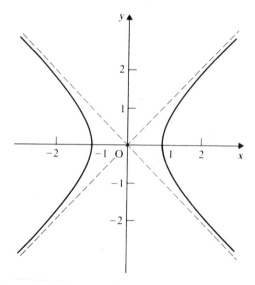

Figure 6.9

The form $y^2 = f(x)$

6.4 If an equation can be expressed in the form $y^2 = f(x)$, then it will have a number of special features. Since y^2 cannot be negative, x must be limited to values for which $f(x)$ is non-negative; for any such value we can write $y = \pm \sqrt{f(x)}$, so the graph will be symmetrical about the x-axis.

Example 10   *Sketch the graph of $y^2 = x(x-2)^2$.*

The factor $(x - 2)^2$ is never negative so the sign of the R.H.S. is determined by the factor x. So x must be greater than, or equal to, zero to obtain real values of y. Also, y is zero at $x = 0$ and $x = 2$.

Consider $y^2 = x(x - 2)^2$.

On differentiating, we obtain

$$2y\frac{dy}{dx} = (x - 2)^2 + 2x(x - 2)$$

$$\frac{dy}{dx} = \frac{(x - 2)(3x - 2)}{\pm 2x^{1/2}(x - 2)}$$

$$= \pm\frac{3x - 2}{2x^{1/2}}$$

From this we can see that $\dfrac{dy}{dx} = 0$ when $x = \dfrac{2}{3}$, and that

as $x \to 0$, $\dfrac{dy}{dx} \to \infty$

as $x \to 2$, $\dfrac{dy}{dx} \to \pm\sqrt{2}$

and

as $x \to \infty$, $\dfrac{dy}{dx} \to \pm\infty$

The graph of $y^2 = x(x - 2)^2$ is shown in Fig. 6.10.

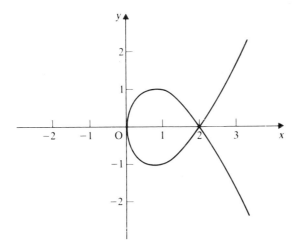

Figure 6.10

[It should be noted that the equation $y^2 = x(x - 2)^2$ cannot be regarded as a rule for expressing y as a *function* of x, because there are *two* values of y for each value of x. However $y = +\sqrt{x(x-2)}$ and $y = -\sqrt{x(x-2)}$ could be regarded as two functions whose graphs could be combined to produce the graph of $y^2 = x(x-2)^2$.]

Qu. 9 Find the gradient of
(a) $y^2 = x(x-2)(x-4)$,   (b) $y^2 = x^2(x+2)$,
at the points where the graphs cut the x-axis.
  Sketch the curves by the method of Example 10.

Simple changes of axes

6.5   The equation of a circle, centre $C(a, b)$ and radius r, is (Book 1, §21.1)

$$(x-a)^2 + (y-b)^2 = r^2$$

and the equation of an equal circle, centre the origin, is

$$x^2 + y^2 = r^2$$

Therefore, if new axes CX and CY were taken parallel to Ox and Oy, the equation of the former would become

$$X^2 + Y^2 = r^2$$

  This is equivalent to making the substitutions

$$X = x - a \qquad Y = y - b$$

or, as is often more convenient,

$$x = X + a \qquad y = Y + b$$

These relationships may easily be verified from a diagram.

  Such a change of axes is sometimes helpful in curve sketching. Thus

$$(y-1)^2 = 4(x+2)$$

becomes

$$Y^2 = 4X$$

referred to parallel axes through $(-2, 1)$ and the curve is now easily drawn, as in Fig. 6.11.
  Note that the equation $y = ax^2 + bx + c$ may be written

$$y + \frac{b^2}{4a} - c = a\left(x + \frac{b}{2a}\right)^2$$

Referred to parallel axes through $\left(-\dfrac{b}{2a}, -\dfrac{b^2 - 4ac}{4a}\right)$ the equation becomes

$$Y = aX^2$$

which is a parabola (see Book 1, §10.4 and §22.6).

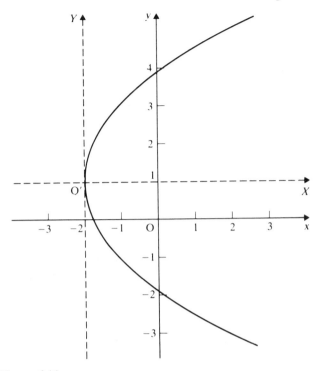

Figure 6.11

Example 11   *Sketch the function* $1 + 2 \sin (\theta + \frac{1}{4}\pi)$ *for values of* θ *from* 0 *to* 2π.

Write $y = 1 + 2 \sin (\theta + \frac{1}{4}\pi)$
$\therefore y - 1 = 2 \sin (\theta + \frac{1}{4}\pi)$

With the substitutions

$$\Theta = \theta + \tfrac{1}{4}\pi \quad \text{and} \quad Y = y - 1 \tag{1}$$

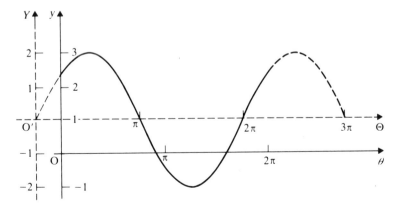

Figure 6.12

the equation becomes

$$Y = 2 \sin \Theta$$

The graph of $Y = 2 \sin \Theta$ has been sketched in Fig. 6.12. Writing $\theta = y = 0$ in equations (1), the origin of the θ, y axes is found to be $(\frac{1}{4}\pi, -1)$, referred to the Θ, Y axes. The θ, y axes were then drawn to pass through this point.

The form $y = 1/f(x)$

6.6   Example 12   *Sketch on the same axes the graphs of*
(a) $y = (x + 1)(2x - 3)$,   (b) $y = 1/\{(x + 1)(2x - 3)\}$.

(a) The graph of $f(x) = (x + 1)(2x - 3)$ is a parabola meeting the x-axis at $(-1, 0)$ and $(1\frac{1}{2}, 0)$. As $x \to \pm\infty$, $y \to +\infty$. See the broken line in Fig. 6.13.

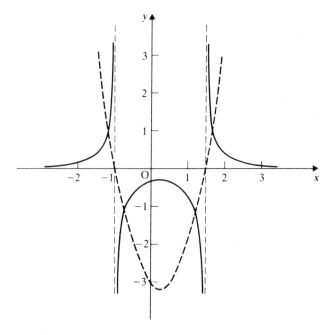

Figure 6.13

(b) The reciprocal of $f(x)$ is sketched as follows:
   (i)   the signs of $f(x)$ and $1/f(x)$ are the same,
   (ii) as $f(x) \to \infty$, $1/f(x) \to 0$ and vice versa,
   (iii) when $f(x) = \pm 1$, $1/f(x)$ has the same value.
  The two graphs are shown in Fig. 6.13.

Exercise 6c

Sketch the following curves:

1 $2x^2 + y^2 = 4$.

2 $2x^2 - y^2 = 4$.

3 $x^2 y = 1$.

4 $x^2 y^2 = 4$.

5 $y^2 = x(x - 2)$.

6 $y^2 = x^3(4 - x)$.

7 $y^2 = x^2(2 - x)$.

8 $y^2 = x^2(x^2 - 1)$.

9 $y^2 = x^2(1 - x^2)$.

10 $y^2 = x^2(2 - x)^3$.

11 $y^2 = (x^2 - 1)(4 - x^2)$.

12 $y = 1/\{(x - 2)(5 - x)\}$.

13 $y = 1/(x^2 - 4x + 3)$.

14 $xy(x^2 - 1) = 1$.

15 $y = (x - 2)^2 + 1$.

16 $(x - 1)(y + 2) = 1$.

17 $y = 1 + \frac{1}{2}\cos(x + \frac{1}{6}\pi)$

18 $y = 3 - 2\sin(2x - \frac{1}{4}\pi)$.

19 $y = 1/(1 + 2\sin x)$, $0 \leqslant x \leqslant 4\pi$.

20 $y = 1/(1 - 2\cos 2x)$, $0 \leqslant x \leqslant 2\pi$.

Exercise 6d (Miscellaneous)

1 Determine the intervals in which

(a) $x^2 - 4x + 3 > 0$, (b) $\dfrac{2x^2 - 4x + 5}{x^2 + 2} > 1$, (c) $\dfrac{x^2 - 4x + 3}{x - 2} > 0$. (L)

2 By completing the square, or otherwise, prove that the inequality

$$x^2 - 2px + q > 0$$

holds for all values of x if and only if $q > p^2$. Find the intervals in which the inequality is broken if $q = p^2 - 1$. (L)

3 Write down conditions that the roots of the equation $ax^2 + bx + c = 0$ may be real and positive.

Prove that if these conditions are satisfied, the roots of the equation

$$a^2 y^2 + a(3b - 2c)y + (2b - c)(b - c) + ac = 0$$

are real and positive. (L)

4 Show that, if λ is positive but not greater than 3, the roots of the equation

$$(\lambda - 2)x^2 - (8 - 2\lambda)x - (8 - 3\lambda) = 0$$

are real.

Find the range of values of λ for which one root is real and positive and the other root is real and negative.† (L)

5 Prove that $ax^2 + bx + c$ is positive for all real values of x if $a > 0$ and $b^2 < 4ac$.

Find the range of values of k for which $x^2 + kx + 3 + k$ is positive for all real values of x. Deduce the range of values of k for which $k(x^2 + kx + 3 + k)$ is positive for all real values of x. (JMB)

---

† It is a common practice, especially in examination questions, to use the phrase 'range of values' when an interval is required: this should not be confused with the use of 'range' to mean the set of images of a function f(x).

6 Show that, if a, b, c are real, $a^2 + b^2 + c^2 - bc - ca - ab$ cannot be negative. Show also that the roots of the equation

$$3x^2 - 2x(a + b + c) + bc + ca + ab = 0$$

are real, and find the relation between a, b, c if one root is three times the other. (L)

7 Find the ranges of values of α in the interval $0 \leqslant \alpha \leqslant 2\pi$ for which the roots of the equation in x

$$x^2 \cos^2 \alpha + ax(\sqrt{3} \cos \alpha + \sin \alpha) + a^2 = 0$$

are real. (C)

8 Prove that $ax^2 + 2bx + c$ has the same sign as a except when the equation $ax^2 + 2bx + c = 0$ has real roots and x lies between them.

By using the substitution $x = x' + 4$, $y = y' - 1$, or otherwise, prove that, for any real values of x and y,

$$x^2 + 2xy + 3y^2 - 6x - 2y \geqslant -11$$ (JMB)

9 Show that the expression $x^2 + 8xy - 5y^2 - k(x^2 + y^2)$ can be put in the form $a(x + by)^2$ when k has either one or other of two values. Find these values and the values of a and b corresponding to each value of k.

Prove that when the variables x and y are restricted by the relation $x^2 + y^2 = 1$ but are otherwise free, then

$$-7 \leqslant x^2 + 8xy - 5y^2 \leqslant 3$$ (O & C)

10 If the roots of the equation $ax^2 + 2bx + c = 0$ are real and unequal, prove that the function

$$(a + c)(ax^2 + 2bx + c) - 2(ac - b^2)(x^2 + 1)$$

is positive for all real values of x. (JMB)

11 Prove, by first squaring, that, if $0 < x < 1$,

$$(1 - x)^{1/2} < \frac{4 - 3x}{4 - x}$$

and deduce that

$$(1 - x)^{1/2} > \frac{(4 - x)(1 - x)}{4 - 3x}$$

Hence prove that $(1 - x)^{1/2}$ differs from $\dfrac{4 - 3x}{4 - x}$ by less than

$$\frac{x^3}{(4 - x)(4 - 3x)}.$$

By putting $x = \tfrac{1}{9}$, show that $\sqrt{2}$ differs from $99/70$ by less than 2×10^{-4}. (O & C)

12 (a) Find what restrictions must be imposed on the values of x and y in order to satisfy both the inequalities

$$x > y \quad \text{and} \quad \frac{x}{x+1} > \frac{y}{y+1}$$

(b) Show that $b^2 + c^2 \geqslant \frac{1}{2}(b+c)^2$, and hence find the range of values of a for which the simultaneous equations

$$a + b + c = 1$$
$$a^2 + b^2 + c^2 = 3$$

may be satisfied for real values of b and c. (JMB)

13 (a) If $pq > 0$, prove that $p/q > 0$.

Find the intervals in which

(i) $\dfrac{1 - 4x}{2x - 3} > 0$, (ii) $\dfrac{2x}{x-1} + \dfrac{x-5}{x-2} > 3$.

(b) In one diagram sketch the three straight lines

$$x + y - 3 = 0, \qquad y - \tfrac{1}{2}x - 5 = 0, \qquad y - 3x + 10 = 0$$

and shade the region in which the following three inequalities are all satisfied:

$$x + y - 3 > 0, \qquad y - \tfrac{1}{2}x - 5 > 0, \qquad y - 3x + 10 > 0$$

In a second diagram shade the region in which none of them is satisfied.

14 Prove that $-\dfrac{1}{4} \leqslant \dfrac{x}{x^2 + 4} \leqslant \dfrac{1}{4}$.

Draw in the same diagram the graphs of

$$y = \frac{x}{x^2 + 4}, \qquad y = x + 1$$

from $x = -4$ to $x = +4$, and deduce that the equation

$$\frac{x}{x^2 + 4} = x + 1$$

has only one real root in this interval. Find the value of this root as accurately as you can. (O & C)

15 Prove that, if x is real, the function $(2x^2 - 5x + 2)/(x - 1)$ can assume all real values.

Sketch the graph of $y = (2x^2 - 5x + 2)/(x - 1)$ from $x = -1$ to $x = +3$, omitting the portion given by the values of x very near to $+1$. (L)

16 For what real values of x is $\left| \dfrac{1}{1 + 2x} \right| = 1$?

Solve the inequality $\left| \dfrac{1}{1 + 2x} \right| < 1$. (O & C)

17 Determine the set of values of x for which

(a) $\dfrac{6}{x+1} < x$, (b) $\dfrac{6}{|x|+1} < |x|$. (O & C)

18 Find the set of values of x for which

(a) $x^2 - 5x + 6 \geqslant 2$,   (b) $1/(x^2 - 5x + 6) \leqslant \frac{1}{2}$.   (O & C)

19 Find the set of values of x for which $\dfrac{x(x + 2)}{x - 3} < x + 1$.   (C)

20 Solve the inequality $\dfrac{2x}{x + 1} > x$, $\quad (x \in \mathbb{R}, x \neq -1)$.   (C)

21 The function f is defined for real values of x (except -1 and 2) by

$$f(x) = \frac{x + 2}{x^2 - x - 2}$$

Find the set of values taken by $f(x)$ and sketch the graph of $y = f(x)$.   (C)

22 Given that $y(x - 1) = x^2 + 3$, where x is real, show that y cannot take any value between -2 and 6.

Find the asymptotes of the curve $y = (x^2 + 3)/(x - 1)$ and sketch the curve, showing the coordinates of the turning points.   (C)

23 Given that $f(x) = x - 1 + 1/(x + 1)$, x real, $x \neq 1$, find the values of x for which $f'(x) = 0$. Sketch the graph of f, showing the coordinates of the turning points and indicating clearly the form of the graph when $|x|$ becomes large.
                    (JMB)

24 Show that the graph of $y = (x^2 - 1)/(x^2 + 4x)$ has no real stationary points and sketch this graph.   (JMB)

25 The domain of the function f is the set $D = \{x: x \in \mathbb{R}, x \neq -1, x \neq 1\}$. The function $f: D \to \mathbb{R}$ is defined by

$$f(x) = \frac{x^2}{(x - 1)(x + 1)}$$

Find the coordinates of the maximum point on the graph of f and state the equation of each of the asymptotes of the graph.

Sketch the graph showing in particular how the curve approaches each of its asymptotes.   (L)

Chapter 7

Further equations and factors

Equations reducing to quadratics

7.1 Certain types of equation can be solved by reducing them to a quadratic equation and, as no new principles are involved, this topic is simply illustrated by examples. There is no need to read all the Examples 1 to 6 before attempting Exercise 7a—some readers will prefer to work the corresponding questions before going on to the next example.

Example 1 *Solve the equation* $\dfrac{x^2 + 4x}{3} + \dfrac{84}{x^2 + 4x} = 11.$

Substitute $y = x^2 + 4x$ in the given equation.

$$\therefore \frac{y}{3} + \frac{84}{y} = 11$$

$$\therefore y^2 - 33y + 252 = 0$$
$$\therefore (y - 12)(y - 21) = 0$$

(a) If $y = 12$,
$$x^2 + 4x - 12 = 0$$
$$\therefore (x + 6)(x - 2) = 0$$
$$\therefore x = -6, 2$$

(b) If $y = 21$,
$$x^2 + 4x - 21 = 0$$
$$\therefore (x + 7)(x - 3) = 0$$
$$\therefore x = -7, 3$$

Therefore the roots of the equation are $-7, -6, 2, 3$.

Example 2 *Solve the equation* $\sqrt{(5x - 25)} - \sqrt{(x - 1)} = 2.$†

The method used is to isolate one square root on one side of the equation and then to square both sides.

---

† By convention the square root sign is always taken to mean the *positive* square root, e.g. $\sqrt{9} = +3$.

$$\sqrt{(5x - 25)} - \sqrt{(x - 1)} = 2$$
$$\therefore \sqrt{(5x - 25)} = 2 + \sqrt{(x - 1)}$$
$$\therefore 5x - 25 = 4 + 4\sqrt{(x - 1)} + x - 1$$
$$\therefore 4x - 28 = 4\sqrt{(x - 1)}$$
$$\therefore x - 7 = \sqrt{(x - 1)}$$

Squaring,

$$x^2 - 14x + 49 = x - 1$$
$$\therefore x^2 - 15x + 50 = 0$$
$$\therefore (x - 5)(x - 10) = 0$$

Now check the values $x = 5$ and $x = 10$ in the original equation.

If $x = 5$,   L.H.S. $= \sqrt{0} - \sqrt{4} = -2$

Therefore 5 is not a root of the equation.

If $x = 10$,   L.H.S. $= \sqrt{25} - \sqrt{9} = 2$

Therefore the only root of the equation is 10.

Example 2 illustrates the need for checking the roots obtained by squaring both sides of an equation.

Example 3 *Find the coordinates of the points of intersection of the circles* $x^2 + y^2 - 6x + 4y - 13 = 0$ *and* $x^2 + y^2 - 10x + 10y - 15 = 0$.

To find the points of intersection, we solve simultaneously the equations

$$x^2 + y^2 - 6x + 4y - 13 = 0 \tag{1}$$
$$x^2 + y^2 - 10x + 10y - 15 = 0$$

Subtracting,

$$4x - 6y + 2 = 0$$

$$\therefore 2x = 3y - 1 \quad \text{and} \quad 4x^2 = 9y^2 - 6y + 1$$

$4 \times (1)$:

$$4x^2 + 4y^2 - 24x + 16y - 52 = 0$$

$$\therefore (9y^2 - 6y + 1) + 4y^2 - 12(3y - 1) + 16y - 52 = 0$$

$$\therefore 13y^2 - 26y - 39 = 0$$
$$\therefore 13(y + 1)(y - 3) = 0$$

Substituting $y = -1$ and $y = 3$ in $2x = 3y - 1$, we obtain $x = -2$ and $x = 4$, respectively.

Therefore the circles meet at $(-2, -1)$, $(4, 3)$.

The next example uses one of the results (Book 1, §9.7) that if α, β are the roots of the equation $ax^2 + bx + c = 0$, then $\alpha + \beta = -b/a$ and $\alpha\beta = c/a$.

Example 4   *Find where the normal at* $(at^2, 2at)$ *to the parabola* $y^2 = 4ax$ *cuts the curve again.*

The gradient at $(at^2, 2at)$ is given by

$$\frac{dy}{dx} = \frac{dy}{dt} \div \frac{dx}{dt} = \frac{2a}{2at} = \frac{1}{t}$$

Therefore the gradient of the normal is $-t$ and its equation is

$$tx + y - at^3 - 2at = 0$$

To solve simultaneously with $y^2 = 4ax$, multiply the former by $4a$:

$$t \times 4ax + 4ay - 4a^2t^3 - 8a^2t = 0$$
$$\therefore ty^2 + 4ay - 4a^2t^3 - 8a^2t = 0$$

Now one root of the equation is $y = 2at$ (since the normal meets the curve at $(at^2, 2at)$). But the sum of the roots is $-4a/t$. Therefore the other root is

$$-\frac{4a}{t} - 2at = -2a(t + 2/t)$$

$$x = y^2/(4a) = a(t + 2/t)^2$$

Therefore the normal meets the parabola again at

$$(a(t + 2/t)^2, \ -2a(t + 2/t))$$

(For another method, see Exercise 9a, No. 7.)

Example 5   *Show that, if the equations*

$$x^2 + ax + 1 = 0 \quad and \quad x^2 + x + b = 0$$

have a common root, then $(b - 1)^2 = (a - 1)(1 - ab)$.

If x_1 is a root of both equations,

$$x_1^2 + ax_1 + 1 = 0 \tag{1}$$
$$x_1^2 + x_1 + b = 0 \tag{2}$$

We must now eliminate x_1 from equations (1) and (2). Subtracting,

$$x_1(a - 1) + 1 - b = 0$$

$$\therefore x_1 = \frac{(b - 1)}{(a - 1)} \qquad (a \neq 1) \tag{3}$$

We may now substitute for x_1 in (1) or (2), or proceed as follows.

$a \times (2) - (1)$:

$$x_1^2(a - 1) + (ab - 1) = 0$$

$$\therefore x_1^2 = \frac{(1 - ab)}{(a - 1)} \qquad (a \neq 1) \tag{4}$$

From (3) and (4),

$$\frac{(b-1)^2}{(a-1)^2} = \frac{1-ab}{a-1}$$

$$\therefore (b-1)^2 = (a-1)(1-ab) \qquad (a \neq 1)$$

The next example shows a method which can be used to solve a quartic equation whose coefficients are arranged symmetrically in the form

$$Ax^4 + Bx^3 + Cx^2 + Bx + A = 0$$

Example 6   *Use the substitution* $y = x + 1/x$ *to solve the equation*

$$2x^4 - 9x^3 + 14x^2 - 9x + 2 = 0$$

Let $y = x + \dfrac{1}{x}$

$$\therefore y^2 = x^2 + 2 + \frac{1}{x^2}$$

Dividing both sides of the given equation by x^2,

$$2x^2 - 9x + 14 - \frac{9}{x} + \frac{2}{x^2} = 0$$

$$\therefore 2\left(x^2 + 2 + \frac{1}{x^2}\right) - 9\left(x + \frac{1}{x}\right) + 10 = 0$$

$$\therefore 2y^2 - 9y + 10 = 0$$

$$\therefore (y - 2)(2y - 5) = 0$$

(a) If $y = 2$,

$$x + \frac{1}{x} = 2$$

$$\therefore x^2 - 2x + 1 = 0$$
$$\therefore (x - 1)^2 = 0$$
$$\therefore x = 1$$

(b) If $y = \frac{5}{2}$,

$$x + \frac{1}{x} = \frac{5}{2}$$

$$\therefore 2x^2 - 5x + 2 = 0$$
$$\therefore (2x - 1)(x - 2) = 0$$
$$\therefore x = \frac{1}{2}, 2$$

Therefore the roots of the equation are $\frac{1}{2}$, 1, 1, 2.

Qu. 1   Show that the substitution $y = x + 1/x$ transforms the quadratic equation $ay^2 + by + c = 0$ to the form $ax^2 + bx + d + b/x + a/x^2 = 0$.

Exercise 7a

Solve the equations in Nos. 1–19.

1 $(x^2 - 2x)^2 + 24 = 11(x^2 - 2x)$.   **2** $x^2 + 2x = 34 + 35/(x^2 + 2x)$.
3 $2 - 5e^{-x} + 2e^{-2x} = 0$.   **4** $4^x - 5 \times 2^x + 4 = 0$.

5 $x^2 + 9/x^2 = 10$.

6 $x^{4/3} + 16x^{-4/3} = 17$.

7 $\sqrt{(x+1)} + \sqrt{(x-2)} = 3$.

8 $\sqrt{(x-5)} + \sqrt{x} = 5$.

9 $\sqrt{(3x-3)} - \sqrt{x} = 1$.

10 $2\sqrt{(x+4)} - \sqrt{(x-1)} = 4$.

11 $2\sqrt{(x-1)} + \sqrt{(x-4)} = x$.

12 $\sqrt{(x-1)} + 2\sqrt{(x-4)} = 4$.

13 $2\sqrt{(2x-12)} - \sqrt{(2x-3)} = 3$.

14 $xy = 4, \quad x - 2y - 2 = 0$.

15 $x^2 + y^2 - 2x - 2y - 23 = 0, \quad x - 7y + 31 = 0$.

16 $3x - 4y - 5 = 0, \quad x^2 + y^2 + 2x + 4y - 20 = 0$.

17 $x^2 + y^2 - 8x + 6y = 0, \quad x^2 + y^2 - 5x + 10y = 0$.

18 $x^2 + y^2 + 8x - 4y + 15 = 0, \quad x^2 + y^2 + 6x + 2y - 15 = 0$.

19 $4x^2 + 25y^2 = 100, \quad xy = 4$.

20 One root of the equation in x

$$bx^2 - x(ab + 2a + 2b) + 2a(a + b) = 0$$

is a. Use the formulae for (a) the sum, (b) the product of the roots of a quadratic equation to find the other root of the equation.

21 Repeat No. 20 for the equations

(a) $cx^2 - acx + dx + cx - ad - ac = 0$,

(b) $ax^2 - bx^2 - a^2x + abx + ax + bx - a^2 - ab = 0$.

22 Find the equation of the normal to $xy = c^2$ at $(ct, c/t)$ and obtain the coordinates of the point where the normal cuts the curve again.

23 A chord of gradient 2 passes through the point $(ap^2, 2ap)$ on the parabola $y^2 = 4ax$. Find the coordinates of the other end of the chord.

24 A line with gradient t cuts the rectangular hyperbola at $(ct, c/t)$. Find the coordinates of the other intersection.

25 For the ellipse $b^2x^2 + a^2y^2 = a^2b^2$, find the coordinates of the other end of the chord through $(a, 0)$ with gradient a/b.

26 Find the condition that the equations

$$x^2 + 2x + a = 0, \qquad x^2 + bx + 3 = 0$$

should have a common root.

27 Show that, if the equations

$$x^2 + 2px + q = 0, \qquad x^2 + 2Px + Q = 0$$

have a common root, then, $(q - Q)^2 + 4(P - p)(Pq - pQ) = 0$.

28 Solve the equations

$$ay + bx + c = 0, \qquad Ay + Bx + C = 0$$

and deduce the condition that the equations

$$ax^2 + bx + c = 0, \qquad Ax^2 + Bx + C = 0$$

should have a common root.

Solve the following equations:

29 $6x^4 - 35x^3 + 62x^2 - 35x + 6 = 0$.

30 $4x^4 + 17x^3 + 8x^2 + 17x + 4 = 0$.

A theorem about ratios

7.2   There is a theorem about ratios which on occasions can greatly simplify algebraic working and which sometimes shortens working in trigonometry:

If $\dfrac{a}{b} = \dfrac{c}{d} = \dfrac{e}{f} = \ldots$, *then for any numbers l, m, n …, not all zero,*

$$\frac{a}{b} = \frac{c}{d} = \frac{e}{f} = \ldots = \frac{la + mc + ne + \ldots}{lb + md + nf + \ldots}$$

To prove this, let $a/b = c/d = e/f = \ldots = k$, say.

$$\therefore a = bk, \qquad c = dk, \qquad e = fk \quad \text{etc.}$$

$$\therefore \frac{la + mc + ne + \ldots}{lb + md + nf + \ldots} = \frac{lbk + mdk + nfk + \ldots}{lb + md + nf + \ldots}$$

$$= \frac{k(lb + md + nf + \ldots)}{lb + md + nf + \ldots}$$

$$= k$$

$$\therefore \frac{a}{b} = \frac{c}{d} = \frac{e}{f} = \ldots = \frac{la + mc + ne + \ldots}{lb + md + nf + \ldots}$$

Qu. 2   With $a = 3$, $b = 5$, $c = 6$, $d = 10$, $e = 12$, $f = 20$, $l = 5$, $m = -2$, $n = -1$, or other suitable numbers, verify that the above ratios are equal.

Express the result of the theorem in your own words.

Example 7   If $\dfrac{a}{b} = \dfrac{c}{d}$, *prove that* $\dfrac{a - c}{a + c} = \dfrac{b - d}{b + d}$.

Let $\dfrac{a}{b} = \dfrac{c}{d} = k$, $\therefore a = bk, c = dk$.

$$\therefore \frac{a - c}{a + c} = \frac{bk - dk}{bk + dk} = \frac{b - d}{b + d}$$

Qu. 3   With $a = 6$, $b = 8$, $c = 3$, $d = 4$, or other suitable numbers, check the result of Example 7.

Qu. 4   If $\dfrac{a}{b} = \dfrac{c}{d}$, prove that $\dfrac{a - b}{a + b} = \dfrac{c - d}{c + d}$.

*Qu. 5   Prove that, if $\dfrac{a}{b} = \dfrac{c}{d}$, then $\dfrac{la + mb}{\lambda a + \mu b} = \dfrac{lc + md}{\lambda c + \mu d}$, where l, m, λ, μ are any numbers such that λ, μ are not both zero.

Example 8   If $\dfrac{a}{b} = \dfrac{c}{d}$, *prove that* $\dfrac{a^2 - b^2}{a^2 + b^2} = \dfrac{c^2 - d^2}{c^2 + d^2}$.

Let $\dfrac{a}{b} = \dfrac{c}{d} = k,$ $\quad \therefore a = bk, \quad c = dk.$

$$\therefore \frac{a^2 - b^2}{a^2 + b^2} = \frac{b^2 k^2 - b^2}{b^2 k^2 + b^2} = \frac{k^2 - 1}{k^2 + 1}$$

Similarly $\dfrac{c^2 - d^2}{c^2 + d^2} = \dfrac{k^2 - 1}{k^2 + 1}$, and the result is proved.

(Nos. 1–10 of Exercise 7b are on the above work.)

Example 9 *Prove that, in triangle* ABC,

$$\frac{b - c}{b + c} = \tan \frac{B - C}{2} \cot \frac{B + C}{2}$$

By the sine formula,

$$\frac{b}{\sin B} = \frac{c}{\sin C}$$

$$\therefore \frac{b - c}{b + c} = \frac{\sin B - \sin C}{\sin B + \sin C}$$

$$= \frac{2 \cos \tfrac{1}{2}(B + C) \sin \tfrac{1}{2}(B - C)}{2 \sin \tfrac{1}{2}(B + C) \cos \tfrac{1}{2}(B - C)}$$

$$\therefore \frac{b - c}{b + c} = \cot \tfrac{1}{2}(B + C) \tan \tfrac{1}{2}(B - C)$$

(Now work Nos. 11–15 of Exercise 7b.)

Example 10 *Solve the simultaneous equations,*

$$2x + 3y + 4z = 8, \qquad 3x - 2y - 3z = -2, \qquad 5x + 4y + 2z = 3.$$

$$2x + 3y + 4z = 8 \tag{1}$$
$$3x - 2y - 3z = -2 \tag{2}$$
$$5x + 4y + 2z = 3 \tag{3}$$

y and z may be eliminated equally easily. We eliminate y as follows:

$2 \times (2) + (3)$: $\qquad 11x - 4z = -1$ $\qquad\qquad$ (4)

$3 \times (3) - 4 \times (1)$: $\qquad 7x - 10z = -23$ $\qquad\qquad$ (5)

$5 \times (4) - 2 \times (5)$: $\qquad 41x \quad\;\; = 41$

$\therefore x = 1, \quad z = 3, \quad y = -2$

We could regard the three equations in Example 10 as the equations of three planes in a three-dimensional space (see Book 1, §15.13), in which case the solution $x = 1, y = -2, z = 3$, gives the coordinates of the point of intersection of the three planes, i.e. $(1, -2, 3)$.

Qu. 6 Solve Example 10 by eliminating z instead of y.

(Now work Nos. 16–20 of Exercise 7b.)

Exercise 7b

In Nos. 1–10 it is given that $\dfrac{a}{b} = \dfrac{c}{d} = \dfrac{e}{f}$.

Complete the statements in Nos. 1–4.

1 $\dfrac{a}{b} = \dfrac{c}{d} = \dfrac{2a - c}{}$.

2 $\dfrac{a}{b} = \dfrac{c}{d} = \dfrac{}{3b - 4d}$.

3 $\dfrac{a - c}{b - d} = \dfrac{2a + 3c}{}$.

4 $\dfrac{a + c}{b + d} = \dfrac{}{3b - d}$.

Prove the results in Nos. 5–10.

5 $\dfrac{a + 2c}{b + 2d} = \dfrac{3a + c}{3b + d}$.

6 $\dfrac{a - c + e}{b - d + f} = \dfrac{a + c - e}{b + d - f}$.

7 $\dfrac{a^2 + c^2}{b^2 + d^2} = \dfrac{c^2 + e^2}{d^2 + f^2}$.

8 $\dfrac{a^2}{b^2} = \dfrac{c^2 - e^2}{d^2 - f^2}$.

9 $\dfrac{a + 2b}{c + 2d} = \dfrac{2a + b}{2c + d}$.

10 $\dfrac{3c + 2e}{3d + 2f} = \dfrac{c + e}{d + f}$.

In Nos. 11–15 use the sine formula to prove that, in triangle ABC,

11 $\sin \frac{1}{2}(B - C) = \dfrac{b - c}{a} \cos \frac{1}{2}A$.

12 $\cos \frac{1}{2}(B - C) = \dfrac{b + c}{a} \sin \frac{1}{2}A$.

13 $\dfrac{a + b - c}{a - b + c} = \tan \frac{1}{2}B \cot \frac{1}{2}C$.

14 $\dfrac{a + b + c}{a + b - c} = \cot \frac{1}{2}A \cot \frac{1}{2}B$.

15 $a \cos 2B + 2b \cos A \cos B = c \cos B - b \cos C$.

Solve the equations in Nos. 16–20.

16 $2a + b + 3c = 11$,
$a + 2b - 2c = 3$,
$4a + 3b + c = 15$.

17 $3p + 2q + 5r = 7$,
$2p - 4q + 9r = 9$,
$6p - 8q + 3r = 4$.

18 $2x + 3y + 4z = -4$,
$4x + 2y + 3z = -11$,
$3x + 4y + 2z = -3$,

19 $a - 3b + 6c = 5$,
$a + 6b + 2c = 4$,
$2a + b + c = 7$.

20 $d - 2e + 3f = 4$,
$5d + 6e - 7f = 8$,
$7d - 5e + 6f = 4$.

Homogeneous expressions

7.3 An expression, or equation, is said to be **homogeneous** if every term is of the same degree. For instance, $x^2z + 3x^3$, $3x + 2y - 4z$, $1/x + 1/y + 1/z$ are homogeneous expressions of degree 3, 1, -1 respectively.

Qu. 7 Which of the following expressions are homogeneous? State the degree of those that are:

(a) $x^3 + y^3 + z^3$,   (b) $3yz + zx - 2xy$,   (c) $x^2 + y^2 + 2x + 2y$,

(d) $\dfrac{1}{yz} + \dfrac{1}{zx} + \dfrac{1}{xy}$,   (e) $x^3 + y^3z + z^3x$,   (f) $x^4 + y^2z^2 + xyz^2$.

Example 11 *Solve, for the ratio $x:y$, the equation $x^2 - 3xy - 40y^2 = 0$.*

$$x^2 - 3xy - 40y^2 = 0$$
$$\therefore (x - 8y)(x + 5y) = 0$$
$$\therefore x - 8y = 0 \quad \text{or} \quad x + 5y = 0$$
$$\therefore x:y = 8 \quad \text{or} \quad -5$$

Example 12 *Solve for $x:y:z$ the equations*

$$6x - 5y - 6z = 0 \qquad 10x + 7y - 33z = 0$$

$$\begin{array}{lll} & 6x - 5y - 6z = 0 & (1) \\ & 10x + 7y - 33z = 0 & (2) \end{array}$$

$$\begin{array}{ll} 5 \times (1): & 30x - 25y - 30z = 0 \\ -3 \times (2): & -30x - 21y + 99z = 0 \end{array}$$

$$\begin{array}{lll} \text{Adding,} & -46y + 69z = 0 & \\ & \therefore -2y + 3z = 0 & (3) \end{array}$$

Substituting $3z = 2y$ in (1),

$$\begin{array}{ll} 6x - 5y - 4y = 0 & \\ \therefore 6x - 9y = 0 & \\ \therefore 2x - 3y = 0 & (4) \end{array}$$

Writing (4), (3) as ratios,

$$x:y = 3:2 \qquad y:z = 3:2$$

Writing these ratios so that the number corresponding to y is the same in each,

$$x:y = 9:6 \qquad y:z = 6:4$$
$$\therefore x:y:z = 9:6:4$$

The two equations in Example 12 could be regarded as the equations of two planes through the origin. The solution, which could be written in the form

$$\frac{x}{9} = \frac{y}{6} = \frac{z}{4}$$

would then be the equation of their common line (see Book 1, §15.14).

It may be noticed that a number of the equations used in coordinate geometry are homogeneous in x, y, a, b, c, e.g. the parabola $y^2 = 4ax$, the ellipse $b^2x^2 + a^2y^2 = a^2b^2$, the rectangular hyperbola $xy = c^2$. One advantage of using homogeneous equations is that equations and expressions derived from them are homogeneous, which provides a method of detecting slips. For example the following equations cannot have been derived correctly from equations homogeneous in x, y, a, b.

$$x^2 + y^2 = ax + b$$

$$x - y - ab = 0$$

$$\frac{a^2x^2 + b^2y^2}{xy} = \frac{x^2 + y^2}{x}$$

$$\frac{x}{a} + y = b$$

The reader who is used to checking dimensions in applied mathematics and physics will see that the expressions, 'A homogeneous expression of degree n', and 'An expression whose terms all have dimensions $[L^n]$', are equivalent. Thus the terms of the equations $ay^2 = x^3$, $x^2 + y^2 = a^2$, $b^2x^2 - a^2y^2 = a^2b^2$ have dimensions $[L^3]$, $[L^2]$, $[L^4]$, and they are homogeneous expressions in x, y, a, b of degree 3, 2, 4, respectively.

Consider now the equation of the tangent to $y^2 = 4ax$ (homogeneous in x, y, a) at $P(at^2, 2at)$. The coordinates of P are lengths, so that t is a ratio, or it may be said to have dimensions $[L^0]$.

$$\frac{dy}{dx} = \frac{dy}{dt} \div \frac{dx}{dt} = \frac{2a}{2at} = \frac{1}{t}$$

Check: $\dfrac{dy}{dx}$ is a ratio, so is $\dfrac{1}{t}$.

Therefore the tangent at P is $x - ty + at^2 = 0$,

Check: the equation is homogeneous in x, y, a.

Symmetrical and cyclic expressions

7.4   Symmetrical functions of α, β were introduced in Book 1, §9.8. They are expressions such as $\alpha + \beta + 2\alpha\beta$, $\alpha/\beta + \beta/\alpha$, which are unchanged when α, β are interchanged. This idea can easily be extended to functions of several variables. For instance, $yz + zx + xy$, $x^3 + y^3 + z^3 + a^3$ are unchanged when any two of the variables included in the expressions are interchanged. These, then, are said to be symmetrical in x, y, z and x, y, z, a respectively.

On the other hand, expressions such as

$$(b - c)(c - a)(a - b) \quad \text{and} \quad a^2(b - c) + b^2(c - a) + c^2(a - b)$$

are changed if two of a, b, c are interchanged, but are unaltered if a is replaced by b, b by c, and c by a according to Fig. 7.1. Such expressions are said to by **cyclic** in a, b, c.

Figure 7.1

When dealing with cyclic expressions in, say, three variables a, b, c, the $\sum$ notation (Book 1, §13.8) gives a convenient shorthand. Thus

$$\sum bc = bc + ca + ab \tag{1}$$

$$\sum a^2(b - c) = a^2(b - c) + b^2(c - a) + c^2(a - b) \tag{2}$$

The term $\sum$ determines the others which are written down according to the method indicated in Fig. 7.1. Note the order of the terms: in (1) the position of a term is determined by the letter it *lacks*; in (2) the letter which is squared is also the letter lacking in the brackets, and it determines the positions of the terms.

Qu. 8 Write in full cyclic expressions in the three variables a, b, c given by

(a) $\sum a^3$,   (b) $\sum a(b + c)$,   (c) $\sum \dfrac{1}{a}$,   (d) $\sum ab^2c^2$.

Some useful identities

7.5 Certain identities will be needed occasionally in this book and for convenience they are grouped together here, which is appropriate because the identities are given in homogeneous forms:

$$(a + b)^3 \equiv a^3 + 3a^2b + 3ab^2 + b^3 \tag{1}$$
$$(a - b)^3 \equiv a^3 - 3a^2b + 3ab^2 - b^3 \tag{2}$$

$$a^3 + b^3 \equiv (a + b)(a^2 - ab + b^2) \tag{3}$$
$$a^3 - b^3 \equiv (a - b)(a^2 + ab + b^2) \tag{4}$$

$$a^3 + b^3 + c^3 - 3abc \equiv (a + b + c)(a^2 + b^2 + c^2 - bc - ca - ab) \tag{5}$$

(1) and (2) follow from the binomial theorem and are easily written down with the help of Pascal's triangle (Book 1, §14.1).

(3) and (4) may be obtained as follows. Consider $a^3 + b^3$ as a function of a. When we substitute $a = -b$ the expression vanishes and so, by the remainder theorem (Book 1, §9.9), $(a + b)$ is a factor. The other factor is found by inspection or long division.

Similarly, $a^3 - b^3$ vanishes when the substitution $b = a$ is made, therefore $(a - b)$ is a factor.

Qu. 9   Verify the identities (3) and (4).

Identity (5) may be verified by long multiplication of the R.H.S., but it is more instructive to proceed as indicated in the table below.

	factors from brackets:		
term in	first	second	result
a^3	a	a^2	a^3
a^2b	$\begin{cases} a \\ b \end{cases}$	$\left.\begin{array}{l} -ab \\ a^2 \end{array}\right\}$	0
abc	$\begin{cases} a \\ b \\ c \end{cases}$	$\left.\begin{array}{l} -bc \\ -ca \\ -ab \end{array}\right\}$	$-3abc$

Not only is the term in a^2b zero, but since the R.H.S. of (5) is symmetrical in a, b, c, the terms in a^2c, b^2c, b^2a, c^2a, c^2b are zero too. Hence the R.H.S. is $a^3 + b^3 + c^3 - 3abc$.

Qu. 10   Verify the following identities by the method above:
(a) $(a + b + c)^2 \equiv \sum a^2 + 2 \sum bc$,
(b) $(a + b + c)^3 \equiv \sum a^3 + 3 \sum a^2(b + c) + 6abc$,
(c) $(a + b + c)^4 \equiv \sum a^4 + 4 \sum a^3(b + c) + 12 \sum a^2bc + 6 \sum b^2c^2$.
Qu. 11   Check the identities in Qu. 10 by the substitutions $a = b = c = 1$. (This check does not prove they are correct but it is worth doing when you have expanded expressions like these.)

Example 13   *Factorise*:
(a) $27a^3b^6 - 8c^3$,   (b) $a^3 + b^3 + c^3 + 3ac(a + c)$.

(a) $27a^3b^6 - 8c^3 \equiv (3ab^2)^3 - (2c)^3$
$ \equiv (3ab^2 - 2c)(9a^2b^4 + 6ab^2c + 4c^2)$

(b) $a^3 + b^3 + c^3 + 3ac(a + c) \equiv b^3 + a^3 + 3a^2c + 3ac^2 + c^3$
$ \equiv b^3 + (a + c)^3$
$ \equiv \{b + (a + c)\}\{b^2 - b(a + c) + (a + c)^2\}$
$ \equiv (a + b + c)(a^2 + b^2 + c^2 - bc + 2ca - ab)$

Exercise 7c

Solve the equations in Nos. 1–5 for the ratio $x:y$.

1 $6x^2 - xy - 12y^2 = 0.$ **2** $2x^2 - 7xy - 30y^2 = 0.$
3 $x^3 - 3x^2y + 4y^3 = 0.$ **4** $6x^3 + 7x^2y - 7xy^2 - 6y^3 = 0.$
5 $4x^4 - 37x^2y^2 + 9y^4 = 0.$

Solve for Nos. 6–10 the ratios $x:y:z$.

6 $2x + 3y - z = 0, \quad 3x - 2y + 4z = 0.$
7 $4x - 5y + 6z = 0, \quad 2x + 3y - 4z = 0.$
8 $ax + by + cz = 0, \quad bx + ay - cz = 0.$
9 $a^2x + ay + z = 0, \quad x + ay + a^2z = 0.$
10 $x \cos \theta - y \sin \theta + z = 0, \quad x \sin \theta + y \cos \theta - z = 0.$

Write in full the cyclic functions in x, y, z given by

11 $\sum x^4.$ **12** $\sum 1/(yz).$ **13** $\sum x^2(y + z).$
14 $\sum x^2y.$ **15** $\sum xy^2.$

Show that:

16 $\sum x(y - z) = 0.$ **17** $\sum x(y + z) = 2 \sum yz.$

Factorise:

18 $1 - t^3.$ **19** $64x^3 + y^3.$ **20** $8 + 27z^3.$
21 $125y^3 - z^6.$ **22** $a^6 - b^6.$
*23 Use the result of No. 22, together with the identity

$$a^6 - b^6 \equiv (a^2 - b^2)(a^4 + a^2b^2 + b^4)$$

to show that $a^4 + a^2b^2 + b^4 \equiv (a^2 + ab + b^2)(a^2 - ab + b^2).$
*24 Find the sum of the geometric progression

$$x^{n-1} + x^{n-2}a + \dots + xa^{n-2} + a^{n-1}$$

Hence, or otherwise, show that

$$x^n - a^n \equiv (x - a)(x^{n-1} + x^{n-2}a + \dots + xa^{n-2} + a^{n-1})$$

*25 Given the polynomial $f(x) \equiv b_nx^n + b_{n-1}x^{n-1} + \dots + b_1x + b_0$ where b_r is a constant, use the result of No. 24 to show that

$$f(x) - f(a) \equiv (x - a)(c_{n-1}x^{n-1} + \dots + c_1x + c_0)$$

where c_r is a constant. (This proves the remainder theorem.)

Roots of cubic equations

7.6 It has been shown (Book 1, §9.7) that if α, β are the roots of the equation $ax^2 + bx + c = 0$, then $\alpha + \beta = -b/a$, $\alpha\beta = c/a$.

We now consider the cubic equation

$$ax^3 + bx^2 + cx + d = 0 \qquad (a \neq 0)$$

Let α, β, γ be the roots of the equation

$$ax^3 + bx^2 + cx + d = 0$$

Now the equation with roots α, β, γ may be written

$$(x - \alpha)(x - \beta)(x - \gamma) = 0$$
$$\therefore x^3 - (\alpha + \beta + \gamma)x^2 + (\beta\gamma + \gamma\alpha + \alpha\beta)x - \alpha\beta\gamma = 0$$

Writing the original equation as

$$x^3 + \frac{b}{a}x^2 + \frac{c}{a}x + \frac{d}{a} = 0$$

and equating coefficients of x^2, x, and the constant terms,

$$\alpha + \beta + \gamma = -\frac{b}{a}$$

$$\beta\gamma + \gamma\alpha + \alpha\beta = \frac{c}{a}$$

$$\alpha\beta\gamma = -\frac{d}{a}$$

Qu. 12 Write down the sum, the sum of the products in pairs, and the product of the roots of the equations:
(a) $3x^3 - 4x^2 + 2x + 5 = 0$,   (b) $x^3 = 1$,
(c) $7x^3 + 6x - 5 = 0$,   (d) $(x + 1)^3 = (x + 2)^2$,
(e) $x^3 - 5x^2 + 2 = 0$,   (f) $x^3 + x^2 + x + 1 = 0$.

Qu. 13 Write down the equations whose roots have the following sums, sums of products in pairs, and products respectively:
(a) 6, 11, 6,   (b) 0, -13, -12,   (c) 14, 0, -288.

Example 14   *The equation $3x^3 + 6x^2 - 4x + 7 = 0$ has roots α, β, γ. Find the equations with roots (a) $1/\alpha$, $1/\beta$, $1/\gamma$, (b) $\beta + \gamma$, $\gamma + \alpha$, $\alpha + \beta$.*

(a) If x is a root of the given equation

$$3x^3 + 6x^2 - 4x + 7 = 0$$

then $y = 1/x$ is a root of the required equation. Substituting $x = 1/y$, it follows that the required equation is

$$\frac{3}{y^3} + \frac{6}{y^2} - \frac{4}{y} + 7 = 0$$

i.e.

$$7y^3 - 4y^2 + 6y + 3 = 0$$

(b) If x is a root of the given equation, then $y = \alpha + \beta + \gamma - x$ is a root of the required equation.

$$\alpha + \beta + \gamma = -\frac{6}{3}$$

$\therefore \ y = -2 - x \quad \text{or} \quad x = -y - 2$
$3x^3 + 6x^2 - 4x + 7 = 0$
$\therefore \ 3(-y^3 - 6y^2 - 12y - 8) + 6(y^2 + 4y + 4) - 4(-y - 2) + 7 = 0$
$\therefore \ -3y^3 - 12y^2 - 8y + 15 = 0$

Therefore the required equation is $3y^3 + 12y^2 + 8y - 15 = 0$.

Qu. 14   What substitutions would have been required in Example 14 to find the equations with roots

(a) $\alpha^2, \beta^2, \gamma^2,$ (b) $\alpha - 2, \beta - 2, \gamma - 2,$

(c) $2\alpha + 1, 2\beta + 1, 2\gamma + 1,$ (d) $\dfrac{1}{\beta\gamma}, \dfrac{1}{\gamma\alpha}, \dfrac{1}{\alpha\beta}$?

Example 15   *Solve the equations*

$$\alpha + \beta + \gamma = 4 \qquad \alpha^2 + \beta^2 + \gamma^2 = 66 \qquad \alpha^3 + \beta^3 + \gamma^3 = 280$$

The method used is to form an equation with roots α, β, γ; in order to find the values of $\sum \beta\gamma$ and $\alpha\beta\gamma$ we use the identities:

$$(\alpha + \beta + \gamma)^2 = \alpha^2 + \beta^2 + \gamma^2 + 2(\beta\gamma + \gamma\alpha + \alpha\beta) \tag{1}$$
$$\alpha^3 + \beta^3 + \gamma^3 - 3\alpha\beta\gamma = (\alpha + \beta + \gamma)(\alpha^2 + \beta^2 + \gamma^2 - \beta\gamma - \gamma\alpha - \alpha\beta) \tag{2}$$

From (1), $16 = 66 + 2 \sum \beta\gamma$.

$$\therefore \ \sum \beta\gamma = -25$$

From (2), $280 - 3\alpha\beta\gamma = 4(66 + 25)$.

$$\therefore \ \alpha\beta\gamma = -28$$

Therefore α, β, γ are the roots of the equation

$$x^3 - 4x^2 - 25x + 28 = 0$$

The L.H.S. vanishes when $x = 1$, therefore $x - 1$ is a factor. Hence

$(x - 1)(x^2 - 3x - 28) = 0$
$\therefore \ (x - 1)(x + 4)(x - 7) = 0$
$\therefore \ x = -4, 1, 7$

Therefore the equations are satisfied by $\alpha = -4, \beta = 1, \gamma = 7$ and the other five permutations of these numbers.

It is worth noting that the product of the degrees of the three given equations is 6 and that six solutions are obtained.

Repeated roots

7.7   For the quadratic equation $ax^2 + bx + c = 0$ with roots α, β,

$$\alpha + \beta = -b/a \qquad \alpha\beta = c/a$$

If the roots are equal, substitute $\beta = \alpha$, then

$$2\alpha = -\frac{b}{a} \qquad \alpha^2 = \frac{c}{a}$$

$$\therefore 4\alpha^2 = \frac{b^2}{a^2} = \frac{4c}{a}$$

$$\therefore b^2 = 4ac$$

The cubic equation $ax^3 + bx^2 + cx + d = 0$ may be treated similarly, but it is more instructive to consider the problem graphically. Figure 7.2 shows two cubic curves of the form $y = ax^3 + bx^2 + cx + d$ $(a > 0)$. The y-axis is not shown: the point to emphasise is that, if the equation $y = 0$ has a repeated root, the x-axis is a tangent.

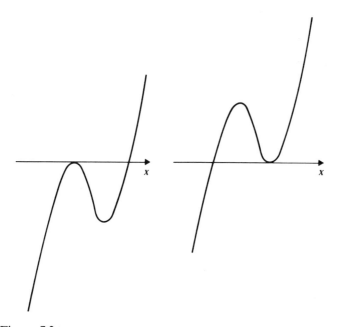

Figure 7.2

In this case, $\dfrac{dy}{dx} = 0$ *has a root in common with* $y = 0$.

Hence, if the equation $ax^3 + bx^2 + cx + d = 0$ has a repeated root, the equations

$$ax^3 + bx^2 + cx + d = 0$$
$$3ax^2 + 2bx + c = 0$$

have a root in common.

Example 16   *Given that the equation $18x^3 + 3x^2 - 88x - 80 = 0$ has a repeated root, solve the equation.*

If $18x^3 + 3x^2 - 88x - 80 = 0$ has a repeated root, then it is also a root of the equation

$$\frac{d}{dx}(18x^3 + 3x^2 - 88x - 80) = 0$$

$$\therefore 54x^2 + 6x - 88 = 0$$

$$\therefore 2(3x + 4)(9x - 11) = 0$$

The sum and product of the roots of $18x^3 + 3x^2 - 88x - 80 = 0$ are respectively $-\frac{1}{6}$ and $\frac{40}{9}$.

If $x = \frac{11}{9}$ is a repeated root, the other is $-\frac{1}{6} - \frac{22}{9} = -\frac{47}{18}$. The product is $(\frac{11}{9})^2(-\frac{47}{18})$, which does not check.

If $x = -\frac{4}{3}$ is a repeated root, the other is $-\frac{1}{6} + \frac{8}{3} = \frac{5}{2}$. The product is $(-\frac{4}{3})^2(\frac{5}{2}) = \frac{40}{9}$, which is correct.

The roots, then, are $-\frac{4}{3}, -\frac{4}{3}, \frac{5}{2}$.

Qu. 15   In Example 16, verify that the sum of the products of the roots in pairs is $-\frac{44}{9}$.

Qu. 16   If $f(x) \equiv (x - a)^2 g(x)$, show that $(x - a)$ is a factor of $f'(x)$.

Qu. 17   If the equation $ax^3 + bx^2 + cx + d = 0$ has three equal roots, find the conditions that must be satisfied by a, b, c, d.

Example 17   *Find the equation of the tangent to $y = x^3$ at the point (t, t^3) and find the coordinates of the point where the tangent meets the curve again.*

$$y = x^3 \qquad \therefore \frac{dy}{dx} = 3x^2$$

Therefore the tangent at (t, t^3) has gradient $3t^2$. Therefore its equation is

$$3t^2 x - y - 2t^3 = 0$$

To find the point of intersection with $y = x^3$, solve the equations simultaneously.

$$3t^2 x - x^3 - 2t^3 = 0$$
$$\therefore x^3 - 3t^2 x + 2t^3 = 0$$

Now, since the line is a tangent to the curve, two of the roots are t, t. The sum of the roots is zero, so the other root is $-2t$. Therefore the line meets the curve again at $(-2t, -8t^3)$.

Qu. 18   Check that $(-2t, -8t^3)$ lies on the tangent in Example 17. Also check that the product of the roots is correct.

Exercise 7d

1 Write down the sum, the sum of the products in pairs, and the product of the roots of the equations
   (a) $3x^3 - 4x^2 - x + 2 = 0$,   (b) $4x^3 + 5x - 6 = 0$,
   (c) $(x + 1)^3 + 2(x + 1)^2 - 3(x + 1) + 4 = 0$,
   (d) $3(x + 1)^3 = 2(x - 1)^2$.

2 Write down the equations whose roots are α, β, γ, when $\sum \alpha$, $\sum \beta\gamma$, $\alpha\beta\gamma$ are, respectively:
   (a) $2, 0, -5$;   (b) $-3, 2, 6$;   (c) $0, -1, 5$.

3 The equation $2x^3 + 3x^2 - 13x - 7 = 0$ has roots α, β, γ. Find the equations with roots
   (a) $\alpha + 1, \beta + 1, \gamma + 1$;   (b) $1/\alpha, 1/\beta, 1/\gamma$;
   (c) $\alpha - 2, \beta - 2, \gamma - 2$;   (d) $\beta\gamma, \gamma\alpha, \alpha\beta$.

4 The equation $x^3 + 3x^2 - 3x - 10 = 0$ has roots α, β, γ. Find the equations with roots
   (a) $-\alpha + \beta + \gamma, \alpha - \beta + \gamma, \alpha + \beta - \gamma$;   (b) $2\alpha + 1, 2\beta + 1, 2\gamma + 1$;

   (c) $\dfrac{1}{\beta\gamma}, \dfrac{1}{\gamma\alpha}, \dfrac{1}{\alpha\beta}$.

5 If the equation $x^3 + 3hx + g = 0$ has roots α, β, γ, find the equations with roots
   (a) $\alpha^2, \beta^2, \gamma^2$;   (b) $1/\alpha^2, 1/\beta^2, 1/\gamma^2$;   (c) $\alpha^3, \beta^3, \gamma^3$.

6 Repeat No. 5 for the equation $ax^3 + bx^2 + cx + d = 0$.

7 For the equation $x^3 + 3hx + g = 0$, with roots α, β, γ, find
   (a) $\sum \alpha^2$,   (b) $\sum \beta^2\gamma^2$,   (c) $\sum \alpha^4$, given that $\sum \alpha^4 = (\sum \alpha^2)^2 - 2\sum \beta^2\gamma^2$.

8 For the equation $x^3 + 3hx + g = 0$, with roots α, β, γ,
   (a) show that $\alpha^3 = -3h\alpha - g$, and use similar expressions for β, γ to deduce that $\sum \alpha^3 = -3h \sum \alpha - 3g$,
   (b) show that $\alpha^4 = -3h\alpha^2 - g\alpha$ and deduce that $\sum \alpha^4 = -3h\sum \alpha^2 - g\sum \alpha$. Find $\sum \alpha^2, \sum \alpha^3, \sum \alpha^4$ in terms of g and h.

9 For the equation $x^3 + 3ax^2 + 3bx + c = 0$, with roots α, β, γ find
   (a) $\sum \alpha^2$,   (b) $\sum \alpha^3$,   (c) $\sum \alpha^4$.

10 Find the relation between a, b, c, d if the roots of the equation

$$ax^3 + bx^2 + cx + d = 0$$

are in (a) arithmetic, (b) geometric progression.
   Find the equation whose roots are the reciprocals of the roots of the given equation and deduce the condition that the roots of the given equation are in harmonic progression.

11 Find the relation between a, b, c, d if one root of the equation

$$ax^3 + bx^2 + cx + d = 0$$

is equal to the sum of the other two.

12 Solve the equation $54x^3 - 111x^2 + 74x - 16 = 0$, given that the roots are in geometric progression.

13 Solve the equation $64x^3 - 240x^2 + 284x - 105 = 0$, given that the roots are in arithmetic progression.

Solve the equations in Nos. 14–16, given that each has a repeated root.

14 $12x^3 - 52x^2 + 35x + 50 = 0$.

15 $18x^3 + 21x^2 - 52x + 20 = 0$.

16 $12x^3 - 20x^2 - 21x + 36 = 0$.

17 Find the condition that the equation $x^3 - 3hx + g = 0$ should have a repeated root. What conditions must be satisfied if all its roots are equal?

18 Sketch the graph of $y^2 = x^3$. Find the point where the tangent at (t^2, t^3) meets the curve again and show that the axes divide the chord in constant ratios.

19 Find the equation of the normal at $(at^2, 2at)$ to the parabola $y^2 = 4ax$. This equation is a cubic in t. Find the condition that it should have two equal roots.

20 Find the equation of the tangent to $x^2y = 1$ at the point $(t, 1/t^2)$. Also find the ratios in which the axes divide the segment of the tangent bounded by the curve.

Solve the following simultaneous equations.

21 $\alpha + \beta + \gamma = -2$, $\quad \alpha^2 + \beta^2 + \gamma^2 = 14$, $\quad \alpha\beta\gamma = 6$.

22 $\alpha + \beta + \gamma = 4$, $\quad \alpha^2 + \beta^2 + \gamma^2 = 38$, $\quad \alpha^3 + \beta^3 + \gamma^3 = 106$.

23 $\alpha + \beta + \gamma = 0$, $\quad \alpha^2 + \beta^2 + \gamma^2 = 42$, $\quad \alpha^3 + \beta^3 + \gamma^3 = -60$.

24 $\alpha + \beta + \gamma = 2$, $\quad \alpha^2 + \beta^2 + \gamma^2 = 14$, $\quad \alpha^3 + \beta^3 + \gamma^3 = 20$.

25 The equation $ax^4 + bx^3 + cx^2 + dx + e = 0$ has roots $\alpha, \beta, \gamma, \delta$. Find, in terms of a, b, c, d, e, the sum of the roots, the sums of the products of the roots in pairs and threes, and the product of the roots. Find also, the equations with roots

(a) $1/\alpha$, $1/\beta$, $1/\gamma$, $1/\delta$; (b) α^2, β^2, γ^2, δ^2.

26 Prove that $\tan 3x = (3 \tan x - \tan^3 x)/(1 - 3 \tan^2 x)$. Hence, or otherwise, solve the equation $t^3 - 6t^2 - 3t + 2 = 0$ correct to two significant figures.

27 Use the identity $\sin 3\theta = 3 \sin \theta - 4 \sin^3 \theta$ to solve the equation

$$8x^3 - 6x + 1 = 0$$

correct to four significant figures.

28 Use the substitution $x = 2 \sin \theta$ to solve the equation $3x^3 - 9x + 2 = 0$ correct to three significant figures.

29 The equation $x^3 + 3x^2 + 3 = 0$ has a root somewhere in the interval $-5 < x < 3$. Find between which two integral values of x in this interval the function $f(x) = x^3 + 3x^2 + 3$ changes sign and find the root of the equation to the nearest integer.

*30 Show that the cubic equation $t^3 + 3at^2 + 3bt + c = 0$ can be reduced to the form $x^3 + 3fx + g = 0$ by means of the substitution $t = x - a$. Obtain f and g in terms of a, b, c and explain how any cubic equation may be solved graphically by drawing a straight line to cut the graph $y = x^3$.

Exercise 7e (Miscellaneous)

1 Solve the equations:
   (a) $9x^{2/3} + 4x^{-2/3} = 37$;
   (b) $x + 2y = 3$, $3x^2 + 4y^2 + 12x = 7$.   (O & C)

2 Solve the equations:

   (a) $\dfrac{1}{x} - \dfrac{1}{x+3} = \dfrac{1}{k} - \dfrac{1}{k+3}$,   (b) $2\sqrt{x} + \sqrt{(2x+1)} = 7$,

   where the *positive* values of the square roots are taken.   (O & C)

3 Solve the equations:
   (a) $3^{2(x+1)} - 10 \times 3^x + 1 = 0$;
   (b) $x^2 + 4y^2 = 25$, $xy + 6 = 0$.   (O & C)

4 (a) If $(x+1)^2$ is a factor of $2x^4 + 7x^3 + 6x^2 + Ax + B$, find the values of A and B.

   (b) Prove that, if the equations $x^2 + ax + b = 0$ and $cx^2 + 2ax - 3b = 0$ have a common root and neither a nor b is zero, then

$$b = \frac{5a^2(c-2)}{(c+3)^2}$$ (O & C)

5 (a) Prove that, if two polynomials P(x) and Q(x) have a common linear factor $x - p$, then $x - p$ is a factor of the polynomial $[P(x) - Q(x)]$. Hence prove that, if the equations

$$ax^3 + 4x^2 - 5x - 10 = 0 \qquad ax^3 - 9x - 2 = 0$$

   have a common root, then $a = 2$ or 11.

   (b) Prove that, if $x + 1/x = y + 1$, then

$$\frac{(x^2 - x + 1)^2}{x(x-1)^2} = \frac{y^2}{y-1}$$

   Hence solve the equation $(x^2 - x + 1)^2 - 4x(x-1)^2 = 0$.   (O & C)

6 (a) Show that the equation $\sqrt{(x^2 + 2)} - \sqrt{(x^2 + 2x + 5)} = 1$ has no solution if it is assumed that the square roots are positive.

   (b) Show that, if $x^3 + 3px + q = 0$ and $x = y - p/y$, then y^3 satisfies a certain quadratic equation.

   By solving the quadratic equation in the case $p = q = 2$, obtain *one* root of the equation $x^3 + 6x + 2 = 0$ leaving your answer in surd form.   (O & C)

7 Factorise completely the expression $a^2(b^3 - c^3) + b^2(c^3 - a^3) + c^2(a^3 - b^3)$.
                                                                        (C)

8 Prove that the remainder when the polynomial P(x) is divided by $(x - a)^2$ is

$$(x - a)P'(a) + P(a)$$

   where P'(x) is the derivative of P(x) with respect to x.
   Given that $x^4 + bx + c$ is divisible by $(x - 2)^2$, find the value of b and of c.
                                                                        (C)

9 The roots of the equation $x^3 - 5x^2 + x + 12 = 0$ are α, β, γ. Calculate the value of $(\alpha + 2)(\beta + 2)(\gamma + 2)$. (C)

10 (a) Given that $a/b = c/d$, prove that each of these ratios equals

$$\frac{ka + lc}{kb + ld}$$

where k, l are any numbers for which $kb + ld \neq 0$.
Solve the simultaneous equations

$$\frac{x}{1} = \frac{x + y}{3} = \frac{x - y + z}{2}$$

$$x^2 + y^2 + z^2 + x + 2y + 4z - 6 = 0$$

(b) Find positive integers a, b for which $x^4 + 2x^2 + 9 \equiv (x^2 + a)^2 - b^2 x^2$ and hence find the quadratic factors of $x^4 + 2x^2 + 9$. (O & C)

11 Solve the equations:
(a) $\sqrt{(2 - x)} + \sqrt{(x + 3)} = 3$,

(b) $\dfrac{x - y}{4} = \dfrac{z - y}{3} = \dfrac{2z - x}{1}$, $x + 3y + 2z = 4$. (O & C)

12 Solve the simultaneous equations

$$a_1 x + b_1 y + c_1 z = 0$$
$$a_2 x + b_2 y + c_2 z = 0$$

for the ratios $x:y:z$.
Hence, or otherwise,
(a) solve the equations

$$x + 4y + 2z = 0$$
$$2x - y + z = 0$$
$$8x + 5y + 6z = 6$$

for x, y, and z; and
(b) find the condition that the quadratic equations

$$a_1 x^2 + b_1 x + c_1 = 0$$
$$a_2 x^2 + b_2 x + c_2 = 0$$

should have a common root. (C)

13 (a) Solve the equation $x^3 - x^2 - 5x + 2 = 0$.
(b) Find the only solution of the equation

$$\sqrt{(4x - 2)} + \sqrt{(x + 1)} - \sqrt{(7 - 5x)} = 0$$ (O & C)

14 (a) Solve $\sqrt{(4x + 13)} - \sqrt{(x + 1)} = \sqrt{(12 - x)}$.
(b) One root of the equation $3x^3 + 14x^2 + 2x - 4 = 0$ is rational. Obtain this root and complete the solution of the equation. (C)

15 (a) One of the roots of the equation $21x^3 - 50x^2 - 37x - 6 = 0$ is a positive integer. Find this root and hence solve the equation completely.

(b) Three numbers α, β, γ are in arithmetical progression and two of them are roots of the equation $x^2 + ax + b = 0$. Prove that the third is either $-\frac{1}{2}a$ or one of the roots of the equation $y^2 + ay + 9b - 2a^2 = 0$.

(O & C)

16 (a) Show that, if the roots of the equation $x^3 - 5x^2 + qx - 8 = 0$ are in geometric progression, then $q = 10$.

(b) If α, β, γ are the roots of the equation $x^3 - x^2 + 4x + 7 = 0$, find the equation whose roots are $\beta + \gamma$, $\gamma + \alpha$, $\alpha + \beta$.

(C)

17 Prove that, if α is a repeated root of the equation $f(x) = 0$, where $f(x)$ is a polynomial, then α is a root of the equation $f'(x) = 0$.

Given that the equation $4x^4 + x^2 + 3x + 1 = 0$ has a repeated root, find its value.

(C)

18 (a) Use the remainder theorem to express $x^3 + 2x^2 + x - 18$ as a product of two factors.

(b) Find the value of the constant p for which the polynomial

$$x^4 + x^3 + px^2 + 5x - 10$$

has $x + 2$ as a factor.

(c) Show that if $y = (x - a)^2 V$ where V is a polynomial in x, then dy/dx is a polynomial with $x - a$ as a factor. Hence or otherwise find the values of the constants k and l for which $x^4 - 2x^3 + 5x^2 + kx + l$ has a factor $(x - 1)^2$.

(JMB)

19 Given that two roots of the equation $x^4 + bx^3 + cx^2 + dx + e = 0$ are such that their sum is zero and also that b, c, d and e are all non-zero, prove that the product of these two roots is d/b and that the product of the other two roots is be/d. Hence, or otherwise, prove that $b^2e + d^2 = bcd$.

Solve the equations

(a) $x^4 + x^3 - 2x^2 - 3x - 3 = 0$

(b) $x^4 - x^3 + 2x^2 - 3x - 3 = 0$

assuming that in each case two of the roots, not necessarily real, are such that their sum is zero.

(JMB)

20 The roots of the equation $x^3 = qx + r$ are α, β, γ. Prove that

(a) $\alpha^2 + \beta^2 + \gamma^2 = 2q$,

(b) $\alpha^3 + \beta^3 + \gamma^3 = 3r$,

(c) $\alpha^5 = q\alpha^3 + r\alpha^2$,

(d) $6(\alpha^5 + \beta^5 + \gamma^5) = 5(\alpha^3 + \beta^3 + \gamma^3)(\alpha^2 + \beta^2 + \gamma^2)$.

(O & C)

Chapter 8

Further matrices and determinants

Introduction

8.1 We have already met the determinant of a 2×2 matrix $\mathbf{M} = \begin{pmatrix} a & b \\ c & d \end{pmatrix}$ (see Book 1, Chapter 11), and we have seen that it plays an important role in finding the inverse of $\mathbf{M}$ (see Book 1, §11.5) and in finding the transformation which corresponds to $\mathbf{M}$ (see Book 1, §11.6). In this chapter, we shall be looking more closely at determinants and their properties. Any 'square' matrix has a corresponding determinant; the determinant of a 3×3 matrix will be defined in the next section. The *ideas* involved in studying an $n \times n$ determinant are no more difficult than those involved in a 2×2 determinant, but since an $n \times n$ matrix contains n^2 entries, the work involved in higher order determinants becomes very laborious.

Before embarking on the study of 3×3 determinants, it is necessary to introduce some standard notation:

(a) the determinant of a given matrix $\mathbf{M}$ is written det($\mathbf{M}$),
(b) when it is necessary to write out the determinant in full, the array of numbers is enclosed in a pair of vertical lines (instead of the round brackets used in matrices),
(c) the matrix formed by interchanging the rows and columns of a matrix $\mathbf{M}$ is called the **transpose** of $\mathbf{M}$ and it is written $\mathbf{M}^{\mathrm{T}}$.

So, if $\mathbf{M} = \begin{pmatrix} 2 & 3 \\ 4 & 7 \end{pmatrix}$, then we write

$$\det(\mathbf{M}) = \begin{vmatrix} 2 & 3 \\ 4 & 7 \end{vmatrix} = 2 \times 7 - 3 \times 4 = 14 - 12 = 2$$

and

$$\mathbf{M}^{\mathrm{T}} = \begin{pmatrix} 2 & 4 \\ 3 & 7 \end{pmatrix}$$

Qu. 1 Given that $\mathbf{M} = \begin{pmatrix} 3 & 4 \\ x & 8 \end{pmatrix}$, find x such that $\det(\mathbf{M}) = 0$.

Qu. 2   Given that $\mathbf{A} = \begin{pmatrix} 3 & 5 \\ 1 & 2 \end{pmatrix}$ and $\mathbf{B} = \begin{pmatrix} 2 & 7 \\ 3 & -1 \end{pmatrix}$, verify that

$\det(\mathbf{AB}) = \det(\mathbf{A})\det(\mathbf{B})$.

*Qu. 3   Prove that, for any 2×2 matrix $\mathbf{M}$, $\det(\mathbf{M^T}) = \det(\mathbf{M})$.

3×3 determinants

8.2   We shall frequently need to refer to a general 3×3 matrix in this chapter, and, when this is necessary, we shall write

$$\mathbf{M} = \begin{pmatrix} a_1 & b_1 & c_1 \\ a_2 & b_2 & c_2 \\ a_3 & b_3 & c_3 \end{pmatrix}$$

Definition

$$\det(\mathbf{M}) = a_1 \begin{vmatrix} b_2 & c_2 \\ b_3 & c_3 \end{vmatrix} - b_1 \begin{vmatrix} a_2 & c_2 \\ a_3 & c_3 \end{vmatrix} + c_1 \begin{vmatrix} a_2 & b_2 \\ a_3 & b_3 \end{vmatrix}.$$

Notice that the three 2×2 matrices in this definition are obtained by deleting the row and column containing the letter by which each is multiplied. This definition can easily be extended to cover determinants of higher order.

When the 2×2 determinants are multiplied out, we obtain

$$\begin{aligned} \det(\mathbf{M}) &= a_1(b_2c_3 - b_3c_2) - b_1(a_2c_3 - a_3c_2) + c_1(a_2b_3 - b_2a_3) \\ &= a_1b_2c_3 - a_1b_3c_2 - b_1a_2c_3 + b_1a_3c_2 + c_1a_2b_3 - c_1b_2a_3 \end{aligned}$$

For convenience, these terms should be re-arranged so that in each term the letters occur in alphabetical order. Putting the terms with + signs first, we have

$$\det(\mathbf{M}) = a_1b_2c_3 + a_2b_3c_1 + a_3b_1c_2 - a_1b_3c_2 - a_3b_2c_1 - a_2b_1c_3$$

When written in this form, we can see that the terms in which the suffixes 1, 2, 3 occur in *clockwise cyclic* order (see Fig. 8.1 (i)) the sign is 'plus', and when the suffixes go the other way round (see Fig. 8.1 (ii)) the sign is 'minus'.

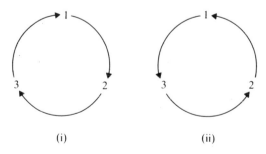

(i)   (ii)

Figure 8.1

Example 1 *Evaluate* $\begin{vmatrix} 7 & 2 & 3 \\ 4 & 1 & 5 \\ 2 & 0 & 3 \end{vmatrix}$

$$\begin{vmatrix} 7 & 2 & 3 \\ 4 & 1 & 5 \\ 2 & 0 & 3 \end{vmatrix} = 7 \begin{vmatrix} 1 & 5 \\ 0 & 3 \end{vmatrix} - 2 \begin{vmatrix} 4 & 5 \\ 2 & 3 \end{vmatrix} + 3 \begin{vmatrix} 4 & 1 \\ 2 & 0 \end{vmatrix}$$

$$= 7(3 - 0) - 2(12 - 10) + 3(0 - 2)$$

$$= 7 \times 3 - 2 \times 2 + 3 \times (-2)$$

$$= 21 - 4 - 6$$

$$= 11$$

Example 2 *Solve the equation* $\begin{vmatrix} x - 3 & 1 & -1 \\ -7 & x + 5 & -1 \\ -6 & 6 & x - 2 \end{vmatrix} = 0.$

$$\begin{vmatrix} x - 3 & 1 & -1 \\ -7 & x + 5 & -1 \\ -6 & 6 & x - 2 \end{vmatrix} = (x - 3) \begin{vmatrix} x + 5 & -1 \\ 6 & x - 2 \end{vmatrix} - \begin{vmatrix} -7 & -1 \\ -6 & x - 2 \end{vmatrix} -$$

$$- \begin{vmatrix} -7 & x + 5 \\ -6 & 6 \end{vmatrix}$$

$$= (x - 3)(x^2 + 3x - 10 + 6) - (-7x + 14 - 6) -$$

$$- (-42 + 6x + 30)$$

$$= (x - 3)(x^2 + 3x - 4) - (-7x + 8) - (6x - 12)$$

$$= x^3 - 13x + 12 + 7x - 8 - 6x + 12$$

$$= x^3 - 12x + 16$$

Hence we must solve the cubic equation

$$x^3 - 12x + 16 = 0$$

Factorising,

$$(x - 2)(x^2 + 2x - 8) = 0$$
$$\therefore (x - 2)(x - 2)(x + 4) = 0$$
$$\therefore x = 2 \quad \text{or} \quad -4$$

Example 3 *Prove that* $\begin{vmatrix} 1 & 1 & 1 \\ x & y & z \\ x^2 & y^2 & z^2 \end{vmatrix} = (x - y)(y - z)(z - x).$

$$\begin{vmatrix} 1 & 1 & 1 \\ x & y & z \\ x^2 & y^2 & z^2 \end{vmatrix} = \begin{vmatrix} y & z \\ y^2 & z^2 \end{vmatrix} - \begin{vmatrix} x & z \\ x^2 & z^2 \end{vmatrix} + \begin{vmatrix} x & y \\ x^2 & y^2 \end{vmatrix}$$

$$= (yz^2 - zy^2) - (xz^2 - zx^2) + (xy^2 - yx^2)$$

$$= yz^2 - zy^2 - xz^2 + zx^2 + xy^2 - yx^2$$

$$= z^2(y - x) - z(y^2 - x^2) + xy(y - x)$$

$$= -z^2(x - y) + z(x + y)(x - y) - xy(x - y)$$

$$= (x - y)(-z^2 + zx + zy - xy)$$

$$= (x - y)\{-z(z - x) + y(z - x)\}$$

$$= (x - y)(z - x)(y - z)$$

$$= (x - y)(y - z)(z - x)$$

[Note, however, that this is not the best way to do this example; we shall shortly be meeting a much more efficient method for tackling problems like this (see Example 6).]

Exercise 8a

1 Evaluate:

(a) $\begin{vmatrix} 5 & 2 \\ 3 & 4 \end{vmatrix}$,   (b) $\begin{vmatrix} 7 & 1 \\ 2 & -1 \end{vmatrix}$,   (c) $\begin{vmatrix} 21 & 14 \\ 15 & 10 \end{vmatrix}$,   (d) $\begin{vmatrix} 91 & 35 \\ 65 & 25 \end{vmatrix}$.

2 Simplify:

(a) $\begin{vmatrix} x & y \\ -y & x \end{vmatrix}$,   (b) $\begin{vmatrix} x^2 & xy \\ xy & y^2 \end{vmatrix}$,

(c) $\begin{vmatrix} \cos\theta & -\sin\theta \\ \sin\theta & \cos\theta \end{vmatrix}$,   (d) $\begin{vmatrix} x+1 & 1 \\ -1 & x-1 \end{vmatrix}$.

3 Prove that $\det(\mathbf{M}) = \det(\mathbf{M}^\mathrm{T})$, where $\mathbf{M} = \begin{pmatrix} a & b \\ c & d \end{pmatrix}$.

4 Solve:

(a) $\begin{vmatrix} x & x \\ 5 & 3x \end{vmatrix} = 0$,   (b) $\begin{vmatrix} x-2 & 1 \\ 2 & x-3 \end{vmatrix} = 0$.

*5 Given that $\mathbf{M} = \begin{pmatrix} a & b \\ c & d \end{pmatrix}$ and $\mathbf{N} = \begin{pmatrix} p & q \\ r & s \end{pmatrix}$, prove that

$$\det(\mathbf{MN}) = \det(\mathbf{M})\det(\mathbf{N})$$

6 Prove that $\det(\mathbf{M}^{-1}) = \dfrac{1}{\det(\mathbf{M})}$, where $\mathbf{M} = \begin{pmatrix} a & b \\ c & d \end{pmatrix}$.

7 Evaluate:

(a) $\begin{vmatrix} 2 & -1 & 0 \\ 3 & 2 & 0 \\ 4 & 7 & 3 \end{vmatrix}$,   (b) $\begin{vmatrix} 3 & 4 & 2 \\ 1 & 2 & 0 \\ -2 & 3 & 5 \end{vmatrix}$,

(c) $\begin{vmatrix} 3 & 6 & 9 \\ 4 & 8 & 12 \\ 5 & 7 & 13 \end{vmatrix}$,   (d) $\begin{vmatrix} 5 & -1 & 1 \\ 5 & -1 & 1 \\ 6 & -1 & 6 \end{vmatrix}$.

8 Solve the equation $\begin{vmatrix} x+3 & 5 & 6 \\ -1 & x-3 & -1 \\ 1 & 1 & x+4 \end{vmatrix} = 0.$

*9 Prove that for any 3×3 matrix $\det(\mathbf{M}) = \det(\mathbf{M}^{\mathrm{T}})$.

10 Given that $\mathbf{A} = \begin{pmatrix} 1 & 3 & 5 \\ 2 & -1 & 0 \\ 4 & 2 & 1 \end{pmatrix}$ and $\mathbf{B} = \begin{pmatrix} 2 & 0 & 1 \\ 1 & -3 & 2 \\ 1 & 1 & -1 \end{pmatrix}$, find $\mathbf{AB}$ and verify that $\det(\mathbf{AB}) = \det(\mathbf{A})\det(\mathbf{B})$.

Properties of determinants

8.3   As in the previous section, $\mathbf{M}$ is the matrix $\begin{pmatrix} a_1 & b_1 & c_1 \\ a_2 & b_2 & c_2 \\ a_3 & b_3 & c_3 \end{pmatrix}$, and

$$\det(\mathbf{M}) = a_1 b_2 c_3 + a_2 b_3 c_1 + a_3 b_1 c_2 - a_1 b_3 c_2 - a_3 b_2 c_1 - a_2 b_1 c_3 \qquad (1)$$

The properties of determinants, (a)–(g) below, are true for matrices of any order, even though in the text we shall only refer to the 3×3 matrix $\mathbf{M}$.

(a) $\det(\mathbf{M}) = \det(\mathbf{M}^{\mathrm{T}})$.

$$\mathbf{M}^{\mathrm{T}} = \begin{vmatrix} a_1 & a_2 & a_3 \\ b_1 & b_2 & b_3 \\ c_1 & c_2 & c_3 \end{vmatrix}$$

$$\begin{aligned}
\therefore \det(\mathbf{M}^{\mathrm{T}}) &= a_1(b_2 c_3 - b_3 c_2) - a_2(b_1 c_3 - b_3 c_1) + a_3(b_1 c_2 - b_2 c_1) \\
&= a_1 b_2 c_3 - a_1 b_3 c_2 - a_2 b_1 c_3 + a_2 b_3 c_1 + a_3 b_1 c_2 - a_3 b_2 c_1 \\
&= a_1 b_2 c_3 + a_2 b_3 c_1 + a_3 b_1 c_2 - a_1 b_3 c_2 - a_3 b_2 c_1 - a_2 b_1 c_3 \\
&= \det(\mathbf{M})
\end{aligned}$$

This property is especially valuable, because it means that whenever one of the statements, below, refers to a *row* of $\mathbf{M}$, the same statement can be made about the corresponding *column* of $\mathbf{M}$, without further proof.

(b) *If $\mathbf{M}$ has a pair of identical rows (or columns) then* $\det(\mathbf{M}) = 0$.
This follows immediately from making the appropriate substitutions in (1).

(c) *If each element in a row (or column) of* **M** *is multiplied by k, then* det(**M**) *is multiplied by k.*†

This may be easily checked by making the appropriate substitutions in (1).

A step like $\begin{vmatrix} a_1 & b_1 & c_1 \\ ka_2 & kb_2 & kc_2 \\ a_3 & b_3 & c_3 \end{vmatrix} = k \begin{vmatrix} a_1 & b_1 & c_1 \\ a_2 & b_2 & c_2 \\ a_3 & b_3 & c_3 \end{vmatrix}$, where a factor k has been

'taken out' of the second row, is very common, when simplifying determinants.

(d) *Interchanging a pair of rows (or columns) of* **M**, *changes the sign of* det(**M**). Once again this is easily verified.

(e) *If the elements of one row (or column) of* **M** *are multiples of the elements of another row (or column), then* det(**M**) $= 0$.
For example, put $b_1 = ka_1$, $b_2 = ka_2$ and $b_3 = ka_3$ in (1); it should be easy to see that det(**M**) $= 0$. (Property (b) could be regarded as a special case of this in which $k = 1$.)

(f) *If each element of one row (or column) of* **M** *is replaced by a new element consisting of the original element plus a multiple of the corresponding element from another row (or column), then the value of* det(**M**) *is unchanged.*
Suppose, for example, that the elements in the first row of **M** are replaced by $a_1 + ka_2$, $b_1 + kb_2$, $c_1 + kc_2$. The new determinant is now

$$\begin{vmatrix} a_1 + ka_2 & b_1 + kb_2 & c_1 + kc_2 \\ a_2 & b_2 & c_2 \\ a_3 & b_3 & c_3 \end{vmatrix}$$

and this equals

$$(a_1 + ka_2)(b_2 c_3 - c_2 b_3) - (b_1 + kb_2)(a_2 c_3 - c_2 a_3) + (c_1 + kc_2)(a_2 b_3 - b_2 a_3)$$
$$= a_1(b_2 c_3 - c_2 b_3) - b_1(a_2 c_3 - c_2 a_3) + c_1(a_2 b_3 - b_2 c_3) +$$
$$+ ka_2(b_2 c_3 - c_2 b_3) - kb_2(a_2 c_3 - c_2 a_3) + kc_2(a_2 b_3 - b_2 a_3)$$

$$= \begin{vmatrix} a_1 & b_1 & c_1 \\ a_2 & b_2 & c_2 \\ a_3 & b_3 & c_3 \end{vmatrix} + k \begin{vmatrix} a_2 & b_2 & c_2 \\ a_2 & b_2 & c_2 \\ a_3 & b_3 & c_3 \end{vmatrix}$$

$= $ det(**M**)   (since, by Property (b), the second determinant is zero)

This property is very useful when simplifying determinants (see Example 5, below). The notation $\mathbf{r}_1' = \mathbf{r}_1 + k\mathbf{r}_2$ is a convenient way of saying that the *new* first row is the *old* first row plus k times the second row. Similarly, $\mathbf{c}_2' = \mathbf{c}_2 + k\mathbf{c}_3$ would mean 'form a new second column by adding k times the third column to the existing second column'.

---

† But remember that when a *matrix* is multiplied by a scalar k, *every* element must be multiplied by k.

(g) $\det(\mathbf{AB}) = \det(\mathbf{A}) \det(\mathbf{B})$.

The proof of this for 2×2 matrices is not difficult (see Exercise 8a, No. 5), but it is rather tedious since there is a lot of rather elementary algebra to be done. For 3×3 matrices the task becomes even more troublesome; if the reader has a great deal of time and patience he or she could try writing out the proof, otherwise it should be taken on trust!

Example 4 *Evaluate* $\begin{vmatrix} 7 & 3 & 35 \\ 3 & 21 & 15 \\ 1 & 5 & 5 \end{vmatrix}$.

This determinant is zero, because $c_3 = 5c_1$ (see Property (e)).

Example 5 *Evaluate* $\begin{vmatrix} 10 & 40 & 56 \\ 1 & 5 & 7 \\ 3 & 4 & 6 \end{vmatrix}$.

$$\begin{vmatrix} 10 & 40 & 56 \\ 1 & 5 & 7 \\ 3 & 4 & 6 \end{vmatrix} = \begin{vmatrix} 2 & 0 & 0 \\ 1 & 5 & 7 \\ 3 & 4 & 6 \end{vmatrix} \qquad (\mathbf{r'_1} = \mathbf{r_1} - 8\mathbf{r_2}, \text{ Property (f)})$$

$$= 2 \begin{vmatrix} 5 & 7 \\ 4 & 6 \end{vmatrix}$$

$$= 2 \times 2$$

$$= 4$$

Example 6 *Factorise* $\begin{vmatrix} 1 & 1 & 1 \\ x & y & z \\ x^2 & y^2 & z^2 \end{vmatrix}$.

If $x = y$, the first two columns would be equal, and hence, by Property (b), the determinant would be zero. Hence, by the remainder theorem (see Book 1, Chapter 9), $(x - y)$ is a factor. Similarly $(y - z)$ and $(z - x)$ are also factors. Since each term of the expansion is of degree three, there are no further *algebraic* factors. However there could be a *numerical* factor k, so, at this stage, we can deduce that

$$\begin{vmatrix} 1 & 1 & 1 \\ x & y & z \\ x^2 & y^2 & z^2 \end{vmatrix} = k(x - y)(y - z)(z - x)$$

However if we look at the first term of the expansion of the determinant, we see that it is $+ yz^2$, and if we look for the same term from expanding the factors, we see that it would be kyz^2. Hence $k = 1$. Therefore, the factorised form of the determinant is $(x - y)(y - z)(z - x)$.

The reader should compare this method with the more elementary method used in Example 3. The method used in Example 6 is especially useful in determinants which are, in some way, symmetrical and the factors can be 'spotted', by exploiting the symmetry and using the remainder theorem, as above.

Example 7 *Factorise* $\begin{vmatrix} a & a^3 & a^4 \\ b & b^3 & b^4 \\ c & c^3 & c^4 \end{vmatrix}$.

Firstly, notice that a is a factor of the first row, b is a factor of the second and c of the third. So, by Property (c), we can write

$$\begin{vmatrix} a & a^3 & a^4 \\ b & b^3 & b^4 \\ c & c^3 & c^4 \end{vmatrix} = abc \begin{vmatrix} 1 & a^2 & a^3 \\ 1 & b^2 & b^3 \\ 1 & c^2 & c^3 \end{vmatrix}$$

Then, as in Example 5, above, $(a - b), (b - c)$ and $(c - a)$ are factors. However, by inspection, we can see that every term in the expansion of the original determinant would have a total degree of *eight*. The factors extracted so far, namely, $abc(a - b)(b - c)(c - a)$, would only yield a total degree of *six*. Consequently, we need a further factor of degree *two*, which is symmetrical in a, b and c. The only possibilities are $(a^2 + b^2 + c^2)$ and $(bc + ca + ab)$, or some combination of these. So we must now consider

$$abc(a - b)(b - c)(c - a)\{\lambda(a^2 + b^2 + c^2) + \mu(bc + ca + ab)\}$$

However, inclusion of the term $(a^2 + b^2 + c^2)$ in the final factor would yield terms in a^5, b^5, c^5 which, by inspection of the determinant, are not required. We can therefore discard this term. We are left with the additional factor $\mu(bc + ca + ab)$. By inspection, e.g. of the term ab^3c^4, we see that $\mu = 1$. Hence the factorised form of the determinant is $abc(a - b)(b - c)(c - a)(bc + ca + ab)$.

Exercise 8b

The numerical exercises are intended to give the reader practice in using the standard properties of determinants; they should not be done by more elementary methods.

1 Evaluate: (a) $\begin{vmatrix} 21 & 31 & 50 \\ 17 & 3 & 35 \\ 22 & 31 & 52 \end{vmatrix}$,   (b) $\begin{vmatrix} 21 & 10 & 30 \\ 9 & -6 & 3 \\ 1 & -1 & 0 \end{vmatrix}$.

2 Evaluate: (a) $\begin{vmatrix} 39 & 91 & 143 \\ 296 & 43 & 151 \\ 51 & 119 & 187 \end{vmatrix}$,   (b) $\begin{vmatrix} 11 & 31 & 41 \\ 1 & 2 & 2 \\ 13 & 36 & 47 \end{vmatrix}$.

3 Evaluate: (a) $\begin{vmatrix} 207 & 52 & 135 \\ 184 & 17 & 120 \\ 69 & 109 & 45 \end{vmatrix}$, (b) $\begin{vmatrix} \frac{1}{2} & \frac{1}{3} & \frac{1}{6} \\ \frac{1}{2} & \frac{1}{4} & \frac{1}{8} \\ \frac{1}{10} & \frac{1}{10} & 0 \end{vmatrix}$.

4 Show that $\begin{vmatrix} a & x & x+a \\ x+a & a & x \\ x & x+a & a \end{vmatrix} = 2(x^3 + a^3).$

5 Solve the equation $\begin{vmatrix} x & 1 & 2 \\ 1 & x & 2 \\ 1 & 2 & x \end{vmatrix} = 0.$

6 Factorise $\begin{vmatrix} p & q & r \\ r & p & q \\ q & r & p \end{vmatrix}.$

7 Factorise $\begin{vmatrix} x & yz & x^2 \\ y & zx & y^2 \\ z & xy & z^2 \end{vmatrix}.$

8 Factorise $\begin{vmatrix} x^2 & x & 1 \\ 1 & x^2 & x \\ x & 1 & x^2 \end{vmatrix}.$

9 Prove that, if (x_1, y_1), (x_2, y_2) and (x_3, y_3) are three collinear points, then $\begin{vmatrix} 1 & 1 & 1 \\ x_1 & x_2 & x_3 \\ y_1 & y_2 & y_3 \end{vmatrix} = 0.$

10 Prove that, if, in three dimensions, (x_1, y_1, z_1), (x_2, y_2, z_2) and (x_3, y_3, z_3) lie in a plane through the origin, then $\begin{vmatrix} x_1 & x_2 & x_3 \\ y_1 & y_2 & y_3 \\ z_1 & z_2 & z_3 \end{vmatrix} = 0.$

Cofactors

8.4 In the following sections it will be convenient to refer to a standard determinant Δ, where

$$\Delta = \begin{vmatrix} a_1 & b_1 & c_1 \\ a_2 & b_2 & c_2 \\ a_3 & b_3 & c_3 \end{vmatrix}$$

Definition

The **cofactor** *of an element of a* 3×3 *determinant is the* 2×2 *determinant obtained by deleting the row and column containing that element and multiplying* *by* +1 *or* −1 *according to the pattern:*

$$+ \quad - \quad +$$
$$- \quad + \quad -$$
$$+ \quad - \quad +$$

A cofactor is always designated by the capital letter corresponding to the element to which it belongs.

No doubt this definition sounds rather complicated; it is not as complicated as it seems if we write out the nine cofactors in full. They are:

$$A_1 = \begin{vmatrix} b_2 & c_2 \\ b_3 & c_3 \end{vmatrix} \qquad B_1 = -\begin{vmatrix} a_2 & c_2 \\ a_3 & c_3 \end{vmatrix} \qquad C_1 = \begin{vmatrix} a_2 & b_2 \\ a_3 & b_3 \end{vmatrix}$$

$$A_2 = -\begin{vmatrix} b_1 & c_1 \\ b_3 & c_3 \end{vmatrix} \qquad B_2 = \begin{vmatrix} a_1 & c_1 \\ a_3 & c_3 \end{vmatrix} \qquad C_2 = -\begin{vmatrix} a_1 & b_1 \\ a_3 & b_3 \end{vmatrix}$$

$$A_3 = \begin{vmatrix} b_1 & c_1 \\ b_2 & c_2 \end{vmatrix} \qquad B_3 = -\begin{vmatrix} a_1 & c_1 \\ a_2 & c_2 \end{vmatrix} \qquad C_3 = \begin{vmatrix} a_1 & b_1 \\ a_2 & b_2 \end{vmatrix}$$

Expanding the determinant in the usual way, we can see that

$$\Delta = a_1 A_1 + b_1 B_1 + c_1 C_1$$

and the reader should verify that

$$a_2 A_2 + b_2 B_2 + c_2 C_2 \quad \text{and} \quad a_3 A_3 + b_3 B_3 + c_3 C_3$$

are also equal to Δ.

Expanding the determinant by columns we can see that $a_1 A_1 + a_2 A_2 + a_3 A_3$, $b_1 B_1 + b_2 B_2 + b_3 B_3$ and $c_1 C_1 + c_2 C_2 + c_3 C_3$ are also all equal to Δ. In other words, whenever the elements of one row (or column) are combined with the cofactors of the *same* row (or column), the sum is equal to Δ. However, notice what happens when the elements of one row are combined with the cofactors of a *different* row. For example, consider

$$a_1 A_2 + b_1 B_2 + c_1 C_2$$

(i.e. the elements of the first row, combined with the cofactors of the second row). Writing the cofactors as 2×2 determinants, we have

$$-a_1 \begin{vmatrix} b_1 & c_1 \\ b_3 & c_3 \end{vmatrix} + b_1 \begin{vmatrix} a_1 & c_1 \\ a_3 & c_3 \end{vmatrix} - c_1 \begin{vmatrix} a_1 & b_1 \\ a_3 & b_3 \end{vmatrix}$$

$$= -\begin{vmatrix} a_1 & b_1 & c_1 \\ a_1 & b_1 & c_1 \\ a_3 & b_3 & c_3 \end{vmatrix}$$

$$= 0 \qquad \text{(because this determinant has a pair of identical rows)}$$

The reader should verify that this will happen whenever the elements of a row (or column) are combined with the cofactors of a different row (or column). (In this case the cofactors are sometimes called *alien* cofactors.)

Qu. 4 Evaluate the determinant below, and find the values of its nine cofactors. Verify that they satisfy the relationships in the preceding section.

$$\begin{vmatrix} 1 & 2 & 3 \\ 3 & 1 & 0 \\ 2 & -1 & 1 \end{vmatrix}$$

These properties of cofactors can be exploited to produce an important method for solving linear equations, known as Cramer's rule. (This was first published by Gabriel Cramer in 1750.)

Cramer's rule

8.5 Consider the following three linear equations, and the associated determinant as discussed in the previous section.

$$a_1 x + b_1 y + c_1 z = d_1$$
$$a_2 x + b_2 y + c_2 z = d_2$$
$$a_3 x + b_3 y + c_3 z = d_3$$

Multiplying the first by A_1, the second by A_2 and the third by A_3 (where A_1, A_2, A_3 are the cofactors corresponding to a_1, a_2 and a_3 respectively) and adding, we obtain

$$(a_1 A_1 + a_2 A_2 + a_3 A_3)x + (b_1 A_1 + b_2 A_2 + b_3 A_3)y + (c_1 A_1 + c_2 A_2 + c_3 A_3)z$$
$$= d_1 A_1 + d_2 A_2 + d_3 A_3$$

Now, using the relationships established in the preceding section, we can see that the coefficient of x is Δ, and the coefficients of y and z are both zero. Hence, we have

$$\Delta x = d_1 A_1 + d_2 A_2 + d_3 A_3$$

$$= \begin{vmatrix} d_1 & b_1 & c_1 \\ d_2 & b_2 & c_2 \\ d_3 & b_3 & c_3 \end{vmatrix}$$

This determinant is the original determinant Δ, with the first column replaced by d_1, d_2 and d_3. It is convenient to abbreviate this determinant to Δ_1. (This abbreviation can be used more generally if we write Δ_i, to mean the determinant formed by replacing the ith column of Δ by d_1, d_2 and d_3.) The result we have just obtained can then be written $\Delta x = \Delta_1$. Proceeding in a similar fashion, it is fairly easy to show that $\Delta y = \Delta_2$ and $\Delta z = \Delta_3$. Hence the solutions to the three linear equations can be expressed in a very neat form, namely

$$\Delta x = \Delta_1, \qquad \Delta y = \Delta_2, \qquad \Delta z = \Delta_3 \qquad (1)$$

and, provided $\Delta \neq 0$, we can divide through by it and obtain,

$$x = \Delta_1/\Delta, \qquad y = \Delta_2/\Delta, \qquad z = \Delta_3/\Delta$$

This is Cramer's rule.

However, if Δ is equal to zero, we must *not* divide by it. In this case the equations (1) read

$$0x = \Delta_1, \qquad 0y = \Delta_2, \qquad 0z = \Delta_3$$

and these have no solution unless Δ_1, Δ_2 and Δ_3 are also zero. If these three determinants are equal to zero, then it *is* possible to find a solution (see Example 9, below).

Example 8   *Use Cramer's rule to solve*

$$\begin{aligned} x - 2y - 3z &= 0 \\ 3x + 5y + 2z &= 0 \\ 2x + 3y - z &= 2 \end{aligned}$$

Using the notation from the previous section,

$$\Delta = \begin{vmatrix} 1 & -2 & -3 \\ 3 & 5 & 2 \\ 2 & 3 & -1 \end{vmatrix}$$

$$= 1 \times (-11) + 2 \times (-7) - 3 \times (-1) = -11 - 14 + 3 = -22$$

$$\Delta_1 = \begin{vmatrix} 0 & -2 & -3 \\ 0 & 5 & 2 \\ 2 & 3 & -1 \end{vmatrix}$$

$$= 2 \begin{vmatrix} -2 & -3 \\ 5 & 2 \end{vmatrix} = 2 \times (+11) = 22$$

$$\Delta_2 = \begin{vmatrix} 1 & 0 & -3 \\ 3 & 0 & 2 \\ 2 & 2 & -1 \end{vmatrix}$$

$$= -2 \begin{vmatrix} 1 & -3 \\ 3 & 2 \end{vmatrix} = -2 \times (+11) = -22$$

$$\Delta_3 = \begin{vmatrix} 1 & -2 & 0 \\ 3 & 5 & 0 \\ 2 & 3 & 2 \end{vmatrix}$$

$$= 2 \begin{vmatrix} 1 & -2 \\ 3 & 5 \end{vmatrix} = 2 \times (+11) = 22$$

Hence, by Cramer's rule, $x = \Delta_1/\Delta = \quad 22/(-22) = -1,$
$$y = \Delta_2/\Delta = -22/(-22) = +1,$$
$$z = \Delta_3/\Delta = \quad 22/(-22) = -1.$$

Example 9 *Solve the equations*

$$\begin{aligned} x + 2y + z &= 7 \\ 3x + y &= -2 \\ 5x + 5y + 2z &= 12 \end{aligned}$$

$$\Delta = \begin{vmatrix} 1 & 2 & 1 \\ 3 & 1 & 0 \\ 5 & 5 & 2 \end{vmatrix}$$

$$= 1 \times (+2) - 2 \times (+6) + 1 \times (+10)$$

$$= 2 - 12 + 10$$

$$= 0$$

Also $\Delta_1 = \Delta_2 = \Delta_3 = 0$. (The detailed working is left to the reader.) Hence, in this case, Cramer's rule gives

$$0 \times x = 0, \qquad 0 \times y = 0, \qquad 0 \times z = 0$$

Now it is certainly *possible* for values of x, y and z to exist which satisfy these equations, but Cramer's rule is not very helpful, in this special case. If we return to the original equations and eliminate z from the first and third equations, we obtain

$$\begin{aligned} 2x + 4y + 2z &= 14 \\ 5x + 5y + 2z &= 12 \end{aligned}$$

and subtracting, we have

$$3x + y = -2$$

This is identical to the second equation. (If the right-hand side had *not* been -2, we would have had to have concluded, at this stage, that there was no solution; if, on the other hand the equations had been *distinct*, we could have solved them to find x and y, but Cramer's rule has already shown us that this is not possible.)

Although we cannot solve $3x + y = -2$ and find a *unique* solution, we can let $x = t$, where t is any real number. Having assigned this value to x, we have no choice in the value of y; it must be $-2 - 3t$. Then, substituting these values for x and y into the first of the original equations, we have

$$\begin{aligned} z &= 7 - t - 2(-2 - 3t) \\ &= 11 + 5t \end{aligned}$$

Hence, we have a *set* of solutions,

$$x = t \qquad y = -2 - 3t \qquad z = 11 + 5t$$

where t is any real number.

(Compare this with Book 1, §15.14.)

Qu. 5 Give a geometrical interpretation of the result of Example 9.

Example 10   *Solve the equations*

$$x + 2y + z = 7$$
$$3x + y = -2$$
$$5x + 5y + 2z = 10$$

(Notice that the first two equations are the same as those in Example 9, and in the third equation only the constant is different. As in Example 9, $\Delta = 0$, but this time $\Delta_1 \neq 0$.)

$$\Delta_1 = \begin{vmatrix} 7 & 2 & 1 \\ -2 & 1 & 0 \\ 10 & 5 & 2 \end{vmatrix}$$

$$= 7 \times (+2) - 2 \times (-4) + 1 \times (-20)$$

$$= 14 + 8 - 20$$

$$= 2$$

Since $\Delta_1 \neq 0$, there is *no* solution.

Qu. 6   Give a geometrical interpretation of the result of Example 10.

Cramer's rule is of considerable interest because of the light it sheds on the behaviour of simultaneous linear equations; in practice, most readers will find that elimination is usually the most satisfactory method to use. (See Chapter 7, Example 10.)

Exercise 8c

The questions in this exercise are intended to give the reader practice in using Cramer's rule.

Solve the simultaneous equations:

1 $x + y + z = 6,$
    $2x + y - z = 1,$
    $x - y + z = 2.$

2 $2x + 3y + z = 1,$
    $x - y + z = 4,$
    $5x + y + 3z = 10.$

3 $7x + y + z = -1,$
    $x - 3y + 2z = 0,$
    $x + 4y - 3z = 4.$

4 $5x - 7y + 3z = 0,$
    $2x - 3y + 5z = -1,$
    $3x - 4y + 2z = 1.$

5 $10x + 20y + 40z = 1,$
    $3x + 7y + 10z = 0,$
    $25x + 12y + 37z = 0.$

6 Find the condition for the simultaneous equations below to have no solution:

$$x + 5y + az = 2$$
$$2x + y + 3z = 1$$
$$7x + 8y + 8z = k$$

In Nos. 7–10, the equations should be regarded as the equations of three planes which

(a) meet in a (unique) point, or
(b) meet in a common line, or
(c) do *not* meet.

The reader should distinguish between these three possibilities. In case (a) the coordinates of the point of intersection should be found and in case (b) the general form of any point on the common line should be found. (In the latter case, the form of the answer is not unique; the form printed in the answers at the back of this book can be obtained by putting $x = t$.)

7   $2x - y + 4z = 0,$
     $3x + 4y + 12z = 0,$
     $7x - 9y + 8z = 0.$

8   $x + y + z = 0,$
     $x - y + z = 0,$
        $y - z = 0.$

9   $x + 2y + 3z = 4,$
     $2x + y + 5z = -2,$
     $4x + 5y + 11z = 6.$

10   $2x + y - z = 5,$
     $x + 2y + z = 4,$
     $3x + 9y + 6z = 7.$

The inverse of a 3×3 matrix

8.6   As in the previous sections, $\mathbf{M} = \begin{pmatrix} a_1 & b_1 & c_1 \\ a_2 & b_2 & c_2 \\ a_3 & b_3 & c_3 \end{pmatrix}$, and the capital letters, A_1, A_2, A_3, etc., are used to represent the cofactors which correspond to the elements a_1, a_2, a_3, etc.

Definition

The **adjoint** *of the matrix* $\mathbf{M}$ *is* $\begin{pmatrix} A_1 & A_2 & A_3 \\ B_1 & B_2 & B_3 \\ C_1 & C_2 & C_3 \end{pmatrix}.$

The standard abbreviation for 'the adjoint of matrix $\mathbf{M}$' is adj($\mathbf{M}$). Notice that adj($\mathbf{M}$) is the transpose of the matrix formed by replacing each element of $\mathbf{M}$ by its cofactor.

Qu. 7   Given that $\mathbf{A} = \begin{pmatrix} 1 & 1 & 1 \\ 3 & 2 & 1 \\ 1 & 2 & 3 \end{pmatrix}$, find adj($\mathbf{A}$) and determine the matrix product $\mathbf{A}$ adj($\mathbf{A}$).

Qu. 8   Repeat Qu. 7, for $\mathbf{A} = \begin{pmatrix} 1 & -1 & 1 \\ 2 & 1 & 2 \\ 0 & 1 & 1 \end{pmatrix}$.

The product $\mathbf{M}$ adj($\mathbf{M}$) is always a **diagonal matrix**, that is, a matrix in which all the elements are zero, except those on the 'leading diagonal' (the diagonal which goes from the top left-hand corner to the bottom right-hand corner). To see the reason for this, we must look at the working in detail:

$$\mathbf{M}\,\mathrm{adj}(\mathbf{M}) = \begin{pmatrix} a_1 & b_1 & c_1 \\ a_2 & b_2 & c_2 \\ a_3 & b_3 & c_3 \end{pmatrix} \begin{pmatrix} A_1 & A_2 & A_3 \\ B_1 & B_2 & B_3 \\ C_1 & C_2 & C_3 \end{pmatrix}$$

$$= \begin{pmatrix} a_1A_1 + b_1B_1 + c_1C_1 & a_1A_2 + b_1B_2 + c_1C_2 & a_1A_3 + b_1B_3 + c_1C_3 \\ a_2A_1 + b_2B_1 + c_2C_1 & a_2A_2 + b_2B_2 + c_2C_2 & a_2A_3 + b_2B_3 + c_2C_3 \\ a_3A_1 + b_3B_1 + c_3C_1 & a_3A_2 + b_3B_2 + c_3C_2 & a_3A_3 + b_3B_3 + c_3C_3 \end{pmatrix}$$

Now, each term which is *on* the leading diagonal, consists of an element of one of the rows of $\mathbf{M}$ combined with its *own* cofactor; this, as we saw in §8.4, is always equal to Δ, the determinant of $\mathbf{M}$. Each term which is *not* on the leading diagonal consists of an element from a row of $\mathbf{M}$, combined with an *alien* cofactor; this, as we also saw in §8.4, is always zero.

Hence,

$$\mathbf{M}\,\mathrm{adj}(\mathbf{M}) = \begin{pmatrix} \Delta & 0 & 0 \\ 0 & \Delta & 0 \\ 0 & 0 & \Delta \end{pmatrix} = \Delta\,\mathbf{I}$$

where $\mathbf{I}$ is the 3×3 unit matrix (see Book 1, §11.5).

The reader should verify that $\mathrm{adj}(\mathbf{M})\,\mathbf{M}$ is also $\Delta\,\mathbf{I}$

Qu. 9   Verify that the answers to Qu. 7 and Qu. 8 are equal to $\Delta\,\mathbf{I}$.

This enables us to tackle a problem which in Book 1, §11.5, we had to postpone. In that section we found the inverse of a general 2×2 matrix, but we did not attempt to find the inverse of any other square matrix. In view of the long and tortuous path we have had to follow in order to arrive at a point where we can tackle this problem, it is not surprising that in the earlier chapter we postponed it! If, now we look at the product $\mathbf{M}\,\mathrm{adj}(\mathbf{M}) = \Delta\,\mathbf{I}$, we can see that the problem is almost solved; only one more step is necessary, namely, divide each side by Δ; we then have

$$\mathbf{M}\,\frac{\mathrm{adj}(\mathbf{M})}{\Delta} = \mathbf{I}$$

In other words $\mathbf{M}^{-1}$, the inverse of $\mathbf{M}$, is $\dfrac{\mathrm{adj}(\mathbf{M})}{\Delta}$.

(Δ must not equal zero; if it does, there is no inverse matrix, i.e. $\mathbf{M}$ is a *singular* matrix.)

Example 11   *Find the inverse of* $\mathbf{A} = \begin{pmatrix} 1 & 1 & 1 \\ 1 & 1 & -1 \\ 1 & -2 & 3 \end{pmatrix}$.

$$\mathrm{adj}(\mathbf{A}) = \begin{pmatrix} 1 & -5 & -2 \\ -4 & 2 & 2 \\ -3 & 3 & 0 \end{pmatrix}$$

$$\mathbf{A}\,\text{adj}(\mathbf{A}) = \begin{pmatrix} 1 & 1 & 1 \\ 1 & 1 & -1 \\ 1 & -2 & 3 \end{pmatrix} \begin{pmatrix} 1 & -5 & -2 \\ -4 & 2 & 2 \\ -3 & 3 & 0 \end{pmatrix}$$

$$= \begin{pmatrix} -6 & 0 & 0 \\ 0 & -6 & 0 \\ 0 & 0 & -6 \end{pmatrix}$$

(Notice that det($\mathbf{A}$) is required, but it is not necessary to work it out separately, because we know it must be the -6 which appears on the leading diagonal.) Hence

$$\mathbf{A}^{-1} = \frac{1}{-6}\begin{pmatrix} 1 & -5 & -2 \\ -4 & 2 & 2 \\ -3 & 3 & 0 \end{pmatrix} = \frac{1}{6}\begin{pmatrix} -1 & 5 & 2 \\ 4 & -2 & -2 \\ 3 & -3 & 0 \end{pmatrix}$$

Example 12 *Find the inverse of the matrix* $\begin{pmatrix} 1 & 2 & 3 \\ 2 & 1 & 1 \\ 3 & 1 & -2 \end{pmatrix}$. *Write the following equations in matrix form and hence find x, y and z:*

$$x + 2y + 3z = 6$$
$$2x + y + z = 5$$
$$3x + y - 2z = 1$$

Let $\mathbf{A} = \begin{pmatrix} 1 & 2 & 3 \\ 2 & 1 & 1 \\ 3 & 1 & -2 \end{pmatrix}$, then $\text{adj}(\mathbf{A}) = \begin{pmatrix} -3 & 7 & -1 \\ 7 & -11 & 5 \\ -1 & 5 & -3 \end{pmatrix}$, and

$$\mathbf{A}\,\text{adj}(\mathbf{A}) = \begin{pmatrix} 1 & 2 & 3 \\ 2 & 1 & 1 \\ 3 & 1 & -2 \end{pmatrix} \begin{pmatrix} -3 & 7 & -1 \\ 7 & -11 & 5 \\ -1 & 5 & -3 \end{pmatrix}$$

$$= \begin{pmatrix} 8 & 0 & 0 \\ 0 & 8 & 0 \\ 0 & 0 & 8 \end{pmatrix}$$

(The reader should note that this step serves two purposes; it checks the accuracy of the arithmetic up to this point, and it evaluates det($\mathbf{A}$). In this case, det($\mathbf{A}$) = 8.) Hence

$$\mathbf{A}^{-1} = \frac{1}{8}\begin{pmatrix} -3 & 7 & -1 \\ 7 & -11 & 5 \\ -1 & 5 & -3 \end{pmatrix}$$

The simultaneous equations can be written

$$\begin{pmatrix} 1 & 2 & 3 \\ 2 & 1 & 1 \\ 3 & 1 & -2 \end{pmatrix} \begin{pmatrix} x \\ y \\ z \end{pmatrix} = \begin{pmatrix} 6 \\ 5 \\ 1 \end{pmatrix}$$

The matrix of the coefficients is the matrix **A**, whose inverse we have found.

Multiplying both sides by $\mathbf{A}^{-1}$, the left-hand side becomes $\begin{pmatrix} x \\ y \\ z \end{pmatrix}$, since

$\mathbf{A}^{-1}\mathbf{A} = \mathbf{I}$, and $\mathbf{I}\begin{pmatrix} x \\ y \\ z \end{pmatrix} = \begin{pmatrix} x \\ y \\ z \end{pmatrix}$.

$$\therefore \begin{pmatrix} x \\ y \\ z \end{pmatrix} = \frac{1}{8}\begin{pmatrix} -3 & 7 & -1 \\ 7 & -11 & 5 \\ -1 & 5 & -3 \end{pmatrix}\begin{pmatrix} 6 \\ 5 \\ 1 \end{pmatrix}$$

$$= \frac{1}{8}\begin{pmatrix} 16 \\ -8 \\ 16 \end{pmatrix}$$

$$= \begin{pmatrix} 2 \\ -1 \\ 2 \end{pmatrix}$$

Hence $x = 2$, $y = -1$ and $z = 2$.

Exercise 8d

Find, where possible, the inverses of the following matrices.

1 $\begin{pmatrix} 1 & -1 & 2 \\ 1 & 2 & 3 \\ 3 & 0 & 1 \end{pmatrix}$. **2** $\begin{pmatrix} 2 & 1 & 7 \\ 3 & 4 & 5 \\ 1 & -2 & 9 \end{pmatrix}$. **3** $\begin{pmatrix} 1 & -10 & 7 \\ 1 & 4 & -3 \\ -1 & 2 & -1 \end{pmatrix}$.

4 $\begin{pmatrix} 2 & 1 & 0 \\ 3 & 7 & 5 \\ 1 & 17 & 15 \end{pmatrix}$. **5** $\begin{pmatrix} 3 & 3 & -1 \\ -6 & 2 & 2 \\ -1 & -1 & 3 \end{pmatrix}$. **6** $\begin{pmatrix} 1 & p & q \\ 0 & 1 & r \\ 0 & 0 & 1 \end{pmatrix}$.

7 $\begin{pmatrix} 1 & 0 & 0 \\ 0 & \cos\alpha & \sin\alpha \\ 0 & \sin\alpha & -\cos\alpha \end{pmatrix}$.

Solve the following simultaneous equations, using matrices (see Example 12).

8 $x - y + 2z = 4,$
$x + 2y + 3z = 2,$
$3x \quad\ + z = 4.$

9 $x - 10y + 7z = 13,$
$x + 4y - 3z = -3,$
$-x + 2y - z = -3.$

10 $2x + y + z = 4,$
$x - y - 2z = 0,$
$5x - 2y - 4z = 3.$

11 Given that $\mathbf{A} = \begin{pmatrix} 1 & 0 & 2 \\ 2 & 1 & 0 \\ 3 & 1 & 1 \end{pmatrix}$ and $\mathbf{B} = \begin{pmatrix} 1 & -1 & 1 \\ 0 & 2 & 1 \\ 1 & 3 & 0 \end{pmatrix}$, find

(a) $\mathbf{A}^{-1}$, (b) $\mathbf{B}^{-1}$, (c) $\mathbf{AB}$, (d) $(\mathbf{AB})^{-1}$, and verify that $(\mathbf{AB})^{-1} = \mathbf{B}^{-1}\mathbf{A}^{-1}$.

12 Repeat No. 11 with $\mathbf{A} = \begin{pmatrix} 2 & 1 & 0 \\ 1 & 5 & 2 \\ 2 & -1 & 1 \end{pmatrix}$ and $\mathbf{B} = \begin{pmatrix} -1 & 2 & 0 \\ 1 & 3 & 2 \\ 2 & 0 & 1 \end{pmatrix}$.

13 Given that $\mathbf{A}$ and $\mathbf{B}$ are non-singular square matrices of the same order, prove that $(\mathbf{AB})^{-1} = \mathbf{B}^{-1}\mathbf{A}^{-1}$. [Hint: consider the product $(\mathbf{AB})(\mathbf{B}^{-1}\mathbf{A}^{-1})$.] (You may assume that matrix multiplication is associative.)
What is the corresponding result for $(\mathbf{ABC})^{-1}$?

Exercise 8e (Miscellaneous)

In Nos. 1–3, evaluate the determinants.

1 (a) $\begin{vmatrix} 2 & 1 & -5 \\ 3 & 7 & 0 \\ 4 & 2 & 1 \end{vmatrix}$, (b) $\begin{vmatrix} 1 & -2 & 3 \\ 2 & 1 & 0 \\ 4 & 7 & 1 \end{vmatrix}$, (c) $\begin{vmatrix} 1 & 0 & -2 \\ 2 & 7 & 3 \\ 1 & 4 & 0 \end{vmatrix}$.

2 (a) $\begin{vmatrix} 1 & 5 & 7 \\ 2 & 3 & 1 \\ 7 & 21 & 23 \end{vmatrix}$, (b) $\begin{vmatrix} 1 & 7 & 11 \\ 2 & 17 & 23 \\ 7 & 54 & 79 \end{vmatrix}$, (c) $\begin{vmatrix} 17 & 34 & 119 \\ 26 & 221 & 91 \\ 7 & 7 & 0 \end{vmatrix}$.

3 (a) $\begin{vmatrix} 1 & 0 & 0 \\ 0 & \cos\alpha & -\sin\alpha \\ 0 & \sin\alpha & \cos\alpha \end{vmatrix}$, (b) $\begin{vmatrix} 1 & x & y \\ 0 & 1 & z \\ 0 & 0 & 1 \end{vmatrix}$.

4 Express in factors:

(a) $\begin{vmatrix} 1 & 1 & 1 \\ yz & zx & xy \\ y+z & z+x & x+y \end{vmatrix}$, (b) $\begin{vmatrix} x & x & x \\ x & y & y \\ x & y & z \end{vmatrix}$.

5 Given that $\mathbf{M} = \begin{pmatrix} \cos\alpha\cos\beta & \cos\alpha\sin\beta & -\sin\alpha \\ -\sin\beta & \cos\beta & 0 \\ \sin\alpha\cos\beta & \sin\alpha\sin\beta & \cos\alpha \end{pmatrix}$, evaluate $\det(\mathbf{M})$ and find $\mathbf{M}^{-1}$.

*6 Prove that, for any two 3×3 matrices $\mathbf{M}$ and $\mathbf{N}$, $(\mathbf{MN})^{\mathrm{T}} = \mathbf{N}^{\mathrm{T}}\mathbf{M}^{\mathrm{T}}$. [Hint: it is sufficient to consider the element in the ith row and jth column.]

7 Prove that if $\mathbf{A}\mathbf{A}^{\mathrm{T}} = \mathbf{I}$, then $\det(\mathbf{A}) = 1$.

8 Given that $N = \dfrac{1}{\sqrt{6}} \begin{pmatrix} \sqrt{2} & \sqrt{3} & 1 \\ \sqrt{2} & 0 & -2 \\ \sqrt{2} & -\sqrt{3} & 1 \end{pmatrix}$, prove that $N N^T = N^T N = I$.

(Any matrix with the property $N N^T = I$, is called an *orthogonal* matrix.)

9 Find the non-trivial solution of the simultaneous equations

$$\begin{aligned} x + y - z &= 0 \\ 5x - y - z &= 0 \\ 4x + y - 2z &= 0 \end{aligned}$$

(The solution $x = 0$, $y = 0$, $z = 0$, is a *trivial* solution; a *non-trivial* solution is one in which x, y and z are *not* all zero.)

*10 Prove that the equations

$$\begin{aligned} a_1 x + b_1 y + c_1 z &= 0 \\ a_2 x + b_2 y + c_2 z &= 0 \\ a_3 x + b_3 y + c_3 z &= 0 \end{aligned}$$

have a non-trivial solution, if and only if $\begin{vmatrix} a_1 & b_1 & c_1 \\ a_2 & b_2 & c_2 \\ a_3 & b_3 & c_3 \end{vmatrix} = 0.$

11 Given that (using the notation of §8.3) a 3×3 matrix has the property, $\mathbf{r}_3 = \alpha \mathbf{r}_1 + \beta \mathbf{r}_2$, where $\alpha, \beta \in \mathbb{R}$, prove that real numbers λ and μ can be found, such that $\mathbf{c}_3 = \lambda \mathbf{c}_1 + \mu \mathbf{c}_2$.

12 A transformation, in three dimensions, is given by

$$\begin{pmatrix} x' \\ y' \\ z' \end{pmatrix} = \begin{pmatrix} a_1 & b_1 & c_1 \\ a_2 & b_2 & c_2 \\ a_3 & b_3 & c_3 \end{pmatrix} \begin{pmatrix} x \\ y \\ z \end{pmatrix}$$

Write down the images of the unit vectors $\begin{pmatrix} 1 \\ 0 \\ 0 \end{pmatrix}$, $\begin{pmatrix} 0 \\ 1 \\ 0 \end{pmatrix}$ and $\begin{pmatrix} 0 \\ 0 \\ 1 \end{pmatrix}$. Hence write down the matrices which give the following transformations:

(a) a reflection in the plane $z = 0$,
(b) a rotation about the z-axis through $90°$,
(c) a reflection in the plane $x = y$.

13 Show that the equation

$$x^2 - y^2 = 1 \tag{1}$$

can be expressed in the form $(x\ y) \begin{pmatrix} a & b \\ b & -a \end{pmatrix} \begin{pmatrix} x \\ y \end{pmatrix} = 1$. Transform this equation by writing $\begin{pmatrix} x \\ y \end{pmatrix} = \dfrac{1}{\sqrt{2}} \begin{pmatrix} 1 & 1 \\ -1 & 1 \end{pmatrix} \begin{pmatrix} X \\ Y \end{pmatrix}$, and show that it can then be written

$$XY = \tfrac{1}{2} \tag{2}$$

What is the relationship between the curves represented by equations (1) and (2)?

14 Express the equation

$$-2x^2 + 4\sqrt{3}xy + 2y^2 = 1$$

in matrix form (as in No. 13), and transform it by writing

$$\begin{pmatrix} x \\ y \end{pmatrix} = \begin{pmatrix} 1 & -\sqrt{3} \\ \sqrt{3} & 1 \end{pmatrix} \begin{pmatrix} X \\ Y \end{pmatrix}$$

What is the relationship between the curves represented by the two equations?

15 Find two values of k such that $\begin{pmatrix} 5 & 6 \\ -2 & -2 \end{pmatrix} \begin{pmatrix} x \\ y \end{pmatrix} = k \begin{pmatrix} x \\ y \end{pmatrix}$. For each value of k, find corresponding values for x and y.

16 Repeat No. 15 for $\begin{pmatrix} 19 & 4 \\ 1 & 16 \end{pmatrix} \begin{pmatrix} x \\ y \end{pmatrix} = k \begin{pmatrix} x \\ y \end{pmatrix}$.

17 Show that if $\mathbf{M} = \begin{pmatrix} 19 & 4 \\ 1 & 16 \end{pmatrix}$ and $\mathbf{X} = \begin{pmatrix} 1 & 4 \\ -1 & 1 \end{pmatrix}$, then

$$\mathbf{X}^{-1}\mathbf{M}\mathbf{X} = \begin{pmatrix} 15 & 0 \\ 0 & 20 \end{pmatrix}$$

By writing $\mathbf{M}$ in the form $\mathbf{X}\mathbf{D}\mathbf{X}^{-1}$, where $\mathbf{D} = \begin{pmatrix} 15 & 0 \\ 0 & 20 \end{pmatrix}$, show that

$$\mathbf{M}^n = \mathbf{X}\mathbf{D}^n\mathbf{X}^{-1}$$

and hence show that

$$\mathbf{M}^n = \frac{1}{5}\begin{pmatrix} 15^n + 4 \times 20^n & -4 \times 15^n + 4 \times 20^n \\ -15^n + 20^n & 4 \times 15^n + 20^n \end{pmatrix}$$

18 Find the inverse of the matrix $\begin{pmatrix} 1 & -3 & 0 \\ 2 & 0 & 1 \\ 4 & 1 & 3 \end{pmatrix}$.

Hence solve the equations

$$x - 3y - a = 0, \quad 2x + z - b = 0 \quad \text{and} \quad 4x + y + 3z - c = 0,$$

for x, y, z in terms of a, b, c.   (O & C)

19 The linear transformation $\begin{pmatrix} x' \\ y' \\ z' \end{pmatrix} = \mathbf{M} \begin{pmatrix} x \\ y \\ z \end{pmatrix}$, where $\mathbf{M}$ is a 3×3 matrix,

maps the points with position vectors $\begin{pmatrix} 1 \\ 0 \\ 0 \end{pmatrix}$, $\begin{pmatrix} 0 \\ 1 \\ 0 \end{pmatrix}$, $\begin{pmatrix} 0 \\ 0 \\ 1 \end{pmatrix}$, to the points with position vectors $\begin{pmatrix} 3 \\ 2 \\ 1 \end{pmatrix}$, $\begin{pmatrix} 2 \\ 1 \\ 0 \end{pmatrix}$, $\begin{pmatrix} 1 \\ 0 \\ 0 \end{pmatrix}$, respectively. Write down the matrix **M** and find the inverse matrix $\mathbf{M}^{-1}$. Show that the transformation with matrix **M** maps points of the plane $x + y + z = 0$, to points of the plane $x = y$, and verify that the inverse transformation with matrix $\mathbf{M}^{-1}$, maps points of the plane $x = y$, to points of the plane $x + y + z = 0$.   (L)

20 Prove that $\begin{vmatrix} (b+c)^2 & a^2 & a^2 \\ b^2 & (a+c)^2 & b^2 \\ c^2 & c^2 & (a+b)^2 \end{vmatrix} = 2abc(a+b+c)^3$.   (JMB)

21 The matrix **A** is $\begin{pmatrix} 3 & 1 & 1 \\ 1 & 0 & -2 \\ 10 & 3 & 1 \end{pmatrix}$. Show that **A** is singular. Find a non-zero matrix **B**, such that $\mathbf{AB} = \mathbf{BA} = \mathbf{O}$.   (JMB)

22 Find the inverse $\mathbf{A}^{-1}$ of the matrix $\mathbf{A} = \begin{pmatrix} 1 & 0 & 0 \\ -1 & 1 & 0 \\ 3 & 2 & 1 \end{pmatrix}$. Find also $\mathbf{B}^{-1}$ and

$(\mathbf{AB})^{-1}$ where $\mathbf{B} = \begin{pmatrix} 1 & 4 & -2 \\ 0 & 1 & 3 \\ 0 & 0 & 1 \end{pmatrix}$.

Given that $\mathbf{AB} \begin{pmatrix} x \\ y \\ z \end{pmatrix} = \begin{pmatrix} 1 \\ -2 \\ 1 \end{pmatrix}$, find x, y and z.   (L)

23 Find the complete solutions of the two systems of equations:

(a) $3x + 4y + z = 5$,   (b) $3x + 4y + z = 5$,
$\quad\ 2x - y - z = 4$, $2x - y - z = 4$,
$\quad\ x + 3y + z = 1$. $5x + 14y + 5z = 7$.   (O)

24 **M** is the matrix $\begin{pmatrix} 3 & 1 & -3 \\ 1 & 2a & 1 \\ 0 & 2 & a \end{pmatrix}$.

(a) Find two values of a for which **M** is singular.

(b) Solve the equation $\mathbf{M} \begin{pmatrix} x \\ y \\ z \end{pmatrix} = \begin{pmatrix} -3\frac{1}{2} \\ 5\frac{1}{2} \\ 5 \end{pmatrix}$ in the case $a = 2$, and determine

whether or not solutions exist for each of the two values of a found in (a).   (C)

25 If a, b and c are real, find the factors of the determinant

$$\Delta = \begin{vmatrix} b+c & c+a & a+b \\ c+a & a+b & b+c \\ a+b & b+c & c+a \end{vmatrix}$$

Show that, if $\Delta = 0$, then either $a + b + c = 0$, or $a = b = c$.   (C)

Chapter 9

Coordinate geometry

Conic sections

9.1 In this chapter we shall be dealing with three curves, the parabola, ellipse, and hyperbola, which are all known as **conic sections** or **conics**. The Greek mathematicians even before Euclid (third century B.C.) were interested in these curves and examined their properties by pure geometry starting from their definitions by means of sections of a cone. From the point of view of coordinate geometry it is better to start from another definition (which can be shown to be equivalent) that a conic is the locus of a point which moves so that its distance from a fixed point bears a constant ratio to its distance from a fixed line (see Fig. 9.1).

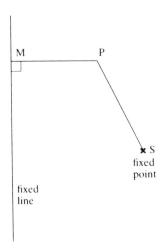

Figure 9.1

The fixed point S is called the **focus**. The fixed line is called the **directrix**. The constant ratio is called the **eccentricity** and is denoted by e. Thus, if P is a point on the locus, M is the foot of the perpendicular from P to the directrix and if

$$\frac{SP}{PM} = e$$

then the locus of P is a conic.

When: $e = 1$, the conic is a **parabola**,
           $e < 1$, the conic is an **ellipse**,
           $e > 1$, the conic is a **hyperbola**.

We shall first take the parabola, which was briefly mentioned in Book 1, Chapter 22.

The parabola

9.2   Given the focus S of the parabola and the directrix, we are at liberty to take what axes we find most convenient. First note that the figure formed by the focus and directrix has an axis of symmetry through S perpendicular to the directrix. This we take as the x-axis, as shown. If we now plot a few points, using the definition of the locus given in the last section,

$$\frac{SP}{PM} = 1$$

an indication of the shape of the curve may be obtained (see Fig. 9.2). (The plotting may be done very simply using squared paper and a pair of compasses.) It now seems reasonable to take the y-axis through the point on the axis of symmetry mid-way between the focus and directrix. This point is called the **vertex** of the parabola. Let the distance from the vertex to the focus be a; then

the focus S is $(a, 0)$
and
the directrix is the line $x = -a$

If $P(x, y)$ is any point on the parabola, and M is the foot of the perpendicular from P to the directrix,

$$SP^2 = (x - a)^2 + y^2$$
and
$$PM = x + a$$

But from the definition,

$$\frac{SP}{PM} = 1, \quad \text{so} \quad SP^2 = PM^2$$

$$\therefore (x - a)^2 + y^2 = (x + a)^2$$
$$\therefore x^2 - 2ax + a^2 + y^2 = x^2 + 2ax + a^2$$
$$\therefore y^2 = \mathbf{4ax}$$

which is the standard equation of a parabola.

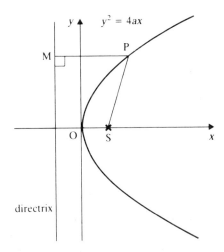

Figure 9.2

Qu. 1 Find the equations of the parabolas:
(a) focus $(-a, 0)$, directrix $x = a$,
(b) focus $(0, b)$, directrix $y = -b$.

Example 1 *Find, in terms of a, m, the value of c which makes the line $y = mx + c$ a tangent to the parabola $y^2 = 4ax$. Also obtain the coordinates of the point of contact.*

$$y = mx + c$$

Multiply both sides by $4a$.

$$4ay = m \times 4ax + 4ac$$

Substituting from $y^2 = 4ax$ and collecting terms,

$$my^2 - 4ay + 4ac = 0 \tag{1}$$

The line will be a tangent if this equation has equal roots (see Book 1, §10.2)

$$\therefore (-4a)^2 = 16mac$$

$$\therefore c = \frac{a}{m}$$

When the roots of equation (1) are equal, they will be given by half the sum of the roots,

$$\therefore y = \frac{1}{2} \times \frac{4a}{m} = \frac{2a}{m}$$

Now

$$x = \frac{y^2}{4a} = \left(\frac{2a}{m}\right)^2 \times \frac{1}{4a} = \frac{a}{m^2}$$

Therefore the point of contact is $\left(\dfrac{a}{m^2}, \dfrac{2a}{m}\right)$

Note that the equation of a general tangent to $y^2 = 4ax$ may be written

$$y = mx + \frac{a}{m} \qquad (m \neq 0)$$

This last result leads us to a very useful way of representing a point on the parabola. Substituting $t = 1/m$, we see that the tangent

$$y = \frac{x}{t} + at$$

touches the parabola at $(at^2, 2at)$. Since the tangent was a general one, we have shown that *any point on the parabola $y^2 = 4ax$ may be written $(at^2, 2at)$*. The equations $x = at^2$, $y = 2at$ are called the **parametric equations** of the parabola $y^2 = 4ax$.

Qu. 2   Verify, by substitution, that $(at^2, 2at)$ always lies on the parabola $y^2 = 4ax$.

We have found the equation of the tangent at $(at^2, 2at)$, but a more direct method follows in Example 2.

Example 2   *Find the equation of the tangent to $y^2 = 4ax$ at $(at^2, 2at)$.*

To find the gradient at $(at^2, 2at)$,

$$\frac{dy}{dx} = \frac{dy}{dt} \bigg/ \frac{dx}{dt}$$

But $y = 2at$, $x = at^2$,

$$\therefore \frac{dy}{dx} = \frac{2a}{2at} = \frac{1}{t}$$

The equation of the tangent is obtained by the method of Book 1, §22.1, which will be used from now on.

$$x - ty = at^2 - t \times 2at$$

i.e.

$$x - ty + at^2 = 0$$

*Qu. 3   Show that the equation of the normal to $y^2 = 4ax$ at $(at^2, 2at)$ is

$$tx + y - at^3 - 2at = 0$$

Example 3   *Show that the equation of the tangent to the parabola $y^2 = 4ax$ at (x_1, y_1) is $yy_1 = 2a(x + x_1)$.*

Differentiating both sides of $y^2 = 4ax$ with respect to x, to find the gradient:

$$2y \frac{dy}{dx} = 4a$$

Therefore at (x_1, y_1), $\dfrac{dy}{dx} = \dfrac{4a}{2y_1} = \dfrac{2a}{y_1}$, and the tangent is

$$2ax - y_1 y = 2ax_1 - y_1{}^2$$
$$\therefore y_1 y = 2ax - 2ax_1 + y_1{}^2$$

Now (x_1, y_1) lies on the parabola, so $y_1{}^2 = 4ax_1$.

$$\therefore y_1 y = 2ax - 2ax_1 + 4ax_1$$
$$\therefore yy_1 = 2a(x + x_1)$$

Qu. 4 Find the equation of the normal to $y^2 = 4ax$ at (x_1, y_1).

Example 4 *Find the equation of the chord joining the points* $(at_1{}^2, 2at_1)$, $(at_2{}^2, 2at_2)$.

The gradient of the chord is

$$\frac{2at_1 - 2at_2}{at_1{}^2 - at_2{}^2} = \frac{2a(t_1 - t_2)}{a(t_1 - t_2)(t_1 + t_2)}$$

$$= \frac{2}{t_1 + t_2}$$

Therefore the equation of the chord is

$$2x - (t_1 + t_2)y = 2at_1{}^2 - (t_1 + t_2) \times 2at_1$$
$$= -2at_1 t_2$$

Therefore the chord is $2x - (t_1 + t_2)y + 2at_1 t_2 = 0$.

*$*$**Qu. 5** As $t_2 \to t_1$, the chord approaches the tangent at t_1. Deduce the equation of the tangent from the equation of the chord.

Definitions

Any chord of a parabola passing through the focus is called a **focal chord**. *The axis of symmetry is usually simply called the* **axis** *of the parabola. The focal chord perpendicular to the axis is called the* **latus rectum**.

To find the length of the latus rectum of the parabola $y^2 = 4ax$, substitute $x = a$;

$$y^2 = 4a^2$$
$$\therefore y = \pm 2a$$

Hence the length of the latus rectum is $4a$.

The reader is advised to work all the questions in Exercise 9a, using the parametric coordinates $(at^2, 2at)$ whenever the opportunity arises. The point $(at^2, 2at)$ is frequently abbreviated to the point t.

Exercise 9a

1 Find the coordinates of the point of intersection of the tangents at the points t_1, t_2 of the parabola $y^2 = 4ax$.

2 Points t_1, t_2 lie on the parabola $y^2 = 4ax$. Find a relation connecting t_1, t_2 if the line joining the points is a focal chord.

3 Prove that the tangents at the ends of a focal chord of a parabola are perpendicular.

4 Find the focus of the parabola $x^2 = 2y$.

5 Find the equation of a parabola whose focus is (2, 0) and directrix $y = -2$.

6 Find the equation of the parabola whose focus is $(-1, 1)$ and directrix $x = y$.

7 Find the gradient of the normal to the parabola $y^2 = 4ax$ at $P(at^2, 2at)$ and the gradient of the chord joining P to $(at_1{}^2, 2at_1)$. Deduce the coordinates of the point where the normal at P cuts the parabola again.

8 Prove that the foot of the perpendicular from the focus of a parabola on to any tangent lies on the tangent at the vertex.

9 Find the points on the parabola $y^2 = 8x$ where (a) the tangent and (b) the normal are parallel to the line $2x + y = 1$.

10 The tangents at the end of a focal chord meet each other at P and the tangent at the vertex at Q, R. Show that the centroid of the triangle PQR lies on the line $3x + a = 0$.

11 Find the point of intersection of the normals at the points t_1, t_2 of the parabola $y^2 = 4ax$.

12 Prove that, in general, from any point (h, k) three normals can be drawn to a parabola.

13 If the normals from a point (h, k) meet the parabola $y^2 = 4ax$ at the three points t_1, t_2, t_3, show that $t_1 + t_2 + t_3 = 0$.

14 PQ is a variable chord of a parabola. If the chords joining the vertex A to P and Q are perpendicular, show that PQ meets the axis of the parabola in a fixed point R, and find the length of AR.

15 Find the equations of the tangents to the parabola $y^2 = 4ax$ from the point $(16a, 17a)$.

16 If the tangents at the end of a focal chord of a parabola meet the tangent at the vertex in C, D, prove that CD subtends a right angle at the focus.

Further examples on the parabola

9.3   Example 5   *Find the focus and directrix of the parabola $y^2 = 2a(x - 4a)$ and give the length of its latus rectum.*

The equation $y^2 = 2a(x - 4a)$ may be written in the form

$$Y^2 = 2aX$$

by the substitutions $y = Y$, $x - 4a = X$. We have thus taken new axes as shown in Fig. 9.3.

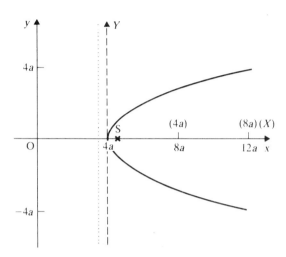

Figure 9.3

The parabola $y^2 = 4bx$ has focus $(b, 0)$, directrix $x = -b$ and latus rectum $4b$. Comparing this with $Y^2 = 2aX$, it follows that the latter has focus $(\frac{1}{2}a, 0)$, directrix $X = -\frac{1}{2}a$ and latus rectum $2a$. Therefore, with the original axes (see Fig. 9.3), the focus is $(9a/2, 0)$, the directrix $x = 7a/2$ and latus rectum $2a$.

Example 6   *Show that the equation $y = 5x - 2x^2$ represents a parabola and find the length of its latus rectum.*

We shall try to express the equation in the form $X^2 = -4aY$. The equation may be written as

$$x^2 - \frac{5}{2}x = -\frac{y}{2}$$

$$\therefore \left(x - \frac{5}{4}\right)^2 = \left(\frac{5}{4}\right)^2 - \frac{y}{2}$$

$$\therefore \left(x - \frac{5}{4}\right)^2 = -\frac{1}{2}\left(y - \frac{25}{8}\right)$$

This is now in the form $X^2 = -4aY$, giving the latus rectum as length $\frac{1}{2}$.

Qu. 6   Find the coordinates of the focus and the equation of the directrix in Example 6.

Example 7   *If the line $lx + my + n = 0$ touches the parabola $y^2 = 4ax$, find the equation connecting l, m, n, a.*

Since any tangent to $y^2 = 4ax$ may be written

$$x - ty + at^2 = 0 \tag{1}$$

let it represent the same tangent as

$$lx + my + n = 0 \tag{2}$$

Comparing coefficients,

$$\frac{1}{l} = -\frac{t}{m} = \frac{at^2}{n}$$

$$\therefore t = -\frac{m}{l} \quad \text{and} \quad t^2 = \frac{n}{al}$$

$$\therefore \frac{m^2}{l^2} = \frac{n}{al}$$

Therefore the condition is $am^2 = ln$.

The next two examples have been chosen to illustrate the use of symmetrical relationships between t_1, t_2. When symmetry exists, the working is made easier and care should be taken to use symmetrical equations and expressions.

Example 8   *A chord of the parabola $y^2 = 4ax$ subtends a right angle at the vertex. Find the locus of the mid-point of the chord.*

Let the ends of the chord be $P_1(at_1{}^2, 2at_1)$, $P_2(at_2{}^2, 2at_2)$. Then the gradient of the line joining the vertex $O(0, 0)$ to P_1 is

$$\frac{2at_1}{at_1{}^2} = \frac{2}{t_1}$$

Similarly the gradient of OP_2 is $\dfrac{2}{t_2}$.

P_1P_2 subtends a right angle at O if OP_1, OP_2 are perpendicular,

$$\therefore \frac{2}{t_1} \times \frac{2}{t_2} = -1$$

$$\therefore t_1 t_2 = -4 \tag{1}$$

The mid-point of P_1P_2 is given by

$$x = \frac{a(t_1{}^2 + t_2{}^2)}{2} \tag{2}$$

$$y = a(t_1 + t_2) \tag{3}$$

[Note that we have three equations, (1), (2), (3), from which to eliminate the two parameters t_1, t_2. Note, also, that these equations are symmetrical in t_1, t_2. Here, as is often the case, we use the following identity.]

$$(t_1 + t_2)^2 = t_1{}^2 + t_2{}^2 + 2t_1 t_2$$

Substituting from equations (3), (2), (1):

$$\frac{y^2}{a^2} = \frac{2x}{a} - 8$$

Therefore the locus is $y^2 = 2a(x - 4a)$.

Example 9   *Show that the equation of the normal to the parabola* $y^2 = 4ax$ *at the point* $(at^2, 2at)$ *is*

$$y + tx = 2at + at^2$$

The normal at a point P$(ap^2, 2ap)$ *meets the x-axis at* G. *Find the coordinates of the point* G. H *is the point on* PG *produced, such that* PG $=$ GH; *find the coordinates of* H *in terms of* p *and show that* H *lies on the parabola* $y^2 = 4a(x - 4a)$.

The gradient of the parabola at the point $(at^2, 2at)$ is given by

$$\frac{dy}{dx} = \frac{dy}{dt} \Big/ \frac{dx}{dt} = \frac{2a}{2at} = \frac{1}{t}$$

Hence the gradient of the normal is $-t$, and consequently the equation of the normal at the point $(at^2, 2at)$ is

$$y + tx = 2at + at^3$$

The equation of the normal at the point $(ap^2, 2ap)$ is

$$y + px = 2ap + ap^3$$

and we obtain the x-coordinate of G (see Fig. 9.4.) by putting $y = 0$. Therefore at G

$$px = 2ap + ap^3$$
i.e. $x = 2a + ap^2$

Hence G is the point $(2a + ap^2, 0)$.

Let H be the point (X, Y), then since G is the mid-point of PH,

$$\tfrac{1}{2}(ap^2 + X) = 2a + ap^2$$
$$\therefore ap^2 + X = 4a + 2ap^2$$
$$X = 4a + ap^2 \tag{1}$$

Similarly,

$$\tfrac{1}{2}(2ap + Y) = 0$$
$$\therefore 2ap + Y = 0$$
$$\therefore Y = -2ap \tag{2}$$

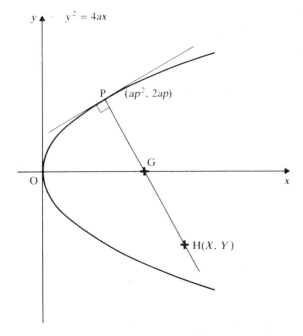

Figure 9.4

Hence H is the point $(4a + ap^2, -2ap)$.

Eliminating p from equations (1) and (2) gives

$$X = 4a + a\left(-\frac{Y}{2a}\right)^2$$

$$= 4a + \frac{Y^2}{4a}$$

$$\therefore Y^2 = 4a(X - 4a)$$

Hence the point H lies on the curve $y^2 = 4a(x - 4a)$. This is the equation of a parabola, with its vertex at $(4a, 0)$.

Example 10   *A variable tangent is drawn to the parabola $y^2 = 4ax$. If the perpendicular from the vertex meets the tangent at P, find the locus of P.*

Let the variable tangent be

$$x - ty + at^2 = 0 \tag{1}$$

Then the perpendicular from the vertex $(0, 0)$ is

$$tx + y = 0 \tag{2}$$

$P(x, y)$ satisfies equations (1), (2) so that the locus of P may be found by eliminating t from these equations. [Note that it is *not* necessary to solve them to find the coordinates of P in terms of t.]

From (2), $t = -y/x$. Substituting in (1),

$$x + \frac{y^2}{x} + a\frac{y^2}{x^2} = 0$$

So the locus of P is $x^3 + xy^2 + ay^2 = 0$.

Exercise 9b

1 Show that the equation $x^2 + 4x - 8y - 4 = 0$ represents a parabola whose focus is at $(-2, 1)$. Find the equation of the tangent at the vertex.

2 Prove that $x = 3t^2 + 1$ and $y = \frac{1}{2}(3t + 1)$ are the parametric equations of a parabola and find its vertex and the length of the latus rectum.

3 Find the focus of the parabola $y = 2x^2 + 3x - 5$.

4 Prove that the line $y = mx + \frac{3}{4}m + 1/m$ touches the parabola $y^2 = 4x + 3$ whatever the value of m.

5 If $ax + by + c = 0$ touches the parabola $x^2 = 4y$, find an equation connecting a, b, c.

6 A parabola, symmetrical about the axis of y, passes through the points $(1, 3)$ and $(2, 0)$. Find its equation and that of the tangent at $(1, 3)$.

7 Prove that the circles which are drawn on a focal chord of a parabola as diameter touch the directrix.

8 A variable chord of the parabola $y^2 = 4ax$ has a fixed gradient k. Find the locus of the mid-point.

9 A chord of the parabola $y^2 = 4ax$ is drawn to pass through the point $(-a, 0)$. Find the locus of the point of intersection of the tangents at the ends of the chord.

10 The difference of the ordinates of two points on the parabola $y^2 = 4ax$ is constant and equal to k. Find the locus of the point of intersection of the tangents at the two points.

11 Find the locus of the mid-points of focal chords of the parabola $y^2 = 4ax$.

12 The tangent at any point P of the parabola $y^2 = 4ax$ meets the tangent at the vertex at the point Q. S is the focus and SQ meets the line through P parallel to the tangent at the vertex at the point R. Find the locus of R.

13 Show that $y = ax^2 + bx + c$ is the equation of a parabola. Find its focus and directrix.

14 Two tangents to the parabola $y^2 = 4ax$ pass through the point (x_1, y_1). Find the equation of their chord of contact.

15 Find the points of contact on the parabola of the tangents common to the circle $(x - a)^2 + y^2 = 4a^2$ and the parabola $y^2 = 4ax$. [Start by writing down the equation of the tangent at $(at^2, 2at)$.]

16 The normal at the point P of the parabola $y^2 = 4ax$ meets the curve again at Q. The circle on PQ as diameter goes through the vertex. Find the x-coordinate of P.

17 Prove that rays of light parallel to the axis of a parabolic mirror are reflected through the focus.

18 A variable chord of the parabola $y^2 = 4ax$ passes through the point (h, k). Find the locus of the orthocentre of the triangle formed by the chord and the tangents at the two ends.

19 A tangent to the parabola $y^2 = 4ax$ meets the parabola $y^2 = 8ax$ at P, Q. Find the locus of the mid-point of PQ.

20 Find the locus of the mid-point of a variable chord through the point $(a, 2a)$ of the parabola $y^2 = 4ax$.

The ellipse

9.4   An ellipse was defined at the beginning of this chapter. Given a fixed point S, the *focus*, and a fixed line, the *directrix*, if P is a point on the locus and M is the foot of the perpendicular from P on to the directrix, then

$$\frac{\text{SP}}{\text{PM}} = e \qquad (e < 1)$$

e is called the *eccentricity* of the ellipse.

Qu. 7   On a sheet of squared paper, rule the directrix along one line near the edge, take the focus 2.7 cm in and plot an ellipse with a pair of compasses, taking $e = 4/5$. Measure the width of the ellipse parallel and perpendicular to the directrix.

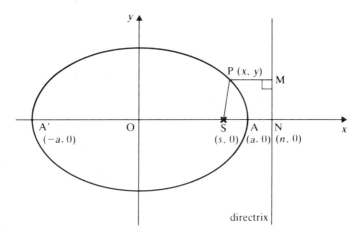

Figure 9.5

The result of Qu. 7 should be like the ellipse in Fig. 9.5, only larger. It follows from the definition that an ellipse is symmetrical about the line through S perpendicular to the directrix, so we take the *x*-axis along this axis of symmetry. Let the *x*-axis cut the ellipse in A′, A, as shown in Fig. 9.5. It appears from our drawing that there may be an axis of symmetry parallel to the directrix, so we shall take the *y*-axis passing through the mid-point of A'A, parallel to the directrix.

Let A be $(a, 0)$ so that A' is $(-a, 0)$; let S be $(s, 0)$ and let the *x*-axis cut the directrix at N$(n, 0)$. We shall now find *s*, *n* in terms of *a*, *e*.

A', A lie on the ellipse and so, by the definition of the locus,

$$\frac{\text{SA}'}{\text{A}'\text{N}} = e \quad \text{and} \quad \frac{\text{SA}}{\text{AN}} = e$$

Hence

$$a + s = e(n + a)$$
$$a - s = e(n - a)$$

Adding,

$$2a = 2en \qquad \therefore n = \frac{a}{e}$$

Subtracting,

$$2s = 2ae \qquad \therefore s = ae$$

Therefore S is the point $(ae, 0)$ and the equation of the directrix is $x = a/e$.

To find the equation of the ellipse, let P(x, y) be any point on the locus, then

$$\frac{\text{SP}}{\text{PM}} = e$$

$$\therefore \text{SP}^2 = e^2 \text{PM}^2$$

But $\text{SP}^2 = (x - ae)^2 + y^2$, and $\text{PM} = a/e - x$.

$$\therefore (x - ae)^2 + y^2 = e^2 \left(\frac{a}{e} - x \right)^2$$

$$\therefore x^2 - 2aex + a^2e^2 + y^2 = a^2 - 2aex + e^2x^2$$
$$\therefore x^2(1 - e^2) + y^2 = a^2(1 - e^2)$$

$$\therefore \frac{x^2}{a^2} + \frac{y^2}{a^2(1 - e^2)} = 1$$

Therefore the equation of the ellipse is

$$\frac{x^2}{a^2} + \frac{y^2}{b^2} = 1 \qquad \text{where} \quad b^2 = a^2(1 - e^2)$$

Note that we have also found that the focus S is $(ae, 0)$ and the directrix is $x = a/e$; but since the equation of the ellipse is unaltered by replacing *x* by $-x$, it follows that there is another focus $(-ae, 0)$ and another directrix $x = -a/e$. Hence

the foci are $(ae, 0)$ and $(-ae, 0)$

the directrices are $x = \dfrac{a}{e}$ and $x = -\dfrac{a}{e}$

The axes of symmetry meet at the **centre** of the ellipse. Any chord passing through the centre is called a **diameter**.

The diameter through the foci is the **major axis** and the perpendicular diameter is called the **minor axis**.

Qu. 8   Show that the lengths of the axes are $2a$, $2b$.
Qu. 9   Find the length of the semi-axes of the ellipse $x^2/16 + y^2/9 = 1$.
Qu. 10   Find the eccentricity of the ellipse $x^2/25 + y^2/16 = 1/4$.
Qu. 11   Find the foci of the ellipse $x^2 + 4y^2 = 9$.

Parametric coordinates for an ellipse

9.5   When dealing with an ellipse

$$\frac{x^2}{a^2} + \frac{y^2}{b^2} = 1$$

working is generally made easier by using a parameter, but the question arises of what parameter to use. Now an equation in the form

$$(\qquad)^2 + (\qquad)^2 = 1$$

suggests the identity

$$\cos^2 \theta + \sin^2 \theta = 1$$

Thus, if we write $x = a \cos \theta$, $y = b \sin \theta$, the equation

$$\frac{x^2}{a^2} + \frac{y^2}{b^2} = 1$$

will always be satisfied. We therefore take as a general point on the ellipse

($a \cos \theta$, $b \sin \theta$)

θ is called the **eccentric angle** of the point.

In Book 1, §22.5, we saw that the parameter θ for a circle in

$$x = a \cos \theta, \qquad y = a \sin \theta$$

could be interpreted in terms of an angle. This is not so simple for an ellipse but it will now be done.

In Fig. 9.6, P is the point $(a \cos \theta, b \sin \theta)$ on the ellipse $x^2/a^2 + y^2/b^2 = 1$ and P′ is a point on the circle $x^2 + y^2 = a^2$ (called the auxiliary circle) such that OP′ makes an angle θ with Ox. Since P, P′ have the same x-coordinate, a $\cos \theta$, PP′ is perpendicular to the major axis of the ellipse. Therefore the eccentric angle θ of any point P may be found as follows: draw the ordinate of P to meet the auxiliary circle at P′, join P′ to the origin, then OP′ makes an angle θ with the positive x-axis.

Qu. 12   Show how to obtain the y-coordinate of the point $(a \cos \theta, b \sin \theta)$ from a circle of radius b. Draw two concentric circles and hence plot an ellipse.

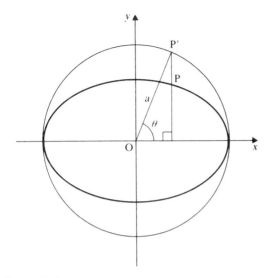

Figure 9.6

Example 11 *Find the equation of the tangent to the ellipse* $\dfrac{x^2}{a^2} + \dfrac{y^2}{b^2} = 1$ *at* $(a \cos \theta, b \sin \theta)$.

The gradient $\dfrac{dy}{dx} = \dfrac{dy}{d\theta} \bigg/ \dfrac{dx}{d\theta}$.

$$x = a \cos \theta, \qquad y = b \sin \theta$$

$$\therefore \frac{dy}{dx} = \frac{b \cos \theta}{-a \sin \theta}.$$

Therefore the equation of the tangent is

$$b \cos \theta \, x + a \sin \theta \, y = b \cos \theta \times a \cos \theta + a \sin \theta \times b \sin \theta$$
$$\therefore bx \cos \theta + ay \sin \theta = ab(\cos^2 \theta + \sin^2 \theta)$$

Therefore the tangent is

$$bx \cos \theta + ay \sin \theta - ab = 0$$

*Qu. 13 Show that the equation of the normal to the ellipse $x^2/a^2 + y^2/b^2 = 1$ at $(a \cos \theta, b \sin \theta)$ is

$$ax \sin \theta - by \cos \theta - (a^2 - b^2) \sin \theta \cos \theta = 0$$

If the general point on the curve is taken to be (x_1, y_1), it is frequently necessary to bring in the extra equation

$$\frac{x_1{}^2}{a^2} + \frac{y_1{}^2}{b^2} = 1$$

This is why working is usually easier when parameters are used.

*Qu. 14　Show that the equation of the tangent at (x_1, y_1) to the ellipse

$$\frac{x^2}{a^2} + \frac{y^2}{b^2} = 1$$

is

$$\frac{xx_1}{a^2} + \frac{yy_1}{b^2} = 1$$

Verify that this gives the equation found in Example 11 for the tangent at $(a \cos \theta, b \sin \theta)$.

Example 12　*Find the equation of the chord of the ellipse*

$$\frac{x^2}{a^2} + \frac{y^2}{b^2} = 1$$

joining the points whose eccentric angles are θ, ϕ.

The ends of the chord are $(a \cos \theta, b \sin \theta)$, $(a \cos \phi, b \sin \phi)$, therefore the gradient of the chord is

$$\frac{b \sin \theta - b \sin \phi}{a \cos \theta - a \cos \phi} = \frac{2b \cos \frac{1}{2}(\theta + \phi) \sin \frac{1}{2}(\theta - \phi)}{-2a \sin \frac{1}{2}(\theta + \phi) \sin \frac{1}{2}(\theta - \phi)}$$

$$= -\frac{b \cos \frac{1}{2}(\theta + \phi)}{a \sin \frac{1}{2}(\theta + \phi)}$$

Therefore the equation of the chord is

$$b \cos \tfrac{1}{2}(\theta + \phi)\, x + a \sin \tfrac{1}{2}(\theta + \phi)\, y$$
$$= b \cos \tfrac{1}{2}(\theta + \phi) \times a \cos \theta + a \sin \tfrac{1}{2}(\theta + \phi) \times b \sin \theta$$

$$\text{R.H.S.} = ab\{\cos \tfrac{1}{2}(\theta + \phi) \cos \theta + \sin \tfrac{1}{2}(\theta + \phi) \sin \theta\}$$
$$= ab \cos \{\tfrac{1}{2}(\theta + \phi) - \theta\}$$
$$= ab \cos \tfrac{1}{2}(\phi - \theta)$$
$$= ab \cos \tfrac{1}{2}(\theta - \phi)$$

Therefore the equation of the chord is

$$\boldsymbol{bx \cos \tfrac{1}{2}(\theta + \phi) + ay \sin \tfrac{1}{2}(\theta + \phi) - ab \cos \tfrac{1}{2}(\theta - \phi) = 0}$$

*Qu. 15　Show, by putting $\phi = \theta$, that the equation of the chord approaches the equation of the tangent at θ as $\phi \to \theta$.

Example 13　*A tangent to the ellipse*

$$\frac{x^2}{a^2} + \frac{y^2}{b^2} = 1$$

at the point P meets the minor axis at L. If the normal at P meets the major axis at M, find the locus of the mid-point of LM.

Let P be the point $(a \cos \theta, b \sin \theta)$, then the tangent at P has equation

$$bx \cos \theta + ay \sin \theta - ab = 0$$

This meets the minor axis $x = 0$ at $L\left(0, \dfrac{b}{\sin \theta}\right)$.

The normal at P is

$$ax \sin \theta - by \cos \theta - (a^2 - b^2) \sin \theta \cos \theta = 0$$

This meets the major axis $y = 0$ at $M\left(\dfrac{a^2 - b^2}{a} \cos \theta, 0\right)$.

The mid-point of LM is given by

$$x = \frac{a^2 - b^2}{2a} \cos \theta \qquad y = \frac{b}{2 \sin \theta}$$

θ can be eliminated from these equations by means of the identity

$$\cos^2 \theta + \sin^2 \theta = 1$$

Therefore the locus of the mid-point of LM is

$$\left(\frac{2ax}{a^2 - b^2}\right)^2 + \left(\frac{b}{2y}\right)^2 = 1$$

Exercise 9c

1 Find the foci and directrices of the ellipse
 (a) $4x^2 + 9y^2 = 36$,   (b) $x^2 + 16y^2 = 25$.
2 Write down the equation of the tangent to

 (a) $\dfrac{x^2}{9} + \dfrac{y^2}{4} = 1$ at $(3 \cos \theta, 2 \sin \theta)$,

 (b) $9x^2 + 16y^2 = 25$ at $(1, 1)$.
3 Find the equation of the normal to
 (a) $9x^2 + 16y^2 = 25$ at $(1, 1)$,
 (b) $x^2 + 2y^2 = 9$ at $(1, -2)$.
4 A point moves so that its distance from $(3, 2)$ is half its distance from the line $2x + 3y = 1$. Why is the locus an ellipse? Find the equation of the major axis.
5 P is any point on an ellipse; S, S′ are the foci. Prove directly from the focus-directrix definition of the ellipse that $SP + S'P = 2a$, where $2a$ is the length of the major axis.
6 Find the relation between the eccentric angles of the points which are at the ends of a focal chord.
7 Prove that the chord joining points of an ellipse whose eccentric angles are $(\alpha + \beta), (\alpha - \beta)$ is parallel to the tangent at the point whose eccentric angle is α.
8 Find the equation of the tangent to the ellipse $x^2/a^2 + y^2/b^2 = 1$ at the end of the latus rectum which lies in the first quadrant.

9 The tangent at P to an ellipse meets a directrix at Q. Prove that lines joining the corresponding focus to P and Q are perpendicular.

10 Find the coordinates of the point of intersection of the tangents to the ellipse $x^2/a^2 + y^2/b^2 = 1$ at the points whose eccentric angles are θ, ϕ.

11 P is any point on an ellipse and S, S' are the foci. Prove that the normal at P bisects the angle S'PS.

12 Find the locus of the mid-point of the line joining the focus $(ae, 0)$ to any point on the ellipse $x^2/a^2 + y^2/b^2 = 1$.

13 The eccentric angles of two points P, Q differ by a constant k. Find the locus of the mid-point of PQ.

14 The normal at the point $(a \cos \theta, b \sin \theta)$ on the ellipse $x^2/a^2 + y^2/b^2 = 1$ meets the axes at L, M. Find the locus of the mid-point of LM.

15 A variable tangent to the ellipse $x^2/a^2 + y^2/b^2 = 1$ meets the axes at R, S. Find the locus of the mid-point of RS.

16 Prove that the tangents to the ellipse $x^2/a^2 + y^2/b^2 = 1$ at points whose eccentric angles differ by a right angle meet on a concentric ellipse and find its equation.

17 Prove that perpendicular tangents to the ellipse $x^2/a^2 + y^2/b^2 = 1$ meet on the circle $x^2 + y^2 = a^2 + b^2$ (called the *director circle*).

18 The tangents to the ellipse $x^2/a^2 + y^2/b^2 = 1$ at P, Q meet at the point (x_1, y_1). Show that the equation of the chord of contact PQ is

$$\frac{xx_1}{a^2} + \frac{yy_1}{b^2} = 1$$

[Use the results of No. 10 and Example 12.]

Further examples on the ellipse

9.6   Example 14   *Find the condition that the line $y = mx + c$ should touch the ellipse $\dfrac{x^2}{a^2} + \dfrac{y^2}{b^2} = 1$.*

The equation of any tangent to the ellipse may be written

$$bx \cos \theta + ay \sin \theta - ab = 0$$

Let this equation represent the same tangent as the given line which we shall write as

$$mx - y + c = 0$$

Comparing coefficients,

$$\frac{b \cos \theta}{m} = \frac{a \sin \theta}{-1} = \frac{-ab}{c}$$

$$\therefore \cos \theta = -\frac{am}{c}, \qquad \sin \theta = \frac{b}{c}$$

But $\cos^2 \theta + \sin^2 \theta = 1$.

$$\therefore \frac{a^2m^2}{c^2} + \frac{b^2}{c^2} = 1$$

Therefore $y = mx + c$ touches the ellipse if

$$c^2 = a^2m^2 + b^2$$

Qu. 16 Work Example 14 by eliminating y between the two equations and writing down the condition that the resulting quadratic in x should have equal roots.

Example 15 *Prove that perpendicular tangents to the ellipse $\dfrac{x^2}{a^2} + \dfrac{y^2}{b^2} = 1$ meet on a circle and find its equation.*

From Example 14 we see that the equation of a general tangent to the ellipse may be written

$$y = mx + (a^2m^2 + b^2)^{1/2}$$
$$\therefore (y - mx)^2 = a^2m^2 + b^2$$
$$\therefore m^2(x^2 - a^2) - 2xym + y^2 - b^2 = 0 \qquad (1)$$

If (x, y) is a point of intersection of two perpendicular tangents to the ellipse, we may regard equation (1) as a quadratic equation for m, the gradient of the tangents. Since the tangents are perpendicular the product of the roots of the equation is -1,

$$\therefore \frac{y^2 - b^2}{x^2 - a^2} = -1$$

$$\therefore y^2 - b^2 = a^2 - x^2$$

Therefore the equation of the locus is

$$x^2 + y^2 = a^2 + b^2$$

This is called the **director circle** of the ellipse.

Example 16 *A variable straight line with constant gradient m meets the ellipse*

$$\frac{x^2}{a^2} + \frac{y^2}{b^2} = 1$$

at Q, R. Find the locus of P, the mid-point of QR.

Let the equation of the line be

$$y = mx + c \qquad (1)$$

To find the coordinates of Q, R, we would solve the equation of the line and the equation of the ellipse

$$b^2x^2 + a^2y^2 = a^2b^2$$

simultaneously:

$$b^2x^2 + a^2(m^2x^2 + 2mxc + c^2) - a^2b^2 = 0$$
$$\therefore x^2(b^2 + a^2m^2) + 2a^2mcx + a^2c^2 - a^2b^2 = 0$$

The x-coordinates of Q, R, say x_1, x_2, are the roots of this equation. But if P is the point (X, Y),

$$X = \tfrac{1}{2}(x_1 + x_2)$$

$$\therefore X = \tfrac{1}{2} \times \frac{-2a^2mc}{b^2 + a^2m^2} \tag{2}$$

Now the coordinates of P satisfy equation (1), so

$$Y = mX + c \tag{3}$$

Therefore we may find the locus of P by eliminating c between the equations (2), (3). Substituting

$$c = Y - mX$$

in equation (2) rearranged as

$$X(b^2 + a^2m^2) = -a^2mc$$

we obtain

$$X(b^2 + a^2m^2) = -a^2m(Y - mX)$$

Therefore the locus of P is

$$b^2x + a^2my = 0$$

which is a diameter of the ellipse.

Exercise 9d

1 Write down the equations of the tangents to
   (a) $x^2/4 + y^2/9 = 1$ with gradient 2,
   (b) $x^2 + 3y^2 = 3$ with gradient -1,
   (c) $4x^2 + 9y^2 = 144$ with gradient $\tfrac{1}{2}$.
2 *Without* solving the equations completely, find the coordinates of the mid-points of the chords formed by the intersection of
   (a) $x - y - 1 = 0$ and $x^2/9 + y^2/4 = 1$,
   (b) $10x - 5y + 6 = 0$ and $4x^2 + 5y^2 = 20$,
   (c) $2x + 3y - 4 = 0$ and $y^2 = 8x$.
3 Prove that the line $x - 2y + 10 = 0$ touches the ellipse $9x^2 + 64y^2 = 576$.
4 Find the equations of the tangents to the ellipse $x^2 + 4y^2 = 4$ which are perpendicular to the line $2x - 3y = 1$.

5 The line $y = x - c$ touches the ellipse $9x^2 + 16y^2 = 144$. Find the value of c and the coordinates of the point of contact.

6 Find the condition for the line $y = mx + c$ to cut the ellipse $x^2/a^2 + y^2/b^2 = 1$ in two distinct points.

7 The line $y = mx + c$ touches the ellipse $x^2/a^2 + y^2/b^2 = 1$.
   Prove that the foot of the perpendicular from a focus on to this line lies on the auxiliary circle $x^2 + y^2 = a^2$.

8 Find the locus of the foot of the perpendicular from the centre of the ellipse $x^2/a^2 + y^2/b^2 = 1$ on to any tangent.

9 Find the equation of the normal at the point (x_1, y_1) on the ellipse $x^2/a^2 + y^2/b^2 = 1$.

10 Find the coordinates of the mid-point of the chord formed by the intersection of
   (a) $y = mx + c$ and $b^2x^2 + a^2y^2 = a^2b^2$,
   (b) $lx + my + n = 0$ and $y^2 = 4ax$.

11 Find the equation of the diameter bisecting the chord $3x + 2y = 1$ of the ellipse $4x^2 + 9y^2 = 16$.

12 Find the equation of the line with gradient m passing through the focus $(ae, 0)$ of the ellipse $b^2x^2 + a^2y^2 = a^2b^2$.
   If the line meets the ellipse in P, Q, find the coordinates of the mid-point of PQ and show that they satisfy the equation

$$a^2my + b^2x = 0$$

   By substituting the value of m obtained from this equation into the equation of PQ, find the locus of the mid-point of PQ.

13 A variable line passes through the point $(a, 0)$. Find the locus of the mid-point of the chord formed by the intersection of this line and the ellipse $b^2x^2 + a^2y^2 = a^2b^2$.

14 Find the locus of points from which the tangents to the ellipse

$$b^2x^2 + a^2y^2 = a^2b^2$$

   are inclined at $45°$.

15 Lines of gradient m are drawn to cut the ellipse $b^2x^2 + a^2y^2 = a^2b^2$.
   Prove that the mid-points of the chords so formed lie on a straight line through the origin with gradient $-b^2/(a^2m)$. Deduce the equation of the chord whose mid-point is (h, k).

16 Show that a general tangent to the circle $x^2 + y^2 - a^2 = 0$ may be written

$$y = mx \pm a\sqrt{(1 + m^2)}$$

   A variable tangent to the circle $x^2 + y^2 - a^2 = 0$ meets the ellipse $b^2x^2 + a^2y^2 = a^2b^2$ $(b > a)$† at P, Q. Find the locus of the mid-point of PQ.

17 A variable tangent to the ellipse $b^2x^2 + a^2y^2 = a^2b^2$ meets the parabola $y^2 = 4ax$ at L, M. Find the locus of the mid-point of LM.

†The major axis lies along the y-axis.

18 The chord of contact of the point (x_1, y_1) with respect to the ellipse

$$\frac{x^2}{a^2} + \frac{y^2}{b^2} = 1$$

cuts the axes at L, M. If the locus of the mid-point of LM is the circle $x^2 + y^2 = 1$, find the locus of (x_1, y_1). [Use the result of Exercise 9c, No. 18.]

The hyperbola

9.7   In §9.4, certain results were obtained for the ellipse. The working is so similar for the hyperbola that it is left to the reader to obtain the corresponding results. Starting with the focus–directrix definition with $e > 1$ the reader should work through the following questions.

Qu. 17   On a sheet of squared paper, rule the directrix along one line near the middle, take the focus 4 cm out towards the nearer edge and plot part of a hyperbola (there are two branches of it) taking $e = 2$.

Qu. 18   Show that, with suitable choice of axes, the equation of a hyperbola may be written

$$\frac{x^2}{a^2} - \frac{y^2}{b^2} = 1$$

where $b^2 = a^2(e^2 - 1),$

the foci are $(ae, 0)$ and $(-ae, 0),$

and the directrices $x = \dfrac{a}{e}$ and $x = -\dfrac{a}{e}.$

Qu. 19   Show that any point on the hyperbola $x^2/a^2 - y^2/b^2 = 1$ may be written

$$(a \sec \theta, b \tan \theta)$$

Qu. 20   Show that at $(a \sec \theta, b \tan \theta)$ on the hyperbola

$$\frac{x^2}{a^2} - \frac{y^2}{b^2} = 1$$

the equation of the tangent is

$$bx - ay \sin \theta - ab \cos \theta = 0$$

and the equation of the normal is

$$ax \sin \theta + by - (a^2 + b^2) \tan \theta = 0$$

Qu. 21   Show that the equation of the tangent at (x_1, y_1) to the hyperbola $x^2/a^2 - y^2/b^2 = 1$ is

$$\frac{xx_1}{a^2} - \frac{yy_1}{b^2} = 1$$

Show that the equation of the tangent in Qu. 20 may be deduced from this.

Asymptotes to a hyperbola

9.8   **Example 17**   *Find c in terms of a, b, m if $y = mx + c$ is a tangent to the hyperbola $\dfrac{x^2}{a^2} - \dfrac{y^2}{b^2} = 1$.*

Solving the two equations simultaneously,

$$b^2x^2 - a^2y^2 = a^2b^2$$
$$\therefore \; b^2x^2 - a^2(m^2x^2 + 2mcx + c^2) - a^2b^2 = 0$$
$$\therefore \; x^2(b^2 - a^2m^2) - 2a^2mcx - a^2(b^2 + c^2) = 0 \tag{1}$$

The line is a tangent if and only if this equation has equal roots,

i.e. if and only if
$$(-2a^2mc)^2 = -4(b^2 - a^2m^2)a^2(b^2 + c^2)$$
$$a^2m^2c^2 = -(b^2 - a^2m^2)(b^2 + c^2)$$
$$a^2m^2c^2 = -b^4 - b^2c^2 + a^2m^2b^2 + a^2m^2c^2$$
$$b^2c^2 = a^2m^2b^2 - b^4$$

Therefore $y = mx + c$ is a tangent to the hyperbola if and only if

$$c^2 = a^2m^2 - b^2$$

[Compare the method of Example 14, §9.6.]

In Example 17, the value of x at the point of contact is given by half the sum of the roots of equation (1) since the roots are equal.

$$\therefore \; x = \frac{a^2mc}{b^2 - a^2m^2}$$

$$= \mp \frac{a^2m\sqrt{(a^2m^2 - b^2)}}{a^2m^2 - b^2}$$

Therefore, at the point of contact,

$$x = \mp \frac{a^2m}{\sqrt{(a^2m^2 - b^2)}}.$$

Hence as $m \to \pm b/a$, $x \to \infty$ and, since $c^2 = a^2m^2 - b^2$, $c \to 0$, so that

$$y = \pm \frac{b}{a}x$$

may be regarded as the limit of a tangent to the hyperbola as the point of contact tends to infinity.

One way of remembering the equation of the asymptotes

$$\frac{x^2}{a^2} - \frac{y^2}{b^2} = 0$$

is that, when x, y are very large, terms other than those of the highest degree may be neglected in comparison.

The rectangular hyperbola

9.9   There is a special case of the hyperbola which has interesting properties and so receives special attention. A **rectangular hyperbola** is one whose asymptotes are perpendicular. The asymptotes of

$$\frac{x^2}{a^2} - \frac{y^2}{b^2} = 1 \quad \text{are} \quad y = \pm \frac{b}{a} x$$

and these are perpendicular when

$$-\frac{b}{a} \times \frac{b}{a} = -1$$

that is when $b = a$. Hence

$$x^2 - y^2 = a^2$$

represents a rectangular hyperbola and its asymptotes are

$$x - y = 0, \qquad x + y = 0$$

The fact that the asymptotes are perpendicular enables us to write the equation of the rectangular hyperbola in a very simple way. Let (x, y) be any point on the curve in Fig. 9.7, then

$$x^2 - y^2 = a^2$$

Note that this equation can be written

$$(x - y)(x + y) = a^2 \tag{1}$$

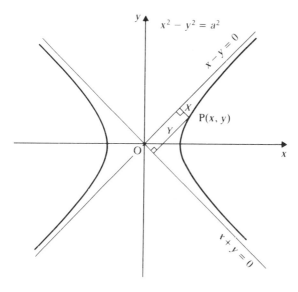

Figure 9.7

If we rotate the plane through 45° the asymptotes will coincide with the axes (remember that under such transformations we always regard the axes as fixed); the two branches of the hyperbola will then occupy the first and third quadrants (see Fig. 9.8).

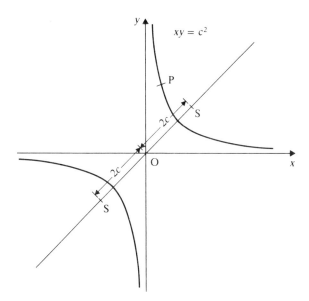

Figure 9.8

Using the matrix method described in Book 1, §11.7 (d), the point (x, y) will be mapped on to point (X, Y), where

$$\begin{pmatrix} X \\ Y \end{pmatrix} = \begin{pmatrix} \cos 45° & -\sin 45° \\ \sin 45° & \cos 45° \end{pmatrix} \begin{pmatrix} x \\ y \end{pmatrix}$$

$$= \frac{1}{\sqrt{2}} \begin{pmatrix} 1 & -1 \\ 1 & 1 \end{pmatrix} \begin{pmatrix} x \\ y \end{pmatrix}$$

$$= \frac{1}{\sqrt{2}} \begin{pmatrix} x - y \\ x + y \end{pmatrix}$$

Hence $X = \frac{1}{\sqrt{2}}(x - y)$ and $Y = \frac{1}{\sqrt{2}}(x + y)$, i.e.,

$$(x - y) = \sqrt{2}X \quad \text{and} \quad (x + y) = \sqrt{2}Y$$

Substituting these expressions into equation (1) gives

$$2XY = a^2$$
$$\therefore XY = \tfrac{1}{2}a^2$$

Hence, with the rectangular hyperbola in its new position, any point on the curve with coordinates (x, y) satisfies the equation

$$xy = c^2 \qquad \text{where } c = \frac{1}{\sqrt{2}} a$$

The eccentricity of $x^2 - y^2 = a^2$ is given by $a^2 = a^2(e^2 - 1)$ from which we find that $e = \sqrt{2}$ and hence the foci are $(\pm\sqrt{2}a, 0)$. Now $a = \sqrt{2}c$, so the foci of $xy = c^2$ are on the major axis at a distance $2c$ from the centre (see Fig. 9.8). Therefore the coordinates of the foci of $xy = c^2$ are $(\sqrt{2}c, \sqrt{2}c)$ and $(-\sqrt{2}c, -\sqrt{2}c)$.

The reader should now work through the following questions which contain very important results for problems on the rectangular hyperbola.

Qu. 22 Show that any point on the rectangular hyperbola $xy = c^2$ may be represented by

$$\left(ct, \frac{c}{t}\right)$$

Qu. 23 Show that the gradient of the hyperbola at $(ct, c/t)$ is

$$-\frac{1}{t^2}$$

Show also that the equation of the tangent is

$$x + t^2 y - 2ct = 0$$

and that the equation of the normal is

$$t^2 x - y - ct^3 + \frac{c}{t} = 0$$

Qu. 24 Show that the gradient of the chord joining the points $(ct_1, c/t_1)$, $(ct_2, c/t_2)$ on the hyperbola $xy = c^2$ is

$$-\frac{1}{t_1 t_2}$$

and that the equation of the chord is

$$x + t_1 t_2 y - c(t_1 + t_2) = 0$$

Qu. 25 Verify that the equation of the chord in Qu. 24 becomes the equation of the tangent in Qu. 23 when $t_1 = t_2 = t$.

Further examples on the hyperbola

9.10 The following examples do not illustrate any new principles but rather serve to show that the same methods that were used for problems about the ellipse may also be used in connection with the hyperbola.

Example 18 *A tangent to a hyperbola at P meets a directrix at Q. If S is the corresponding focus, prove that PQ subtends a right angle at S. (Fig. 9.9.)*

Let P be the point $(a \sec \theta, b \tan \theta)$ on the hyperbola

$$\frac{x^2}{a^2} - \frac{y^2}{b^2} = 1$$

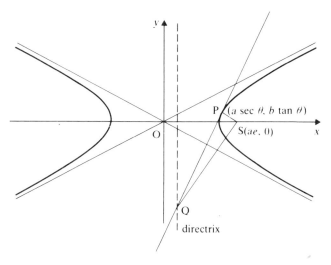

P $(a \sec \theta, b \tan \theta)$

S$(ae, 0)$

Q

directrix

Figure 9.9

The tangent at P is

$$bx - ay \sin \theta - ab \cos \theta = 0$$

This meets the directrix $x = a/e$ at a point given by

$$\frac{ba}{e} - ay \sin \theta - ab \cos \theta = 0$$

$$\therefore y \sin \theta = \frac{b(1 - e \cos \theta)}{e}$$

$$\therefore \text{Q is } \left(\frac{a}{e}, \frac{b(1 - e \cos \theta)}{e \sin \theta} \right)$$

Therefore the gradient of QS

$$m_1 = \frac{\dfrac{b(1 - e \cos \theta)}{e \sin \theta}}{\dfrac{a}{e} - ae} = \frac{b(1 - e \cos \theta)}{a(1 - e^2) \sin \theta}$$

The gradient of PS

$$m_2 = \frac{b \tan \theta}{a \sec \theta - ae} = \frac{b \sin \theta}{a(1 - e \cos \theta)}$$

$$\therefore m_1 m_2 = \frac{b(1 - e\cos\theta)}{a(1 - e^2)\sin\theta} \times \frac{b\sin\theta}{a(1 - e\cos\theta)}$$

$$= \frac{b^2}{a^2(1 - e^2)} = -1$$

since $b^2 = a^2(e^2 - 1)$. (See §9.7.)

Therefore SQ and SP are perpendicular and so PQ subtends a right angle at S.

Example 19   S *is a focus of the hyperbola*

$$\frac{x^2}{a^2} - \frac{y^2}{b^2} = 1$$

and P *is the foot of the perpendicular from* S *on to a variable tangent. Find the locus of* P.

$$y = mx \pm (a^2 m^2 - b^2)^{1/2} \tag{1}$$

is the equation of a general tangent to the hyperbola.

$$x + my - ae = 0 \tag{2}$$

is the equation of the perpendicular from S to the tangent.

The coordinates of P satisfy equations (1) and (2), and so we may find the locus by eliminating m between these equations.

From (1),

$$y^2 - 2mxy + m^2 x^2 = a^2 m^2 - b^2$$

From (2),

$$m^2 y^2 + 2mxy + x^2 = a^2 e^2$$

Adding,

$$y^2(1 + m^2) + x^2(1 + m^2) = a^2 m^2 - b^2 + a^2 e^2$$
$$\text{R.H.S.} = a^2 m^2 + a^2 \quad \text{since} \quad b^2 = a^2(e^2 - 1)$$

$$\therefore y^2(1 + m^2) + x^2(1 + m^2) = a^2(1 + m^2)$$

Therefore the locus of P is $x^2 + y^2 = a^2$, which is the **auxiliary circle**.

Example 20   PQ *is a chord of the rectangular hyperbola* $xy = c^2$ *and* R *is its mid-point. If* PQ *has a constant length* k, *find the locus of* R.

Let P be $(cp, c/p)$ and Q be $(cq, c/q)$. Then, if R is (x, y),

$$x = \tfrac{1}{2}c(p + q) \tag{1}$$

$$y = \frac{c(p + q)}{2pq} \tag{2}$$

Since the length of PQ is given to be k,

$$PQ^2 = (cp - cq)^2 + \left(\frac{c}{p} - \frac{c}{q}\right)^2 = k^2$$

$$\therefore c^2(p - q)^2 + c^2 \frac{(q - p)^2}{p^2 q^2} = k^2$$

$$\therefore c^2(p - q)^2(1 + p^2 q^2) = k^2 p^2 q^2 \tag{3}$$

From (1) and (2), $p + q = 2x/c$ and $pq = x/y$.

Now $(p - q)^2 \equiv (p + q)^2 - 4pq$, so that (3) becomes

$$c^2\{(p + q)^2 - 4pq\}(1 + p^2 q^2) = k^2 p^2 q^2$$

Substituting for $p + q$ and pq,

$$c^2 \left(\frac{4x^2}{c^2} - \frac{4x}{y}\right)\left(1 + \frac{x^2}{y^2}\right) = k^2 \frac{x^2}{y^2}$$

$$\therefore \left(4x^2 - \frac{4c^2 x}{y}\right)(y^2 + x^2) = k^2 x^2$$

Therefore the locus of P is $4(xy - c^2)(x^2 + y^2) = k^2 xy$.

Exercise 9e

1 P is any point on the rectangular hyperbola $xy = c^2$. Show that the line joining P to the centre, and the tangent at P, are equally inclined to the asymptotes.

2 P is any point on the hyperbola $x^2/a^2 - y^2/b^2 = 1$ and Q is the point (a, b). Find the locus of the point dividing PQ in the ratio 2:1.

3 Prove that the product of the lengths of the perpendiculars from any point of a hyperbola to its asymptotes is constant.

4 The normal at any point of a hyperbola meets the axes at E, F. Find the locus of the mid-point of EF.

5 Find the coordinates of the point at which the normal at $(ct, c/t)$ meets the rectangular hyperbola $xy = c^2$ again.

6 Any tangent to the rectangular hyperbola $xy = c^2$ meets the asymptotes at L and M. Find the locus of the mid-point of LM.

7 The normal at any point on the hyperbola $xy = c^2$ meets the x-axis at A, and the tangent meets the y-axis at B. Find the locus of the mid-point of AB.

8 Find the equation of the chord of the hyperbola $xy = c^2$ whose mid-point is (x_1, y_1).

9 Show that, in general, four normals can be drawn from any point to the rectangular hyperbola $xy = c^2$.

10 The normal at any point P of the rectangular hyperbola $xy = c^2$ meets the y-axis at A, and the tangent meets the x-axis at B. Find the coordinates of the fourth vertex Q of the rectangle APBQ in terms of t, the parameter of P.

11 Find the locus of the foot of the perpendicular from the origin on to a tangent to the rectangular hyperbola $xy = c^2$.

12 Find the condition that the line $lx + my + n = 0$ should touch the rectangular hyperbola $xy = c^2$.

13 Prove that the locus of middle points of parallel chords of the rectangular hyperbola $xy = c^2$ is a diameter.

14 Find the locus of the point of intersection of perpendicular tangents to the hyperbola $x^2/a^2 - y^2/b^2 = 1$.

15 Find the locus of the foot of the perpendicular from the origin to a tangent of the hyperbola $x^2/a^2 - y^2/b^2 = 1$.

16 PQ is a variable chord of the hyperbola $x^2/a^2 - y^2/b^2 = 1$ with constant gradient m_1. Show that the locus of the mid-point of PQ is a diameter with gradient m_2 such that $m_1 m_2 = b^2/a^2$.

17 The chord AB of a hyperbola meets the asymptotes at M, N. Prove that AM = BN. [Show that AB and MN have the same mid-point.]

18 Find the equation of the chord joining the points $(a \sec \theta, b \tan \theta)$, $(a \sec \phi, b \tan \phi)$ on the hyperbola $x^2/a^2 - y^2/b^2 = 1$.

19 The tangent at any point P on the hyperbola $x^2/a^2 - y^2/b^2 = 1$ meets the asymptotes at Q and Q'. Prove that PQ = PQ'.

20 P, Q, R are three points on a rectangular hyperbola such that PQ subtends a right angle at R. Show that PQ is perpendicular to the tangent at R.

Exercise 9f (Miscellaneous)

1 The distance between the foci of an ellipse is 8 and between the directrices is 18. Find its equation in the simplest form.

2 R, S are two fixed points a distance 5 units apart and the point P moves so that PR + PS is constant and equal to 12 units. Find the locus of P.

3 P is any point on a parabola and S is the focus. Prove that the circle on SP as diameter touches the tangent at the vertex.

4 The equation of a parabola is $y^2 = 12x - 12$. Find the equations of the straight lines that pass through the origin and cut the parabola where $x = 4$. Find also the equations of the tangents to the parabola that are parallel to these lines.

5 Show that the line $5y - 4x = 25$ touches the ellipse $x^2/25 + y^2/9 = 1$ and find the equation of the normal to the ellipse at the point of contact. What is the eccentricity of the ellipse?

6 Find the equations of the tangents to the hyperbola $3x^2 - 4y^2 = 1$ which make equal angles with the axes.

7 The perpendiculars from the foci of the hyperbola $b^2 x^2 - a^2 y^2 = a^2 b^2$ on to any tangent are of length p_1, p_2. Prove that $|p_1 p_2| = b^2$.

8 The gradients of the tangents to the parabola $y^2 = 4ax$ and the rectangular hyperbola $xy = c^2$ at the point at which they cut are m_1 and m_2 respectively. Prove that $m_2 = -2m_1$.

9 Find the coordinates of the mid-point of the chord $x + y - 1 = 0$ of the parabola $y^2 = 6x$.

10 Show that the equation $2x^2 + y^2 = 6y$ represents an ellipse with eccentricity $\frac{1}{2}\sqrt{2}$. Find the coordinates of the centre and the length of the minor axis.

11 Show that $x^2 + 4x - 8y - 4 = 0$ represents a parabola whose focus is at $(-2, 1)$. Find the equation of the tangent at the vertex.

12 A man stands on a ladder which rests on smooth horizontal ground against a smooth vertical wall. Prove that his feet will describe part of an ellipse as the ladder falls.

13 Find the eccentricity and focus of the curve $y^2 - 4y + 2x + 2 = 0$ and write down the equations of the tangent and normal at the point $(\frac{1}{2}, 1)$.

14 P $(a \sec \theta, b \tan \theta)$ is any point on the hyperbola $x^2/a^2 - y^2/b^2 = 1$ and N is the foot of the perpendicular from P to the x-axis. NT is drawn to touch the auxiliary circle at T. Prove that the line joining T to the centre of the circle makes an angle with the x-axis equal to θ.

15 Show that, if the chord joining the points $P(ap^2, 2ap)$, $Q(aq^2, 2aq)$ on the parabola $y^2 = 4ax$ passes through $(a, 0)$, then $pq = -1$.

　　Further, the tangent at P meets the line through Q parallel to the axis of the parabola at R. Prove that the line $x + a = 0$ bisects PR.　　　　(O & C)

16 Show that the tangent at the point P, with parameter t, on the curve $x = ct$, $y = c/t$ has the equation $x + t^2 y = 2ct$

This tangent meets the x-axis in a point Q and the line through P parallel to the x-axis cuts the y-axis in a point R. Show that, for any position of P on the curve, QR is a tangent to the curve with parametric equations $x = ct$, $y = c/(2t)$.　　　　(L)

17 The chord PQ of a parabola $y^2 = 4ax$, where P is the point $(ap^2, 2ap)$ and Q is the point $(aq^2, 2aq)$ subtends an angle of $90°$ at the origin. Show that $pq = -4$.

　　As p varies:

(a) show that the tangents to the parabola at P and Q meet on a fixed straight line and find the equation of this line,

(b) show that the chord PQ passes through a fixed point and find the coordinates of this point.　　　　(L)

18 Prove that the equation of the tangent to the ellipse

$$\frac{x^2}{a^2} + \frac{y^2}{b^2} = 1 \qquad (a > 0, b > 0)$$

at the point P $(a \cos \theta, b \sin \theta)$ is $\dfrac{x}{a} \cos \theta + \dfrac{y}{b} \sin \theta = 1$

The tangent at P meets the axes Ox and Oy at X and Y respectively. Find the area of the triangle OXY.

　　The points A and B have coordinates $(a, 0)$ and $(0, b)$ respectively. Show that the area of triangle APB is $\frac{1}{2}ab(\cos \theta + \sin \theta - 1)$.

　　Prove that, as θ varies in the interval $0 < \theta < \frac{1}{2}\pi$, the area of the triangle APB is a maximum when the tangent to the ellipse at P is parallel to AB. Prove also that triangle OXY has its minimum area when triangle APB has its maximum area.　　　　(C)

19 Find the equation of the tangent at the point $(ct, c/t)$ on the rectangular hyperbola $xy = c^2$.

Tangents drawn from the point $R(h, k)$ touch the hyperbola at points $P(cp, c/p)$ and $Q(cq, c/q)$. Find h and k in terms of p and q.

The mid-point M of the chord PQ is (x_M, y_M).

(a) If R lies on the straight line $y = mx$, prove that M also lies on this line.

(b) If R lies on the rectangular hyperbola $xy = \frac{1}{2}c^2$, prove that M also lies on a certain rectangular hyperbola. State the least distance of M from the origin. (O & C)

20 If a particle is projected under gravity from a point on a level plane with velocity V at an angle of elevation α, the range

$$R = \frac{V^2 \sin 2\alpha}{g}$$

and the greatest height

$$H = \frac{V^2 \sin^2 \alpha}{2g}$$

With axes through the point of projection, the equation of the parabolic trajectory is

$$y = x \tan \alpha - \frac{gx^2}{2V^2 \cos^2 \alpha}$$

Show that this equation may be written

$$(x - \tfrac{1}{2}R)^2 = \frac{2V^2 \cos^2 \alpha}{g}(H - y)$$

and determine the coordinates of the focus, and the equation of the directrix.

Chapter 10

Series for e^x and $\ln(1 + x)$

Introduction

10.1 The expansion of functions of a variable as series has considerable theoretical and practical importance. There are some problems that are most easily tackled by means of series, for instance estimating the value of the constant e, and there are problems in science and engineering which have no practicable solution except by series. Further, the development of computers has considerably added to the practical importance of approximate numerical solutions to problems. So far in this book only the function $(1 + x)^n$ has been expanded in a series and in this chapter two more functions, e^x and $\ln(1 + x)$ will be considered.

The exponential series

10.2 The fundamental property of the function e^x is that

$$\frac{d}{dx}(e^x) = e^x$$

If two assumptions are made:

   (a) that e^x can be expanded as a series of ascending powers of x and
   (b) that the nth derivative of such a series is the sum to infinity of the nth derivatives of the individual terms,

it is easy to find the coefficients of the terms in the series.
   Suppose that

$$e^x = a_0 + a_1 x + a_2 x^2 + a_3 x^3 + \dots + a_n x^n \dots \tag{1}$$

Differentiating (1) once, twice, and three times respectively,

$$e^x = \quad a_1 + 2a_2 x + 3a_3 x^2 + \dots + na_n x^{n-1} + \dots \tag{2}$$

$$e^x = \quad\quad 2a_2 + 3 \times 2a_3 x + \dots + n(n-1)a_n x^{n-2} + \dots \tag{3}$$

$$e^x = \quad\quad\quad 3 \times 2a_3 + \dots + n(n-1)(n-2)a_n x^{n-3} + \dots \tag{4}$$

Differentiating (1) n times,

$$e^x = n!a_n + \ldots \tag{5}$$

Now substituting $x = 0$ in (1), (2), (3), (4), (5),

$$1 = a_0$$
$$1 = a_1$$
$$1 = 2a_2$$
$$1 = 3 \times 2a_3$$
$$1 = n!a_n$$

Substituting the values we have just found for a_0, a_1, a_2, a_3, a_n into equation (1),

$$\mathbf{e^x = 1 + x + \frac{x^2}{2!} + \frac{x^3}{3!} + \ldots + \frac{x^n}{n!} + \ldots}$$

This series is often denoted by **exp x** and it is valid for all values of x (see below).

Qu. 1   Write down the first four terms and the general terms in the expansions of: (a) e^{-x}, (b) e^{x^2}, (c) e^{3x} in ascending powers of x; (d) $e^{1/x}$, (e) e^{-1/x^2} in descending powers of x.

Qu. 2   (Another method of proof.) Find the coefficients of the terms in the expansion of e^x by equating coefficients in equations (1) and (2) above.

Alternatively, the expansion of e^x can be obtained by integration. The assumption is made that $x^n \to 0$ as $n \to \infty$ when $|x| < 1$. This has already been assumed in connection with infinite geometrical progressions (Book 1, §13.9). Most readers, however, will prefer to leave this proof until the second reading, in which case they should proceed to Example 1, p. 185.

Let the variable x lie in the range of values from 0 to c, where c is any positive constant, thus

$$0 < x < c$$

Now $e^0 = 1$,

$$\therefore 1 < e^x < e^c$$

Integrating from 0 to x,

$$x < e^x - 1 < x\,e^c$$

Again integrating from 0 to x,

$$\tfrac{1}{2}x^2 < e^x - 1 - x < \tfrac{1}{2}x^2\,e^c$$

Integrating a further $n - 2$ times,

$$\frac{x^n}{n!} < e^x - 1 - \frac{x^2}{2!} - \ldots - \frac{x^{n-1}}{(n-1)!} < \frac{x^n}{n!}\,e^c$$

When $n \to \infty$, $x^n/n! \to 0$ (proved below),

$$\therefore e^x - 1 - x - \frac{x^2}{2!} - \ldots - \frac{x^{n-1}}{(n-1)!} \to 0$$

Therefore the difference between e^x and the series

$$1 + x + \frac{x^2}{2!} + \ldots + \frac{x^r}{r!} + \ldots + \frac{x^{n-1}}{(n-1)!}$$

approaches zero as $n \to \infty$.

$$\therefore \text{ e}^x = 1 + x + \frac{x^2}{2!} + \ldots + \frac{x^r}{r!} + \ldots$$

To prove the series for negative values of x, take

$$a < x < 0$$
$$\therefore \text{ e}^a < \text{e}^x < 1$$

Integrating from x to 0,

$$-x\,\text{e}^a < 1 - \text{e}^x < -x$$

Again integrating from x to 0,

$$\tfrac{1}{2}x^2\,\text{e}^a < \text{e}^x - 1 - x < \tfrac{1}{2}x^2$$

It is left to the reader to complete the proof.

Note that the expansion of e^x is valid for all values of x. It is clear that exp x must have a finite sum for any value of x, since it has been shown that the sum to infinity is e^x, which is finite.

––––––

To show that $x^n/n! \to 0$ as $n \to \infty$, let $u_r = x^r/r!$

$$\therefore \frac{u_{r+1}}{u_r} = \frac{x^{r+1}}{(r+1)!} \times \frac{r!}{x^r} = \frac{x}{r+1}$$

Let k be the first integer greater than or equal to $2x$, then if $r > k$,

$$\frac{u_{r+1}}{u_r} < \frac{1}{2}$$

$$\therefore u_{k+1} < \tfrac{1}{2}u_k \qquad u_{k+2} < (\tfrac{1}{2})^2 u_k, \ldots, u_n < (\tfrac{1}{2})^{n-k} u_k$$

$$\therefore u_n < \frac{x^k}{k!}\left(\frac{1}{2}\right)^{n-k}$$

But $x^k/k!$ is finite and $(\tfrac{1}{2})^{n-k} \to 0$ as $n \to \infty$, therefore $x^n/n! \to 0$ as $n \to \infty$.

––––––

Example 1 *Find the value of* e *correct to four places of decimals.*

Substituting $x = 1$ in the series for e^x,

$$\text{e}^1 = 1 + 1 + \frac{1}{2!} + \frac{1}{3!} + \ldots + \frac{1}{r!} + \ldots$$

The working is shown, although readers may prefer to use a calculator, if available. Each term in the series, after the first, is obtained from the previous one by dividing by 1, 2, 3,..., 9,... respectively. The working has been taken to five places of decimals. The value obtained for e is 2.7183, correct to four places of decimals.

1.00000
1.00000
0.50000
0.16667
0.04167
0.00833
0.00139
0.00020
0.00002 (5)
0.00000
———
2.71828

It can be shown that e is irrational, and it can also be shown that e is *transcendental*, that is, e satisfies no algebraic equation in the form

$$a_0 + a_1 x + \ldots + a_n x^n = 0$$

where the coefficients $a_0, a_1, \ldots, a_n$ are integers.

Example 2   *Find the first four terms in the expansions in ascending powers of x of* (a) e^{1-x^2}, (b) e^{x-x^2}, *giving the general term in* (a).

(a) $e^{1-x^2} = e^1 \times e^{-x^2}$

$$= e\left\{1 + (-x^2) + \frac{(-x^2)^2}{2!} + \frac{(-x^2)^3}{3!} + \ldots + \frac{(-x^2)^r}{r!} + \ldots\right\}$$

$$\therefore e^{1-x^2} = e\{1 - x^2 + \tfrac{1}{2}x^4 - \tfrac{1}{6}x^6 + \ldots + (-1)^r x^{2r}/r! + \ldots\}$$

(b) $e^{x-x^2} = 1 + (x - x^2) + \frac{(x - x^2)^2}{2!} + \frac{(x - x^2)^3}{3!} + \ldots$

$$= 1 + x - x^2 + \tfrac{1}{2}x^2 - x^3 + \ldots + \tfrac{1}{6}x^3 + \ldots$$

$$\therefore e^{x-x^2} = 1 + x - \tfrac{1}{2}x^2 - \tfrac{5}{6}x^3 + \ldots$$

Example 3   *Find the sum to infinity of the series*

$$1 + \frac{3x}{1!} + \frac{5x^2}{2!} + \frac{7x^3}{3!} + \ldots$$

The general term is $\dfrac{1 + 2n}{n!} x^n$. We aim to find terms in the form $x^r/r!$, so the general term is split up as

$$\frac{x^n}{n!} + \frac{2n}{n!} x^n = \frac{x^n}{n!} + 2x \times \frac{x^{n-1}}{(n-1)!} \qquad (n \geqslant 1)$$

Therefore the series may be written:

$$1 + (x + 2x \times 1) + \left(\frac{x^2}{2!} + 2x \times x\right) + \left(\frac{x^3}{3!} + 2x \times \frac{x^2}{2!}\right) + \dots$$

$$= 1 + x + \frac{x^2}{2!} + \frac{x^3}{3!} + \frac{x^4}{4!} + \dots + 2x \times 1 + 2x \times x + 2x \times \frac{x^2}{2!} + 2x \times \frac{x^3}{3!} + \dots$$

$$= e^x + 2x\,e^x = (1+2x)\,e^x$$

Exercise 10a

1 Use the expansion exp x to find the values of (a) $e^{0.1}$, (b) $1/e$, (c) $\sqrt{e}$, giving your answers correct to four places of decimals.

In Nos. 2–10, expand the functions of x as far as the fourth non-zero terms and give the general terms.

2 e^{x^3}.

3 $\sqrt[3]{(e^x)}$.

4 $(1/e^x)^2$.

5 e^{2+x}.

6 $1/\sqrt{(e^x)}$.

7 $(1+x)e^x$.

8 $(1+2x)\,e^{-2x}$.

9 $\dfrac{e^{3x} \times e^{2x}}{e^x}$.

10 $\dfrac{e^{3x} + e^{2x}}{e^x}$.

11 Find the greatest terms in the expansion of e^x when $x = 10$.

In Nos. 12–15 expand the functions in ascending powers of x as far as the term in x^3.

12 e^{x^2+2x}.   **13** e^{x^2-3x+1}.   **14** $\dfrac{e^x}{1+x}$.   **15** $\dfrac{1-e^{-x}}{e^x-1}$.

16 Find the limits of the following functions as x approaches zero:

(a) $\dfrac{e^x - (1+x)}{e^{2x} - (1+2x)}$,   (b) $\dfrac{e^{2x} - (1+4x)^{1/2}}{e^{-x} - (1-3x)^{1/3}}$,   (c) $\dfrac{e^x + e^{-x} - 2}{e^{x^2} - 1}$.

Find the sums to infinity of the following series:

17 $1 + \dfrac{2x}{1!} + \dfrac{3x^2}{2!} + \dfrac{4x^3}{3!} + \dots$

18 $1 + \dfrac{3x}{2!} + \dfrac{9x^2}{3!} + \dfrac{27x^3}{4!} + \dots$

19 $1 + \dfrac{x^2}{2!} + \dfrac{x^4}{4!} + \dots$ [Start by writing down the series for e^x and e^{-x}.]

20 $x + \dfrac{x^3}{3!} + \dfrac{x^5}{5!} + \dots$

The logarithmic series

10.3   The geometric series $1 - u + u^2 - u^3 + \dots$ has a sum to infinity (Book 1, §13.9) of $1/(1 + u)$. So we may write

$$\frac{1}{1 + u} = 1 - u + u^2 - u^3 + \dots$$

Assuming that the integral of the sum of an infinite series is the sum of the integrals of its terms, integrate between 0 and x:

$$\ln(1 + x) = x - \frac{x^2}{2} + \frac{x^3}{3} - \frac{x^4}{4} + \dots$$

The nth term of the geometric series is $(-1)^{n-1}u^{n-1}$ so that the nth term of the logarithm series is $(-1)^{n-1}x^n/n$. Since the geometric series only has a sum if $|u| < 1$, we should expect that the logarithmic series would be valid when $|x| < 1$ but it can also be shown that the series has a sum when $x = 1$ (see Exercise 10b, No. 25; see also Fig. 16.3, p. 316). Thus

$$\mathbf{ln\,(1 + x) = x - \frac{x^2}{2} + \frac{x^3}{3} - \dots + (-1)^{n-1}\frac{x^n}{n} + \dots}$$

provided $-1 < x \leqslant 1$.

Note that if x is replaced by $-x$ in this series,

$$\mathbf{ln\,(1 - x) = -x - \frac{x^2}{2} - \frac{x^3}{3} - \dots - \frac{x^n}{n} - \dots}$$

provided $-1 \leqslant x < 1$.

We can, however, prove the expansion of $\ln(1 + x)$ without making the assumption about integrating an infinite series which was made at the beginning of this section. It is suggested that most readers should omit the following proof on first reading and proceed to Example 4.

Consider the sum of n terms of the geometric progression

$$1 - u + u^2 - \dots + (-1)^{n-1}u^{n-1} = \frac{1 - (-u)^n}{1 + u}$$

$$\therefore \frac{1}{1 + u} = 1 - u + u^2 - \dots + (-1)^{n-1}u^{n-1} + \frac{(-1)^n u^n}{1 + u}$$

Integrating from 0 to x,

$$\ln(1 + x) = x - \frac{x^2}{2} + \frac{x^3}{3} - \dots + (-1)^{n-1}\frac{x^n}{n} + R_n$$

where

$$R_n = \int_0^x \frac{(-1)^n u^n}{1 + u}\,du = (-1)^n \int_0^x \frac{u^n}{1 + u}\,du$$

We now examine what happens to R_n as $n \to \infty$.

Case 1: $0 < x \leqslant 1$. Consider the function $u^n/(1+u)$, where u lies in the range $0 \leqslant u \leqslant x$. The least value of the denominator is 1, so

$$0 \leqslant \frac{u^n}{1+u} \leqslant u^n$$

Integrating with respect to u from 0 to x,

$$0 < \int_0^x \frac{u^n}{1+u}\,du < \int_0^x u^n\,du$$

$$\therefore\ 0 < |R_n| < \left[\frac{u^{n+1}}{n+1}\right]_0^x = \frac{x^{n+1}}{n+1}$$

$$\therefore |R_n| < \frac{x^{n+1}}{n+1} \leqslant \frac{1}{n+1}$$

Hence if $0 < x \leqslant 1$, $R_n \to 0$ as $n \to \infty$.

Case 2: $-1 < x < 0$. Consider the function $(-1)^n u^n/(1+u)$, where u lies in the range $x \leqslant u \leqslant 0$. We now have

$$0 \leqslant \frac{(-1)^n u^n}{1+u} \leqslant \frac{(-1)^n u^n}{1+x}$$

Integrating with respect to u from x to 0,

$$0 < \int_x^0 \frac{(-1)^n u^n}{1+u}\,du < \int_x^0 \frac{(-1)^n u^n}{1+x}\,du$$

$$\therefore\ 0 < |R_n| < \left[\frac{(-1)^n u^{n+1}}{(1+x)(n+1)}\right]_x^0$$

$$\therefore |R_n| < \frac{(-1)^{n+1} x^{n+1}}{(1+x)(n+1)} < \frac{1}{(1+x)(n+1)}$$

Hence if $-1 < x < 0$ (but *not* for $x = -1$), $R_n \to 0$ as $n \to \infty$.

Case 3: $x = 0$. Both sides of the expansion are zero. Therefore

$$\ln(1+x) = x - \frac{x^2}{2} + \frac{x^3}{3} - \ldots + (-1)^{n-1}\frac{x^n}{n} + \ldots$$

provided $-1 < x \leqslant 1$.

Example 4 *Expand as series in ascending powers of* x:

(a) $\ln(2+x)$,   (b) $\ln(2+x)^3$,   (c) $\ln(x^2 - 3x + 2)$.

(a) $\ln(2+x) = \ln\{2(1+\tfrac{1}{2}x)\}$

$$= \ln 2 + \ln(1+\tfrac{1}{2}x)$$

$$\ln(2+x) = \ln 2 + \tfrac{1}{2}x - \frac{(\tfrac{1}{2}x)^2}{2} + \ldots + (-1)^{n-1}\frac{(\tfrac{1}{2}x)^n}{n} + \ldots$$

$$\therefore \ln(2+x) = \ln 2 + \frac{x}{2} - \frac{x^2}{8} + \ldots + (-1)^{n-1}\frac{x^n}{2^n \times n} + \ldots$$

The expansion is valid if $-1 < \tfrac{1}{2}x \leqslant 1$, i.e. if $-2 < x \leqslant 2$.

(b) $\ln(2+x)^3 = 3\ln(2+x)$.

Therefore, using the result of part (a),

$$\ln(2+x)^3 = 3\ln 2 + \frac{3x}{2} - \frac{3x^2}{8} + \ldots + (-1)^{n-1}\frac{3x^n}{2^n \times n} + \ldots$$

The expansion is again valid if $-2 < x \leqslant 2$.

(c) $\ln(x^2 - 3x + 2) = \ln\{(1-x)(2-x)\}$
$$= \ln(1-x) + \ln(2-x)$$

$$\ln(1-x) = -x - \frac{x^2}{2} - \ldots - \frac{x^n}{n} - \ldots$$

From (a),

$$\ln(2-x) = \ln 2 - \frac{x}{2} - \frac{x^2}{8} - \ldots - \frac{x^n}{2^n \times n} - \ldots$$

Adding,

$$\ln(x^2 - 3x + 2) = \ln 2 - \frac{3}{2}x - \frac{5}{8}x^2 - \ldots - \frac{x^n}{n}\{1 + (\tfrac{1}{2})^n\} - \ldots$$

For the expansions to be valid, x must satisfy both $-1 \leqslant x < 1$ and $-2 \leqslant x < 2$,
i.e. $-1 \leqslant x < 1$.

Qu. 3   Expand in ascending powers of x:
(a) $\ln(1 + \tfrac{1}{4}x)$,   (b) $\ln(3-x)$,   (c) $\ln(x^2 - 2x + 1)$.
Give the first three terms and the general term and state the ranges of values of x
for which the expansions are valid.

Example 5   *If $|x| > 1$, show that*

$$\ln(1+x) = \ln x + \frac{1}{x} - \frac{1}{2x^2} + \ldots + \frac{(-1)^{n-1}}{nx^n} + \ldots$$

[We are told that $|x| > 1$, so we express the series in terms of $\dfrac{1}{x}$, which is
numerically *smaller* than 1.]

$$\ln(1+x) = \ln\{x(1+1/x)\}$$
$$= \ln x + \ln(1 + 1/x)$$

$$\therefore \ln(1+x) = \ln x + \frac{1}{x} - \frac{1}{2x^2} + \ldots + \frac{(-1)^{n-1}}{nx^n} + \ldots$$

Other series have been devised for the calculation of logarithms and one of these will now be obtained.

$$\ln(1+x) = \quad x - \frac{x^2}{2} + \frac{x^3}{3} - \frac{x^4}{4} + \ldots + \frac{x^{2n-1}}{2n-1} - \frac{x^{2n}}{2n} + \ldots$$

$$\ln(1-x) = -x - \frac{x^2}{2} - \frac{x^3}{3} - \frac{x^4}{4} - \ldots - \frac{x^{2n-1}}{2n-1} - \frac{x^{2n}}{2n} - \ldots$$

The expansions are valid if $-1 < x \leqslant 1$, $-1 \leqslant x < 1$, respectively, so for both to be valid, $-1 < x < 1$.

Subtracting,

$$\ln\left(\frac{1+x}{1-x}\right) = 2\left(x + \frac{x^3}{3} + \ldots + \frac{x^{2n-1}}{2n-1} + \ldots\right)$$

Dividing by 2 and writing

$$\frac{1}{2}\ln\left(\frac{1+x}{1-x}\right) = \ln\sqrt{\left(\frac{1+x}{1-x}\right)}$$

we obtain

$$\ln\sqrt{\left(\frac{1+x}{1-x}\right)} = x + \frac{x^3}{3} + \ldots + \frac{x^{2n-1}}{2n-1} + \ldots$$

provided $-1 < x < +1$.

The advantage of this series may be seen by attempting to calculate, say, $\ln 1.5$ by two methods.

(a) Substitute $x = \frac{1}{2}$ in

$$\ln(1+x) = x - \tfrac{1}{2}x^2 + \tfrac{1}{3}x^3 - \ldots + (-1)^{n-1}x^n/n + \ldots$$

$$\ln 1.5 \quad = \tfrac{1}{2} - \tfrac{1}{8} + \tfrac{1}{24} - \tfrac{1}{64} + \tfrac{1}{160} - \tfrac{1}{384} + \tfrac{1}{896} - \tfrac{1}{2048} + \ldots$$

(b) Substitute $x = \frac{1}{5}$ in

$$\ln\left(\frac{1+x}{1-x}\right) = 2\left\{x + \frac{1}{3}x^3 + \frac{1}{5}x^5 + \ldots + x^{2n-1}/(2n-1) + \ldots\right\}.$$

$$\ln 1.5 \quad = 2(\tfrac{1}{5} + \tfrac{1}{375} + \tfrac{1}{15625} + \ldots)$$
$$= 0.4055 \quad \text{to four places of decimals}$$

It is clear that the value correct to four places of decimals can be obtained far more rapidly by the second series.

Note that, using $\log_{10} 1.5 = \log_{10} e \times \ln 1.5$, $\log_{10} 1.5$ can be obtained. (The abbreviation $\lg x$ is sometimes used for $\log_{10} x$; with this notation we could write $\lg 1.5 = \lg e \times \ln 1.5$.)

Example 6 *Find the first three terms in the expansion of*

$$\frac{\ln (1 + x)}{\ln (1 - x)}$$

in ascending powers of x.

Let $\dfrac{\ln (1 + x)}{\ln (1 - x)} = a_0 + a_1 x + a_2 x^2 + ...,$ where a_0, a_1, a_2 are constants to be determined.

$$\therefore \ \ln (1 + x) = \ln (1 - x) (a_0 + a_1 x + a_2 x^2 + ...)$$

$$\therefore \ x - \frac{x^2}{2} + \frac{x^3}{3} - ... = \left(- x - \frac{x^2}{2} - \frac{x^3}{3} - ... \right) (a_0 + a_1 x + a_2 x^2 + ...)$$

Equating coefficients of x, x^2, x^3:

$$1 = -a_0$$
$$-\tfrac{1}{2} = -\tfrac{1}{2}a_0 - a_1$$
$$\tfrac{1}{3} = -\tfrac{1}{3}a_0 - \tfrac{1}{2}a_1 - a_2$$

from which we obtain

$$a_0 = -1, \qquad a_1 = 1, \qquad a_2 = -\tfrac{1}{2}$$

$$\therefore \ \frac{\ln (1 + x)}{\ln (1 - x)} = -1 + x - \frac{1}{2}x^2 + ...$$

Qu. 4 Write down the first three terms of the expansion of $\dfrac{\lg (1 + x)}{\lg (1 - x)}$ in ascending powers of x.

Example 7 *Find the sum to infinity of the series*

$$\frac{1}{1 \times 2} \times \frac{1}{3} + \frac{1}{2 \times 3} \times \frac{1}{3^2} + \frac{1}{3 \times 4} \times \frac{1}{3^3} + ...$$

The general term is

$$\frac{1}{n(n + 1)} \times \frac{1}{3^n}$$

which may be expressed in partial fractions as

$$\left(\frac{1}{n} - \frac{1}{n + 1} \right) \frac{1}{3^n}$$

Therefore the series may be written

$$1 \times \frac{1}{3} + \frac{1}{2} \times \frac{1}{3^2} + \frac{1}{3} \times \frac{1}{3^3} + \ldots + \frac{1}{n} \times \frac{1}{3^n} + \ldots$$

$$-\frac{1}{2} \times \frac{1}{3} - \frac{1}{3} \times \frac{1}{3^2} - \frac{1}{4} \times \frac{1}{3^3} - \ldots - \frac{1}{n+1} \times \frac{1}{3^n} - \ldots$$

$$= -\ln(1 - \tfrac{1}{3}) + S = -\ln\tfrac{2}{3} + S \tag{1}$$

where $S = -\dfrac{1}{2} \times \dfrac{1}{3} - \dfrac{1}{3} \times \dfrac{1}{3^2} - \dfrac{1}{4} \times \dfrac{1}{3^3} - \ldots - \dfrac{1}{n+1} \times \dfrac{1}{3^n} - \ldots$

$$\therefore -\frac{1}{3} + \frac{1}{3} S = -\frac{1}{3} - \frac{1}{2} \times \frac{1}{3^2} - \frac{1}{3} \times \frac{1}{3^3} - \ldots - \frac{1}{n+1} \times \frac{1}{3^{n+1}} - \ldots$$

$$= \ln(1 - \tfrac{1}{3})$$

$$\therefore S = 3 \ln \tfrac{2}{3} + 1$$

Therefore, from (1), the sum of the series is $2 \ln \frac{2}{3} + 1$.

Exercise 10b

1 Expand the following functions in ascending powers of x, giving the first three or four terms, as indicated, and the general term. State the ranges of values of x for which the expansions are valid.

(a) $\ln(3 + x)$, (4),   (b) $\ln(1 - \frac{1}{2}x)$, (4),   (c) $\ln(2 - 5x)$, (4),

(d) $\ln(1 - x^2)$, (4),   (e) $\ln\left(\dfrac{3 + x}{3 - x}\right)$, (3),   (f) $\ln\left(\dfrac{4 - 3x}{4 + 3x}\right)$, (3).

Find the first three terms and the general terms in the expansions of the functions in Nos. 2–8. State the necessary restrictions on the values of x.

2 $\ln\left(\dfrac{2 - x}{3 - x}\right)$.   **3** $\ln\dfrac{1}{3 - 4x - 4x^2}$.

4 $\ln\left\{\dfrac{(1 + 4x)^3}{(1 + 3x)^4}\right\}$.   **5** $\ln\sqrt{(x^2 + 3x + 2)}$.

6 $\ln(1 + x + x^2)$ [Hint: $(1 - x^3) = (1 - x)(1 + x + x^2)$]

7 $\ln\{(1 + x)^{1/x}\}$.   **8** $\ln(1 - x + x^2)$.

Expand the following functions in ascending powers of x as far as the terms indicated. State the ranges of values of x for which the expansions are valid.

9 $\dfrac{\ln(1 + x)}{1 - x}$, (x^3).   **10** $e^x \ln(1 + x)$, (x^3).

11 $\dfrac{x + x^2}{\ln(1 + x)}$, (x^2).   **12** $\{\ln(1 - x)\}^2$, (x^4).

13 By substituting $x = \frac{1}{3}$ in the expansion of $\ln\{(1 + x)/(1 - x)\}$ in ascending powers of x, find the value of $\ln 2$ correct to four significant figures. Taking $\ln 1.5 = 0.4055$, estimate the value of $\ln 3$.

In Nos. 14–16, take $\ln 2 = 0.693\ 147$ and $\ln 3 = 1.098\ 612$.

14 Find $\ln 10$ correct to four places of decimals by substituting $x = \frac{1}{9}$ in the expansion of $\ln(1 + x)$. Deduce an approximate value of $\lg e$.

15 Find the value of $\ln 7$ by substituting $x = \frac{1}{8}$ in the expansion of $\ln\{(1 + x)/(1 - x)\}$. Give your answer correct to four places of decimals.

16 Find the value of $\lg 11$ correct to four places of decimals. Use the expansion of $\ln\{(1 + x)/(1 - x)\}$ with $x = 0.1$. Take $\lg e = 0.434\ 29$.

17 Find the limits of the following functions as x approaches zero:

(a) $\ln\{(1 - x^2)^{1/x^2}\}$,

(b) $\dfrac{\ln(1 + x) - x}{\ln(1 - x) + x}$,

(c) $\dfrac{\ln\{(1 + x)^2\} + x^2 - 2x}{\ln(1 - x^3)}$,

(d) $\dfrac{\ln(1 - x) + x\sqrt{(1 + x)}}{\ln(1 + x^2)}$.

Find the sums to infinity of the following series.

18 $\dfrac{1}{3} - \dfrac{1}{2} \times \dfrac{1}{9} + \dfrac{1}{3} \times \dfrac{1}{27} - \dfrac{1}{4} \times \dfrac{1}{81} + \dots$

19 $\dfrac{1}{2} + \dfrac{1}{2} \times \dfrac{1}{2^2} + \dfrac{1}{3} \times \dfrac{1}{2^3} + \dfrac{1}{4} \times \dfrac{1}{2^4} + \dots$

20 $\dfrac{1}{4} + \dfrac{1}{3} \times \dfrac{1}{4^3} + \dfrac{1}{5} \times \dfrac{1}{4^5} + \dfrac{1}{7} \times \dfrac{1}{4^7} + \dots$

21 $1 - \dfrac{1}{2} \times \dfrac{2}{5} + \dfrac{1}{3} \times \dfrac{2^2}{5^2} - \dfrac{1}{4} \times \dfrac{2^3}{5^3} + \dots$

22 $1 + \dfrac{1}{3 \times 2^2} + \dfrac{1}{5 \times 2^4} + \dfrac{1}{7 \times 2^6} + \dots$

*23 Integrate the inequalities

$$\frac{1}{(1 + t)^2} < \frac{1}{1 + t} < 1 \qquad (t > 0)$$

from 0 to u and deduce that

$$\frac{u}{1 + u} < \ln(1 + u) < u \qquad (u > 0)$$

Sketch the graph of $y = 1/x$ and illustrate the latter inequalities graphically. Also prove that, if $-1 < u < 0$,

$$\frac{u}{1 + u} < \ln(1 + u) < u$$

24 Sketch the graph of $y = 1/x$ and show that, when n is a positive integer greater than 1,

$$\frac{1}{2} + \frac{1}{3} + \ldots + \frac{1}{n} < \ln n < 1 + \frac{1}{2} + \ldots + \frac{1}{n-1}$$

25 Let s_n denote the sum of n terms of the series

$$1 - \tfrac{1}{2} + \tfrac{1}{3} - \tfrac{1}{4} + \tfrac{1}{5} - \tfrac{1}{6} + \ldots$$

By considering the terms of the series in pairs, show that s_{2n} increases as $n \to \infty$. By considering the terms of the series after 1 in pairs, show that s_{2n+1} is less than 1 and decreases as $n \to \infty$. Show that $|s_{2n}| - |s_{2n+1}| \to 0$ as $n \to \infty$. What can you conclude about s_n as $n \to \infty$?

Exercise 10c (Miscellaneous)

1 By expanding the integrand of

$$\int_0^x \frac{1}{1+x} \, dx$$

as a series of powers of x and integrating term by term, find the series for ln $(1 + x)$, assuming your method to be valid provided that $|x| < 1$.

Write down the series for ln $(1 - x)$, obtain the series for $\ln \dfrac{1+x}{1-x}$ and deduce a series for $\ln \dfrac{m}{n}$ in terms of $\dfrac{m-n}{m+n}$.

Hence calculate ln 8 correct to five places of decimals, given that ln 7 = 1.945 910.   (JMB)

2 Assuming that $|x| < 1$, write down
(a) the sum of the infinite geometric series $1 + x^2 + x^4 + \ldots$,
(b) the first three terms of the series for ln $(1 + x)$.
Obtain the first two terms of the series for

$$\frac{1}{2} \ln \left(\frac{1+x}{1-x} \right)$$

Assuming also that x is positive, show that the sum of the remaining terms of this series is less than

$$\frac{x^5}{5(1 - x^2)} \qquad\qquad \text{(JMB)}$$

3 Write down the expansions of ln $(1 + x)$ and ln $(1 - x)$, stating for what values of x they are valid. Prove that,
(a) if $-\tfrac{1}{2} < x \leqslant \tfrac{1}{2}$, then

$$\ln (1 + x - 2x^2) = x - \tfrac{5}{2}x^2 + \tfrac{7}{3}x^3 - \tfrac{17}{4}x^4 \ldots$$

(b) if m/n is positive,

$$\ln \frac{m}{n} = 2 \left\{ \left(\frac{m-n}{m+n}\right) + \frac{1}{3}\left(\frac{m-n}{m+n}\right)^3 + \frac{1}{5}\left(\frac{m-n}{m+n}\right)^5 + \dots \right\} \tag{L}$$

4 Prove that, when $a > 0$,

(a) $\ln (a + x) = \ln a + \ln \left(1 + \dfrac{x}{a}\right)$,

(b) $a^x = e^{x \ln a}$.

Write down the first three terms of the expansions of $\ln (1 + y)$ and e^y.

Prove that the expansion of $a^x - 1 - x \ln (a + x)$ as a series of ascending powers of x begins with a term in x^2, and find the coefficient of x^2 in this term.   (O & C)

5 (a) Write down the expansion of $\ln (1 + x)$ in ascending powers of x, giving the first three terms and the coefficient of x^m; state the limitations on the value of x.

Prove that

$$2 \ln n - \ln (n + 1) - \ln (n - 1) = \frac{1}{n^2} + \frac{1}{2n^4} + \frac{1}{3n^6} + \dots$$

stating the necessary restriction on the value of n.

Given that $\ln 10 = 2.302\,59$ and $\ln 3 = 1.098\,61$, calculate the value of $\ln 11$ correct to four places of decimals.

(b) Find the coefficient of x^n in the expansion of $(1 + 3x)e^{-3x}$ as a series of ascending powers of x.   (O & C)

6 Find the sum of the first n terms of a geometric progression of which the first term is a and the common ratio is r.

If p is any odd positive integer and q any even positive integer, and if $x > 0$, prove that

$$1 - x + x^2 - \dots - x^p < \frac{1}{1+x} < 1 - x + x^2 - \dots + x^q$$

Deduce that

$$\frac{x}{1} - \frac{x^2}{2} + \frac{x^3}{3} - \dots - \frac{x^{p+1}}{p+1} < \ln (1 + x) < \frac{x}{1} - \frac{x^2}{2} + \frac{x^3}{3} - \dots + \frac{x^{q+1}}{q+1}$$

By taking $p = 5$, $q = 4$, $x = 0.1$, calculate the value of $\ln (1.1)$ correct to six places of decimals.   (JMB)

7 (a) Write down the expansions of e^x and $\ln (1 + x)$ in series of ascending powers of x, giving in each case the first three terms and the nth term.

(b) By considering the factors of $(1 - x^3)$, obtain the coefficients of x^{3n}, x^{3n+1}, x^{3n+2} in the expansion of $\ln (1 + x + x^2)$.

(c) Obtain the first two non-vanishing terms in the expansion of

$$(x - 1)(1 - e^x) - \ln (1 + x)$$

in ascending powers of x.   (O & C)

8 (a) Prove that if $x^p = (xy)^q = (xy^2)^r$ for all values of x and y, then $2pr = q(p + r)$.

(b) Assuming the expansion for $\ln (1 + x)$ in ascending powers of x, prove that

$$\ln \sqrt{\left(\frac{1 + x}{1 - x}\right)} = x + \frac{1}{3}x^3 + \frac{1}{5}x^5 + \dots$$

and, when $0 < \theta < \frac{1}{2}\pi$, deduce that

$$\sin \theta + \tfrac{1}{3} \sin^3 \theta + \tfrac{1}{5} \sin^5 \theta + \dots = \ln (\tan \theta + \sec \theta)$$

(c) Establish the identity

$$n^2 \equiv (n + 2)(n + 1) - 3(n + 2) + 4$$

and hence find the sum of the series

$$\frac{1^2}{3!} + \frac{2^2}{4!} + \frac{3^2}{5!} + \dots \qquad\qquad \text{(O \& C)}$$

9 (a) Prove the identity

$$1 + 2x + 2x^2 + x^3 \equiv \frac{(1 + x)(1 - x^3)}{1 - x}$$

and hence expand $\ln (1 + 2x + 2x^2 + x^3)$ in ascending powers of x as far as the term in x^6, stating the necessary restrictions on the values of x.

(b) Write down the series for e^x and e^{-x} in ascending powers of x. Prove that, if x^4 and higher powers of x are neglected, then

$$\frac{x}{1 - e^{-2x}} - \frac{1}{2} = \frac{1}{2}x + \frac{1}{6}x^2 \qquad\qquad \text{(O \& C)}$$

10 Assuming that x is sufficiently small, find the values of p and q, other than zero, for which

$$(1 + x)^p - \ln (1 + qx) = 1 + ax^3 + \dots,$$

where the terms omitted contain powers of x higher than the third. Determine the value of the coefficient a. (JMB)

11 Write down the expansion of $\ln (1 + x)$ in ascending powers of x, giving the general term and stating for what real values of x the expansion is valid. Determine a and b so that the expansion of

$$\frac{1 + ax}{1 + bx} \ln (1 + x)$$

may contain no term in x^2 or x^3, and show that with these values

$$\frac{1 + bx}{1 + ax} = 1 - \frac{x}{2} + \frac{x^2}{3} - \frac{2x^3}{9}$$

neglecting powers of x above the third. (O \& C)

12 Give the expansion of $\ln(1 + x)$ in ascending powers of x and state for what range of values of x the expansion is valid.

By taking logarithms or otherwise, verify that, when n is large, an approximate value of $(1 + 1/n)^n$ is

$$e\left(1 - \frac{1}{2n} + \frac{11}{24n^2} - \frac{7}{16n^3}\right)$$ (O & C)

13 (a) Expand $\ln\dfrac{2 + x}{1 - x}$ in ascending powers of x, giving the first four terms and the general term.

(b) Show that the first non-zero coefficient in the expansion of

$$e^{-x} - \frac{1 - x}{(1 - x^2)^{1/2}(1 - x^3)^{1/3}}$$

in ascending powers of x is that of x^5.   (L)

14 Write down the series for $\ln(1 + x)$ in ascending powers of x and state the range of values of x for which it is valid.

Prove that, if $n > 1$,

$$\ln\frac{n}{n - 1} > \frac{1}{n} > \ln\frac{n + 1}{n}$$

and deduce that, if n is a positive integer,

$$1 + \ln n > 1 + \frac{1}{2} + \frac{1}{3} + \ldots + \frac{1}{n} > \ln(n + 1)$$ (C)

15 Write down the expansion in ascending powers of x of $\ln\left(\dfrac{1 + x}{1 - x}\right)$ where $-1 < x < 1$.

By using partial fractions, obtain the sum of the series

$$\sum_{n=1}^{\infty} \frac{x^{2n}}{(2n - 1)(2n + 1)}$$

when $0 < x < 1$.

Find the sum of the first N terms of the series when $x = 1$ and deduce that

$$\sum_{n=1}^{\infty} \frac{1}{(2n - 1)(2n + 1)} = \frac{1}{2}$$ (C)

16 State the first four terms in the series expansions of $(1 + x)^n$, $\ln(1 + x)$.

Find the sum of the infinite series

$$\frac{1}{2} + \frac{1}{2} \times \frac{1}{2^2} + \frac{1}{3} \times \frac{1}{2^3} + \ldots + \frac{1}{n} \times \frac{1}{2^n} + \ldots$$ (JMB)

17 (a) If $0 < x < 1$ and $f(x)$ is the sum of the infinite series

$$1 + \frac{x^2}{3} + \frac{x^4}{5} + \frac{x^6}{7} + \ldots$$

show that, for x in this range,

$$f\left(\frac{2x}{1 + x^2}\right) = (1 + x^2)\, f(x)$$

(b) Sum to infinity the series

$$\frac{3^2}{1!} + \frac{4^2}{2!} + \frac{5^2}{3!} + \dots \tag{L}$$

18 Sum to infinity each of the following series:

(a) $1 + \dfrac{2x}{1!} + \dfrac{3x^2}{2!} + \dfrac{4x^3}{3!} + \dots,$

(b) $\dfrac{x}{1 \times 2} + \dfrac{x^2}{2 \times 3} + \dfrac{x^3}{3 \times 4} + \dots,$ if $|x| < 1$,

(c) $1 + \dfrac{5}{3} + \dfrac{5 \times 7}{3 \times 6} + \dfrac{5 \times 7 \times 9}{3 \times 6 \times 9} + \dots.$ (L)

19 The function $f(t)$ is defined, for non-zero values of t, by the relation

$$f(t) = \frac{1}{2}t\left(\frac{e^t + 1}{e^t - 1}\right)$$

Prove that (a) $f(t) = f(-t)$,   (b) $f(2t) = f(t) + \dfrac{t^2}{4f(t)}.$

Using the expansion $e^t = 1 + t + \dfrac{t^2}{2!} + \dfrac{t^3}{3!} + \dots,$ show that, if t is small enough for t^3 to be neglected, then

$$f(t) = 1 + \tfrac{1}{12}t^2 \tag{C}$$

20 Show that, if x is so small that x^6 and higher powers of x may be neglected,

$$\ln(1 + x) = x - \tfrac{1}{2}x^2 + \tfrac{1}{3}x^3 - \tfrac{1}{4}x^4 + \tfrac{1}{5}x^5$$

Deduce that, for such values of x,

$$\ln\left(\frac{1 + x}{1 - x}\right) = 2\left(x + \frac{1}{3}x^3 + \frac{1}{5}x^5\right)$$

By giving a suitable value to x in this last result, prove that, for large N,

$$\ln\left(\frac{N + 1}{N - 1}\right) \approx \frac{2}{N} + \frac{2}{3N^3} + \frac{2}{5N^5}$$

Hence show that, for large N,

$$\left(\frac{N + 1}{N - 1}\right)^N \approx e^2 \tag{C}$$

Chapter 11

Further differentiation

Logarithmic differentiation

11.1 The object of the first three sections of this chapter is to extend the reader's powers of differentiation and to revise earlier work. In the course of this we shall also discuss how to integrate certain functions.

Logarithmic differentiation is a powerful method which can considerably simplify the differentiation of

(a) products (and quotients) of a number of functions,
(b) certain exponential functions.

It is best introduced by examples but first it is advisable to revise some of the properties of logarithms and how to differentiate functions of y with respect to x.

Qu. 1 $\ln (a^3 \sqrt{b}/c^2) = \ln a^3 + \ln \sqrt{b} - \ln c^2,$
$$= 3 \ln a + \tfrac{1}{2} \ln b - 2 \ln c.$$
(See §2.7.) Write in a similar form:
(a) $\ln (a^2 b)$,   (b) $\ln (a^3/b^3)$,   (c) $\ln \sqrt{(abc)}$,
(d) $\ln (a \sqrt[3]{b}/c^3)$,   (e) $\ln (1/c^4)$,   (f) $\ln (a^b)$.

Qu. 2 $\log_{10} 10\,000 = \log_{10} 10^4 = 4$. Simplify in a similar manner:
(a) $\log_{10} 1000$,   (b) $\log_{10} (1/100)$,   (c) $\log_2 (2^4)$,
(d) $\ln (e^2)$,   (e) $\ln (e^{2x})$,   (f) $\ln (e^{3x^2})$.

Qu. 3 Differentiate with respect to x:
(a) $\ln x$,   (b) $\ln (1 + 2x)$,   (c) $\ln (1 - x)$,
(d) $\ln 4x^3$,   (e) $\ln \sin x$,   (f) $\ln \tan x$.

When differentiating functions of y with respect to x we can, if need be, use the chain rule,

$$\frac{dz}{dx} = \frac{dz}{dy} \times \frac{dy}{dx}$$

Thus if $z = y^4$,

$$\frac{dz}{dy} = 4y^3$$

$$\therefore \frac{dz}{dx} = 4y^3 \times \frac{dy}{dx}$$

Qu. 4 Differentiate with respect to x:
(a) $3y^2$,   (b) y^3,   (c) $\cos y$,   (d) $\ln y$.

Express, in your own words, a rule which will help you to differentiate any function of y with respect to x. Use this rule to differentiate with respect to x:
(e) $5y^4$,   (f) $3/y^2$,   (g) $\sqrt{y}$,   (h) $\tan y$.

Example 1 *Differentiate* $\dfrac{e^{x^2} \sqrt{(\sin x)}}{(2x + 1)^3}$.

Let $y = \dfrac{e^{x^2} \sqrt{(\sin x)}}{(2x + 1)^3}$.

$$\therefore \ln y = \ln (e^{x^2}) + \ln \sqrt{(\sin x)} - \ln (2x + 1)^3$$
$$= x^2 + \tfrac{1}{2} \ln \sin x - 3 \ln (2x + 1)$$

Differentiating with respect to x,

$$\frac{1}{y}\frac{dy}{dx} = 2x + \frac{\cos x}{2 \sin x} - \frac{6}{2x + 1}$$

$$\therefore \frac{dy}{dx} = \frac{e^{x^2} \sqrt{(\sin x)}}{(2x + 1)^3} \left\{ 2x + \frac{\cos x}{2 \sin x} - \frac{6}{2x + 1} \right\}$$

(There are occasions when this is the most convenient form in which to use the derivative, but here we shall go on to simplify the expression in brackets.)

$$2x + \frac{\cos x}{2 \sin x} - \frac{6}{2x + 1} = \frac{4x^2 + 2x - 6}{2x + 1} + \frac{\cos x}{2 \sin x}$$

$$\therefore \frac{dy}{dx} = \frac{e^{x^2}}{2\sqrt{(\sin x)}(2x + 1)^4} \{(8x^2 + 4x - 12) \sin x + (2x + 1) \cos x\}$$

Qu. 5 Use the method of Example 1 to differentiate with respect to x:

(a) $\sqrt[3]{\dfrac{x + 1}{x - 1}}$,   (b) $\dfrac{\sqrt{(x^2 + 1)}}{(2x - 1)^2}$,   (c) $\dfrac{x^2 e^x}{(x - 1)^3}$.

Example 2 *Differentiate* 10^x *with respect to* x.

Let $y = 10^x$.
$$\therefore \ln y = \ln 10^x$$
$$= x \ln 10$$

Differentiating with respect to x,

$$\frac{1}{y}\frac{dy}{dx} = \ln 10$$

$$\therefore \frac{dy}{dx} = 10^x \ln 10$$

Example 3   *Differentiate, with respect to x, (a) 2^{x^2}, (b) x^x.*

(a) Let $y = 2^{x^2}$.
$$\therefore \ln y = \ln 2^{x^2}$$
$$= x^2 \ln 2$$

Differentiating with respect to x,

$$\frac{1}{y}\frac{dy}{dx} = 2x \ln 2$$

$$\therefore \frac{dy}{dx} = 2^{x^2} 2x \ln 2$$

(b) Let $y = x^x$.
$$\therefore \ln y = \ln x^x$$
$$= x \ln x$$

Differentiating with respect to x,

$$\frac{1}{y}\frac{dy}{dx} = x \times \frac{1}{x} + 1 \times \ln x$$

$$= 1 + \ln x$$

$$\therefore \frac{dy}{dx} = (1 + \ln x)x^x$$

Qu. 6   We have shown in Example 2 that the derivative of 10^x is $10^x \ln 10$. Write down a function whose derivative is 10^x. What is $\int 10^x \, dx$?
Qu. 7   Differentiate with respect to x:
(a) 2^x,   (b) 3^x,   (c) $(\frac{1}{2})^x$,   (d) 10^{5x},   (e) 10^{x^2}.
Qu. 8   From your answers to Qu. 7, write down:
(a) $\int 2^x \, dx$,   (b) $\int 3^x \, dx$,   (c) $\int (\frac{1}{2})^x \, dx$,   (d) $\int 10^{5x} \, dx$.

Integration by trial

11.2   In Qu. 6 we had an example of what may be called 'integration by trial'. This procedure was discussed in §1.2 and §2.10; its stages are shown in the next two examples.

Example 4 *Integrate 2^{-x} with respect to x.*

(*Stage* 1: make a guess. From the last section it is to be expected that the integral involves 2^{-x}.)

Let $y = 2^{-x}$
$\therefore \ln y = -x \ln 2$

(*Stage* 2: differentiate.)

$$\therefore \frac{1}{y} \frac{dy}{dx} = -\ln 2$$

$$\therefore \frac{dy}{dx} = -2^{-x} \ln 2$$

(*Stage* 3: compare with the given functions.) We have an extra constant factor of $-\ln 2$.

(*Stage* 4: alter the guessed function.)

$$\frac{d}{dx} \left(\frac{2^{-x}}{-\ln 2} \right) = 2^{-x}$$

$$\therefore \int 2^{-x} dx = -\frac{2^{-x}}{\ln 2} + c$$

Example 5 *Integrate $x \ln x$ with respect to x.*

(*Stage* 1: make a guess.) When we *differentiate* a product, we differentiate each function in turn and multiply by the other; so, to integrate $x \ln x$, try integrating one factor and multiply by the other. As we do not know how to integrate $\ln x$, we had better try $\frac{1}{2}x^2 \ln x$.

Let $y = \frac{1}{2}x^2 \ln x$.

(*Stage* 2: differentiate.)

$$\therefore \frac{dy}{dx} = x \ln x + \frac{1}{2}x^2 \times \frac{1}{x}$$

$$\therefore \frac{d}{dx} \left(\frac{1}{2}x^2 \ln x \right) = x \ln x + \frac{1}{2}x$$

(*Stage* 3: compare with the given function.) We have an extra term of $\frac{1}{2}x$ on the right-hand side.

(*Stage* 4: alter the guessed function.)

$$\frac{d}{dx} \left(\frac{1}{2}x^2 \ln x - \frac{1}{4}x^2 \right) = x \ln x + \frac{1}{2}x - \frac{1}{2}x = x \ln x$$

$$\therefore \int x \ln x \, dx = \frac{1}{2}x^2 \ln x - \frac{1}{4}x^2 + c$$

Qu. 9   Integrate the following functions with respect to x by trial:

(a) $(3x + 1)^{1/2}$,   (b) $\sin x \cos^5 x$,   (c) $\dfrac{\sin x}{(1 + \cos x)^3}$,

(d) 5^x,   (e) 2^{2x},   (f) $\ln x$.

Inverse trigonometrical functions

11.3   The functions $\sin^{-1} x$, $\tan^{-1} x$ (or arcsin x, arctan x) have been introduced in Book 1, §18.7. We will now turn to the problem of differentiating such inverse trigonometrical functions. This will be illustrated by examples but, for some readers, a little revision may be advisable. Remember that, in this context, radians *must* be used.

Qu. 10 $y = \sin^{-1} x$ means 'y is the angle (or the number) whose sine is x' so that $\sin y = x$. Rewrite:
(a) $y = \tan^{-1} x$,   (b) $\sec^{-1} x = y$,   (c) $\cos^{-1} p = q$.

Qu. 11   Differentiate with respect to x:
(a) y^2,   (b) $\sin y$,   (c) $\tan y$,   (d) $\sec y$.

Example 6   *Differentiate with respect to x:*
(a) $\sin^{-1} x$,   (b) $\tan^{-1} (x^2 + 1)$.

(a) Let $y = \sin^{-1} x$.
∴ $\sin y = x$

Differentiating with respect to x,

$$\cos y \frac{dy}{dx} = 1$$

$$1 - \sin^2 y = 1 - x^2.$$
$$\therefore \cos^2 y = 1 - x^2.$$

(y was our own introduction, so we must get $\dfrac{dy}{dx}$ in terms of x.)

$$\therefore \sqrt{(1 - x^2)} \frac{dy}{dx} = 1$$

$$\therefore \frac{dy}{dx} = \frac{1}{\sqrt{(1 - x^2)}}$$

(b) Let $y = \tan^{-1} (x^2 + 1)$.
∴ $\tan y = x^2 + 1$

Differentiating with respect to x,

$$\sec^2 y \frac{dy}{dx} = 2x$$

$$1 + \tan^2 y = 1 + (x^2 + 1)^2.$$
$$\therefore \sec^2 y = x^4 + 2x^2 + 2.$$

(We must again express $\dfrac{dy}{dx}$ in terms of x.)

$$\therefore (x^4 + 2x^2 + 2)\frac{dy}{dx} = 2x$$

$$\therefore \frac{dy}{dx} = \frac{2x}{x^4 + 2x^2 + 2}$$

Qu. 12 Differentiate with respect to x:
(a) $\cos^{-1} x$,   (b) $\cot^{-1} x$,   (c) $\sin^{-1}(2x + 1)$.

Exercise 11a

1 Express in the form $p \ln a + q \ln b + r \ln c$:
   (a) $\ln(a^3 b^4)$,   (b) $\ln(a/b)$,   (c) $\ln \sqrt{(a^3/b)}$,
   (d) $\ln(a^2 b/\sqrt{c})$,   (e) $\ln \sqrt{(ab/c)}$,   (f) $\ln \{1/\sqrt{(abc)}\}$.
2 Write the following in a form which does not use the logarithm notation:
   (a) $\lg 100\,000$,   (b) $\log_2 8$,   (c) $\ln e^4$,
   (d) $\ln \sqrt{e}$,   (e) $\ln e^{x^3}$,   (f) $\ln(1/e^{2x})$.

Differentiate Nos. 3–10 with respect to x, using logarithmic differentiation:

3 $\sqrt{\dfrac{(2x + 3)^3}{1 - 2x}}$.

4 $\dfrac{e^{x/2} \sin x}{x^4}$.

5 $\dfrac{1}{\sqrt{(x^2 + 1)}\sqrt[3]{(x^2 - 1)}}$.

6 $\dfrac{1}{x\, e^x \cos x}$.

7 7^x.   8 10^{3x}.   9 $10^{-x/2}$.   10 $1/10^x$.

Integrate with respect to x:

11 5^x.   12 8^x.   13 $(\tfrac{1}{3})^x$.   14 3^{2x}.
15 Convince yourself that $e^{\ln a} = a$ (see §2.7). Write a^x in the form $e^{x \ln a}$ and hence find $\dfrac{d}{dx}(a^x)$.
16 Find $\int a^x \, dx$ by writing $a^x = e^{x \ln a}$.

Differentiate with respect to x:

17 $\tan^{-1} x$.   18 $\sec^{-1} x$.   19 $\sin^{-1}(x + 1)$.
20 $\cos^{-1}(2x - 1)$.   21 $\tan^{-1}(1/x^2)$.   22 $2\cos^{-1} 5x$.

23 Find: (a) $\dfrac{d}{dx}(\sin^{-1} x + \cos^{-1} x)$,   (b) $\dfrac{d}{dx}(\tan^{-1} x + \cot^{-1} x)$.

Explain these answers.

24 Find $\dfrac{d}{dx}(\sin^{-1} x)$ and hence write down

(a) $\dfrac{d}{dx}(\sin^{-1} 2x)$,   (b) $\dfrac{d}{dx}(\sin^{-1} x^2)$,   (c) $\dfrac{d}{dx}\{\cos^{-1} \sqrt{(1 - x^2)}\}$.

Integrate with respect to x by trial:

25 $\sqrt{(4x+3)}$.   **26** $x(2x^2+1)^3$.   **27** $\ln x$.

28 $\sin^{-1} x$. [Find $\dfrac{d}{dx}(x \sin^{-1} x)$.]

Differentiate with respect to x:

29 x^{-x}.   **30** $x^{\sin x}$.

Local maxima and minima; the first derivative test

11.4   We have already met the first derivative test in Book 1, Chapter 5. The diagrams in Fig. 11.1 illustrate (i) a local maximum and (ii) a local minimum. (The $+$ and $-$ signs indicate the sign of the gradient.)

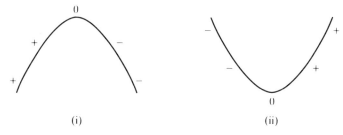

Figure 11.1

At a turning point, $\dfrac{dy}{dx} = 0$ *and it changes sign*; in the case of a maximum (see diagram (i)) it changes from $+$ to $-$ as x increases, whereas at a minimum it changes from $-$ to $+$ (see diagram (ii)).

If however $\dfrac{dy}{dx} = 0$ but does *not* change sign, then we have a **stationary point of inflexion** (see Fig. 11.2).

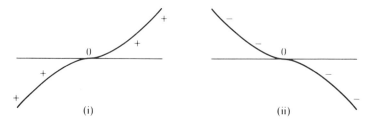

Figure 11.2

The first derivative test is very easy to apply, especially if one makes a habit of factorising the derived function, as the next two examples show.

Example 7   *Investigate the stationary points on the graph of*

$$y = x^2 e^{-x}$$

and sketch the curve.

$$y = x^2 e^{-x}$$

so

$$\frac{dy}{dx} = 2xe^{-x} - x^2 e^{-x}$$
$$= (2x - x^2)e^{-x}$$
$$= x(2 - x)e^{-x}$$

From this we can see that $\dfrac{dy}{dx}$ is zero when $x = 0$ and when $x = 2$. We know that e^{-x} is always positive, so the sign of the gradient is determined by the other factors. By inspection, we can see that $\dfrac{dy}{dx}$ is negative when $x < 0$, and that between 0 and 2, $\dfrac{dy}{dx}$ is positive. When $x > 2$, $\dfrac{dy}{dx}$ is negative again.

Therefore there is a local minimum at $(0,0)$, and a local maximum at $(2, 4e^{-2})$. The curve can now be sketched (see Fig. 11.3).

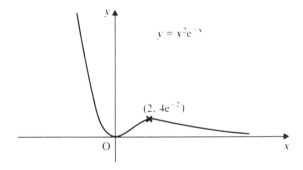

Figure 11.3

Example 8   *Investigate the stationary values of the function*

$$f(x) = x^3 - 3x^2 + 3x$$

and sketch the graph of $y = f(x)$.

In this case,

$$f'(x) = 3x^2 - 6x + 3$$
$$= 3(x^2 - 2x + 1)$$
$$= 3(x - 1)^2$$

We can see that f'(1) is zero, but as $(x-1)^2$ is a square, it can never be negative. In other words, the gradient of $y = f(x)$ is zero at $x = 1$, but everywhere else it is positive. Therefore there is a point of inflexion at $(1,1)$ (see Fig. 11.4).

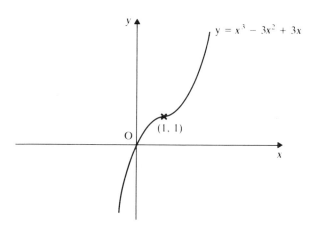

$y = x^3 - 3x^2 + 3x$

$(1, 1)$

Figure 11.4

Qu. 13 Investigate the stationary values of the function xe^{-x} and sketch the graph of $y = xe^{-x}$.

Local maxima and minima; the second derivative test

11.5 The second derivative test depends on the fact that if the gradient of a curve $y = f(x)$ is *increasing* with x, the rate of change of the gradient is *positive*; if the gradient is *decreasing*, its rate of change is *negative*. To put it another way, let us consider the graph of $\dfrac{dy}{dx}$ plotted against x, bearing in mind that the gradient of *this* curve is given by $\dfrac{d^2y}{dx^2}$; if the ordinate, $\dfrac{dy}{dx}$, is increasing with x, the gradient, $\dfrac{d^2y}{dx^2}$, is positive; if $\dfrac{dy}{dx}$ is decreasing, $\dfrac{d^2y}{dx^2}$ is negative.

Looking back to Fig. 11.1, we see that at a local maximum, the gradient is decreasing (it is changing from a positive value, through zero, to a negative value as x increases); so at such a point the derivative *of the gradient function* is negative. In other words, if $y = f(x)$, and at $x = a$

$\dfrac{dy}{dx}$ **is zero and** $\dfrac{d^2y}{dx^2}$ **is *negative*, then y has a (local) maximum at $x = a$;**

on the other hand, if

$\dfrac{dy}{dx}$ **is zero and** $\dfrac{d^2y}{dx^2}$ **is *positive*, then y has a (local) minimum at $x = a$.**

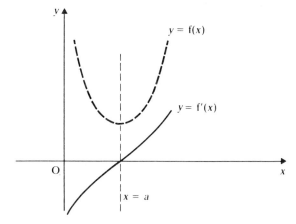

Figure 11.5

Fig. 11.5 shows (as a continuous curve) a graph representing $y = f'(x)$, with positive gradient at $x = a$, i.e. $f''(a) > 0$, and $f'(a) = 0$. The dashed curve represents the corresponding graph of $y = f(x)$, showing a minimum when $x = a$.

Qu. 14   Draw a diagram, like Fig. 11.5, illustrating the graphs of $y = f'(x)$ and $y = f(x)$, with $f''(a) < 0$ and $f'(a) = 0$.

Example 9   *Use the second derivative test to investigate the stationary values of the function xe^{-x}.*

Let $y = xe^{-x}$.

$$\therefore \frac{dy}{dx} = e^{-x} - xe^{-x}$$

$$= (1 - x)e^{-x}$$

From this we can see that there is a stationary value of $1/e$ when $x = 1$.

$$\frac{d^2y}{dx^2} = -e^{-x} - e^{-x} + xe^{-x}$$

$$= -2e^{-x} + xe^{-x}$$

When $x = 1$, $\dfrac{d^2y}{dx^2} = -2e^{-1} + e^{-1} = -e^{-1}$. This is negative, so by the second derivative test, there is a local maximum of $1/e$ when $x = 1$.

It is important to understand that no conclusion can be drawn from the second derivative test when $\dfrac{d^2y}{dx^2}$ is zero. (It is a common mistake to think that there is always a point of inflexion in this case. If the reader has any doubts on this point, consider $y = x^4$; both $\dfrac{dy}{dx}$ and $\dfrac{d^2y}{dx^2}$ are zero at $x = 0$, but the function clearly has a minimum at this point.)

Although examination papers frequently direct the candidate to use the second derivative test (and it is a foolish candidate who ignores the examiner's instructions), if you are free to choose your own method, the first derivative test is often the simpler one to use.

Points of inflexion

11.6 We have already met *stationary* points of inflexion, see Fig. 11.2. In Fig. 11.2(i), the gradient is positive everywhere except at the stationary point, where it is zero, i.e. at this point the gradient has a minimum value. In Fig. 11.2(ii), the gradient is negative everywhere except at the stationary point, where it is zero, so at this point the gradient has a maximum value. In general, a point of inflexion is a point where the gradient has a local maximum or minimum value. Fig. 11.6 shows some points of inflexion for which $\dfrac{dy}{dx} \neq 0$.

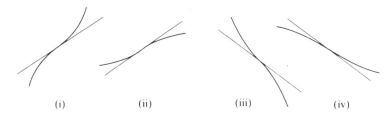

(i)   (ii)   (iii)   (iv)

Figure 11.6

Looking at Fig. 11.6(i), we see that the gradient is always positive; it is decreasing as it approaches the point of inflexion, and after that it increases again, i.e. the *gradient* has a *minimum* value at this point of inflexion. The reader should analyse the other diagrams similarly. (On a graph, a point of inflexion is easily recognised, because the graph 'crosses its own tangent' at such a point.)

Points of inflexion are easily located by applying the first derivative test to the gradient function, i.e.

at a point of inflexion $\dfrac{d^2y}{dx^2}$ is zero *and it changes sign*

(In the case of $y = x^4$, mentioned at the end of §11.5, the second derivative, namely $12x^2$, does not change sign at $x = 0$.)

Example 10   *Find the points of inflexion of the function* $y = \dfrac{48}{12 + x^2}$ *and sketch its graph.*

We are given $y = \dfrac{48}{12 + x^2}$, so in this case,

$$\frac{dy}{dx} = \frac{-96x}{(12 + x^2)^2}$$

From this we can see that $\frac{dy}{dx}$ is zero when $x = 0$, and that its sign changes from positive to negative as x passes through zero, so there is a local maximum at this point. The maximum value of the function is 4.

To find the points of inflexion, we differentiate again:

$$\frac{d^2y}{dx^2} = \frac{-96}{(12 + x^2)^2} + \frac{96 \times 4x^2}{(12 + x^2)^3}$$

$$= 96\left(\frac{-(12 + x^2) + 4x^2}{(12 + x^2)^3}\right)$$

$$= 96\left(\frac{3x^2 - 12}{(12 + x^2)^3}\right)$$

$$= \frac{288(x - 2)(x + 2)}{(12 + x^2)^3}$$

From this we can see that $\frac{d^2y}{dx^2}$ is zero at $x = \pm 2$, and that it changes sign at these points. Hence there are points of inflexion at $(-2, 3)$ and $(+2, 3)$.

In order to sketch the curve, notice that y is always positive and that it tends to zero as x tends to infinity. Also, this is an even function, so its graph is symmetrical about the y-axis (see Fig. 11.7).

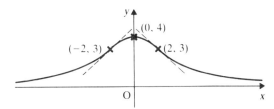

Figure 11.7

Qu. 15 Find the point of inflexion on $y = 2x^3 - 18x^2 + 12x + 80$.

Exercise 11b

Find the nature of the stationary points of

1 $x(x - 3)^2$.   **2** $x + \dfrac{4}{x^2}$.   **3** $x - 2 + \dfrac{1}{x - 3}$.   **4** $x - \ln x$.

Find the points of inflexion in Nos. 5 and 6.

5 $y = x^4 - 54x^2$.   **6** $y = x^4 - 4x^3 + 6x^2 - 4x$.

7 Sketch the graphs of Nos. 1–6.

In Nos. 8–11, find the maxima and minima of the functions of θ in the interval $0 \leqslant \theta \leqslant 2\pi$:

8 $\sin\theta + \frac{1}{2}\sin 2\theta$. Sketch the graph.

9 $(\sin\theta)/(1+\sin\theta)$. Sketch the graph.

10 $\ln\cos\theta - \cos\theta$.

11 $\cos\theta - \frac{1}{3}\cos 3\theta$. Sketch the graph.

12 Find the turning point of the function $x\,e^{-x}$ and determine its nature. Show that there is a point of inflexion when $x = 2$ and sketch the curve.

13 Find the turning point of the function $10\arctan x - \frac{1}{2}x^2$ and sketch the curve.

14 Show that the function $x\ln x$ has a minimum at $(1/e, -1/e)$. Given that $x\ln x \to 0$ as $x \to 0$, sketch the graph of the function.

15 Show that $e^x\cos x$ has turning points at intervals of π in x. Distinguish between maxima and minima and show that these values are in a geometrical progression with common ratio $-e^\pi$.

16 A right circular cylinder is inscribed in a sphere of given radius a. Show that the volume of the cylinder is $\pi h(a^2 - \frac{1}{4}h^2)$, where h is the height of the cylinder. Find the ratio of the height to the radius of the cylinder when its volume is greatest.

17 A right circular cylinder is inscribed in a given sphere. Show that, when the area of the curved surface is greatest, the height of the cylinder is equal to its diameter.

18 A funnel is in the form of a right circular cone. If the funnel is to hold a given quantity of fluid, find the ratio of the height to the radius when the area of the curved surface is a minimum.

19 A right circular cone of vertical angle 2θ is inscribed in a sphere of radius a. Show that the area of the curved surface of the cone is $\pi a^2(\sin 3\theta + \sin\theta)$ and prove that its greatest area is $8\pi a^2/(3\sqrt{3})$.

20 An open box has a square horizontal cross-section. If the box is to hold a given amount of material and the internal surface area of the box is to be a minimum, find the ratio of height to the length of the sides.

The *n*th derivative

11.7 Although we have found the first and second derivatives of given functions many, many times in this book, we have hardly ever looked at the third, fourth or even higher derivatives. However there are many occasions when these are required and it is very helpful if we can find a general form of the *n*th derivative (as in other contexts, it should be assumed that n is used here to represent a positive integer). It is convenient to use the notation y_n to represent $\dfrac{d^n y}{dx^n}$.

First let us consider some very simple cases. The simplest of all is $y = e^x$, because this function does not change when it is differentiated, so $y_n = e^x$. Also if $y = e^{ax}$, then $y_n = a^n e^{ax}$.

Simple powers of x are also straight forward. For example if $y = x^4$, then

$$y_1 = 4x^3, \qquad y_2 = 12x^2, \qquad y_3 = 24x, \qquad y_4 = 24, \quad \text{and} \quad y_n = 0 \text{ if } n > 4$$

More generally, if we are given $y = x^N$, then

$$y_1 = Nx^{N-1}, \qquad y_2 = N(N-1)x^{N-2}, \qquad y_3 = N(N-1)(N-2)x^{N-3},$$

and

$$y_n = N(N-1)(N-2)\ldots(N-n+1)x^{N-n}$$

If N is a positive integer then we can write

$$y_n = \frac{N!}{(N-n)!} x^{N-n}, \quad \text{if } n \leqslant N, \text{ and}$$

$$y_n = 0, \quad \text{if } n > N$$

In the case of $y = \sin x$, the successive derivatives are $\pm\cos x$ or $\pm\sin x$, depending on the number of times we have differentiated. To be precise, we can write

$$y_{2n} = (-1)^n \sin x$$

and

$$y_{2n+1} = (-1)^n \cos x, \quad \text{where } n \in \mathbb{N}$$

The reader should be able to work out the corresponding result for $\cos x$.

Example 11   *Given that $y = x^2 e^x$, prove that $y_n = e^x[x^2 + 2nx + n(n-1)]$, for all positive integers n.*

As this result has to be proved for all positive integers, the method of mathematical induction is clearly the best approach.

We are given that $y = x^2 e^x$, so on differentiating we obtain

$$y_1 = x^2 e^x + 2xe^x$$
$$= e^x(x^2 + 2x)$$

Hence the proposition is true for $n = 1$. Now we suppose it is true when $n = N$, i.e.

$$y_N = e^x(x^2 + 2Nx + N^2 - N)$$

Differentiating this by the product rule gives

$$y_{N+1} = e^x(x^2 + 2Nx + N^2 - N) + e^x(2x + 2N)$$
$$= e^x(x^2 + 2Nx + 2x + N^2 + N)$$
$$= e^x[x^2 + 2(N+1)x + N(N+1)]$$

This is the original proposition, but with $n = N + 1$. Hence, by the principle of mathematical induction, the proposition is true for all positive integers.

Leibnitz's theorem

11.8 Example 11 was a particular case of a general rule, known as Leibnitz's theorem, which enables us to write down the nth derivative of a product. In other words, it generalises the product rule.

We shall write, as in the product rule itself,

$$y = uv, \quad \text{where } u \text{ and } v \text{ are functions of } x$$

Then, differentiating once by the product rule, gives

$$y_1 = u_1 v + u v_1$$

Differentiating again gives

$$y_2 = (u_2 v + u_1 v_1) + (u_1 v_1 + u v_2)$$
$$= u_2 v + 2u_1 v_1 + u v_2$$

Differentiating for the third time gives

$$y_3 = u_3 v + 3u_2 v_1 + 3u_1 v_2 + u v_3$$

Now these coefficients should look familiar to the reader; they are the same numbers which appear in the expansion of $(a + b)^3$, i.e.

$$(a + b)^3 = a^3 + 3a^2 b + 3ab^2 + b^3$$

The reader should have no difficulty in verifying that y_4 has the same coefficients as the expansion of $(a + b)^4$. This suggests that in general, the nth derivative will be

$$y_n = u_n v + \binom{n}{1} u_{n-1} v_1 + \binom{n}{2} u_{n-2} v_2 + \ldots + \binom{n}{r} u_{n-r} v_r + \ldots + u v_n$$

where $\binom{n}{r} = \dfrac{n!}{(n-r)! \, r!}$ (see §4.1).

It is not difficult to prove this by mathematical induction; the proof is left as an exercise for the reader.

Qu. 16 Prove Leibnitz's theorem by induction. $\left[\text{Hint: remember that} \right.$

$$\binom{n}{r} + \binom{n}{r-1} = \binom{n+1}{r}. \Bigg]$$

(*Historical note*: Gottfried Wilhelm Leibnitz (1646–1716) was a contemporary of Sir Isaac Newton. They both discovered the subject we now call calculus at about the same time. Leibnitz in particular invented the notation $\dfrac{dy}{dx}$, etc.)

Leibnitz's theorem could have been used to give the result of Example 11. Starting from $y = x^2 e^x$ and applying Leibnitz's theorem, we have

$$y_n = x^2 e^x + \frac{n}{1} \times (2x) \times e^x + \frac{n}{1} \times \frac{n-1}{2} \times 2 \times e^x$$

$$= e^x [x^2 + 2nx + n(n-1)]$$

Example 12 *Given that* $y = f(x) = (\sin^{-1} x)^2$, *show that*

(a) $(1 - x^2)y_1{}^2 = 4y$,

(b) $(1 - x^2)y_2 = xy_1 + 2$,

(c) $(1 - x^2)y_{n+2} - (2n + 1)xy_{n+1} - n^2 y_n = 0$,

(d) $f^{(n+2)}(0) = n^2 f^{(n)}(0)$,

(e) $f^{(2n)}(0) = [(n - 1)!]^2 \times 2^{2n-1}$.

(a) $y = (\sin^{-1} x)^2$,

$$\therefore y_1 = \frac{dy}{dx} = 2(\sin^{-1} x)\frac{1}{\sqrt{(1 - x^2)}}$$

$$\sqrt{(1 - x^2)}y_1 = 2\sin^{-1} x$$

Squaring,

$$(1 - x^2)y_1{}^2 = 4(\sin^{-1} x)^2$$
$$= 4y$$

(b) Differentiating the result of (a) gives

$$2(1 - x^2)y_1 y_2 - 2xy_1{}^2 = 4y_1$$

i.e. $(1 - x^2)y_2 \quad - xy_1 \quad = 2$

$$\therefore (1 - x^2)y_2 = xy_1 + 2$$

(c) We now use Leibnitz's theorem to differentiate the result of (b) n times.

$$(1 - x^2)y_{n+2} + \frac{n}{1}(-2x)y_{n+1} + \frac{n}{1} \times \frac{(n-1)}{2}(-2)y_n = xy_{n+1} + \frac{n}{1} \times y_n$$

$$\therefore (1 - x^2)y_{n+2} - 2nxy_{n+1} - n(n-1)y_n - xy_{n+1} - ny_n = 0$$
$$\therefore (1 - x^2)y_{n+2} - (2n + 1)xy_{n+1} - n^2 y_n = 0$$

(d) Putting $x = 0$, and writing $f^{(n)}(0)$ for the value of y_n when $x = 0$, we have

$$f^{(n+2)}(0) - n^2 f^{(n)}(0) = 0$$

i.e. $f^{(n+2)}(0) \qquad = n^2 f^{(n)}(0)$

(e) We can see from part (a), that $f(0) = 0$, $f^{(1)}(0) = 0$, and from part (b) that $f^{(2)}(0) = 2$. If we now use $f^{(n+2)}(0) = n^2 f^{(n)}(0)$ and start at $f^{(1)}(0) = 0$, we can see that all *odd* derivatives are zero. The even ones will read

$$f^{(2)}(0) = 2$$
$$f^{(4)}(0) = 2^2 \times f^{(2)}(0)$$
$$= 2^2 \times 2$$
$$f^{(6)}(0) = 4^2 f^{(4)}(0)$$
$$= 4^2 \times 2^2 \times 2$$
$$f^{(8)}(0) = 6^2 \times 4^2 \times 2^2 \times 2$$

In general,

$$f^{(2n)}(0) = (2n - 2)^2 \times \dots \times 6^2 \times 4^2 \times 2^2 \times 2$$
$$= (2^{n-1})^2 \times [(n-1)^2 \times \dots \times 3^2 \times 2^2 \times 1^2] \times 2$$
$$= (2^{n-1})^2 \times [(n-1)!]^2 \times 2$$
$$= [(n-1)!]^2 \times 2^{2n-1}$$

Exercise 11c (Miscellaneous)

Differentiate the functions in Nos. 1–7 with respect to x.

1 (a) $\dfrac{x^2}{1 - x^2}$,   (b) $\sin^{-1} \dfrac{1}{x}$,   (c) $e^{2x} \cos 3x$.

2 (a) $\dfrac{x + 1}{\sqrt{(x - 1)}}$,   (b) $\sin^3 x \cos^2 x$,   (c) $x \ln x - x$.

3 (a) $\dfrac{1}{(2x + 1)(3x - 2)}$,   (b) $e^{\cos 2x}$,   (c) $\ln \sin 2x$.

4 (a) $\sqrt{\dfrac{2x + 1}{2x - 1}}$,   (b) $\tan^{-1} \left(\dfrac{1 + x}{1 - x} \right)$,   (c) $\dfrac{e^x}{e^{2x} e^{3x}}$.

5 (a) $(x^2 + 1)^2 (x^3 + 1)^3$,   (b) $\cos^{-1}(\tan x)$,   (c) $\dfrac{\ln x}{x^2}$.

6 (a) $\dfrac{\sin^4 x}{\cos^5 x}$,   (b) $\sin^{-1} \dfrac{x}{\sqrt{(1 + x^2)}}$.

7 (a) 2^{x^2},   (b) x^{2x}.

8 If $y = e^{2x} \cos 3x$, show that $\dfrac{d^2 y}{dx^2} - 4 \dfrac{dy}{dx} + 13y = 0$.

9 If $y = x e^{-x}$, show that

$$\frac{d^2 y}{dx^2} + 2 \frac{dy}{dx} + y = 0 \quad \text{and that} \quad \frac{d^3 y}{dx^3} = 3 \frac{dy}{dx} + 2y$$

10 (a) Show that the gradient of the ellipse $b^2 x^2 + a^2 y^2 = a^2 b^2$ at the point $(a \cos \theta, b \sin \theta)$ is $(-b/a) \cot \theta$ and find an expression for $\dfrac{d^2 y}{dx^2}$ at that point. [For the method, see Book 1, §7.8.]

(b) The equation of a curve is given parametrically by the equations

$$x = \frac{t^2}{1 + t^3}, \qquad y = \frac{t^3}{1 + t^3}$$

Show that $\dfrac{dy}{dx} = \dfrac{3t}{2 - t^3}$, and that $\dfrac{d^2 y}{dx^2} = 48$ at the point $(\tfrac{1}{2}, \tfrac{1}{2})$.

11 Find $\dfrac{d^2 y}{dx^2}$ in terms of the parameter when

(a) $x = t^2 - 4, \quad y = t^3 - 4t,$

(b) $x = \cos^3 t, \quad y = \sin^3 t,$

(c) $x = a(\theta - \sin \theta), \quad y = a(1 - \cos \theta).$

12 (a) Differentiate sin x from first principles.

   (b) Prove that $\dfrac{d}{dx}(uv) = v\dfrac{du}{dx} + u\dfrac{dv}{dx}.$

13 (a) Differentiate tan x from first principles.

   (b) Prove that $\dfrac{d}{dx}\left(\dfrac{u}{v}\right) = \left(v\dfrac{du}{dx} - u\dfrac{dv}{dx}\right)\bigg/ v^2.$

14 Differentiate $\arcsin x + x\sqrt{(1 - x^2)}$ with respect to x and hence find

$$\int \sqrt{(1 - x^2)}\, dx$$

15 Differentiate $x \arctan x - \frac{1}{2}\ln(1 + x^2)$ with respect to x. Hence write down $\int \arctan x\, dx$.

16 Find the maximum and minimum values of the function

$$(a + b \sin x)/(b + a \sin x)$$

where $b > a > 0$ in the interval $0 \leqslant x \leqslant 2\pi$. Sketch the graph when $a = 4$, $b = 5$.

17 Find the maximum and minimum values of y given by the equation

$$x^3 + y^3 - 3xy = 0$$

18 Investigate the stationary values of $\dfrac{x^3}{1 + x^2}$ and sketch the graph of

$$y = \frac{x^3}{1 + x^2}.$$

19 Given that $y = \dfrac{x}{1 - x}$, prove that $y_n = \dfrac{n!}{(1 - x)^{n+1}}.$

20 Show that

(a) $\dfrac{d}{dx}(\sin x) = \sin(x + \pi/2),$ (b) $\dfrac{d^n}{dx^n}(\sin x) = \sin(x + n\pi/2),$

and find similar expressions for $\dfrac{d}{dx}(\cos x)$ and $\dfrac{d^n}{dx^n}(\cos x).$

21 Prove that

(a) $\dfrac{d^n}{dx^n}(e^x \cos x) - 2^{n/2} e^x \cos(x + n\pi/4),$

(b) $\dfrac{d^n}{dx^n}(e^{ax} \sin bx) = (a^2 + b^2)^{n/2} e^{ax} \sin(bx + n\alpha),$ where $\tan \alpha = (b/a).$

22 Show that the function $y = \tan x - 8 \sin x$ has two stationary values between $x = 0$ and $x = 2\pi$. Draw a rough graph of the function between these values of x and show that, if the equation

$$\tan x - 8 \sin x = b$$

has four real roots between 0 and 2π, then $-3\sqrt{3} < b < 3\sqrt{3}.$ (O)

23 By first putting the expression into partial fractions, or otherwise, find the first and second derivatives of the function

$$\frac{3x - 1}{(4x - 1)(x + 5)}$$

Find the coordinates of any maxima, minima and points of inflexion that the function may have, and draw a rough sketch of its graph.   (O)

24 Prove that, if $f'(a) = 0$ and $f''(a)$ is negative, then the graph of the function $f(x)$ has a maximum at the point whose abscissa is a.

Prove that the function

$$y = \frac{\sin x \cos x}{1 + 2 \sin x + 2 \cos x}$$

has turning points in the range $0 \leqslant x \leqslant 2\pi$ when $x = \frac{1}{4}\pi$ and $x = \frac{5}{4}\pi$, distinguishing between maximum and minimum values.

Prove that the tangents at the origin and at the point $(\frac{1}{2}\pi, 0)$ meet at a point whose abscissa is $\frac{1}{4}\pi$.   (O & C)

25 Find the shortest distance between two points, one of which lies on the parabola $y^2 = 4ax$, and the other on the circle $x^2 + y^2 - 24ay + 128a^2 = 0$.

26 Chords of the hyperbola $xy = c^2$ cut both the branches and pass through the point $(2\sqrt{3}c, 0)$. Find the length of the shortest of these chords.

27 Given the four points $P_1 (x_1, y_1)$, $P_2 (x_2, y_2)$, $P_3 (x_3, y_3)$, $P_4 (x_4, y_4)$, show that the variable point $P (x, y)$, whose parametric form is given by

$$\begin{pmatrix} x \\ y \end{pmatrix} = \begin{pmatrix} x_1 & x_2 & x_3 & x_4 \\ y_1 & y_2 & y_3 & y_4 \end{pmatrix} \begin{pmatrix} 1 - 3t + 3t^2 - t^3 \\ 3t - 6t^2 + 3t^3 \\ 3t^2 - 3t^3 \\ t^3 \end{pmatrix} \qquad \text{where } 0 \leqslant t \leqslant 1$$

has the following properties:

(a) P_1 is an end point of the curve (when $t = 0$),
(b) P_4 is the other end point (when $t = 1$),
(c) the gradient at P_1 equals the gradient of $\overrightarrow{P_1 P_2}$,
(d) the gradient at P_4 equals the gradient of $\overrightarrow{P_3 P_4}$.

(Such a curve is called a Bezier curve; it is widely used in computer-aided design.)

Find the parametric form of the curve when P_1, P_2, P_3 and P_4 are the points $(0, 1)$, $(1, 2)$, $(4, 0)$ and $(5, 1)$ respectively. Draw a diagram showing P_1, P_2, P_3, P_4, the line segments $P_1 P_2$ and $P_3 P_4$, and a sketch of the curve.

28 Given that $(x + 1)y = x$, prove that $(x + 1)\dfrac{dy}{dx} + y = 1$ and use induction to prove that

$$(x + 1)\frac{d^n y}{dx^n} + n \frac{d^{n-1} y}{dx^{n-1}} = 0 \qquad (n \geqslant 2) \qquad\qquad \text{(O \& C)}$$

29 Given that $y = \dfrac{1}{\sqrt{(1 + x^2)}}$, prove that $(1 + x^2)y_1 + xy - 0$ and hence show that

$$(1 + x^2)y_n + (2n - 1)xy_{n-1} + (n - 1)^2 y_{n-2} = 0$$

30 Use Leibnitz's theorem to find

(a) $\dfrac{d^3}{dx^3}(x^3 e^x)$,   (b) $\dfrac{d^4}{dx^4}(x \cos x)$,   (c) $\dfrac{d^n}{dx^n}(x^2 e^{2x})$,

(d) $\dfrac{d^n}{dx^n}\{y(1 - x^2)\}$.

Chapter 12

Further trigonometry

General solutions of trigonometrical equations

12.1 The purpose of this chapter is to revise Chapters 16–19 of Book 1, for the benefit of readers who may need it. Since no new work is involved, it consists solely of examples and exercises. At this level of the subject, it should be assumed that radians are required, unless the question explicitly refers to degrees. In this chapter there is more emphasis on *general* solutions; not all examining boards insist on these, but they are not difficult and can be instructive. They are most easily obtained by referring to the graph of the appropriate function.

If $x = \alpha$ is a solution of the equation

$$\sin x = k, \quad \text{where } |k| \leqslant 1$$

then so is $\pi - \alpha$ (see Fig. 12.1, which is not drawn to scale), and since the function $\sin x$ is periodic (see Book 1, §2.15), with a period of 2π, any multiple of 2π may be added to these solutions. So the general solution can be expressed

$$x = \alpha + 2n\pi$$
$$\text{or} \quad x = (\pi - \alpha) + 2n\pi, \quad \text{where } n \in \mathbb{Z}$$

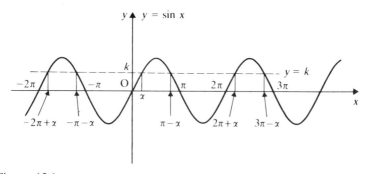

Figure 12.1

Similarly, if α is a solution of $\cos x = k$, then so is $-\alpha$ (see Fig. 12.2, which is not drawn to scale), so the general solution of this equation can be expressed

$$x = \pm\alpha + 2n\pi$$

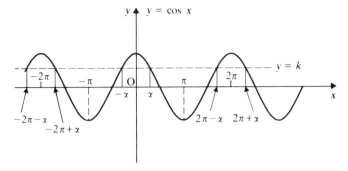

Figure 12.2

The equation tan $x = k$ can be solved for *any* real value of k. However, unlike sin x and cos x, the period of tan x is π (not 2π), so if $x = \alpha$ is a solution, we can add any multiple of π to it, hence the general solution (see Fig. 12.3) is

$$x = \alpha + n\pi$$

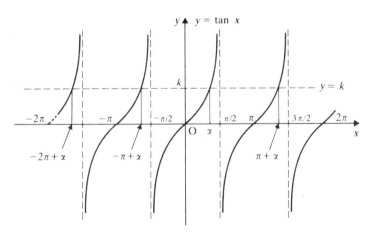

Figure 12.3

Qu. 1 Write down the general solutions of the following equations:
(a) sin $\theta = 0$,   (b) cos $\theta - -1$,   (c) tan $\theta = 1$,
(d) sin $\theta = 1$,   (e) cos $\theta = 0$,   (f) sin $\theta = \frac{1}{2}$,
(g) tan $\theta = -1$,   (h) sin $\theta = 1/\sqrt{2}$,   (i) cos $\theta = -1/\sqrt{2}$.

Example 1 *Find the general solution of the equation* cos $2\theta + \sin \theta = 0$.

$$\cos 2\theta + \sin \theta = 0$$
$$\therefore\ 1 - 2\sin^2 \theta + \sin \theta = 0$$
$$\therefore\ 2\sin^2 \theta - \sin \theta - 1 = 0$$
$$\therefore\ (\sin \theta - 1)(2\sin \theta + 1) = 0$$

(a) If $\sin\theta = 1$, then $\theta = (2n + \frac{1}{2})\pi = (4n + 1)\pi/2$.

(b) If $\sin\theta = -\frac{1}{2}$, then $\theta = n\pi - (-1)^n\frac{1}{6}\pi$.

These may be combined as $\theta = \frac{1}{2}\pi + \frac{2}{3}n\pi$.

Example 2   *Find the general solution of the equation* $\sin 3\theta + \sin 2\theta = 0$.

$\sin 3\theta + \sin 2\theta = 0$

$\therefore\ 2\sin\frac{5}{2}\theta\cos\frac{1}{2}\theta = 0$

(a) If $\sin\frac{5}{2}\theta = 0$, then $\frac{5}{2}\theta = n\pi$.

$\therefore\ \theta = 2n\pi/5$

(b) If $\cos\frac{1}{2}\theta = 0$, then $\frac{1}{2}\theta = n\pi + \frac{1}{2}\pi$.

$\therefore\ \theta = (2n + 1)\pi$

Therefore the general solution is $\theta = 2n\pi/5$ or $(2n + 1)\pi$.

Example 3   *Solve the equation* $4\cos x - 6\sin x = 5$, *for values of x between* $0°$ *and* $360°$ *correct to* $0.1°$.

$4\cos x - 6\sin x = 5$

Divide both sides by $\sqrt{(4^2 + 6^2)} = \sqrt{52} = 2\sqrt{13}$.

$$\therefore\ \frac{2}{\sqrt{13}}\cos x - \frac{3}{\sqrt{13}}\sin x = \frac{5}{2\sqrt{13}} \qquad (1)$$

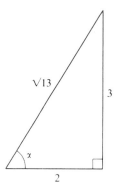

Figure 12.4

From Fig. 12.4, $\tan\alpha = 3/2$. From tables or a calculator $\alpha = 56.31°$, correct to two decimal places, but we shall delay substituting the numerical value of α as long as possible. Equation (1) can now be written

$$\cos x \cos \alpha - \sin x \sin \alpha = \frac{5}{2\sqrt{13}}$$

$$\therefore \cos (x + \alpha) = \frac{5}{2\sqrt{13}}$$

$$\therefore (x + \alpha) = 46.10°, 313.90°, 406.10°,\dots$$

Now, subtracting α $(= 56.31)$ from both sides, we obtain

$$x = -10.21°, 257.59°, 349.79°,\dots$$

So the solution, correct to 0.1°, within the required range of values is

$$x = 257.6° \quad \text{or} \quad 349.8°$$

Exercise 12a

Solve the following equations. Solutions are given in radians unless tables (or a calculator) have been used, in which case the answers are given in degrees.

1 $\cos 2\theta = \frac{1}{2}$.
2 $\tan \theta = \cos 130°$.
3 $\cos \theta = -\tan 146°$.
4 $\sin 2\theta = \sin \theta$.
5 $\cos 2\theta = \sin \theta$.
6 $\sin 2\theta = \cos \theta$.
7 $\cos \theta = 3 \tan \theta$.
8 $2 \tan \theta = \tan 2\theta$.
9 $\cos \theta = \cos 3\theta$.
10 $\sin 3\theta + \sin \theta = 0$.
11 $\sin \theta + \sin 3\theta + \sin 5\theta = 0$.
12 $\cos \theta + \sin 2\theta - \cos 3\theta = 0$.
13 $\cos 4\theta + 1 = 2 \cos^2 \theta$.
14 $1 - \sin \theta = 2 \cos^2 \theta$.
15 $2 \cos^2 \theta = 1 + \sin \theta$.
16 $2 \tan \theta = 1 - \tan^2 \theta$.
17 $3 \tan^2 \theta = 2 \sin \theta$.
18 $2 \sin^2 \theta + 3 \cos \theta = 3$.
19 $3 \cot^2 \theta + 5 = 7 \csc \theta$.
20 $\sin 2\theta = \cot \theta$.
21 $2 \csc^2 \theta = 5(\cot \theta + 13)$.
22 $\cos 2\theta = 5 \sin \theta + 3$.
23 $\cos \theta + \sqrt{3} \sin \theta = 1$.
24 $7 \cos \theta = 5 + \sin \theta$.
25 $7 \sin \theta - 24 \cos \theta = 25$.

Qu. 2 Write down the values of
(a) $\tan^{-1}(-1)$,   (b) $\sin^{-1}(-\frac{1}{2})$,   (c) $\cos^{-1}(-1)$,
(d) $\tan^{-1}(\sqrt{3})$,   (e) $\sin^{-1} 1$,   (f) $\cos^{-1} 1$.

Remember that the notation arcsin x, etc., is an alternative to $\sin^{-1} x$, etc. Both forms are in common use, and readers should be prepared to see either in examination papers.

Qu. 3 Find the values of
(a) arctan $\frac{1}{3}$,   (b) arccos $(-\frac{1}{4})$,   (c) arcsin (0.01),
(d) arcsin $(-\frac{1}{5})$,   (e) arctan (-3),   (f) arccos $\frac{3}{5}$.

Example 4 *Find the general solution of the equation* $4 \cos 3\theta + 3 \cos \theta = 0$.

$$4 \cos 3\theta + 3 \cos \theta = 0$$
$$\therefore\ 4(4 \cos^3 \theta - 3 \cos \theta) + 3 \cos \theta = 0$$
$$\therefore\ 16 \cos^3 \theta - 9 \cos \theta = 0$$
$$\therefore\ \cos \theta(16 \cos^2 \theta - 9) = 0$$

$$\therefore\ \cos \theta = 0, \quad \pm \tfrac{3}{4}$$
$$\therefore\ \theta = n\pi + \tfrac{1}{2}\pi, \quad n\pi \pm \arccos \tfrac{3}{4}$$

Example 5 *Solve the equation* $\sin 3\theta = \sin \theta$.

$$\sin 3\theta = \sin \theta$$
$$\therefore\ 3\theta = 2n\pi + \theta \quad \text{or} \quad 3\theta = (2n+1)\pi - \theta$$
$$\therefore\ 2\theta = 2n\pi \quad\quad \text{or} \quad 4\theta = (2n+1)\pi$$
$$\therefore\ \theta = n\pi, \quad (2n+1)\pi/4$$

Qu. 4 Solve the equation $\sin 3\theta = \sin \theta$ by means of the identity
$$\sin 3\theta = 3 \sin \theta - 4 \sin^3 \theta$$

Example 6 *Solve the equation* $\cos 5\theta = \sin 4\theta$.

The identity $\sin \phi = \cos (\tfrac{1}{2}\pi - \phi)$ is used.

$$\cos 5\theta = \sin 4\theta$$
$$\therefore\ \cos 5\theta = \cos (\tfrac{1}{2}\pi - 4\theta)$$
$$\therefore\ 5\theta = 2n\pi \pm (\tfrac{1}{2}\pi - 4\theta)$$

$$\therefore\ 5\theta = 2n\pi + \tfrac{1}{2}\pi - 4\theta \quad \text{or} \quad 5\theta = 2n\pi - \tfrac{1}{2}\pi + 4\theta$$
$$\therefore\ 9\theta = (4n+1)\pi/2 \quad\quad \text{or} \quad \theta = (4n-1)\pi/2$$

$$\therefore\ \theta = (4n+1)\pi/18, \quad (4n-1)\pi/2$$

Qu. 5 Solve Example 6 by writing $\cos 5\theta = \sin (\tfrac{1}{2}\pi - 5\theta)$.

Example 7 *Show that* $2 \tan^{-1} 2 + \tan^{-1} 3 = \pi + \tan^{-1} \tfrac{1}{3}$.

Let $A = \tan^{-1} 2$, $B = \tan^{-1} 3$, then $\tan A = 2$, $\tan B = 3$, and the left-hand side of the equation is $2A + B$. We shall find $\tan (2A + B)$.

$$\tan 2A = \frac{2 \tan A}{1 - \tan^2 A} = \frac{4}{1 - 4} = -\frac{4}{3}$$

$$\tan (2A + B) = \frac{\tan 2A + \tan B}{1 - \tan 2A \tan B} = \frac{-\frac{4}{3} + 3}{1 + 4} = \frac{1}{3}$$

$$\therefore\ 2A + B = n\pi + \tan^{-1} \tfrac{1}{3} \quad \text{for an appropriate value of } n$$

Now $\frac{1}{4}\pi < A < \frac{1}{2}\pi$, $\frac{1}{4}\pi < B < \frac{1}{2}\pi$.

$\therefore \frac{3}{4}\pi < 2A + B < \frac{3}{2}\pi$

$\therefore 2A + B = \pi + \tan^{-1}\frac{1}{3}$

$\therefore 2\tan^{-1}2 + \tan^{-1}3 = \pi + \tan^{-1}\frac{1}{3}$

Exercise 12b

1 Write down the values of
   (a) $\sin^{-1}(\frac{1}{2}\sqrt{3})$, (b) $\cos^{-1}(\frac{1}{2}\sqrt{2})$, (c) $\tan^{-1}(1/\sqrt{3})$,
   (d) $\cos^{-1}0$, (e) $\tan^{-1}(-\sqrt{3})$, (f) $\sin^{-1}(-1)$.
2 Use tables or a calculator to evaluate:
   (a) $\arctan 2$, (b) $\arcsin\frac{3}{5}$, (c) $\arccos\frac{2}{5}$,
   (d) $\arcsin(-\frac{1}{3})$, (e) $\arctan(-\frac{1}{2})$, (f) $\arccos\frac{1}{4}$.

Find the general solutions of the following equations in θ:

3 $3\sin 2\theta = \sin\theta$. 4 $3\cos 2\theta + 2\cos\theta = 1$.
5 $9\sin 3\theta = 2\sin\theta$. 6 $\tan 2\theta = 4\tan\theta$.
7 $\cos\theta + \sin\theta = \sqrt{2}$. 8 $\cos\theta - \sqrt{3}\sin\theta = 1$.
9 $\sin(\theta + \alpha) = 2\sin(\theta - \alpha)$. 10 $3\cos(\theta - \alpha) = 4\sin(\alpha - \theta)$.
11 $1 + 2\cos 2\theta = \cos 2\alpha + 2\cos\theta\cos\alpha$.
12 $4\cos\theta - 7\sin\theta + 8 = 0$. 13 $3\cos\theta = 7(\sin\theta - 1)$.
14 $\cos 3\theta = \cos\theta$. 15 $\sin 3\theta = \sin 2\theta$.
16 $\tan 4\theta = \tan\theta$. 17 $\sin 3\theta = \cos 2\theta$.
18 $\cos 4\theta = \sin 3\theta$. 19 $\tan 4\theta + \tan 2\theta = 0$.
20 $\sin 2\theta + \cos 3\theta = 0$.
21 Find the general solution of the equation $\cos 3\theta = \cos 2\theta$. Also express the
   equation as an equation for $\cos\theta$ and hence show that $\cos\frac{2}{5}\pi = \frac{1}{4}(-1 + \sqrt{5})$.
   Find a similar expression for $\cos\frac{4}{5}\pi$.

Prove the relations in Nos. 22–25.

22 $2\tan^{-1}\frac{1}{2} = \tan^{-1}\frac{4}{3}$. 23 $\tan^{-1}\frac{1}{2} + \tan^{-1}\frac{1}{3} = \frac{1}{4}\pi$.
24 $\frac{1}{4}\pi - \tan^{-1}\frac{2}{3} = \tan^{-1}\frac{1}{5}$. 25 $3\tan^{-1}2 - \pi = \tan^{-1}(\frac{2}{11})$.

Exercise 12c (Miscellaneous)

1 Find all the angles between $0°$ and $360°$ which satisfy the equations:
   (a) $\sin\theta = \cos 127°$, (b) $3\cot^2\theta = 2\cos\theta$,
   (c) $3\sin\theta - 4\cos\theta = 2$. (O & C)
2 Find all the values of x between $0°$ and $180°$ inclusive for which
   (a) $\sin 3x = \sin x$, (b) $2\cos^2 x - \sin^2 x = 1$,
   (c) $\sin 2x + \cos x = 0$.
3 Find all the values of x between $0°$ and $360°$ inclusive which satisfy the
   equations:
   (a) $\sin x + \cos x = \sin 18° + \cos 18°$,
   (b) $\sin 2x = \cos x \sin 3x$. (L)

4 Find the general solutions of the equations:
(a) $\cos x + \cos 3x = \cos 2x$,   (b) $4 \cos x - 3 \sin x = 2$.   (C)

5 Find all the values of x (in radian measure) which satisfy the following equations:
(a) $\sin x + \cos 4x = 0$,   (b) $\cos 2x - 5 \cos x + 3 = 0$,
(c) $\tan 2x + \tan 4x = 0$.   (O & C)

6 Give the general solutions of the following equations:
(a) $2 \sin 3\theta - 7 \cos 2\theta + \sin \theta + 1 = 0$,
(b) $\cos \theta - \sin 2\theta + \cos 3\theta - \sin 4\theta = 0$.   (C)

7 (a) Express $\cos 5\theta$ in terms of $\cos \theta$. Hence prove that

$$\cos 18° = \tfrac{1}{4}\sqrt{(10 + 2\sqrt{5})}$$

and evaluate $\cos 54°$, leaving your answer in surd form.
(b) Solve the equation $\sin 4x - \sin 3x + \sin 2x = 0$ completely, expressing your answers in radian measure.   (O & C)

8 Write down and solve the quadratic equation in x whose roots p and q are given by the relations

$$p + q = \tfrac{1}{2}, \qquad pq = -\tfrac{1}{2}$$

Use your result to find the simultaneous values of θ and ϕ which satisfy the equations

$$\cos \theta + \cos \phi = \tfrac{1}{2}, \qquad \cos \theta \cos \phi = -\tfrac{1}{2}$$

and lie in the interval 0 to π inclusive.   (JMB)

9 (a) Find all the solutions of the equation

$$\sin 60° + \sin (60° + x) + \sin (60° + 2x) = 0$$

which lie between $0°$ and $360°$.
(b) Prove that $(\sin 3\theta)/(1 + 2 \cos 2\theta) = \sin \theta$ and hence show that $\sin 15° = (\sqrt{3} - 1)/(2\sqrt{2})$.   (L)

10 Prove the formula $\tan 3\theta = \dfrac{3 \tan \theta - \tan^3 \theta}{1 - 3 \tan^2 \theta}$.

Find the general solution of the equation $\tan 3\theta + \tan 2\theta = 0$ and show that $\tan^2 \tfrac{1}{5}\pi$, $\tan^2 \tfrac{2}{5}\pi$ are the roots of the quadratic equation

$$x^2 - 10x + 5 = 0$$ (O & C)

11 Find all the solutions in the interval $0° \leqslant x \leqslant 360°$ of the equations
(a) $\tan 3x + 1 = 0$,
(b) $2 \cos^2 x + 3 \sin x = 0$.   (O & C)

12 Find all the solutions of the equation $1 + \cos 2A = 2 \sin 2A$ in the interval $0° \leqslant A \leqslant 360°$.   (O & C: SMP)

13 By expressing the equations in the form $\sin (A + B) = R$, or otherwise, find all the angles which satisfy
(a) $\sin x + \sqrt{3} \cos x = 1$,   in $0 < x < 2\pi$,
(b) $\sin x (\sin x + \cos x) = 1$,   in $\pi < x < 2\pi$.

Find the set of values of k for which the equation

$$\sin x(\sin x + \cos x) = k$$

has real roots. (L)

14 (a) Express $\sqrt{3} \sin x + \cos x$ in the form $r \sin(x + \theta)$, where $r > 0$ and $0 < \theta < \pi/2$. Hence, or otherwise, find the general solution for x of the equation

$$\sqrt{3} \sin x + \cos x = \sin \alpha + \sqrt{3} \cos \alpha$$

(b) Express the products $\cos A \cos B$ and $\sin A \sin B$ each in terms of $\cos(A + B)$ and $\cos(A - B)$. Hence, or otherwise, find the values of θ, in the interval $\pi/2 < \theta < 3\pi/2$, which satisfy the equation

$$\cot \theta \cot(\theta - \pi/6) = 2 + \sqrt{3}$$ (L)

15 Given that $7 \sin^2 x - 5 \sin x + \cos^2 x = 0$, find all the possible values of $\sin x$. (L)

16 (a) Show that, if $\sin 2x = 0$, then $\sin 5x = \sin x$.
   Is it also true that, if $\sin 5x = \sin x$, then $\sin 2x = 0$? Justify your answer.
   (b) Find all values of x, to the nearest degree, between $0°$ and $360°$ for which $2 \tan 3x + 1 = 0$. (O & C)

17 Express $\tan(45° + x)$ in terms of $\tan x$. Hence, or otherwise,
   (a) express $\tan 75°$ in the form $a + b\sqrt{3}$, where a and b are integers;
   (b) express $\tan(45° + x) + \cot(45° + x)$ in terms of $\cos 2x$. (JMB)

18 In the triangle ABC, AB is of one unit length and $BC = CA = p$. The point P lies in AB at a distance x from A and is such that $\angle ACP = \theta$ and $\angle BCP = 2\theta$. By using the sine rule, or otherwise, show that

$$\cos \theta = \frac{1 - x}{2x}$$

State the possible values of θ as p varies, and deduce that $\frac{1}{3} < x < \frac{1}{2}$. Express $\cos 3\theta$ in terms of p. Hence, or otherwise, find the value of x correct to two decimal places, when $p = 1/\sqrt{2}$. (JMB)

19 Using the expressions for $\sin(A + B)$ and $\cos(A + B)$, derive the following results:
   (a) $\sin 2A = 2 \sin A \cos A$, (b) $\cos 2A = \cos^2 A - \sin^2 A$,
   (c) $\sin 3A = 3 \sin A \cos^2 A - \sin^3 A$,
   (d) $\cos 3A = \cos^3 A - 3 \cos A \sin^2 A$.
   From results (c) and (d) above, obtain an expression for $\tan 3A$ in terms of $\tan A$.
   Find all the values of x in the interval $0° \leqslant x \leqslant 180°$, for which

$$\tan 3x + 2 \tan x = 0$$

giving your answers to the nearest $0.1°$ where necessary. (C)

20 Find the sets of values of x that satisfy the following inequalities:
   (a) $8x^3 - 4x^2 - 6x + 3 > 0$,
   (b) $8 \sin^3 x - 4 \sin^2 x - 6 \sin x + 3 > 0$,
   (c) $8 \sec^6 x - 4 \sec^4 x - 6 \sec^2 x + 3 > 0$. (C)

Chapter 13

Further integration

Integration by parts

13.1 We have learnt the importance of recognising such integrals as

$$\int x \, e^{x^2} \, dx = \tfrac{1}{2}e^{x^2} + c \quad \text{and} \quad \int 2x \cos(x^2 + 2) \, dx = \sin(x^2 + 2) + c$$

When, however, the integrand is the product of two functions of x but is not susceptible to this treatment, e.g. $\int x \, e^x \, dx$, $\int x \cos x \, dx$, we may often successfully apply a technique known as *integration by parts*; this is based upon the idea of differentiating the product of two functions of x.

If u and v are two functions of x,

$$\frac{d}{dx}(uv) = v\frac{du}{dx} + u\frac{dv}{dx}$$

Integrating each side with respect to x,

$$uv = \int v\frac{du}{dx}\,dx + \int u\frac{dv}{dx}\,dx$$

$$\therefore \int u\frac{dv}{dx}\,dx = uv - \int v\frac{du}{dx}\,dx$$

Example 1 *Find $\int x \cos x \, dx$.*

$$\int u\frac{dv}{dx}\,dx = uv - \int v\frac{du}{dx}\,dx \qquad\qquad \text{Let } u = x.$$

$$\int x \cos x \, dx = x \sin x - \int \sin x \times 1 \, dx \qquad \text{Let } \frac{dv}{dx} = \cos x,$$

$$= x \sin x + \cos x + c \qquad\qquad\qquad \therefore v = \sin x.$$

This method can of course only be attempted if the factor chosen as $\dfrac{dv}{dx}$ can be integrated; Example 1 illustrates the fact that its successful application usually depends upon the *correct choice of u*, since it is this which determines whether $\int v \dfrac{du}{dx} dx$ is easier to tackle than the original integral.

Qu. 1 Check the answer to Example 1 by differentiation.
Qu. 2 Attempt Example 1 taking cos x as u.
Qu. 3 Find the following integrals:
(a) $\int x \sin x \, dx$, (b) $\int x \cos 2x \, dx$, (c) $\int x \ln x \, dx$, (d) $\int x \, e^x \, dx$.

Qu. 4 Find $\dfrac{d}{dx} (e^{x^2})$, and deduce $\int x^3 \, e^{x^2} \, dx$.

The integral $\int \tan^{-1} x \, dx$ does not at first sight appear to be susceptible to the method under discussion. However, this is one of a small group of integrals which may be found by taking $\dfrac{dv}{dx}$ as 1.

Example 2 *Find* $\int \tan^{-1} x \, dx$.

$$\int u \frac{dv}{dx} dx = uv - \int v \frac{du}{dx} dx$$

$$\int \tan^{-1} x \times 1 \, dx = \tan^{-1} x \times x - \int x \times \frac{1}{1+x^2} dx$$

$$= x \tan^{-1} x - \tfrac{1}{2} \ln (1 + x^2) + c$$

Let $u = \tan^{-1} x$.
Let $\dfrac{dv}{dx} = 1$,
$\therefore v = x$.

Qu. 5 Find $\int \ln x \, dx$.

Qu. 6 (a) Find $\dfrac{d}{dx} (\sin^{-1} x)$, (b) find $\int \sin^{-1} x \, dx$.

To some integrals it is necessary to apply the method of integration by parts more than once, as is illustrated in the next example.

Example 3 *Find* $\int x^2 \sin x \, dx$.

$$\int x^2 \sin x \, dx = x^2(-\cos x) - \int -\cos x \times 2x \, dx$$

$$= -x^2 \cos x + \int 2x \cos x \, dx$$

$$\int 2x \cos x \, dx = 2x \sin x - \int \sin x \times 2 \, dx$$

$$= 2x \sin x + 2 \cos x + c$$

$$\therefore \int x^2 \sin x \, dx = -x^2 \cos x + 2x \sin x + 2 \cos x + c$$

$$= 2x \sin x + (2 - x^2) \cos x + c$$

Let $u = x^2$.
Let $\dfrac{dv}{dx} = \sin x$,
$\therefore v = -\cos x$.

Let $u = 2x$.
Let $\dfrac{dv}{dx} = \cos x$,
$v = \sin x$.

Qu. 7 Check the answer to Example 3 by differentiation.
Qu. 8 Find the following integrals:
(a) $\int x^2 \cos x \, dx$, (b) $\int x^2 \, e^x \, dx$.

Exercise 13a

1 Find the following integrals, and check by differentiation:

(a) $\int 2x \sin x \, dx$,   (b) $\int \frac{1}{2}x \, e^x \, dx$,   (c) $\int x \sin 2x \, dx$,

(d) $\int x^2 \ln x \, dx$,   (e) $\int x \cos (x + 2) \, dx$,   (f) $\int x(1 + x)^7 \, dx$,

(g) $\int x \, e^{2x} \, dx$,   (h) $\int x \, e^{x^2} \, dx$,   (i) $\int \dfrac{\ln x}{x^2} \, dx$,

(j) $\int x \sec^2 x \, dx$,   (k) $\int x^n \ln x \, dx$,   (l) $\int x \, 3^x \, dx$.

2 Find the following integrals, and check by differentiation:

(a) $\int \ln 2x \, dx$,   (b) $\int \sin^{-1} 3x \, dx$,   (c) $\int \ln y^2 \, dy$,

(d) $\int \tan^{-1} \dfrac{\theta}{2} \, d\theta$,   (e) $\int \cos^{-1} t \, dt$,   (f) $\int e^{\sqrt{x}} \, dx$.

3 Find the following integrals (see Qu. 4 on p. 229):

(a) $\int x^5 \, e^{x^3} \, dx$,   (b) $\int x \, e^{-x^2} \, dx$,   (c) $\int x^3 \, e^{-x^2} \, dx$,

(d) $\int x^3 \cos x^2 \, dx$,   (e) $\int x^3 \sec^2 (x^2) \, dx$.

4 Find the following integrals:

(a) $\int x^2 \cos 3x \, dx$,   (b) $\int x^3 \, e^x \, dx$,

(c) $\int x^2 \sin x \cos x \, dx$,   (d) $\int x^2 \, e^{-x} \, dx$,

(e) $\int (x \cos x)^2 \, dx$,   (f) $\int x \, (\ln x)^2 \, dx$.

5 Find the following integrals:

(a) $\int x \sin x \cos x \, dx$,   (b) $\int \dfrac{x}{e^x} \, dx$,   (c) $\int x(1 + 2x)^5 \, dx$,

(d) $\int \dfrac{\ln y}{y} \, dy$,   (e) $\int u \tan^{-1} u \, du$,   (f) $\int x \, e^{-x^2} \, dx$,

(g) $\int x^3 \, e^{-x} \, dx$,   (h) $\int x(1 - x^2)^6 \, dx$,   (i) $\int t \sin^2 t \, dt$,

(j) $\int v \, e^{3v} \, dv$.

6 (a) Find $\int x \tan^2 x \, dx$.

   (b) Show that $\int x \sin^{-1} x \, dx = \frac{1}{4}(2x^2 - 1) \sin^{-1} x + \frac{1}{4}x\sqrt{(1 - x^2)} + c$.

7 Evaluate:

(a) $\displaystyle\int_0^{\pi/2} x \cos x \, dx$,   (b) $\displaystyle\int_0^1 x^2 \, e^x \, dx$,   (c) $\displaystyle\int_1^{e^2} \ln x \, dx$,

(d) $\displaystyle\int_0^1 \sin^{-1} y \, dy$,   (e) $\displaystyle\int_0^{\pi} t \sin^2 t \, dt$,   (f) $\displaystyle\int_1^{10} x \log_{10} x \, dx$.

Involving inverse trigonometrical functions

13.2   In §11.3 we dealt with the differentiation of inverse trigonometrical functions. The frequency with which we meet inverse sine and inverse tangent functions in integration is just one good reason why we should be adept at differentiating these functions on sight.

   If $y = \sin^{-1} u$, where u is a function of x,

$$\frac{dy}{dx} = \frac{dy}{du} \times \frac{du}{dx} = \frac{1}{\sqrt{(1-u^2)}} \times \frac{du}{dx}$$

Thus $\dfrac{d}{dx}\left\{3\sin^{-1}\dfrac{x}{2}\right\} = 3\,\dfrac{1}{\sqrt{\left(1-\dfrac{x^2}{4}\right)}} \times \dfrac{1}{2} = \dfrac{3}{\sqrt{(4-x^2)}}$

and $\dfrac{d}{dx}\{2\tan^{-1}5x\} = 2 \times \dfrac{1}{1+25x^2} \times 5 = \dfrac{10}{1+25x^2}$

Qu. 9 Write down, and simplify where necessary, the derivatives of the following functions:

(a) $\sin^{-1}3x$,　　(b) $\tan^{-1}2x$,　　(c) $\sin^{-1}\dfrac{x}{3}$,　　　(d) $\cos^{-1}2x$,

(e) $\tfrac{1}{2}\tan^{-1}3x$,　(f) $3\tan^{-1}\dfrac{x}{2}$,　(g) $\tfrac{1}{2}\sin^{-1}(x-1)$,　(h) $2\tan^{-1}\left(\dfrac{x+1}{2}\right)$.

The reader should now be able to write down certain integrals, hitherto obtained by the change of variable $x = k\sin u$ or $x = k\tan u$.

For example, $\displaystyle\int \dfrac{2}{3+4x^2}\,dx$, written as $\displaystyle\int \dfrac{\frac{2}{3}}{1+\frac{4}{3}x^2}\,dx$, is seen to be of the form

$k\tan^{-1}\dfrac{2x}{\sqrt{3}} + c$. Now,

$$\frac{d}{dx}\left(k\tan^{-1}\frac{2x}{\sqrt{3}}\right) = \frac{1}{1+\frac{4}{3}x^2} \times \frac{2k}{\sqrt{3}}$$

Comparing this with the integrand we find $k = 1/\sqrt{3}$,

$$\therefore \int \frac{2}{3+4x^2}\,dx = \frac{1}{\sqrt{3}}\tan^{-1}\left(\frac{2x}{\sqrt{3}}\right) + c$$

Qu. 10　Find the following integrals:

(a) $\displaystyle\int \dfrac{2}{4+x^2}\,dx$,　　(b) $\displaystyle\int \dfrac{3}{1+4x^2}\,dx$,　　(c) $\displaystyle\int \dfrac{4}{\sqrt{(9-x^2)}}\,dx$,

(d) $\displaystyle\int \dfrac{1}{\sqrt{(1-9x^2)}}\,dx$,　(e) $\displaystyle\int \dfrac{1}{2+25x^2}\,dx$,　(f) $\displaystyle\int \dfrac{2}{\sqrt{(3-4x^2)}}\,dx$,

(g) $\displaystyle\int \dfrac{1}{3-2x+x^2}\,dx$,　(h) $\displaystyle\int \dfrac{5}{\sqrt{\{9-(x+2)^2\}}}\,dx$.

The change of variable $t = \tan\dfrac{x}{2}$

13.3　Of the trigonometrical ratios, two have not yet been integrated in this book, $\sec x$ and $\operatorname{cosec} x$.

Now

$$\csc x = \cfrac{1}{2 \sin \dfrac{x}{2} \cos \dfrac{x}{2}} = \cfrac{\sec^2 \dfrac{x}{2}}{2 \tan \dfrac{x}{2}}$$

$$\left(\text{dividing numerator and denominator by } \cos^2 \dfrac{x}{2} \right).$$

Thus

$$\int \csc x \, dx = \int \cfrac{\dfrac{1}{2} \sec^2 \dfrac{x}{2}}{\tan \dfrac{x}{2}} \, dx$$

$$\therefore \int \csc x \, dx = \ln \tan \frac{x}{2} + c\dagger$$

Furthermore,

$$\sin\left(x + \frac{\pi}{2} \right) = \sin x \cos \frac{\pi}{2} + \cos x \sin \frac{\pi}{2} = \cos x$$

$$\therefore \csc\left(x + \frac{\pi}{2} \right) = \sec x$$

Thus

$$\int \sec x \, dx = \int \csc\left(x + \frac{\pi}{2} \right) dx$$

$$= \ln \tan\left(\frac{x}{2} + \frac{\pi}{4} \right) + c$$

The above working suggests a change of variable which is of considerable importance. If we write $\tan \dfrac{x}{2}$ as t, since $\tan 2A = \dfrac{2 \tan A}{1 - \tan^2 A}$,

$$\tan x = \frac{2t}{1 - t^2}$$

It is also possible (see Book 1, §17.4) to express $\sin x$ and $\cos x$ in terms of t.

$$\sin x = \frac{2t}{1 + t^2} \quad \text{and} \quad \cos x = \frac{1 - t^2}{1 + t^2}$$

Fig. 13.1 provides a useful mnemonic for these identities. Starting with the fact that $\tan x = \dfrac{2t}{1 - t^2}$, one can at once deduce that $AC = 1 + t^2$.

†See footnote to p. 233.

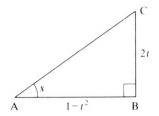

Figure 13.1

When we make the change of variable $t = \tan \dfrac{x}{2}$,

$$\frac{dt}{dx} = \frac{1}{2} \sec^2 \frac{x}{2}$$

$$\frac{dx}{dt} = \frac{2}{\sec^2 \dfrac{x}{2}} = \frac{2}{1 + \tan^2 \dfrac{x}{2}}$$

$$\therefore \frac{dx}{dt} = \frac{2}{1 + t^2}$$

Qu. 11   Find $\int \operatorname{cosec} x \, dx$ using the change of variable $t = \tan \dfrac{x}{2}$.

Qu. 12   Find $\displaystyle\int \frac{\sin \theta}{1 + \cos \theta} \, d\theta$ (a) by expressing the integrand in terms of ratios of $\dfrac{\theta}{2}$, (b) by the change of variable $t = \tan \dfrac{\theta}{2}$.

Qu. 13   Use the change of variable $t = \tan \dfrac{x}{2}$ to show that

$$\int \sec x \, dx = \ln \frac{1 + t}{1 - t} + c$$

Compare this form of the integrand with that obtained earlier and deduce that

$$\int \sec x \, dx = \ln (\sec x + \tan x) + c\dagger$$

This change of variable is best thought of in more general terms as '$t = \tan$ (half angle)'. For example, when applied to $\int \operatorname{cosec} 4x \, dx$ it is $t = \tan 2x$; then $1 = 2 \sec^2 2x \dfrac{dx}{dt}$, giving $\dfrac{dx}{dt} = \dfrac{1}{2(1 + t^2)}$. Care must be taken to establish the correct numerical factor in the expression for $\dfrac{dx}{dt}$.

---

†If $\tan \dfrac{x}{2}$ is negative $\displaystyle\int \operatorname{cosec} x \, dx = \ln \left(-\tan \dfrac{x}{2} \right) + c.$

If $\sec x + \tan x$ is negative $\displaystyle\int \sec x \, dx = \ln (-\sec x - \tan x) + c.$ (See §2.12.)

Example 4 *Find* $\displaystyle\int \frac{1}{3 + 5 \cos \frac{1}{2}x}\, dx.$

$$\int \frac{1}{3 + 5 \cos \frac{1}{2}x} \frac{dx}{dt}\, dt = \int \frac{1}{3 + 5 \times \dfrac{1 - t^2}{1 + t^2}} \times \frac{4}{1 + t^2}\, dt$$

$$= \int \frac{4}{3(1 + t^2) + 5(1 - t^2)}\, dt$$

$$= \int \frac{2}{4 - t^2}\, dt$$

$$= \int \left\{ \frac{1}{2(2 + t)} + \frac{1}{2(2 - t)} \right\} dt$$

$$= \tfrac{1}{2} \ln (2 + t) - \tfrac{1}{2} \ln (2 - t) + c$$

$$= \ln \frac{k\sqrt{(2 + \tan \frac{1}{4}x)}}{\sqrt{(2 - \tan \frac{1}{4}x)}}$$

Let $t = \tan \dfrac{x}{4}$.

$1 = \tfrac{1}{4} \sec^2 \dfrac{x}{4} \times \dfrac{dx}{dt}$

$\dfrac{dx}{dt} = \dfrac{4}{1 + t^2}.$

Qu. 14 Find $\dfrac{dx}{dt}$ in terms of t if

(a) $t = \tan x$,   (b) $t = \tan 4x$,   (c) $t = \tan \tfrac{3}{2}x.$

Qu. 15 Find

(a) $\int \operatorname{cosec} 2x\, dx$,   (b) $\displaystyle\int \frac{1}{1 + \sin 3\theta}\, d\theta$,   (c) $\displaystyle\int \frac{1}{\sqrt{(x^2 - 1)}}\, dx$ (use $x = \sec u$).

The change of variable $t = \tan x$

13.4   An integrand containing $\sin x$ and $\cos x$, particularly even powers of these, may often be expressed as a function of $\tan x$ and $\sec x$. In such a case the change of variable $t = \tan x$ is worth trying.

Example 5 *Find* $\displaystyle\int \frac{1}{1 + \sin^2 x}\, dx.$

[In this case we divide the numerator and denominator by $\cos^2 x$.]

$$\int \frac{1}{1 + \sin^2 x}\, dx = \int \frac{\sec^2 x}{\sec^2 x + \tan^2 x}\, dx$$

$$= \int \frac{\sec^2 x}{1 + 2 \tan^2 x} \times \frac{dx}{dt}\, dt$$

Let $t = \tan x.$

$\dfrac{dx}{dt} = \dfrac{1}{1 + t^2}.$

$$= \int \frac{1+t^2}{1+2t^2} \times \frac{1}{1+t^2} \, dt$$

$$= \int \frac{1}{1+2t^2} \, dt$$

$$= \frac{1}{\sqrt{2}} \tan^{-1}(\sqrt{2}\,t) + c$$

$$= \frac{1}{\sqrt{2}} \tan^{-1}(\sqrt{2}\tan x) + c$$

Qu. 16   Find (a) $\displaystyle\int \frac{1}{1+\cos^2 x} \, dx,$ (b) $\displaystyle\int \frac{2\tan x}{\cos 2x} \, dx.$

Splitting the numerator

13.5 When a fractional integrand with a quadratic denominator cannot be written in simple partial fractions, it may often be usefully expressed as two fractions by splitting the numerator. To take a simple example, such as the reader has already met in Exercises 1d and 1f,

$$\int \frac{1+x}{1+x^2} \, dx = \int \left(\frac{1}{1+x^2} + \frac{x}{1+x^2} \right) dx$$

$$= \tan^{-1} x + \ln\sqrt{(1+x^2)} + c$$

The key to a more general application of this method is to express the numerator in two parts, one of which is *a multiple of the derivative of the denominator*.

Example 6   *Find* $\displaystyle\int \frac{5x+7}{x^2+4x+8} \, dx.$

Since $\dfrac{d}{dx}(x^2+4x+8) = 2x+4,$

let $\quad 5x+7 \equiv A(2x+4) + B;$ whence $A = \frac{5}{2},\ B = -3.$

$$\therefore \int \frac{5x+7}{x^2+4x+8} \, dx = \int \left\{ \frac{\frac{5}{2}(2x+4)}{x^2+4x+8} - \frac{3}{x^2+4x+8} \right\} dx$$

$$= \frac{5}{2}\ln(x^2+4x+8) - 3\int \frac{1}{(x+2)^2+4} \, dx$$

$$= \frac{5}{2}\ln(x^2+4x+8) - \frac{3}{2}\tan^{-1}\left(\frac{x+2}{2}\right) + c$$

This method is also appropriate for integrands of the form

$$\frac{a \cos x + b \sin x}{\alpha \cos x + \beta \sin x}$$

since the numerator may be expressed in the form

$$A \,(derivative \ of \ denominator) + B \,(denominator)$$

Example 7   *Find* $\displaystyle\int \frac{2 \cos x + 3 \sin x}{\cos x + \sin x}\,dx.$

Let $2 \cos x + 3 \sin x \equiv A(-\sin x + \cos x) + B(\cos x + \sin x)$; whence $A = -\frac{1}{2}$, $B = \frac{5}{2}$.

$$\therefore \int \frac{2 \cos x + 3 \sin x}{\cos x + \sin x}\,dx = \int \left\{ \frac{-\frac{1}{2}(-\sin x + \cos x)}{\cos x + \sin x} + \frac{\frac{5}{2}(\cos x + \sin x)}{\cos x + \sin x} \right\} dx$$

$$= -\tfrac{1}{2} \ln (\cos x + \sin x) + \tfrac{5}{2}x + c$$

Qu. 17   Find   (a) $\displaystyle\int \frac{2x + 3}{x^2 + 2x + 10}\,dx,$ (b) $\displaystyle\int \frac{1 - 2x}{\sqrt{\{9 - (x + 2)^2\}}}\,dx,$

(c) $\displaystyle\int \frac{\sin x}{\cos x + \sin x}\,dx,$ (d) $\displaystyle\int \frac{2 \cos x + 9 \sin x}{3 \cos x + \sin x}\,dx.$

Improper integrals

13.6   There are two types of integrals to be discussed under this heading, and we shall consider them in terms of the area under a curve.

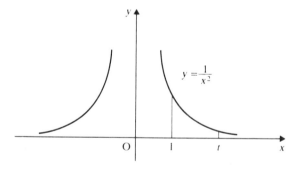

Figure 13.2

Fig. 13.2 shows part of the curve $y = 1/x^2$, to which the x-axis is an asymptote. The area under this curve from $x = 1$ to $x = t$ $(t > 1)$ is

$$\int_1^t \frac{1}{x^2}\,dx = \left[-\frac{1}{x} \right]_1^t = 1 - \frac{1}{t}$$

As $t \to \infty$, this area $\to 1$. Thus although the area 'enclosed' by $y = 1/x^2$, $x = 1$ and the x-axis is not in fact a finite enclosed area, we see that it can be evaluated as the limiting value of the area $\int_1^t \frac{1}{x^2} \, dx$ as $t \to \infty$.

For brevity it is permissible to write

$$\int_1^\infty \frac{1}{x^2} \, dx = \left[-\frac{1}{x} \right]_1^\infty = 1$$

(Integrals like this are usually called 'improper integrals of the first kind'.)

We are faced with a similar situation when we consider 'the area under the curve $y = 1/\sqrt{(1 - x^2)}$ from $x = 0$ to $x = 1$' (Fig. 13.3), since $x = 1$ is an asymptote to the curve. (Integrals like this are usually called 'improper integrals of the second kind'.)

The area under this curve from $x = 0$ to $x = t$ $(0 < t < 1)$ is

$$\int_0^t \frac{1}{\sqrt{(1 - x^2)}} \, dx = \left[\sin^{-1} x \right]_0^t = \sin^{-1} t$$

As $t \to 1$, this area $\to \pi/2$.

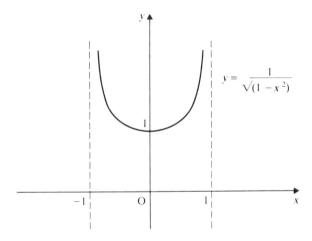

Figure 13.3

Thus, although the integrand $1/\sqrt{(1 - x^2)}$ is meaningless when $x = 1$, the limiting process is implied when we write

$$\int_0^1 \frac{1}{\sqrt{(1 - x^2)}} \, dx = \left[\sin^{-1} x \right]_0^1 = \frac{\pi}{2}$$

Qu. 18 Evaluate:

(a) $\int_1^\infty \frac{1}{x^2} \, dx$, using the change of variable $x = \frac{1}{u}$,

(b) $\displaystyle\int_0^1 \frac{1}{\sqrt{(1-x^2)}}\,dx$, using the change of variable $x = \sin u$.

Qu. 19 Evaluate the following integrals where possible, otherwise show that they are meaningless. Illustrate with a sketch.

(a) $\displaystyle\int_0^1 \frac{1}{x^2}\,dx$,   (b) $\displaystyle\int_0^\infty \frac{1}{1+x^2}\,dx$,   (c) $\displaystyle\int_1^\infty \frac{1}{x}\,dx$,

(d) $\displaystyle\int_0^3 \frac{1}{(x-1)^2}\,dx$,   (e) $\displaystyle\int_0^\infty e^{-x}\,dx$,   (f) $\displaystyle\int_1^2 \frac{1}{\sqrt{(4-x^2)}}\,dx$.

Exercise 13b

1 Differentiate the following with respect to x:

(a) $\sin^{-1} 2x$,   (b) $\tan^{-1}(3x+1)$,   (c) $\frac{1}{3}\cos^{-1} 2x$,

(d) $2\sin^{-1}\left(\dfrac{x-1}{3}\right)$,   (e) $\dfrac{1}{2}\tan^{-1}\dfrac{x}{2}$,   (f) $\dfrac{2}{3}\sin^{-1}\dfrac{3x}{2}$,

(g) $\cot^{-1} x$,   (h) $\sec^{-1} x$,   (i) $x^2\tan^{-1} x^2$,

(j) $\cot^{-1} x + \tan^{-1} x$.

2 Find the following integrals:

(a) $\displaystyle\int \frac{1}{9+x^2}\,dx$,   (b) $\displaystyle\int \frac{3}{\sqrt{(4-y^2)}}\,dy$,   (c) $\displaystyle\int \frac{2}{1+9u^2}\,du$,

(d) $\displaystyle\int \frac{2}{\sqrt{(1-16x^2)}}\,dx$,   (e) $\displaystyle\int \frac{2}{3+4t^2}\,dt$,   (f) $\displaystyle\int \frac{1}{\sqrt{(5-4x^2)}}\,dx$,

(g) $\displaystyle\int \frac{1}{2+3y^2}\,dy$,   (h) $\displaystyle\int \frac{1}{3\sqrt{(3-6x^2)}}\,dx$,   (i) $\displaystyle\int \frac{1}{2y^2-8y+17}\,dy$,

(j) $\displaystyle\int \frac{2}{\sqrt{(1+6x-3x^2)}}\,dx$.

3 Find the following integrals:

(a) $\displaystyle\int \operatorname{cosec}\frac{x}{2}\,dx$,   (b) $\displaystyle\int \sec 2\theta\,d\theta$,   (c) $\displaystyle\int \operatorname{cosec} 3x\,dx$,

(d) $\displaystyle\int \sec 4\phi\,d\phi$,   (e) $\displaystyle\int \sec x \operatorname{cosec} x\,dx$,   (f) $\displaystyle\int \frac{1}{1+\cos y}\,dy$,

(g) $\displaystyle\int \frac{1}{1+\sin 2x}\,dx$,   (h) $\displaystyle\int \frac{\sin\theta}{1-\cos\theta}\,d\theta$,   (i) $\displaystyle\int \frac{1}{4+5\cos x}\,dx$,

(j) $\displaystyle\int \frac{1}{5+3\cos\frac{1}{2}\theta}\,d\theta$.

4 Use the change of variable $\tan x = t$ to find the following integrals:

(a) $\displaystyle\int \frac{1}{1 + 2\sin^2 x}\, dx,$ (b) $\displaystyle\int \frac{1}{\cos 2x - 3\sin^2 x}\, dx,$

(c) $\displaystyle\int \frac{\sin^2 x}{1 + \cos^2 x}\, dx,$ (d) $\displaystyle\int \frac{1}{1 - 10\sin^2 x}\, dx.$

5 Find the following integrals:

(a) $\displaystyle\int \frac{x + 5}{x^2 + 3}\, dx,$ (b) $\displaystyle\int \frac{y + 4}{y^2 + 6y + 9}\, dy,$

(c) $\displaystyle\int \frac{3u + 8}{u^2 + 2u + 5}\, du,$ (d) $\displaystyle\int \frac{3 - 7x}{\sqrt{(4x - x^2)}}\, dx,$

(e) $\displaystyle\int \frac{\cos\theta}{\cos\theta + \sin\theta}\, d\theta,$ (f) $\displaystyle\int \frac{3\cos x - 2\sin x}{\cos x + \sin x}\, dx.$

6 Evaluate:

(a) $\displaystyle\int_3^\infty \frac{1}{(x - 2)^2}\, dx,$ using the change of variable $x - 2 = \dfrac{1}{u}$,

(b) $\displaystyle\int_0^{2/3} \frac{1}{\sqrt{(4 - 9x^2)}}\, dx,$ using the change of variable $x = \dfrac{2}{3}\sin u.$

7 Evaluate the following integrals where possible, otherwise show that they are meaningless. Illustrate with a sketch.

(a) $\displaystyle\int_1^2 \frac{1}{x - 1}\, dx,$ (b) $\displaystyle\int_2^3 \frac{1}{\sqrt{(x - 2)}}\, dx,$ (c) $\displaystyle\int_0^3 \frac{1}{(x - 3)^2}\, dx,$

(d) $\displaystyle\int_1^4 \frac{1}{(x - 2)^2}\, dx,$ (e) $\displaystyle\int_3^\infty \frac{1}{(x - 1)^2}\, dx,$ (f) $\displaystyle\int_{-\infty}^0 e^x\, dx,$

(g) $\displaystyle\int_0^{1/2} \ln x\, dx,$ (h) $\displaystyle\int_{-\infty}^0 x\, e^x\, dx,$ (i) $\displaystyle\int_1^{3/2} \frac{1}{\sqrt{(9 - 4x^2)}}\, dx.$

(j) $\displaystyle\int_0^\infty \frac{1}{4 + 25x^2}\, dx.$

8 The area enclosed by the x-axis, $x = 1$, $x = t$, and the curve $y = 1/x$ is rotated through 2π radians about the x-axis. What may be said about the volume of the solid so generated (a) as $t \to \infty$, (b) as $t \to 0$?

*9 Find the area of the ellipse given by the parametric equations

$$x = 5\cos\theta, \qquad y = 3\sin\theta$$

$$\left(\text{Use the fact that } \int y\, dx = \int y\, \frac{dx}{d\theta}\, d\theta.\right)$$

10 Find the area of the segment cut off by $x = 8$ from the parabola given by the parametric equations $x = 2t^2$, $y = 4t$.

11 If $S \equiv \displaystyle\int_0^{\pi/2} \dfrac{\sin \theta}{\cos \theta + \sin \theta} \, d\theta$, and $C \equiv \displaystyle\int_0^{\pi/2} \dfrac{\cos \theta}{\cos \theta + \sin \theta} \, d\theta$, prove that $S = C = \pi/4$.

Further integration by parts

13.7   The purpose of this section is to consolidate the method of integration by parts, and to introduce an interesting development in its application to certain integrals in which the original integral appears again. This gives us the opportunity to consider two integrals of great importance in physics,

$$\int e^{ax} \cos bx \, dx \quad \text{and} \quad \int e^{ax} \sin bx \, dx$$

Example 8   *Find* $\int e^{ax} \cos bx \, dx$.

$$\int u \frac{dv}{dx} \, dx = uv - \int v \frac{du}{dx} \, dx \qquad\qquad \text{Let } u = \cos bx.$$

Let $I = \int e^{ax} \cos bx \, dx$

$$\qquad\qquad \text{Let } \frac{dv}{dx} = e^{ax},$$

$$= \frac{1}{a} e^{ax} \cos bx - \int \frac{1}{a} e^{ax} \, (-b \sin bx) \, dx \qquad\qquad \therefore v = \frac{1}{a} e^{ax}.$$

$$= \frac{1}{a} e^{ax} \cos bx + \frac{b}{a} \int e^{ax} \sin bx \, dx \qquad\qquad\qquad (1)$$

But

$$\int e^{ax} \sin bx \, dx = \frac{1}{a} e^{ax} \sin bx - \int \frac{1}{a} e^{ax} \, b \cos bx \, dx \qquad \text{Let } u = \sin bx.$$

$$\qquad\qquad \text{Let } \frac{dv}{dx} = e^{ax},$$

$$= \frac{1}{a} e^{ax} \sin bx - \frac{b}{a} I \qquad\qquad \therefore v = \frac{1}{a} e^{ax}.$$

Substituting in (1),

$$I = \frac{1}{a} e^{ax} \cos bx + \frac{b}{a^2} e^{ax} \sin bx - \frac{b^2}{a^2} I$$

$$\therefore a^2 I = a \, e^{ax} \cos bx + b \, e^{ax} \sin bx - b^2 I$$

$$\therefore I(a^2 + b^2) = e^{ax} \, (a \cos bx + b \sin bx) + k$$

$$\therefore I = \int e^{ax} \cos bx \, dx = \frac{e^{ax}}{a^2 + b^2} \, (a \cos bx + b \sin bx) + c$$

Qu. 20   Find $\int e^{2x} \sin 3x \, dx$.

Qu. 21   Find $\int e^x \cos 2x \, dx$,

(a) taking e^x as u throughout,

(b) taking $\cos 2x$ as u in the first step, and $\sin 2x$ as u in the second.

(c) Can we usefully take $\cos 2x$ as u in the first step, and e^x as u in the second?

Exercise 13c

1 Use the method of Example 8 to find the following integrals:

(a) $\int e^{3x} \cos 2x \, dx$,   (b) $\int e^{4x} \sin 3x \, dx$,   (c) $\int e^{-t} \cos \dfrac{t}{2} \, dt$,

(d) $\int e^x \sin (2x + 1) \, dx$,   (e) $\int e^{2\theta} \cos^2 \theta \, d\theta$.

2 Find $\int \sec^3 x \, dx$ by first proving it equal to $\frac{1}{2} \sec x \tan x + \frac{1}{2} \int \sec x \, dx$.

3 Find the following integrals:

(a) $\int x^3 \ln x \, dx$,   (b) $\int \tan^{-1} 2y \, dy$,   (c) $\int \dfrac{x}{e^{x^2}} \, dx$,

(d) $\int x \sin 3x \, dx$,   (e) $\int x^2 \sin 2x \, dx$,   (f) $\int e^{3x} \sin 2x \, dx$.

(g) $\int \frac{1}{2} u^3 e^{u^2} \, du$,   (h) $\int x(2x-1)^5 \, dx$,   (i) $\int x \ln \sqrt{(x-1)} \, dx$,

(j) $\int \ln (3x) \, dx$,   (k) $\int x^2 e^{2x} \, dx$,   (l) $\int e^{-y} \cos \dfrac{y}{2} \, dy$,

(m) $\int x^{-3} \ln x \, dx$,   (n) $\int \sin^{-1} \dfrac{t}{3} \, dt$,   (o) $\int \ln x^3 \, dx$,

(p) $\int y^2 \cos^2 y \, dy$,   (q) $\int x \cos x^2 \, dx$,   (r) $\int x \ln (x^2) \, dx$,

(s) $\int \theta^3 \sin (\theta^2) \, d\theta$,   (t) $\int x^3 \cos 2x \, dx$.

4 If $C = \int e^{ax} \cos bx \, dx$, and $S = \int e^{ax} \sin bx \, dx$, prove that

$$aC - bS = e^{ax} \cos bx \quad \text{and} \quad aS + bC = e^{ax} \sin bx$$

Hence find C and S.

5 Prove that $\displaystyle\int_0^\infty e^{-2x} \sin 3x \, dx = 3/13$.

6 Find the area enclosed by the x-axis and the curve $y = x(2-x)^5$.

7 A uniform lamina is enclosed by the axes and the curve $y = \cos x$ from $x = 0$ to $x = \pi/2$. Find the coordinates of its centre of gravity.

8 The area under $y = \cos x$ from $x = 0$ to $x = \pi/2$ is rotated through four right angles about the x-axis. Find the centre of gravity of the uniform solid so generated.

9 Prove that $\int \cos^4 x \, dx = \frac{1}{4} \sin x \cos^3 x + \frac{3}{4} \int \cos^2 x \, dx$.

10 Find the area bounded by the x-axis and that part of the curve $y = e^{3x} \sin x$ from $x = 0$ to $x = \pi$.

Reduction formulae

13.8 The normal method of finding $\int \cos^2 x \, dx$ is to use the fact that $\cos^2 x = \frac{1}{2}(1 + \cos 2x)$; $\int \cos^4 x \, dx$ may be tackled in the same way by expressing the integrand in terms of cos 2x and cos 4x, but for the integrals of higher even powers of cos x the working becomes tedious.

It is instructive to find ∫ cos² x dx using integration by parts. Once again we find that in the process the original integral reappears; this special aspect of integration by parts is found to have a most powerful application, not only in finding the integrals of high powers of cos x and sin x, but also in establishing general formulae for dealing with integrands of high power.

$$I \equiv \int \cos^2 x \, dx = \cos x \sin x - \int \sin x \, (-\sin x) \, dx \qquad \text{Let } u = \cos x.$$

$$= \cos x \sin x + \int (1 - \cos^2 x) \, dx \qquad \text{Let } \frac{dv}{dx} = \cos x,$$

$$\therefore I = \cos x \sin x + x - I$$

$$\therefore 2I = \cos x \sin x + x + k \qquad \qquad \therefore v = \sin x.$$

$$\therefore I = \tfrac{1}{2} \cos x \sin x + \tfrac{1}{2} x + c$$

Qu. 22   Find ∫ cos⁴ x dx by finding it first in terms of ∫ cos² x dx using integration by parts, and then using the above result.

Qu. 23   Show that

$$\int \cos^6 x \, dx = \frac{1}{6} \cos^5 x \sin x + \frac{5}{24} \cos^3 x \sin x +$$

$$+ \frac{5}{16} \cos x \sin x + \frac{5 \times 3 \times 1}{6 \times 4 \times 2} x + c$$

Qu. 24   Find ∫ cos³ x dx (a) by finding it first in terms of ∫ cos x dx using integration by parts, (b) by another method, giving it as a function of sin x only.

Now the real value, in terms of economy, of the results we are beginning to establish is apparent when we come to consider definite integrals between certain limits. For example, using the result obtained in Qu. 23,

$$\int_0^{\pi/2} \cos^6 x \, dx = \left[f(x) \right]_0^{\pi/2} + \left[\frac{5 \times 3 \times 1}{6 \times 4 \times 2} x \right]_0^{\pi/2}$$

where each term of f(x) contains both cos x and sin x, and therefore vanishes at each limit.

Hence

$$\int_0^{\pi/2} \cos^6 x \, dx = \frac{5 \times 3 \times 1}{6 \times 4 \times 2} \times \frac{\pi}{2} = \frac{5\pi}{32}$$

Qu. 25   Show that

$$\int \cos^5 x \, dx = \frac{1}{5} \cos^4 x \sin x + \frac{4}{15} \cos^2 x \sin x + \frac{4 \times 2}{5 \times 3} \sin x + c$$

and evaluate $\int_0^{\pi/2} \cos^5 x \, dx$.

A pattern is emerging in the last but one term in Qu. 23 and in Qu. 25. We shall now consider the general treatment of this form.

Suppose $I_n = \displaystyle\int_0^{\pi/2} \cos^n x \, dx$ $(n \geqslant 2)$.

Using integration by parts,

$$I_n = \left[\cos^{n-1} x \sin x \right]_0^{\pi/2} -$$

$$- \int_0^{\pi/2} \sin x \,(n-1)\cos^{n-2} x \,(-\sin x)\, dx$$

Let $u = \cos^{n-1} x$.

Let $\dfrac{dv}{dx} = \cos x$,

$\therefore v = \sin x$.

$$= 0 + (n-1) \int_0^{\pi/2} (1 - \cos^2 x) \cos^{n-2} x \, dx$$

$$= (n-1) \int_0^{\pi/2} \cos^{n-2} x \, dx - (n-1) \int_0^{\pi/2} \cos^n x \, dx$$

$\therefore I_n = (n-1)I_{n-2} - (n-1)I_n$
$\therefore nI_n = (n-1)I_{n-2}$

$$\therefore I_n = \frac{n-1}{n} I_{n-2} \qquad (n \geqslant 2) \tag{1}$$

Since this relationship reduces by 2 the power of $\cos x$ in the integrand it is called a **reduction formula**.

Replacing n by $(n-2)$ in (1),

$$I_{n-2} = \frac{n-3}{n-2} I_{n-4} \qquad (n \geqslant 4)$$

Similarly

$$I_{n-4} = \frac{n-5}{n-4} I_{n-6} \qquad (n \geqslant 6)$$

Thus

$$I_n = \frac{n-1}{n} I_{n-2} = \frac{(n-1)(n-3)}{n(n-2)} I_{n-4}$$

$$= \frac{(n-1)(n-3)(n-5)}{n(n-2)(n-4)} I_{n-6} \qquad (n \geqslant 6)$$

If n is *odd*, e.g. $n = 7$, we obtain a multiple of I_1, which is

$$\int_0^{\pi/2} \cos x \, dx = 1$$

If n is *even*, e.g. $n = 6$, we obtain a multiple of I_0, which is

$$\int_0^{\pi/2} 1 \, dx = \frac{\pi}{2}$$

Thus

$$\int_0^{\pi/2} \cos^7 x \, dx = I_7 = \frac{6 \times 4 \times 2}{7 \times 5 \times 3} \times 1 = \frac{16}{35}$$

and

$$\int_0^{\pi/2} \cos^6 x \, dx = I_6 = \frac{5 \times 3 \times 1}{6 \times 4 \times 2} \times \frac{\pi}{2} = \frac{5\pi}{32}$$

Qu. 26   Evaluate:

(a) $\displaystyle\int_0^{\pi/2} \cos^8 x \, dx,$ (b) $\displaystyle\int_0^{\pi/2} \cos^9 x \, dx,$ (c) $\displaystyle\int_0^{\pi/2} \cos^{10} x \, dx.$

Qu. 27   If $I_n = \displaystyle\int_0^{\pi/2} \sin^n x \, dx$, use integration by parts to show that

$$I_n = \frac{n-1}{n} I_{n-2}$$

Qu. 28   Use the change of variable $x = \pi/2 - y$ to show that

$$\int_0^{\pi/2} \sin^n x \, dx = \int_0^{\pi/2} \cos^n x \, dx$$

We can now state a reduction formula which the reader should memorise.

If $\displaystyle I_n = \int_0^{\pi/2} \cos^n x \, dx$ **or** $\displaystyle\int_0^{\pi/2} \sin^n x \, dx,$

$$I_n = \frac{n-1}{n} I_{n-2}$$

Hence, when n is *odd*,

$$I_n = \frac{(n-1)(n-3)\dots 4 \times 2}{n(n-2)\dots 5 \times 3}$$

and when n is *even*,

$$I_n = \frac{(n-1)(n-3)\dots 3 \times 1}{n(n-2)\dots 4 \times 2} \times \frac{\pi}{2}$$

(These formulae are often called Wallis' formulae)

A thorough treatment of reduction formulae is beyond the scope of this book, but as an introduction to this topic the above ideas are developed more fully in Exercise 13d; of particular interest is No. 6, from which the basic formula quoted above may be deduced as a special case.

Exercise 13d

1 Use integration by parts to show that

$\int \sin^2 x \, dx = -\frac{1}{2} \cos x \sin x + \frac{1}{2}x + c$

Assuming this result, find $\int \sin^4 x \, dx$ by the same method, and evaluate

$\int_0^{\pi/2} \sin^4 x \, dx$

2 Use integration by parts to show that

$\int \sin^3 \theta \, d\theta = -\frac{1}{3} \cos \theta \sin^2 \theta - \frac{2}{3} \cos \theta + c$

Assuming this result, find $\int \sin^5 \theta \, d\theta$ by the same method, and evaluate

$\int_0^{\pi/2} \sin^5 \theta \, d\theta$

3 Evaluate the following:

(a) $\int_0^{\pi/2} \sin^3 x \, dx,$ (b) $\int_0^{\pi/2} \sin^6 x \, dx,$ (c) $\int_0^{\pi/2} \sin^9 x \, dx,$

(d) $\int_0^{\pi/2} \cos^4 x \, dx,$ (e) $\int_0^{\pi/2} \sin^{10} x \, dx,$ (f) $\int_0^{\pi/2} \sin^7 x \, dx.$

4 Use the change of variable $u = 2\theta$ to prove that

$\int_0^{\pi} \sin^8 \frac{u}{2} \, du = \frac{35\pi}{128}$

Evaluate the following:

(a) $\int_0^{\pi/4} \cos^7 2y \, dy,$ (b) $\int_0^{3\pi/2} \sin^5 \frac{t}{3} \, dt,$ (c) $\int_0^{\pi/6} \cos^6 3x \, dx.$

*5 Demonstrate graphically that

(i) $\int_0^{\pi} \cos^3 \theta \, d\theta = 0,$ (ii) $\int_0^{\pi} \sin^3 \theta \, d\theta = 2 \int_0^{\pi/2} \sin^3 \theta \, d\theta,$

(iii) $\int_0^{\pi} \cos^4 \theta \, d\theta = 2 \int_0^{\pi/2} \cos^4 \theta \, d\theta.$

Evaluate the following:

(a) $\int_0^{\pi} \sin^7 \theta \, d\theta,$ (b) $\int_{-\pi/2}^{\pi/2} \cos^4 \theta \, d\theta,$ (c) $\int_{-\pi/2}^{\pi/2} \sin^6 \theta \, d\theta,$

(d) $\int_{-\pi/2}^{\pi/2} \sin^7 \theta \, d\theta,$ (e) $\int_0^{\pi} \cos^5 \theta \, d\theta,$ (f) $\int_{-\pi/2}^{\pi/2} \cos^9 \theta \, d\theta,$

(g) $\int_0^{\pi} \sin^{10} \theta \, d\theta,$ (h) $\int_0^{\pi} \cos^8 \theta \, d\theta.$

*6 (a) Writing $\int_0^{\pi/2} \cos^m \theta \sin^n \theta \, d\theta$ as $\int_0^{\pi/2} c^m s^n \, d\theta$, or $I_{m,\,n}$, use integration by parts (taking u as c^{m-1}) to prove that

$$I_{m,\,n} = \frac{m-1}{m+n} I_{m-2,\,n} \qquad (m \geqslant 2),$$

and write down $I_{m-2,\,n}$, and hence $I_{m,\,n}$, in terms of $I_{m-4,\,n}$ ($m \geqslant 4$).

(b) Use the change of variable $x = \dfrac{\pi}{2} - y$ to prove that $I_{m,\,n} = I_{n,\,m}$, and reduce $I_{m-4,\,n}$ to the form $kI_{m-4,\,n-6}$ ($n \geqslant 6$).

(c) Show that $\int_0^{\pi/2} \cos^5 \theta \sin^6 \theta \, d\theta = \dfrac{8}{99} \int_0^{\pi/2} \cos \theta \sin^6 \theta \, d\theta$, and proceed to evaluate this

(i) by reduction to the form $kI_{1,\,0}$,

(ii) by writing the latter integral as a function of $\sin \theta$.

(d) Evaluate the following:

(i) $\displaystyle\int_0^{\pi/2} \cos^8 \theta \sin^5 \theta \, d\theta$,   (ii) $\displaystyle\int_0^{\pi/2} \cos^6 \theta \sin^8 \theta \, d\theta$,

(iii) $\displaystyle\int_0^{\pi/2} \cos^7 \theta \sin^6 \theta \, d\theta$,   (iv) $\displaystyle\int_0^{\pi/2} \cos^5 \theta \sin^7 \theta \, d\theta$.

7 Use a suitable change of variable and the method of No. 6 to evaluate the following:

(a) $\displaystyle\int_0^1 x^5 (1 - x^2)^6 \, dx$,   (b) $\displaystyle\int_0^1 x^4 (1 - x^2)^{7/2} \, dx$.

8 If $I_n = \displaystyle\int_0^\infty x^n e^{-ax} \, dx$, (a) obtain a reduction formula for I_n in terms of I_{n-1}, and (b) evaluate $\displaystyle\int_0^\infty x^9 e^{-2x} \, dx$.

9 If $I_n = \displaystyle\int_0^{\pi/4} \tan^n \theta \, d\theta$, obtain a reduction formula relating I_n and I_{n-2}, and use it to evaluate the following correct to two significant figures:

(a) $\displaystyle\int_0^{\pi/4} \tan^7 \theta \, d\theta$,   (b) $\displaystyle\int_0^{\pi/4} \tan^8 \theta \, d\theta$.

10 (a) If $I_n = \int \sec^n x \, dx$, prove that

$$I_n = \frac{1}{n-1} \tan x \sec^{n-2} x + \frac{n-2}{n-1} I_{n-2} \qquad (n \geqslant 2)$$

and use this reduction formula to write down an expression relating I_n to I_{n-6} ($n \geqslant 6$).

(b) Use the result obtained in (a) to find $\int_0^{\pi/6} \sec^8 x\, dx$, and check your answer by expressing the integral as a function of $\tan x$ only.

(c) Prove that $\int_0^{\pi/3} \sec^7 x\, dx = \dfrac{61}{8}\sqrt{3} + \dfrac{5}{16}\ln(2 + \sqrt{3})$.

The mean value of a function

13.9 In elementary arithmetic the idea of the mean (often loosely called the average) of a set of numbers is a familiar concept. If we have n numbers $y_1, y_2, y_3,\ldots,y_n$, their mean $\bar{y}$ is given by

$$\bar{y} = \frac{1}{n}(y_1 + y_2 + y_3 + \ldots + y_n)$$

$$= \frac{1}{n}\sum_{r=1}^{r=n} y_r$$

Notice that if each of the n values of y_r were replaced by $\bar{y}$, the total would be unaltered, because $n\bar{y} = \sum_{r=1}^{r=n} y_r$.

In more advanced work it is frequently necessary to find the mean of a continuous function; for example we may wish to find the mean value of the temperature over a period of 24 hours, or the mean value of an alternating voltage.

The mean value of a function $f(x)$ over an interval $x = a$ to $x = b$ is defined as

$$\frac{1}{(b-a)}\int_a^b f(x)\, dx$$

Like the elementary idea of the mean in arithmetic, the mean value of $f(x)$ is the (constant) value, $\bar{y}$, which should replace $f(x)$ throughout the interval, so that

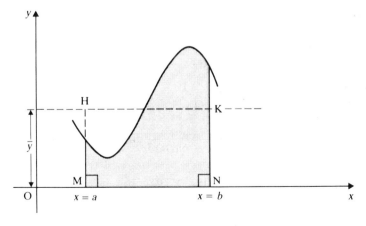

Figure 13.4

the area under $y = \bar{y}$ (i.e. the area of the rectangle HKNM in Fig. 13.4) equals the area under the curve $y = f(x)$ from $x = a$ to $x = b$.

Qu. 29 Find the mean value of
(a) $y = x^2$, from $x = 1$ to $x = 4$,   (b) $y = \sin x$, from $x = 0$ to $x = \pi$,
(c) $y = \sqrt{x}$, from $x = 0$ to $x = 4$,   (d) $y = \sin^2 x$, from $x =$ to $x = \pi/2$.

Exercise 13e (Revision)

No list of 'standard integrals' is given in this chapter, in the belief that the recognition of form is more important than the learning of formulae (see Nos. 13–17).
In this exercise,
   Nos. 1–7 summarise the main methods dealt with in chapters 1, 2, 3, 13.
   Nos. 8–12 gather together the integrals of some trigonometrical functions and inverse functions, to enable the reader to take stock of his or her power of handling these integrals.
   Nos. 13–17 are designed to develop discrimination in choice of method. These questions test the essential skill, recognition of form, and the more experienced reader may confine his or her attention to these questions, together with some of the less obvious integrals in Nos. 8–12.

Find the integrals in Nos. 1–6.

1 (a) $\int x\sqrt{(x^2 + 1)}\,dx$,   (b) $\int \dfrac{x^2 + 1}{\sqrt{(x^3 + 3x - 4)}}\,dx$,

   (c) $\int \cos^5 u\,du$,   (d) $\int \sec^6 \theta\,d\theta$,

   (e) $\int \sec x \tan^5 x\,dx$,   (f) $\int x \sin x^2\,dx$,

   (g) $\int \dfrac{\sec^2 \sqrt{x}}{\sqrt{x}}\,dx$,   (h) $\int x(2x^2 + 3)^{-1}\,dx$,

   (i) $\int \dfrac{x}{e^{x^2}}\,dx$,   (j) $\int \tan \dfrac{\theta}{2}\,d\theta$.

2 *Change of variable.*
   (a) $\int x\sqrt{(2x - 3)}\,dx$,   (b) $\int 2x(3x - 1)^7\,dx$,

   (c) $\int \dfrac{y(y - 8)}{(y - 4)^2}\,dy$,   (d) $\int \dfrac{1}{\sqrt{(4 - 5y^2)}}\,dy$,

   (e) $\int \dfrac{1}{3 + 9u^2}\,du$,   (f) $\int \dfrac{1}{u^2 - 6u + 17}\,du$,

   (g) $\int \dfrac{1}{\sqrt{(7 + 4x - 2x^2)}}\,dx$,   (h) $\int \sqrt{(4 - y^2)}\,dy$,

   (i) $\int \dfrac{1}{x\sqrt{(9x^2 - 1)}}\,dx$,   (j) $\int \dfrac{1}{5 + 4\cos \theta}\,d\theta$.

3 *Involving exponential and logarithmic functions.*

(a) $\int e^{3x} dx$,   (b) $\int 10^y dy$,   (c) $\int \frac{x^2}{e^{x^3}} dx$,

(d) $\int \frac{1}{3x} dx$,   (e) $\int \frac{1}{3x+4} dx$,   (f) $\int \frac{1}{3-2x} dx \quad (x > \frac{3}{2})$,

(g) $\int \frac{1}{3x+9} dx$,   (h) $\int \frac{1}{1-x^2} dx$,   (i) $\int \ln x \, dx$,

(j) $\int e^{\sqrt{x}} dx$ (write as $\int x^{1/2} x^{-1/2} e^{x^{1/2}} dx$).

4 *Partial fractions.*

(a) $\int \frac{2}{9-x^2} dx$,   (b) $\int \frac{1}{y(y-3)} dy$,   (c) $\int \frac{1}{x^3 - x^2} dx$,

(d) $\int \frac{x}{(4-x)^2} dx$,   (e) $\int \frac{2-x^2}{(x+1)^3} dx$,   (f) $\int \frac{(x-2)^2}{x^3+1} dx$.

5 *Integration by parts.*

(a) $\int x \cos \frac{x}{2} dx$,   (b) $\int \frac{x}{2} e^x dx$,   (c) $\int y \operatorname{cosec}^2 y \, dy$,

(d) $\int 2y(1-3y)^6 dy$,   (e) $\int x \, 3^x dx$,   (f) $\int x \ln 2x \, dx$,

(g) $\int \ln t \, dt$,   (h) $\int \tan^{-1} 3x \, dx$,   (i) $\int 4^x dx$,

(j) $\int x^3 \sin x \, dx$,

(k)   Prove $\int \cos^4 \frac{x}{2} dx = \frac{1}{2} \sin \frac{x}{2} \cos^3 \frac{x}{2} + \frac{3}{4} \int \cos^2 \frac{x}{2} dx$.

6 *Splitting the numerator.*

(a) $\int \frac{2x-1}{4x^2+3} dx$,   (b) $\int \frac{1-4y}{\sqrt{(1+2y-y^2)}} dy$,

(c) $\int \frac{\cos \theta}{2 \cos \theta - \sin \theta} d\theta$,   (d) $\int \frac{\cos x - 2 \sin x}{3 \cos x + 4 \sin x} dx$.

7 Evaluate the following:

(a) $\int_{1/3}^{2/3} \frac{1}{\sqrt{(4-9x^2)}} dx$,   (b) $\int_{1}^{\sqrt{2}} \frac{1}{8+y^2} dy$,   (c) $\int_{5}^{\infty} \frac{1}{(x-3)^2} dx$,

(d) $\int_{0}^{\pi/2} \cos^{11} x \, dx$,   (e) $\int_{0}^{\pi/2} \sin^{12} \theta \, d\theta$,   (f) $\int_{0}^{\pi/8} \cos^6 4y \, dy$,

(g) $\displaystyle\int_{-\pi/2}^{\pi/2} \sin^8 u \, du,$ (h) $\displaystyle\int_0^{\pi} \cos^7 x \, dx,$ (i) $\displaystyle\int_0^{\pi/2} \cos^9 \theta \sin^{10} \theta \, d\theta,$

(j) $\displaystyle\int_{-1}^{+1} \frac{1}{2x-3} \, dx.$

---

8 Find the following integrals:

(a) $\int \sin 5x \, dx,$ (b) $\displaystyle\int \cos \frac{x}{3} \, dx,$ (c) $\int \tan 5x \, dx,$ (d) $\int \cot \tfrac{1}{2}x \, dx,$

(e) $\displaystyle\int \operatorname{cosec} x \, dx \left(\text{use } \tan \frac{x}{2} = t \right),$ (f) $\displaystyle\int \sec x \, dx \left(\text{use } \tan \frac{x}{2} = t \right).$

9 Find the following integrals:

(a) $\displaystyle\int \sec^2 \frac{x}{3} \, dx,$ (b) $\int \operatorname{cosec}^2 4x \, dx,$ (c) $\int \sin^2 x \, dx,$

(d) $\int \cos^2 x \, dx,$ (e) $\int \tan^2 x \, dx,$ (f) $\int \cot^2 x \, dx.$

10 Find the following integrals:
(a) $\int \sin^3 x \, dx,$ †(b) $\int \cos^3 x \, dx,$
(c) $\int \tan^3 x \, dx$ (use Pythagoras' theorem),
†(d) $\int \cot^3 x \, dx$ (use Pythagoras' theorem),
(e) $\int \sec^3 x \, dx$ (by reduction),   †(f) $\int \operatorname{cosec}^3 x \, dx$ (by reduction).

11 Find the following integrals ((a) and (b) by expressing the integrands in terms of $\cos 2x$, $\cos 4x$, or by reduction, the remainder by using Pythagoras' theorem):
(a) $\int \sin^4 x \, dx,$ (b) $\int \cos^4 x \, dx,$ (c) $\int \tan^4 x \, dx,$
(d) $\int \operatorname{cosec}^4 x \, dx,$ (e) $\int \sec^4 x \, dx,$ (f) $\int \cot^4 x \, dx.$

12 Find the following integrals using integration by parts (in (e) and (f) continue by using the change of variable $x = \sec u$):
(a) $\int \sin^{-1} x \, dx,$ (b) $\int \cos^{-1} x \, dx,$ (c) $\int \tan^{-1} x \, dx,$
(d) $\int \cot^{-1} x \, dx,$ (e) $\int \sec^{-1} x \, dx,$ (f) $\int \operatorname{cosec}^{-1} x \, dx.$

Find the integrals in Nos. 13–17.

13 (a) $\displaystyle\int \frac{1}{3 + 4x^2} \, dx,$ (b) $\displaystyle\int \frac{x}{\sqrt{(5 + 8x^2)}} \, dx,$

(c) $\displaystyle\int \frac{1}{\sqrt{(1 + x^2)}} \, dx,$ (d) $\displaystyle\int \frac{x}{2 + 3x^2} \, dx,$

(e) $\int x\sqrt{(3 + x^2)} \, dx,$ (f) $\displaystyle\int \frac{x + 1}{3 + 2x^2} \, dx,$

†The change of variable $y = \tfrac{1}{2}\pi - x$ may be used.

(g) $\int \dfrac{x-2}{x^2-4x+7}\,dx,$ (h) $\int \sqrt{(2+x^2)}\,dx,$

(i) $\int \dfrac{3x-11}{x^2-4x+5}\,dx,$ (j) $\int x\sqrt{(2+3x)}\,dx.$

14 (a) $\int \dfrac{1}{\sqrt{(4-5x^2)}}\,dx,$ (b) $\int \dfrac{x}{\sqrt{(1-3x)}}\,dx,$

(c) $\int \dfrac{2}{9-x^2}\,dx,$ (d) $\int \dfrac{3}{(16-x)^2}\,dx,$

(e) $\int x\sqrt{(6-x^2)}\,dx,$ (f) $\int \dfrac{3x}{4-x^2}\,dx,$

(g) $\int \sqrt{(4-x^2)}\,dx,$ (h) $\int \dfrac{x}{\sqrt{(7-2x^2)}}\,dx,$

(i) $\int \dfrac{x-2}{\sqrt{(3-4x^2)}}\,dx,$ (j) $\int \dfrac{1}{\sqrt{(x^2-9)}}\,dx.$

15 (a) $\int \cos x^\circ\,dx,$ (b) $\int x \sin 2x \cos 2x\,dx,$

(c) $\int \sec\dfrac{\theta}{2}\,\text{cosec}\,\dfrac{\theta}{2}\,d\theta,$ (d) $\int \cos^6 x \sin^5 x\,dx,$

(e) $\int y \sec^2 y\,dy,$ (f) $\int x \sin x\,dx,$

(g) $\int x \sin x^2\,dx,$ (h) $\int u^2 \cos u\,du,$

(i) $\int \sin^2 y \cos^2 y\,dy,$ (j) $\int \sin 5x \cos 2x\,dx.$

16 (a) $\int \dfrac{1}{1+\cos\theta}\,d\theta,$ (b) $\int \dfrac{1}{1-5\sin^2\theta}\,d\theta,$

(c) $\int \dfrac{1}{1+\sin x}\,dx,$ (d) $\int \dfrac{2\cos\theta-4\sin\theta}{\cos\theta+3\sin\theta}\,d\theta,$

(e) $\int \dfrac{1}{1-\cos\frac{1}{2}x}\,dx,$ (f) $\int \dfrac{4}{\cos^2 x+9\sin^2 x}\,dx,$

(g) $\int \dfrac{1}{\cos^2 2y-\sin^2 2y}\,dy,$ (h) $\int \dfrac{1+\sin x}{\cos^2 x}\,dx,$

(i) $\int \dfrac{1}{\sin\theta+2\cos\theta+1}\,d\theta,$ (j) $\int \dfrac{1}{1+\tan x}\,dx.$

17 (a) $\int x^3\,e^{-x}\,dx,$ (b) $\int \ln(x+2)\,dx,$

(c) $\int \dfrac{e^{\sqrt{y}}}{\sqrt{y}}\,dy,$ (d) $\int \dfrac{1}{t\sqrt{\ln t}}\,dt,$

(e) $\int x \tan^{-1} 3x \, dx$,

(f) $\int \dfrac{\sin^{-1} x}{\sqrt{(1 - x^2)}} \, dx$,

(g) $\int 4^x \, dx$,

(h) $\int x \, 10^x \, dx$,

(i) $\int x^3 \ln (2x) \, dx$,

(j) $\int x^3 \, e^{x^2} \, dx$.

Exercise 13f (Miscellaneous)

1 Integrate the following functions with respect to x:

(a) $\left(x^2 - \dfrac{2}{x} \right)^2$,   (b) $\sin 3x \cos 5x$,   (c) $\dfrac{2}{x(1 + x^2)}$.

Prove by means of the substitution $t = \tan x$ that

$$\int_0^{\pi/4} \frac{dx}{1 + \sin 2x} = \frac{1}{2}$$

and find the value of $\displaystyle \int_0^{\pi/4} \frac{dx}{(1 + \sin 2x)^2}$.   (O & C)

2 Integrate the following functions with respect to x:

(a) $x(1 + x^2)^{3/2}$,   (b) $\dfrac{3 + 2x}{1 - 4x^2}$,   (c) $x^2 \ln x$.

Evaluate $\displaystyle \int_0^{\pi/2} x \cos^2 3x \, dx$.   (O & C)

3 Express $\dfrac{1}{1 - x^4}$ in partial fractions and hence show that

$$\int_0^{1/2} \frac{dx}{1 - x^4} = \frac{1}{4} \ln 3 + \frac{1}{2} \arctan \left(\frac{1}{2} \right).$$ (L)

4 Integrate the following functions with respect to x:

(a) $\dfrac{1}{x^2 - 3x + 2}$,   (b) $x \arctan x$,   (c) $\dfrac{x^2}{x^2 + 2x + 2}$.

By means of the substitution $x = 1 - 1/u^4$, show that

$$\int_5^{2^{1/4}} \frac{du}{u(2u^4 - 1)^{1/2}} = \frac{\pi}{24}$$ (O & C)

5 Integrate the following functions with respect to x:

(a) $\dfrac{x + 6}{x^2 + 6x + 8}$,   (b) $x \ln x$,   (c) $\arcsin x$.

Evaluate $\displaystyle \int_{1/2}^1 \frac{(1 + x^{3/2})}{x \sqrt{\{x(1 - x)\}}} \, dx$, by means of the substitution $x = \cos^2 \phi$.

(O & C)

6 Use the substitution $u = +\sqrt{(1 + x^2)}$ to evaluate

$$\int_{4/3}^{12/5} \frac{\mathrm{d}x}{x(1 + x^2)^{3/2}}$$ (O & C)

7 By making the substitution $x = \pi - y$, or otherwise, prove that

$$\int_0^\pi x \sin^3 x \, \mathrm{d}x = \frac{2\pi}{3}$$ (O & C)

8 Integrate with respect to x:

(a) $\cos x \operatorname{cosec}^3 x$,   (b) $\dfrac{x + 3}{\sqrt{(7 - 6x - x^2)}}$.

By making the substitution $x = a \cos^2 \theta + b \sin^2 \theta$, prove that

$$\int_a^b \frac{x \, \mathrm{d}x}{\sqrt{\{(x - a)(b - x)\}}} = \frac{1}{2}\pi(a + b)$$ (O & C)

9 Integrate with respect to x:

(a) $\dfrac{1}{\sqrt{(5 - 4x - x^2)}}$,   (b) $x^3 \, \mathrm{e}^{-x^2}$.

If $S = \displaystyle\int \frac{\sin x}{a \sin x + b \cos x} \, \mathrm{d}x$ and $C = \displaystyle\int \frac{\cos x}{a \sin x + b \cos x} \, \mathrm{d}x$,

find $aS + bC$ and $aC - bS$. Hence, or otherwise, prove that

$$\int_0^{\pi/2} \frac{\sin x}{3 \sin x + 4 \cos x} \, \mathrm{d}x = \frac{3\pi}{50} + \frac{4}{25} \ln\left(\frac{4}{3}\right).$$ (O & C)

10 Evaluate $\displaystyle\int_1^2 \frac{\mathrm{d}x}{x^2\sqrt{(x - 1)}}$ by means of the substitution $x = \sec^2 \phi$.   (O & C)

11 By means of the substitution $x^2 = 1/u$, evaluate the integral

$$\int_1^2 \frac{\mathrm{d}x}{x^2\sqrt{(5x^2 - 1)}}$$ (O & C)

12 Integrate with respect to x:

(a) $\dfrac{x^2}{x^2 - 4}$,   (b) $x^3 \, \mathrm{e}^{-x^2/2}$,   (c) $\dfrac{1}{\sqrt{(5 + 4x - x^2)}}$.

By means of the substitution $u = \tan x$, or otherwise, evaluate the integral

$$\int_0^{\pi/2} \frac{\mathrm{d}x}{2 + \cos^2 x}$$ (O & C)

13 Prove that, if $C = \int \mathrm{e}^{ax} \cos bx \, \mathrm{d}x$ and $S = \int \mathrm{e}^{ax} \sin bx \, \mathrm{d}x$,

$$aC - bS = \mathrm{e}^{ax} \cos bx \qquad \text{and} \qquad aS + bC = \mathrm{e}^{ax} \sin bx$$

Evaluate $\displaystyle\int_0^{\pi/2} \mathrm{e}^{2x} \sin 3x \, \mathrm{d}x$.   (O & C)

14 Integrate with respect to x:

(a) $\int (2x - 3)^{-3/2} \, dx$,   (b) $\int x \sin x \, dx$,   (c) $\int x \sqrt{(x - 4)} \, dx$.

By means of the substitution $x = 3 \cos^2 \theta + 6 \sin^2 \theta$, or otherwise, evaluate

$$\int_3^6 \frac{dx}{\sqrt{\{(x - 3)(6 - x)\}}} \qquad\qquad \text{(O \& C)}$$

15 Give a geometrical interpretation of the integral

$$I = \int_a^b f(x) \, dx \qquad (b > a)$$

Without attempting to evaluate them, determine whether the following integrals are positive, negative, or zero:

(a) $\displaystyle\int_0^1 x^3(1 - x^2)^2 \, dx$,   (b) $\displaystyle\int_0^\pi \sin^3 x \cos^3 x \, dx$,   (c) $\displaystyle\int_{1/2}^1 e^{-x} \ln x \, dx$.

$$\text{(O \& C)}$$

16 (a) Let $\displaystyle I(Z) = \int_1^Z \frac{(x - 1)^p(2 - x)^p}{x^{p+1}} \, dx \quad (p > 0)$.

By writing $x = 2/y$, prove that $2I(\sqrt{2}) = I(2)$.

(b) Without attempting to evaluate them, determine whether the following integrals are positive, negative, or zero:

$$\int_0^1 x^3(1 - x)^3 \, dx, \qquad \int_0^\pi \sin^2 x \cos^3 x \, dx, \qquad \int_0^\pi e^{-x} \sin x \, dx$$

$$\text{(O \& C)}$$

17 By means of the substitution $y = a - x$ or otherwise, prove that

$$\int_0^a f(x) \, dx = \int_0^a f(a - x) \, dx$$

Hence prove that

$$\int_0^\pi \frac{x \sin x}{1 + \cos^2 x} \, dx = \int_0^\pi \frac{(\pi - x) \sin x}{1 + \cos^2 x} \, dx = \frac{\pi^2}{4} \qquad \text{(O \& C)}$$

18 (a) Evaluate the integral $\displaystyle\int_0^1 \tan^{-1} x \, dx$.

(b) If $\displaystyle I_n = \int_0^1 \frac{dx}{(1 + x^2)^n}$ show that $2nI_{n+1} = 2^{-n} + (2n - 1)I_n$.

Deduce the value of $\displaystyle\int_0^1 \frac{dx}{(1 + x^2)^3}$. $\qquad\qquad$ (JMB)

19 Prove that, if $u_n = \int_0^{\pi/2} \dfrac{\sin 2n\theta}{\sin \theta} \, d\theta$, where n is a positive integer,

$$u_n - u_{n-1} = \frac{2(-1)^{n-1}}{2n-1}$$

Hence prove that $u_n = 2\left\{ 1 - \dfrac{1}{3} + \dfrac{1}{5} - \dots + \dfrac{(-1)^{n-1}}{2n-1} \right\}$.   (O & C)

20 (a) Prove that, if $I_n = \int \sec^{2n} \theta \, d\theta$, and $n > 1$,

$$(2n-1)I_n = 2(n-1)I_{n-1} + \sec^{2n-2} \theta \tan \theta$$

(b) Using the result of (a) prove that

$$\int_0^{\pi/4} \sec^{10} \theta \, d\theta = \frac{1328}{315}$$ (O & C)

Chapter 14

Differential equations

The general problem

14.1   An equation containing any *differential coefficients* such as $\dfrac{dy}{dx}$, $\dfrac{d^2y}{dx^2}$, is

called a *differential equation*; a solution of such an equation is an equation relating x and y and containing no differential coefficients.

Given the differential equation $\dfrac{dy}{dx} = 3$, we obtain the *general solution*

$y = 3x + c$, which is the equation of all straight lines of gradient 3. If the data also includes the fact that $y = 5$ when $x = 1$, we can determine that $c = 2$, and we obtain the *particular solution* $y = 3x + 2$.

Thus, in simple graphical terms,

  (a) a **differential equation** defines some property common to a family of curves,
  (b) the **general solution**, involving one or more arbitrary constants, is the equation of *any* member of the family,
  (c) a **particular solution** is the equation of *one* member of the family.

Consider the differential equation $\dfrac{dy}{dx} = x$. We can easily solve this with our

existing knowledge, but before we do so, consider for a moment what this differential equation tells us: it says that, for any value of x, the gradient is equal to x. This information is illustrated in Fig. 14.1.

The solution of the differential equation is

$$y = \tfrac{1}{2}x^2 + c$$

(in this context, the constant of integration, c, is usually called 'an **arbitrary constant**'). Equations of this form represent parabolas with the y-axis as the axis of symmetry (see Fig. 14.2).

As with the family of lines $y = 3x + c$, above, if we are given some further information, say when $x = 0$, $y = 2$, then we can find c and identify the particular parabola with this property, in this case $y = \tfrac{1}{2}x^2 + 2$.

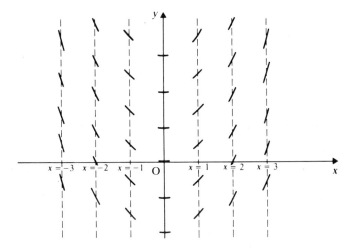

Figure 14.1

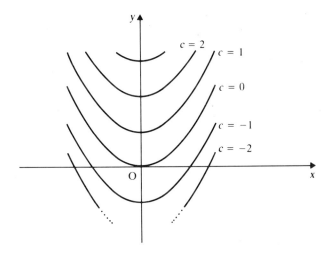

Figure 14.2

Qu. 1 Find the general solution of $\dfrac{d^2y}{dx^2} = 0$. What is the particular solution given by $\dfrac{dy}{dx} = 3$, and $y = -2$ when $x = 1$?

Qu. 2 Find the general solution of $\dfrac{dy}{dx} = 3x^2$. Illustrate with a sketch.

Qu. 3 For any circle centre the origin $\dfrac{dy}{dx} = -\dfrac{x}{y}$. Solve this equation by writing it as $y\dfrac{dy}{dx} = -x$. $\left(\text{What is } \dfrac{d}{dx}(y^2)?\right)$

Qu. 4 Find the general solution of $\dfrac{d^2s}{dt^2} = a$, where a is a constant. What does

this become, given the initial conditions $s = 0$ and $\dfrac{ds}{dt} = u$ when $t = 0$?

Definition

*The **order** of a differential equation is determined by the highest differential coefficient present.*

Thus the equations in Qu. 1 and 4 are of the second order, whereas those in Qu. 2 and 3 are of the first order.

Since each step of integration introduces one arbitrary constant, it is in general true that the order of a differential equation gives us the number of arbitrary constants in the general solution.

This suggests that from an equation involving x, y, and n arbitrary constants there may be formed (by differentiating n times and eliminating the constants) a differential equation of order n.

Qu. 5 Form differential equations by differentiating and eliminating the constants A, B from the following:

(a) $y = Ax + B$,   (b) $y = Ax$,   (c) $r = A \cos \theta$,
(d) $xy = A$,   (e) $y = A e^x$,   (f) $y = e^{Bx}$,
(g) $y = A e^{Bx}$,   (h) $y = A \ln x$, (i) $x = \tan (Ay)$.

Qu. 6 Confirm the given general solution of each of the following differential equations:

(a) $\dfrac{d^2y}{dx^2} - \dfrac{dy}{dx} - 2y = 0$, $y = A e^{2x} + B e^{-x}$,

(b) $\dfrac{d^2x}{dt^2} - 4\dfrac{dx}{dt} + 4x = 0$, $x = e^{2t}(A + Bt)$.

We must now classify some of the simpler forms of differential equations.

First order — separating the variables

14.2 The solutions of $\dfrac{dy}{dx} = f(x)$ and $\dfrac{dy}{dx} = f(y)$ $\left(\text{which may be written}\right.$

$\dfrac{dx}{dy} = \dfrac{1}{f(y)}\right)$ depend upon the integrals $\int f(x)\,dx$ and $\displaystyle\int \dfrac{1}{f(y)}\,dy$. There are other differential equations equally susceptible to direct integration once they have been written in a suitable form.

Consider

$$\frac{dy}{dx} = xy \tag{1}$$

We write this as

$$\frac{1}{y}\frac{dy}{dx} = x$$

then integrating each side with respect to x,

$$\int \frac{1}{y}\frac{dy}{dx}\,dx = \int x\,dx$$

$$\left[\text{But from §1.5 we know that } \int f(y)\,dy = \int f(y)\frac{dy}{dx}\,dx. \right]$$

$$\therefore \int \frac{1}{y}\,dy = \int x\,dx \qquad (2)$$

$$\therefore \ln y + c = \tfrac{1}{2}x^2$$
$$\therefore \ln(ky) = \tfrac{1}{2}x^2, \quad \text{or} \quad y = A\,e^{x^2/2}$$

Note how the arbitrary constant of integration appears in different forms; we have written c as $\ln k$, and A as $1/k$.

Now let us look back at (1) and (2) in the above working. The symbols dx, dy have as yet no meaning for us in isolation; they have been used only in composite symbols such as $\frac{dy}{dx}$, $\frac{d}{dx}f(x)$, $\int f(x)\,dx$. However, in the present circumstances it is convenient to think of dx as an 'x-factor', and dy as a 'y-factor', and proceed direct from (1) to (2) by 'separating the variables' and adding the integral sign. The intervening lines provide the justification for this.

Example 1 *Solve* $x^2\dfrac{dy}{dx} = y(y-1)$.

$$x^2\frac{dy}{dx} = y(y-1)$$

Separating the variables,

$$\int \frac{1}{y(y-1)}\,dy = \int \frac{1}{x^2}\,dx$$

$$\therefore \int \left\{ \frac{1}{y-1} - \frac{1}{y} \right\}\,dy = \int \frac{1}{x^2}\,dx$$

$$\therefore \ln \frac{k(y-1)}{y} = -\frac{1}{x}$$

or $\qquad k(y-1) = y\,e^{-1/x}$

Qu. 7 Solve the following differential equations, and check solutions by differentiation and elimination of arbitrary constants:

(a) $\dfrac{dy}{dx} = \dfrac{x}{y}$,   (b) $\dfrac{dy}{dx} = \dfrac{y}{x}$,   (c) $\dfrac{dx}{dy} = xy$,

(d) $x\dfrac{dy}{dx} = \tan y,$ (e) $e^{-x}\dfrac{dy}{dx} = y^2 - 1,$ (f) $\sqrt{(x^2 + 1)}\dfrac{dy}{dx} = \dfrac{x}{y}.$

Qu. 8 $v\dfrac{dv}{ds} = a,$ where a is a constant. Solve this equation given that $v = u$ when $s = 0.$

Since Newton's time, many physical problems have been expressed in terms of differential equations (readers who are studying applied mathematics or physics have probably met some already). Solving, or at least attempting to solve, a differential equation is a very common task for a scientist, and nowadays the problem frequently arises in other disciplines, such as economics. What follows is an important application of the subject in physics.

It is known that radioactive substances decay at a rate which is proportional to the amount of the radioactive substance present. If we use x to represent the amount present at time t, we can express this in the form of a differential equation, namely,

$$\frac{dx}{dt} = -kx \qquad \text{where } k \text{ is a positive constant.}$$

For different substances, the rate of decay is different; it is usual to quote the 'half-life' of the substance, that is, the time it takes for half of the original quantity to decay. For radium the half-life is about 1600 years. We shall now solve the differential equation, that is, express x as a function of t, and hence find the value of k. We shall then use the solution to find the percentage of a given sample of radium which would still exist after a lapse of 200 years in storage. [Remember, in the following working, to distinguish between the arbitrary constant of integration A (or x_0), and k, a constant which is determined by the half-life of radium.]

$$\frac{dx}{dt} = -kx$$

Separating the variables gives

$$\int \frac{1}{x}\, dx = \int -k\, dt$$

and integrating,

$$\ln x = -kt + c$$
$$\therefore x = e^{-kt + c}$$
$$= e^c \times e^{-kt}$$

and replacing e^c by A, we can write

$$x = Ae^{-kt}$$

This is the general solution of the differential equation. (This particular differential equation is extremely common, and, unless specific instructions to the contrary are given, the solution $x = Ae^{-kt}$ may be quoted.) We now continue with the solution.

When $t = 0$, $x = A$, in other words A is the original value of x. It is convenient to write this as x_0, so

$$x = x_0 e^{-kt}$$

Now, we are told that when $t = 1600$, $x = \frac{1}{2}x_0$, consequently

$$\frac{1}{2}x_0 = x_0 e^{-1600k}$$
$$\therefore \frac{1}{2} = e^{-1600k}$$

i.e. $e^{1600k} = 2$

Taking natural logarithms,

$$1600k = \ln 2,$$

$$k = \frac{\ln 2}{1600}$$

Hence the solution can be expressed

$$x = x_0 e^{-(\ln 2/1600)t}$$

This, in turn, can be written

$$x = x_0(e^{\ln 2})^{-t/1600}$$

But $e^{\ln 2} = 2$, (see §2.8; this step is extremely common in this topic), hence

$$x = x_0 2^{-t/1600}$$

(We can verify by inspection that when $t = 1600$, $x = \frac{1}{2}x_0$. It is important to check your work like this whenever possible.)

Finally, when $t = 200$,

$$x = x_0 2^{-1/8}$$
$$= 0.917 x_0$$

In other words, after 200 years, 91.7% of the original radioactive radium still exists.

Exercise 14a

1 By differentiating and eliminating the constants A and B from the following equations, form differential equations, and illustrate the geometrical significance of each:

(a) $3x - 2y + A = 0$,　(b) $Ax + 2y + 1 = 0$,　(c) $Ax + By = 0$,
(d) $x^2 + y^2 = A$,　(e) $y = Ax^{-1}$,　(f) $y = A(x - 4)$.

2 By differentiating and eliminating the constants A and B from the following equations, form differential equations:

(a) $y = A \cos(3t + B)$,　(b) $y = A + B e^{3t}$,　(c) $y = A e^{3x} + B e^{-3x}$,
(d) $y = A e^{3x} + B e^{-2x}$　(first multiply each side by e^{2x}),

(e) $y = e^{4x}(A + Bx)$ $\left(\text{first show that } \dfrac{dy}{dx} = 4y + B\,e^{4x}\right).$

3 Obtain the equation of the straight line of gradient $\frac{3}{10}$, which passes through $(5, -2)$, by finding a particular solution of the differential equation $\dfrac{dy}{dx} = \frac{3}{10}.$

4 A family of parabolas has the differential equation $\dfrac{dy}{dx} = 2x - 3$. Find the equation of that member of the family which passes through $(4, 5)$.

5 Find the general solution of the differential equation $6t\,\dfrac{dt}{ds} + 1 = 0$, and the particular solution given by the condition $s = 0$ when $t = -2$.

6 Find the particular solutions of the differential equation

$$\operatorname{cosec} x\,\frac{dy}{dx} = e^x \operatorname{cosec} x + 3x$$

given by the conditions (a) $y = 0$ when $x = 0$, (b) $y = 3$ when $x = \pi/2$.

7 Find the general solutions of the following differential equations:

(a) $\dfrac{dy}{dx} = y,$

(b) $\dfrac{1}{x}\dfrac{dy}{dx} = \sqrt{(x - 1)},$

(c) $(x + 2)\dfrac{dy}{dx} = y,$

(d) $\dfrac{dy}{dx} = \sec^2 y,$

(e) $\dfrac{dv}{du} = v(v - 1),$

(f) $\ln x\,\dfrac{dx}{dy} = 1,$

(g) $\dfrac{dy}{dx} = \tan y,$

(h) $\tan^{-1} y\,\dfrac{dy}{dx} = 1,$

(i) $y\dfrac{dy}{dx} = x - 1,$

(j) $(x^2 - 1)\dfrac{dy}{dx} = y,$

(k) $\dfrac{d\theta}{dr} = \sin \theta,$

(l) $x^2\dfrac{dy}{dx} = y + 3,$

(m) $x\dfrac{dy}{dx} = y + xy,$

(n) $\dfrac{d\phi}{d\theta} = \tan \phi \tan \theta,$

(o) $\theta\dfrac{d\theta}{dr} = \cos^2 \theta,$

(p) $\dfrac{y}{x}\dfrac{dy}{dx} = \ln x,$

(q) $2 \sin \theta\,\dfrac{d\theta}{dr} = \cos \theta - \sin \theta,$

(r) $x\dfrac{dy}{dx} - 3 = 2\left(y + \dfrac{dy}{dx}\right),$

(s) $e^t\dfrac{dx}{dt} = \sin t,$

(t) $e^x\dfrac{dy}{dx} + y^2 + 4 = 0.$

8 Find the particular solutions of the following differential equations which

satisfy the given conditions:

(a) $(1 + \cos 2\theta)\dfrac{dy}{d\theta} = 2, \quad y = 1 \quad$ when $\theta = \pi/4$,

(b) $\dfrac{dy}{dx} = x(y - 2), \quad y = 5 \quad$ when $x = 0$,

(c) $(1 + x^2)\dfrac{dy}{dx} = 1 + y^2, \quad y = 3 \quad$ when $x = 2$,

(d) $\dfrac{dy}{dx} = \sqrt{(1 - y^2)}, \quad y = 0 \quad$ when $x = \pi/6$.

9 According to Newton's law of cooling, the rate at which the temperature of a body falls is proportional to the amount by which its temperature exceeds that of its surroundings. Suppose the temperature of an object falls from 200° to 100° in 40 minutes, in a surrounding temperature of 10°. Prove that after t minutes, the temperature, T degrees, of the body is given by

$$T = 10 + 190e^{-kt} \qquad \text{where } k = \tfrac{1}{40}\ln\left(\tfrac{19}{9}\right)$$

Calculate the time it takes to reach 50°.

10 A tank contains a solution of salt in water. Initially the tank contains 1000 litres of water with 10 kg of salt dissolved in it. The mixture is poured off at a rate of 20 litres per minute, and simultaneously pure water is added at a rate of 20 litres per minute. All the time the tank is stirred to keep the mixture uniform. Find the mass of salt in the tank after 5 minutes. The tank must be topped up by adding more salt when the mass of salt in the tank falls to 5 kg; after how many minutes will it need topping up?

First order exact equations

14.3 The equation

$$2xy\frac{dy}{dx} + y^2 = e^{2x}$$

is not one in which the variables may be separated. However, the L.H.S. is $\dfrac{d}{dx}(xy^2)$ and the equation may be solved by integrating each side with respect to x; it is called an **exact equation** and the solution is

$$xy^2 = \tfrac{1}{2}e^{2x} + A$$

Qu. 9 Solve the following exact equations:

(a) $x^2\dfrac{dy}{dx} + 2xy = 1,$

(b) $\dfrac{t^2}{x}\dfrac{dx}{dt} + 2t \ln x = 3 \cos t,$

(c) $x^2 \cos u \dfrac{du}{dx} + 2x \sin u = \dfrac{1}{x},$

(d) $e^y + x e^y \dfrac{dy}{dx} = 2.$

Integrating factors

14.4   There are some differential equations which are not *exact* as they stand, but which may be made so by multiplying each side by an **integrating factor**.

Example 2   *Solve* $xy\dfrac{dy}{dx} + y^2 = 3x$.

> We cannot separate the variables. Can we find a function whose derivative is the L.H.S. as in §14.3? No. Then can we find a function whose derivative is $f(x) \times$ L.H.S.?
>
> $$\frac{d}{dx}(xy^2) = y^2 + 2xy\frac{dy}{dx}; \quad \text{this is no good.}$$
>
> $$\frac{d}{dx}(x^2y^2) = 2xy^2 + 2x^2y \times \frac{dy}{dx} = 2x\left(y^2 + xy\frac{dy}{dx}\right) = 2x \times \text{L.H.S.}$$
>
> The required integrating factor is $2x$.

$$xy\frac{dy}{dx} + y^2 = 3x$$

Multiplying each side by $2x$,

$$2x^2y\frac{dy}{dx} + 2xy^2 = 6x^2$$

$$\therefore x^2y^2 = 2x^3 + A$$

Qu. 10   Find the integrating factors required to make the following differential equations into exact equations, and solve them:

(a) $x\dfrac{dy}{dx} + 2y = e^{x^2}$,   (b) $x\,e^y\dfrac{dy}{dx} + 2e^y = x$,

(c) $2x^2y\dfrac{dy}{dx} + xy^2 = 1$,   (d) $r\sec^2\theta + 2\tan\theta\dfrac{dr}{d\theta} = 2r^{-1}$.

First order linear equations

14.5   A differential equation is *linear in y* if it is of the form

$$\frac{d^n y}{dx^n} + P_1\frac{d^{n-1}y}{dx^{n-1}} + P_2\frac{d^{n-2}y}{dx^{n-2}} + \ldots + P_{n-1}\frac{dy}{dx} + P_n y = Q$$

where $P_1, P_2, \ldots, P_n, Q$ are functions of x, or constants; it is of the nth order. Thus a *first order linear equation* is of the form

$$\frac{dy}{dx} + Py = Q$$

where P, Q are functions of x or constants. This type of differential equation deserves special attention because an integrating factor, when required and if obtainable, is of a standard form.

Let us assume that the general first order linear equation given above can be made into an exact equation by using the integrating factor R, a function of x. If this is so,

$$R\frac{dy}{dx} + RPy = RQ \qquad (1)$$

is an exact equation, and it is apparent from the first term that the L.H.S. of (1) is $\frac{d}{dx}(Ry) = R\frac{dy}{dx} + y\frac{dR}{dx}$. Thus (1) may also be written

$$R\frac{dy}{dx} + y\frac{dR}{dx} = RQ \qquad (2)$$

Equating the second terms on the L.H.S. of (1) and (2),

$$y\frac{dR}{dx} = RPy$$

$$\therefore \frac{dR}{dx} = RP$$

Separating the variables,

$$\int \frac{1}{R}\,dR = \int P\,dx$$

$$\therefore \ln R = \int P\,dx$$

$$\therefore R = e^{\int P\,dx}$$

Thus the required **integrating factor is** $e^{\int P\,dx}$. The initial assumption that an integrating factor exists is therefore justified provided that it is possible to find $\int P\,dx$.

Example 3 *Solve the differential equation* $\dfrac{dy}{dx} + 3y = e^{2x}$, *given that* $y = \frac{6}{5}$ *when* $x = 0$.

The integrating factor is $e^{\int 3\,dx} = e^{3x}$. Multiplying each side of the given equation by e^{3x},

$$e^{3x}\frac{dy}{dx} + 3e^{3x}\,y = e^{5x}$$

$$\therefore e^{3x}\,y = \tfrac{1}{5}e^{5x} + A$$

Therefore the general solution is

$$y = \tfrac{1}{5}e^{2x} + A\,e^{-3x}$$

But $y = \tfrac{6}{5}$ when $x = 0$, $\therefore \tfrac{6}{5} = \tfrac{1}{5} + A$, $\therefore A = 1$.
  Therefore the particular solution is

$$y = \tfrac{1}{5}e^{2x} + e^{-3x}$$

Example 4   *Solve* $\dfrac{dy}{dx} + y \cot x = \cos x$.

  The integrating factor is

$$e^{\int \cot x \, dx} = e^{\ln \sin x} = \sin x \quad \text{(see §2.8)}$$

Multiplying each side of the given equation by $\sin x$,

$$\sin x \frac{dy}{dx} + y \cos x = \cos x \sin x$$

$$\therefore \ y \sin x = \tfrac{1}{2}\sin^2 x + A$$

Therefore the general solution is

$$y = \tfrac{1}{2}\sin x + A \operatorname{cosec} x$$

Qu. 11   Find the general solution of $\dfrac{dy}{dx} + 2xy = x$. What is the particular
solution given by $y = -\tfrac{1}{2}$ when $x = 0$?
Qu. 12   Show that the equation in Qu. 10 (a) is of the type under discussion, and
find the required integrating factor as $e^{\int P \, dx}$.
Qu. 13   Solve: (a) $\dfrac{dy}{dx} - y \tan x = x$,   (b) $\dfrac{dy}{dx} + y + 3 = x$.

First order homogeneous equations

14.6   In a homogeneous differential equation all the terms are of the same
dimensions. To obtain a clear picture of what is meant by this, suppose x and y
to measure units of length. The term $x^2 y$ is of dimensions (length)^3, or L^3; the
term $\dfrac{(x^2 + y^2)^2}{x}$ is of dimensions $\dfrac{L^4}{L} = L^3$.

  The dimensions of some other terms are given below:

$$\frac{y}{x} \qquad\qquad \frac{L}{L} = L^0$$

$$\frac{y}{x^2} \qquad\qquad \frac{L}{L^2} = L^{-1}$$

$$\frac{dy}{dx} = \lim_{\delta x \to 0} \frac{\delta y}{\delta x} \qquad\qquad \frac{L}{L} = L^0$$

$$\frac{d^2y}{dx^2} = \lim_{\delta x \to 0} \frac{\delta\left(\frac{dy}{dx}\right)}{\delta x} \qquad \frac{L^0}{L} = L^{-1}$$

$$x\frac{dy}{dx} \qquad\qquad L \times L^0 = L^1$$

$$\frac{\left(\frac{y^2}{x}\right)^2}{\frac{dy}{dx}} \qquad\qquad \frac{L^2}{L^0} = L^2$$

Qu. 14 Pick out that member of each of the following groups of terms and expressions which is not of the same dimensions as the rest:

(a) $xy\dfrac{dy}{dx}, \quad y^3\dfrac{d^2y}{dx^2}, \quad \left(\dfrac{y}{x}\right)^2, \quad x^2 + y^2,$

(b) $(x + y)^2\dfrac{dy}{dx}, \quad x^2\left(1 + \dfrac{y}{x}\right), \quad \left(\dfrac{dy}{dx}\right)^2 xy, \quad \dfrac{d^2y}{dx^2} + xy,$

(c) $(y + 2x)\dfrac{dy}{dx}, \quad (y^2 - x^2)\dfrac{d^2y}{dx^2}, \quad x\sqrt{(x^2 + y^2)}, \quad 2x + \dfrac{y^2}{x}.$

Qu. 15 Which of the following equations are homogeneous?

(a) $x^2\dfrac{dy}{dx} = y^2,$ (b) $xy\dfrac{dy}{dx} = x^2 + y^2,$ (c) $x^2\dfrac{dy}{dx} = 1 + xy,$

(d) $x^2\dfrac{d^2y}{dx^2} = y\dfrac{dy}{dx},$ (e) $(x^2 - y^2)\dfrac{dy}{dx} = 2xy,$ (f) $(1 + y^2)\dfrac{dy}{dx} = x^2.$

A first order homogeneous equation is of the form

$$P\frac{dy}{dx} = Q$$

Since $\dfrac{dy}{dx}$ is of dimensions 0, P and Q are homogeneous functions of x and y of the same dimensions, i.e. of the same degree. The significant point to note is that, if P and Q are of degree n, we may divide each side of the equation by x^n and thereby obtain

$$P'\frac{dy}{dx} = Q'$$

where P' and Q' are functions of y/x.
  For example, the equation

$$xy\frac{dy}{dx} = x^2 + y^2$$

when each side is divided by x^2, becomes

$$\frac{y}{x}\frac{dy}{dx} = 1 + \left(\frac{y}{x}\right)^2$$

This suggests the substitutions

$$\frac{y}{x} = u \quad \text{and, since } y = ux, \quad \frac{dy}{dx} = u + x\frac{du}{dx}$$

Example 5   *Solve* $xy\dfrac{dy}{dx} = x^2 + y^2.$

Dividing each side by x^2,

$$\frac{y}{x}\frac{dy}{dx} = 1 + \left(\frac{y}{x}\right)^2$$

Let $y = ux$,   then $\dfrac{dy}{dx} = u + x\dfrac{du}{dx}.$

$$\therefore u\left(u + x\frac{du}{dx}\right) = 1 + u^2$$

$$\therefore ux\frac{du}{dx} = 1$$

Separating the variables,

$$\int u\, du = \int \frac{1}{x}\, dx$$

$$\therefore \tfrac{1}{2}u^2 = \ln(Bx)$$

$$\therefore \left(\frac{y}{x}\right)^2 = 2\ln(Bx)$$

$$\therefore \left(\frac{y}{x}\right)^2 = \ln(Ax^2) \qquad \text{where } A = B^2$$

Therefore the general solution is

$$y^2 = x^2 \ln(Ax^2)$$

Qu. 16   Solve the following equations by the method of Example 5,

(a) $x^2\dfrac{dy}{dx} = y^2 + xy,$ (b) $x\dfrac{dy}{dx} = x - y,$ (c) $x^2\dfrac{dy}{dx} = 2y^2.$

Qu. 17   Solve the equation $x\dfrac{dy}{dx} = 2x + y$ (a) by the method of Example 5,
(b) by the method of Example 4.

Qu. 18 Solve the equations in Qu. 16 (b) and (c) *not* using the method of Example 5.

The above questions serve not only to illustrate the method under discussion but also to stress that the types of equations given in this chapter are not all mutually exclusive.

Exercise 14b

1 Solve the following exact differential equations:

(a) $y^2 + 2xy \dfrac{dy}{dx} = \dfrac{1}{x^2}$,

(b) $xy^2 + x^2y \dfrac{dy}{dx} = \sec^2 2x$,

(c) $\ln y + \dfrac{x}{y} \dfrac{dy}{dx} = \sec x \tan x$,

(d) $(1 - 2x) e^y \dfrac{dy}{dx} - 2e^y = \sec^2 x$,

(e) $2t\, e^s + t^2\, e^s \dfrac{ds}{dt} = \sin t + t \cos t$,

(f) $e^u r^2 + 2r\, e^u \dfrac{dr}{du} = -\operatorname{cosec}^2 u$.

2 Find, by inspection, the integrating factors required to make the following differential equations into exact equations, and solve them:

(a) $\sin y + \dfrac{1}{2}x \cos y \dfrac{dy}{dx} = 3$,

(b) $\dfrac{dy}{dx} + \dfrac{y}{x} = \dfrac{e^x}{x}$,

(c) $\dfrac{1}{x} \tan y + \sec^2 y \dfrac{dy}{dx} = 2e^{x^2}$,

(d) $y\, e^x + y^2\, e^x \dfrac{dx}{dy} = 1$.

3 Solve the following first order linear equations:

(a) $\dfrac{dy}{dx} + 2y = e^{-2x} \cos x$,

(b) $\dfrac{1}{t} \dfrac{ds}{dt} = 1 - 2s$,

(c) $\dfrac{dy}{dx} + (2x + 1)y - e^{-x^2} = 0$,

(d) $\dfrac{dr}{d\theta} + 2r \cot \theta = \operatorname{cosec}^2 \theta$,

(e) $\dfrac{dr}{d\theta} + r \tan \theta = \cos \theta$,

(f) $x \dfrac{dy}{dx} + 2y = \dfrac{\cos x}{x}$,

(g) $x \dfrac{dy}{dx} - y = \dfrac{x}{x-1}$,

(h) $2x \dfrac{dy}{dx} = x - y + 3$,

(i) $\sin x \dfrac{dy}{dx} + y = \sin^2 x$,

(j) $3y + (x - 2) \dfrac{dy}{dx} = \dfrac{2}{x-2}$.

4 Solve the following homogeneous equations:

(a) $x^2 \dfrac{dy}{dx} = 3x^2 + xy$,

(b) $xy \dfrac{dy}{dx} = x^2 - y^2$,

(c) $x^2 \dfrac{dy}{dx} = x^2 + xy + y^2$,

(d) $3x^2 \dfrac{dy}{dx} = y^2$,

(e) $(x^2 + y^2)\dfrac{dy}{dx} = xy,$ (f) $(4x - y)\dfrac{dy}{dx} = 4x,$

(g) $x\dfrac{dy}{dx} = y + \sqrt{(x^2 - y^2)},$ (h) $x\dfrac{dy}{dx} = x + 2y,$

(i) $y\dfrac{dy}{dx} = 2x + y,$ (j) $x^2\dfrac{dy}{dx} = x^2 + y^2.$

*5 Solve the equation $\dfrac{dy}{dx} = \dfrac{x - y + 2}{x + y}$, reducing it to a homogeneous equation

by the change of variables $x = X - 1,\ y = Y + 1.$ $\Bigg($ Note that this implies

a change of origin to $(-1, 1)$ the point of intersection of the straight lines
$x - y + 2 = 0$ and $x + y = 0$; see §6.5. The new axes are parallel to the old so
$\dfrac{dy}{dx} = \dfrac{dY}{dX}.\Bigg)$

6 Solve the following equations by the method indicated in No. 5:

(a) $\dfrac{dy}{dx} = \dfrac{y - 2}{x + y - 5},$ (b) $2y\dfrac{dy}{dx} = x + y - 3.$

*7 State why the equation $\dfrac{dy}{dx} = \dfrac{y - x + 2}{y - x - 4}$ may not be reduced to a homo-

geneous equation by the method of No. 5. Solve it by the change of variable
$y - x = z.$

8 Solve the following equations by the method indicated in No. 7:

(a) $\dfrac{dy}{dx} = \dfrac{2x + y - 2}{2x + y + 1},$ (b) $(x + y)\dfrac{dy}{dx} = x + y - 2.$

9 Solve the following differential equations:

(a) $(x + 3)\dfrac{dy}{dx} - 2y = (x + 3)^3,$

(b) $x^2\dfrac{dy}{dx} = x^2 - xy + y^2,$

(c) $\dfrac{dy}{dx} + (y + 3)\cot x = e^{-2x}\operatorname{cosec} x,$

(d) $\sin y + (x + 3)\cos y\dfrac{dy}{dx} = \dfrac{1}{x^2},$

(e) $(x - 4y + 2)\dfrac{dy}{dx} = x + y - 3,$

(f) $\dfrac{dy}{dx} = y + 2 + e^{2x}(x + 1),$

(g) $2y \ln y + x\dfrac{dy}{dx} = \dfrac{y}{x} \cot x,$

(h) $2 \tan \theta \dfrac{dr}{d\theta} + (2r + 3) \tan^2 \theta + 2r = 0,$

(i) $x(y - x)\dfrac{dy}{dx} = y(x + y),$

(j) $\dfrac{dy}{dx} = \dfrac{x - y + 1}{x - y + 3}.$

10 Find the particular solutions of the following differential equations which satisfy the given conditions:

(a) $(x + 1)\dfrac{dy}{dx} - 3y = (x + 1)^4, \quad y = 16$ when $x = 1,$

(b) $\dfrac{du}{d\theta} + u \cot \theta = 2 \cos \theta, \quad u = 3$ when $\theta = \dfrac{\pi}{2},$

(c) $(x + y)\dfrac{dy}{dx} = x - y, \quad y = -2$ when $x = 3,$

(d) $(x^2 - y^2)\dfrac{dy}{dx} = xy, \quad y = 2$ when $x = 4,$

(e) $x - 1 + \dfrac{dx}{dt} = e^{-t} t^{-2}, \quad x = 1$ when $t = 1.$

Second order equations reducible to first order form

14.7   In this section and the next, we shall consider some rather special second order differential equations which can be reduced to first order form. In the following chapter we shall tackle more general second order differential equations.

To the form $\dfrac{d^2 y}{dx^2} = f(x)$ we may apply direct integration twice, as also to an equation such as

$$x\frac{d^2 y}{dx^2} + \frac{dy}{dx} = 2x$$

which is exact, giving

$$x\frac{dy}{dx} = x^2 + A \quad \text{and} \quad y = \frac{1}{2}x^2 + A \ln x + B$$

Of wide application to other forms of second order equations is the substitution $\dfrac{dy}{dx} = p$, from which we obtain

$$\frac{d^2y}{dx^2} = \frac{dp}{dx} = \frac{dp}{dy} \times \frac{dy}{dx} = p\frac{dp}{dy}$$

Thus

(a) the equation $\dfrac{d^2y}{dx^2} = f(y)$ becomes $p\dfrac{dp}{dy} = f(y)$,

(b) an equation containing $\dfrac{d^2y}{dx^2}, \dfrac{dy}{dx}$, y but *with x absent*, becomes a *first order* equation containing $p\dfrac{dp}{dy}$, p, y,

(c) an equation containing $\dfrac{d^2y}{dx^2}, \dfrac{dy}{dx}$, x but *with y absent*, becomes a *first order* equation containing $\dfrac{dp}{dx}$, p, x.

Example 6   *Solve* $(1 + x^2)\dfrac{d^2y}{dx^2} = 2x\dfrac{dy}{dx}$.

Let $\dfrac{dy}{dx} = p$, and since y is absent, write $\dfrac{d^2y}{dx^2}$ as $\dfrac{dp}{dx}$.

$$(1 + x^2)\frac{dp}{dx} = 2xp$$

Separating the variables,

$$\int \frac{1}{p}\, dp = \int \frac{2x}{1 + x^2}\, dx$$

$$\therefore\ \ln p = \ln\{C(1 + x^2)\}$$

$$\therefore\ p = \frac{dy}{dx} = C + Cx^2$$

$$\therefore\ y = Cx + \tfrac{1}{3}Cx^3 + B$$

Therefore, writing $3A$ for C, the general solution is

$$y = Ax^3 + 3Ax + B \tag{1}$$

This equation contains two arbitrary constants A and B. When we considered the solution of $\dfrac{dy}{dx} = 2x$ in §14.1, we saw that the solution $y = x^2 + c$ represented a set of curves (see Fig. 14.2), and that if we were given some further information, for instance a point through which a particular curve passes, we could find the value of c which gives the equation of this particular solution. The solution (1) above represents a set of cubic curves. To identify one particular member of this set, we must be given sufficient information to find the values of both constants. This could be done either by giving *two* points through which the curve passes, or by giving *one* point and the gradient at that point (or indeed at some other point). For example if we are told that the curve passes through $(0, -5)$ and $(1, 3)$, then substituting these coordinates into (1), we obtain

$$-5 = A \times 0^3 + 3 \times A \times 0 + B$$
$$\therefore B = -5$$

and

$$3 = A \times 1^3 + 3 \times A \times 1 + B$$
$$\therefore 4A + B = 3$$
$$4A = 8$$
$$A = 2$$

So the particular member of the set of curves represented by (1) which passes through the two given points is

$$y = 2x^3 + 6x - 5$$

Qu. 19   Solve: (a) $x\dfrac{d^2y}{dx^2} = 2,$ (b) $\dfrac{d^2y}{dx^2} = x\cos x,$

(c) $x\dfrac{d^2y}{dx^2} + \dfrac{dy}{dx} = 9x^2,$ (d) $y\dfrac{d^2y}{dx^2} + \left(\dfrac{dy}{dx}\right)^2 = \cos x.$

Qu. 20   Solve: (a) $y\dfrac{d^2y}{dx^2} = \left(\dfrac{dy}{dx}\right)^2,$ giving the general solution.

(b) $(2x - 1)\dfrac{d^2y}{dx^2} - 2\dfrac{dy}{dx} = 0,$ given that when $x = 0,$

$$y = 2 \text{ and } \dfrac{dy}{dx} = 3.$$

Qu. 21   Write the differential equation $2\dfrac{dy}{dx} + x\dfrac{d^2y}{dx^2} = \dfrac{2}{x},$ by means of the

substitution $\dfrac{dy}{dx} = p,$ as a differential equation linear in $p,$ and proceed as in §14.4.

Simple harmonic motion

14.8   The substitutions mentioned in §14.7 arise in a less abstract form in mechanics. With the usual notation, the velocity

$$v = \dfrac{dx}{dt} \quad \left(\text{compare with } p = \dfrac{dy}{dx}\right)$$

and the acceleration is

$$\frac{\mathrm{d}^2x}{\mathrm{d}t^2} = \frac{\mathbf{d}v}{\mathbf{d}t} = \frac{\mathrm{d}v}{\mathrm{d}x} \times \frac{\mathrm{d}x}{\mathrm{d}t} = v\frac{\mathbf{d}v}{\mathbf{d}x}$$

The reader may already appreciate that in dealing with variable forces, Newton's Second Law of Motion may be usefully written

$$P = m\frac{\mathrm{d}v}{\mathrm{d}t}, \quad \text{if } P \text{ is a function of } t$$

or $$P = mv\frac{\mathrm{d}v}{\mathrm{d}x}, \quad \text{if } P \text{ is a function of } x$$

A particular case of motion under the action of a force varying with displacement is Simple Harmonic Motion (S.H.M.).

Definition

A body moves in Simple Harmonic Motion in a straight line when its acceleration is proportional to its distance from a given point on the line, and is directed always towards that point.

Before studying this section the reader should have some knowledge of this topic. We shall not confine our attention only to finding the general solution of the typical S.H.M. equation

$$\frac{\mathrm{d}^2x}{\mathrm{d}t^2} = -n^2x$$

We must discuss in some detail the constants which arise in the solution. Now the constant n in the above equation is determined by the physical situation which gives rise to S.H.M. For example, if a body of given mass hangs at rest from a given spring attached to a fixed point, and is then displaced vertically and released, it will oscillate in S.H.M.; in this case the mass of the body, and the natural length and elasticity of the spring, together determine the constant n, and the **periodic time** $2\pi/n$. But n must not be confused with the two *arbitrary constants* which will arise in the general solution of the above second order differential equation.

  (a) Quite independent of the periodic time is an *arbitrary* **amplitude** a, the maximum displacement from the centre of oscillation (dependent in the above example upon how far we displace the body from its equilibrium position before releasing it).
  (b) The general solution of the S.H.M. equation will give the displacement x from the centre of oscillation at time t; here is the second *arbitrary* choice, the instant at which we take t to be zero.

Example 7   *Find the general solution of the Simple Harmonic Motion equation*
$$\frac{\mathrm{d}^2x}{\mathrm{d}t^2} = -n^2x.$$

$$\left[\text{Since } t \text{ is absent, we write } \frac{d^2x}{dt^2} \text{ as } v\frac{dv}{dx}.\right]$$

$$v\frac{dv}{dx} = -n^2x$$

$$\therefore \tfrac{1}{2}v^2 = -\tfrac{1}{2}n^2x^2 + c$$

[At this stage we prefer to express the arbitrary c in terms of the arbitrary amplitude a.]

If the amplitude is a, $v = 0$ when $x = a$,

$$\therefore 0 = -\tfrac{1}{2}n^2a^2 + c \quad \text{whence} \quad c = \tfrac{1}{2}n^2a^2$$

$$\therefore v^2 = n^2(a^2 - x^2)$$

We must now deal separately with the positive and negative velocities which occur in any position (other than the extreme positions when $x = \pm a$). Thus

$$\frac{dx}{dt} = +n\sqrt{(a^2 - x^2)} \quad \text{or} \quad \frac{dx}{dt} = -n\sqrt{(a^2 - x^2)}$$

Separating the variables,

$$+\int \frac{1}{\sqrt{(a^2 - x^2)}}\,dx = \int n\,dt \quad \text{or} \quad -\int \frac{1}{\sqrt{(a^2 - x^2)}}\,dx = \int n\,dt$$

[Here it is preferable to use the change of variable $x = a\cos u$ on the L.H.S. of each equation, rather than the more usual $x = a\sin u$, since it enables one to handle more easily the remaining arbitrary constant.]

The solution of these equations may be written

$$-\cos^{-1}\frac{x}{a} = nt + \varepsilon' \qquad \text{and} \qquad \cos^{-1}\frac{x}{a} = nt + \varepsilon,$$

$$\therefore x = a\cos(-nt - \varepsilon') \quad \text{and} \qquad x = a\cos(nt + \varepsilon).$$

But $\cos(-\theta) = \cos\theta$, so we may write

$$x = a\cos(nt + \varepsilon') \qquad \text{and} \qquad x = a\cos(nt + \varepsilon)$$

for motion in the

$$\text{\textit{positive}} \qquad \text{and} \qquad \text{\textit{negative}}$$

directions respectively.

At an extreme position when $x = a$, and $t = t_1$ say, the motion is changing from positive to negative direction, both solutions are valid, and

$$\cos(nt_1 + \varepsilon') = \cos(nt_1 + \varepsilon) = 1$$
$$\therefore nt_1 + \varepsilon' \text{ and } nt_1 + \varepsilon \text{ are multiples of } 2\pi$$
$$\therefore \varepsilon' = \varepsilon + 2k\pi \qquad \text{(where } k \text{ is an integer or zero)}$$

Hence $x = a \cos (nt + \varepsilon') = a \cos (nt + \varepsilon + 2k\pi) = a \cos (nt + \varepsilon)$. Therefore the motion is fully defined by the general solution

$$x = a \cos (nt + \varepsilon) \tag{1}$$

Qu. 22 Write down the general solutions of the following equations:

(a) $\dfrac{d^2x}{dt^2} = -4x$,   (b) $\dfrac{d^2y}{dx^2} + 9y = 0$,   (c) $\dfrac{d^2y}{dx^2} = -16x$.

Qu. 23 In Example 7 what integrating factor will enable you to obtain the first order equation

$$\left(\frac{dx}{dt}\right)^2 = -n^2x^2 + k?$$

Qu. 24 A Simple Harmonic Motion of amplitude 2 cm has the equation $\dfrac{d^2x}{dt^2} = -\tfrac{9}{4}x$. Write down the solution of this equation given that $x = 2$ when $t = 0$. Find expressions for $\dfrac{dx}{dt}$ (a) in terms of x, (b) in terms of t.

Qu. 25 What special form is taken by the general solution $x = a \cos (nt + \varepsilon)$ if the motion is timed (a) from an extreme position (i.e. $x = a$ when $t = 0$), (b) from the centre of oscillation?

The reader is no doubt familiar with the fact that, if a radius OP of a circle centre O radius a rotates about O with constant angular velocity n rad/sec, and Q is the projection of P on a diameter AB, Q moves with S.H.M. along AB.

Let us take $t = 0$ at Q_1, where $\angle AOP_1 = \varepsilon$ radians, and suppose that Q moves directly to position Q_2 in time t. Then $\angle P_1OP_2 = nt$ radians, and $\angle AOP_2 = (nt + \varepsilon)$ radians (Fig. 14.3). Thus if x is the displacement of Q from O at time t,

$$x = a \cos (nt + \varepsilon)$$

When $t = 0$, $x = a \cos \varepsilon$, and so we see the significance of this constant ε, which is called the **initial phase**.

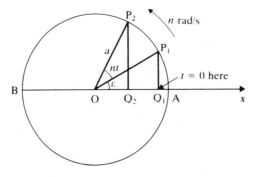

Figure 14.3

Qu. 26 What does the general solution $x = a \cos(nt + \varepsilon)$ become if the initial phase is (a) 0, (b) $-\pi/2$? Illustrate each case with a sketch.

Qu. 27 A Simple Harmonic Motion has amplitude 3 mm. If $t = 0$ when the body is $+1.5$ mm from the centre of oscillation, what is the initial phase?

The two arbitrary constants which appear in the general solution (1) are the amplitude a, and the initial phase ε. However, the general solution is often given in a form in which these are not explicitly stated.

Expanding the R.H.S. of (1)

$$x = a \cos nt \cos \varepsilon - a \sin nt \sin \varepsilon$$

or $\quad x = A \cos nt + B \sin nt$ $\hspace{3cm}$ (2)

where

$$A = a \cos \varepsilon \quad \text{and} \quad B = -a \sin \varepsilon$$

In this form we see that the amplitude

$$a = \sqrt{(A^2 + B^2)}$$

and the initial phase

$$\varepsilon = \tan^{-1}\left(-\frac{B}{A}\right)$$

Reduction to the form $\dfrac{d^2x}{dt^2} = -n^2 x$

14.9 In tackling a problem on S.H.M. the reader may inadvertently choose to measure displacement from a point other than the centre of oscillation. How an equation thus obtained may be reduced to the standard form is illustrated in the following example.

Example 8 *Solve* $\dfrac{d^2x}{dt^2} + 9x - 18 = 0.$

This equation may be written

$$\frac{d^2x}{dt^2} = -9(x - 2)$$

Let $x - 2 = u$, then $\dfrac{dx}{dt} = \dfrac{du}{dt}$ and $\dfrac{d^2x}{dt^2} = \dfrac{d^2u}{dt^2}$,

$$\therefore \frac{d^2u}{dt^2} = -9u$$

The general solution of this equation is

$$u = a \cos(3t + \varepsilon)$$

But $x = u + 2$,

$\therefore x = a \cos(3t + \varepsilon) + 2$

Qu. 28   Solve the following differential equations:

(a) $\dfrac{d^2 y}{dx^2} + 4y + 4 = 0$,   (b) $\dfrac{d^2 \theta}{dt^2} + 2\theta - 6 = 0$,   (c) $\dfrac{d^2 x}{dt^2} + \frac{9}{4}t = -1$.

Exercise 14c

1 Solve the following differential equations:

(a) $x \dfrac{d^2 y}{dx^2} - 1 = 0$,

(b) $\cos\theta \dfrac{d^2 y}{d\theta^2} - \sin\theta \dfrac{dy}{d\theta} = \cos\theta$,

(c) $e^x \dfrac{d^2 y}{dx^2} = 2$,

(d) $(2x + 1)\dfrac{d^2 y}{dx^2} + 2\dfrac{dy}{dx} = 0$,

(e) $\dfrac{1}{x}\dfrac{d^2 y}{dx^2} - \dfrac{1}{x^2}\dfrac{dy}{dx} = e^x$.

2 Use the substitution $\dfrac{dy}{dx} = p$ to solve the following differential equations:

(a) $x \dfrac{d^2 y}{dx^2} = 3\dfrac{dy}{dx}$,

(b) $\dfrac{d^2 y}{dx^2} = 2\left(\dfrac{dy}{dx}\right)^2$,

(c) $\dfrac{d^2 y}{dx^2} + y\left(\dfrac{dy}{dx}\right)^3 = 0$,

(d) $(1 + x^2)\dfrac{d^2 y}{dx^2} + 2x\dfrac{dy}{dx} = 0$,

(e) $(1 - x^2)\dfrac{d^2 y}{dx^2} = x\dfrac{dy}{dx}$.

3 Solve the differential equation $\dfrac{d^2 s}{dt^2} + \dfrac{1}{10}\left(\dfrac{ds}{dt}\right)^2 = 0$, using the substitution $\dfrac{ds}{dt} = v$, and writing $\dfrac{d^2 s}{dt^2}$ (a) as $\dfrac{dv}{dt}$, (b) as $v\dfrac{dv}{ds}$.

4 Use the substitution $\dfrac{dy}{dx} = p$ to write the following as differential equations linear in p, and proceed as in §14.5:

(a) $\dfrac{d^2 y}{dx^2} + \cot x \dfrac{dy}{dx} = 1$,

(b) $\dfrac{d^2 y}{dx^2} + \dfrac{1}{x + 2}\dfrac{dy}{dx} = x$,

(c) $x \dfrac{d^2 y}{dx^2} + 2\dfrac{dy}{dx} = x \ln x$.

5 Find the particular solution of the differential equation

$$\dfrac{dy}{dx} \times \dfrac{d^2 y}{dx^2} + x = 0$$

which satisfies the condition that y and $\dfrac{dy}{dx}$ are both zero when $x = 1$.

6 Write down the general solution of each of the following differential equations:

(a) $\dfrac{d^2s}{dt^2} = -25s,$ (b) $\dfrac{d^2y}{dx^2} + \tfrac{9}{4}y = 0,$

(c) $9\dfrac{d^2s}{d\theta^2} = -2\theta,$ (d) $4\dfrac{d^2y}{dt^2} + 3y = 0.$

7 Find the solution of the differential equation

$$16\dfrac{d^2s}{dt^2} + 9s = 0$$

given that $s = 4$ and $\dfrac{ds}{dt} = 0$ when $t = 0$.

8 A body moves in a straight line so that when it is x cm from a point O on the line its acceleration is $9x$ cm/s² towards O. Write down the differential equation which describes this motion, and then present a complete solution of it (see Example 7) given that the body is at rest when 2 cm from O, and its distance from O is $+\sqrt{3}$ cm at the instant from which time is measured.

9 A body moves in S.H.M. of amplitude 4 cm and has initial phase $-\pi/2$ s. It takes 1 s to travel 2 cm from the centre of oscillation, O. What was its initial position, and what is its periodic time?

10 A body moving in S.H.M. is timed from an extreme position, and is found to take 2 s to reach a point mid-way between the centre of oscillation and the other end of its path. State the initial phase, and calculate the periodic time.

11 A body moves in a straight line so that it is x m from a fixed point on the line at time t s, where $x = \cos 2t + \sin 2t$. Write this in the form $x = a \cos (nt + \varepsilon)$ and state the amplitude, initial phase, and periodic time of the motion.

12 Repeat No. 11 for $x = 3 \cos \tfrac{1}{2}t - 4 \sin \tfrac{1}{2}t$, giving the initial phase correct to three significant figures.

13 The two simple harmonic motions defined by $x = a \cos (nt + \varepsilon_1)$ and $x = a \cos (nt + \varepsilon_2)$ are said to have a *phase difference* of $\varepsilon_1 - \varepsilon_2$. Find the phase difference between the following pairs of S.H.M.:

(a) $x = a \cos nt, \quad x = a \sin nt,$

(b) $x = 2 \cos \left(3t + \dfrac{\pi}{6} \right), \quad x = \sqrt{2}(\cos 3t - \sin 3t),$

(c) $x = \dfrac{3\sqrt{2}}{2} (\cos nt - \sin nt), \quad x = \dfrac{3\sqrt{2}}{2} (\cos nt + \sin nt),$

(d) $x = a \sin nt, \quad x = -a \sin nt,$
(e) $x = 5 \cos nt - 5\sqrt{3} \sin nt, \quad x = 5\sqrt{2} \cos nt + 5\sqrt{2} \sin nt.$

14 Solve the following differential equations:

(a) $\dfrac{d^2y}{dx^2} = -4(y+3),$ (b) $2\dfrac{d^2\theta}{dt^2} + 9\theta = 3,$

(c) $3\dfrac{d^2s}{dt^2} + 4t = 1,$ (d) $\dfrac{d^2x}{dt^2} + 4x + 8 = 0.$

given that in (d), $x = -1$ when $t = 0$, and $x = -3$ when $t = \pi/4$.

15 Solve the following differential equations:

(a) $\dfrac{d^2y}{dx^2} - 2\dfrac{dy}{dx} - 2 = 0,$ (b) $\dfrac{d^2y}{dx^2} = \dfrac{dy}{dx},$ (c) $\dfrac{d^2x}{dt^2} + x = 0,$

(d) $x\dfrac{d^2y}{dx^2} + \dfrac{dy}{dx} = 0,$ (e) $(3y-1)\dfrac{d^2y}{dx^2} = \left(\dfrac{dy}{dx}\right)^2.$

Exercise 14d (Miscellaneous)

Solve the differential equations in Nos. 1–14.

1 $\cos t \dfrac{dx}{dt} = x.$

2 $x\dfrac{dy}{dx} - y = x^2 \ln x.$

3 $\sec x \dfrac{d^2y}{dx^2} = e^x.$

4 $(1 + \cos \theta) = (1 - \cos \theta)\dfrac{d\theta}{dr}.$

5 $\ln(y+1) + \dfrac{x}{y+1}\dfrac{dy}{dx} = \dfrac{1}{x(x+1)}.$

6 $3\dfrac{d^2y}{dx^2} + 4y = 0.$

7 $(2x-1)\dfrac{dy}{dx} + 8y = 4(2x-1)^{-2}.$

8 $y\dfrac{d^2y}{dx^2} + \left(\dfrac{dy}{dx}\right)^2 - 3\dfrac{dy}{dx} = 0.$

9 $x(x+y)\dfrac{dy}{dx} = x^2 + xy - 3y^2.$

10 $u\dfrac{dv}{du} = \ln u.$

11 $(1-x^2)\dfrac{dy}{dx} - xy = 1.$

12 $9\dfrac{d^2x}{dt^2} + 4x = 0$, given that $x = 2$ when $t = 0$, and $x = -4$ when $t = \pi.$

13 $y\dfrac{d^2y}{dx^2} + 25 = \left(\dfrac{dy}{dx}\right)^2$, given that $\dfrac{dy}{dx} = 4$ when $y = 1$, and $y = \tfrac{5}{3}$ when $x = 0.$

14 $\dfrac{d^2s}{dt^2} + 9(s-1) = 0$, given that $s = 2$ when $t = 0$, and $\dfrac{d^2s}{dt^2} = -9\sqrt{3}$ when $t = \pi/6.$

15 Solve the differential equation

$$x\dfrac{dy}{dx} + 2y = e^x \qquad (x > 0)$$

given that $y = 1$ when $x = 1.$ (C)

16 Find the solution of the differential equation

$$\frac{dy}{dx} = xy \ln x$$

which satisfies the initial conditions $x = 1$, $y = 1$, giving $\ln y$ in terms of x.
(O & C)

17 Find the solution of the differential equation

$$ye^{2x} \frac{dy}{dx} = x$$

which satisfies the boundary condition $y = 2$ when $x = 0$; give y in terms of x.
(O & C)

18 (a) Find the general solution of the differential equation

$$\frac{dy}{dx} + 2y = e^{2x}$$

Find also the particular solution for which $y = \frac{1}{2}$ when $x = 0$.

(b) Given that $y = Ae^{mx} + Be^{-mx}$, where A, B and m are constants, find $\frac{dy}{dx}$

and $\frac{d^2y}{dx^2}$. Hence form a differential equation relating y and x which does

not contain the constants A and B. (L)

19 Prove that if $y = f(x)$ satisfies the differential equation

$$\frac{dy}{dx} = -2xy,$$

then so does $y = kf(x)$, where k is any real number.
    Find the equation of the solution curve through the point $(0, 1)$ and sketch
a graph showing several members of the family of solutions.
(O & C: SMP)

20 Find the solution of the differential equation

$$\frac{dy}{dx} + y \cot x = \cos 3x,$$

for which $y = 1$, when $x = \pi/6$. (JMB)

21 The temperature y degrees of a body, t minutes after being placed in a certain
room, satisfies the differential equation

$$6\frac{d^2y}{dt^2} + \frac{dy}{dt} = 0$$

By using the substitution $z = \frac{dy}{dt}$, find y in terms of t, given that $y = 63$ when

$t = 0$ and $y = 36$ when $t = 6 \ln 4$.
    Find after how many minutes the rate of cooling of the body will have
fallen below one degree per minute, giving your answer correct to the nearest
minute. How cool does the body get? (O & C)

22 A particle moves along the x-axis so that its displacement x from the origin O at time t satisfies the differential equation

$$\frac{d^2x}{dt^2} + 4x = k$$

where k is a constant. At times $t = 0, \frac{1}{4}\pi, \frac{1}{2}\pi$ the values of x are 8, 9, 2 units respectively. Find the value of k, and find x in terms of t.

Find the greatest speed of the particle.

Give a sketch of $\dfrac{dx}{dt}$ against t for values of t in the range $0 \leqslant t \leqslant \pi$.

If the motion starts at $t = 0$, calculate the times at which the particle is (a) first at rest, (b) next moving in the positive direction of the x-axis with its initial speed.

(The general solution of the differential equation may be quoted.)

(O & C)

23 In established forest fires, the proportion of the total area of the forest which has been destroyed is denoted by x, and the rate of change of x with respect to time, t hours, is called the destruction rate. Investigations show that the destruction rate is directly proportional to the product of x and $(1 - x)$. A particular fire is initially noticed when one half of the forest is destroyed, and it is found that the destruction rate at this time is such that, if it remained constant thereafter, the forest would be destroyed completely in a further 24 hours. Show that

$$12\frac{dx}{dt} = x(1 - x)$$

and deduce that approximately 73% of the forest is destroyed 12 hours after it is first noticed. (L)

24 A chemical substance X decays, at a rate equal to twice the quantity of X present, so that $\dfrac{dx}{dt} = -2x$ where x is the quantity of X present at time t. Given that initially $x = a$, find an expression for x in terms of a and t.

The quantity, y, of another substance Y changes so that its rate of increase is equal to $2ae^{-2t} - \frac{1}{2}y$. Given that initially $y = 0$, find an expression for y at time t. (L)

25 At any instant, a spherical meteorite is gaining mass because of two effects: (a) mass is condensing onto it at a rate which is proportional to the surface area of the meteorite at that instant, (b) the gravitational field of the meteorite attracts mass onto itself, the rate being proportional to the mass of the meteorite mass at that instant. Assuming that the two effects can be added together and that the meteorite remains spherical and of constant density, show that the radius r at time t satisfies the differential equation

$$\frac{dr}{dt} = A + Br,$$ where A and B are constants. If $r = r_0$ at $t = 0$, show that

$$r = r_0 e^{Bt} + \frac{A}{B}(e^{Bt} - 1)$$

(L)

Chapter 15

Second order linear differential equations with constant coefficients

Introduction; the auxiliary quadratic equation

15.1 This chapter title may seem a rather long and forbidding one but it is not as complicated as it may seem at first reading. 'Second order' means that the *second* derivative will appear, but not derivatives of higher order; 'linear' means that none of the terms containing y will be squared, or cubed, or raised to any power but one; and 'with constant coefficients' means that the coefficients will be constants (not variables). In other words we shall be concerned with equations of the form

$$a\frac{d^2y}{dx^2} + b\frac{dy}{dx} + cy = 0$$

where $a, b, c \in \mathbb{R}$, and $a \neq 0$.

At first, the R.H.S. of the equation will always be zero, but later we shall consider equations in which the 0 is replaced by a function of x.

We shall see that the nature of the solution will depend upon the relative magnitudes of the constants a, b and c, and we shall frequently need to refer to the auxiliary quadratic equation (A.Q.E. for short)

$$am^2 + bm + c = 0$$

The general solution of the A.Q.E. is

$$m = \frac{-b \pm \sqrt{(b^2 - 4ac)}}{2a}$$

As we know (see Book 1, §10.2) there are three cases which can arise from this:

(a) if $b^2 > 4ac$, the A.Q.E. has two real, distinct roots,
(b) if $b^2 = 4ac$, the A.Q.E. has identical, real roots,
(c) if $b^2 < 4ac$, the A.Q.E. has a pair of conjugate complex roots.

Each case gives rise to a distinct type of solution to the original differential equation, and we shall consider each of these in turn. But first, note that we have already solved one important type of second order differential equation, namely the S.H.M. equation

$$\frac{d^2y}{dx^2} + n^2y = 0$$

The A.Q.E. for this is

$$m^2 + n^2 = 0$$

which has roots

$$m = \pm in$$

In other words it is an example of case (c) above, and we already know from §14.8 that one form of the general solution of this equation is

$$y = A \cos nx + B \sin nx$$

Second order differential equations frequently arise in mechanics and physics; see, for example, Exercise 15b, Nos. 15–21.

Type I — the A.Q.E. with real, distinct roots

15.2   Suppose the roots of the A.Q.E. are α and β. It can then be written

$$m^2 - (\alpha + \beta)m + \alpha\beta = 0$$

and the corresponding differential equation is

$$\frac{d^2y}{dx^2} - (\alpha + \beta)\frac{dy}{dx} + \alpha\beta y = 0$$

This differential equation can be rearranged so that it reads

$$\frac{d^2y}{dx^2} - \beta\frac{dy}{dx} - \alpha\left(\frac{dy}{dx} - \beta y\right) = 0$$

and this, in turn, can be written

$$\frac{d}{dx}\left(\frac{dy}{dx} - \beta y\right) - \alpha\left(\frac{dy}{dx} - \beta y\right) = 0$$

So far, we appear to have made the equation even more complicated than it was when we started, but now we can simplify it by substituting

$$v = \frac{dy}{dx} - \beta y \tag{1}$$

The differential equation can then be written

$$\frac{dv}{dx} - \alpha v = 0 \tag{2}$$

Not only is this simpler, but it is a *first order differential equation*, and we already have a large battery of techniques for solving first order equations.

The solution of (2) is

$$v = Ce^{\alpha x} \qquad \text{(see §14.2)}$$

where C is an arbitrary constant. Substituting this in equation (1) gives

$$\frac{dy}{dx} - \beta y = Ce^{\alpha x}$$

Once again we have a first order differential equation, and this one can be solved by using the integrating factor $e^{\int -\beta dx} = e^{-\beta x}$. Multiplying through by this gives

$$e^{-\beta x}\frac{dy}{dx} - \beta e^{-\beta x} y = Ce^{(\alpha - \beta)x}$$

and integrating, this becomes

$$e^{-\beta x} y = \frac{C}{\alpha - \beta} e^{(\alpha - \beta)x} + B$$

where B is an arbitrary constant. Multiplying through by $e^{\beta x}$, we have

$$y = \frac{C}{\alpha - \beta} e^{\alpha x} + Be^{\beta x}$$

Since C was an arbitrary constant, there is nothing to be gained from writing $C/(\alpha - \beta)$, and so we replace this by another arbitrary constant A. Hence the general solution is

$$y = Ae^{\alpha x} + Be^{\beta x}$$

where α and β are the (real) roots of the A.Q.E.

Example 1   *Solve the differential equation*

$$\frac{d^2y}{dx^2} + 2\frac{dy}{dx} - 15y = 0$$

Find the values of the arbitrary constants such that when $x = 0$, $y = 5$ and $\frac{dy}{dx} = 23$.

The A.Q.E. is

$$m^2 + 2m - 15 = 0$$

Hence

$$(m + 5)(m - 3) = 0$$

$$\therefore m = -5 \quad \text{or} \quad +3$$

Since the roots are real and distinct, this is a 'Type I' differential equation and its general solution is

$$y = Ae^{-5x} + Be^{+3x}$$

When $x = 0$, $y = 5$, so

$$5 = A + B$$

Also, when $x = 0$, $\dfrac{dy}{dx} = 23$, and since $\dfrac{dy}{dx} = -5Ae^{-5x} + 3Be^{3x}$, we have

$$23 = -5A + 3B$$

Solving this pair of simultaneous equations for A and B gives $A = -1$, $B = 6$. Hence the solution which fits the given conditions is

$$y = -e^{-5x} + 6e^{3x}$$

In this example we were given the values of y and $\dfrac{dy}{dx}$, when $x = 0$. When differential equations are formed in applied mathematics it is quite common for the independent variable to represent t, the time, and frequently the values of the dependent variable, and its derivative, will be given for $t = 0$, since these determine the initial state of the system; consequently these values are usually called the *initial conditions*.

Qu. 1 Find the general solutions of the following differential equations:

(a) $\dfrac{d^2y}{dx^2} = y$,

(b) $\dfrac{d^2y}{dx^2} - 12\dfrac{dy}{dx} + 20y = 0$,

(c) $2\dfrac{d^2y}{dx^2} - 5\dfrac{dy}{dx} - 3y = 0$,

(d) $15\dfrac{d^2y}{dx^2} - 8\dfrac{dy}{dx} + y = 0$.

The reader should not assume that all differential equations are expressed in terms of x (for the independent variable) and y (for the dependent variable). In Qu. 2, z and t will be used.

Qu. 2 Find the general solution of the differential equations:

(a) $\dfrac{d^2z}{dt^2} - 25z = 0$,

(b) $6\dfrac{d^2z}{dt^2} - \dfrac{dz}{dt} - z = 0$.

Qu. 3 Find the solutions of the differential equations in Qu. 2 which satisfy the initial conditions, $z = 0$, and $\dfrac{dz}{dt} = 10$, when $t = 0$.

Qu. 4 Solve the differential equation $f''(x) - 6f'(x) + 5f(x) = 0$, given that $f(0) = 1$ and $f'(0) = 9$.

Type II — the A.Q.E. with identical roots

15.3 In this type, the A.Q.E. can be written in the form

$$m^2 - 2pm + p^2 = 0$$

This can be factorised to give

$$(m - p)^2 = 0$$

and so the solution of the A.Q.E. is $m = p$. The corresponding differential equation is

$$\frac{d^2 y}{dx^2} - 2p \frac{dy}{dx} + p^2 y = 0$$

Before we tackle this, notice the following very useful transformation. If we write

$$y = e^{px} v$$

where v is a function of x, and p is a constant, then

$$v = e^{-px} y$$

differentiating once,

$$\frac{dv}{dx} = e^{-px} \frac{dy}{dx} - pe^{-px} y$$

and differentiating again,

$$\frac{d^2 v}{dx^2} = e^{-px} \frac{d^2 y}{dx^2} - 2pe^{-px} \frac{dy}{dx} + p^2 e^{-px} y$$

$$= e^{-px} \left(\frac{d^2 y}{dx^2} - 2p \frac{dy}{dx} + p^2 y \right)$$

Hence

$$\frac{d^2 y}{dx^2} - 2p \frac{dy}{dx} + p^2 y = e^{px} \frac{d^2 v}{dx^2}$$

Using this transformation, the Type II second order differential equation

$$\frac{d^2 y}{dx^2} - 2p \frac{dy}{dx} + p^2 y = 0$$

can be written

$$e^{px} \frac{d^2 v}{dx^2} = 0$$

$$\therefore \frac{d^2 v}{dx^2} = 0$$

Integrating once gives

$$\frac{dv}{dx} = A$$

and integrating again,

$$v = Ax + B$$

where A and B are arbitrary constants. If we now replace v by $e^{-px}y$, we have

$$e^{-px}y = Ax + B$$
$$\therefore y = (Ax + B)e^{px}$$

This is the general solution of the Type II equation.

Example 2   *Find the general solution of the differential equation*

$$4\frac{d^2u}{d\theta^2} - 12\frac{du}{d\theta} + 9u = 0$$

The A.Q.E. is

$$4m^2 - 12m + 9 = 0$$
$$(2m - 3)^2 = 0$$
$$2m = 3$$
$$m = 3/2$$

The A.Q.E. has real identical roots, so this is a Type II differential equation, and its general solution is

$$u = e^{(3/2)\theta}(A\theta + B)$$

Qu. 5   Solve the differential equations:

(a) $\dfrac{d^2V}{dt^2} + 6\dfrac{dV}{dt} + 9V = 0,$ (b) $100\dfrac{d^2r}{dt^2} - 60\dfrac{dr}{dt} + 9r = 0.$

Qu. 6   Find the solution of the differential equation

$$\frac{d^2y}{dx^2} - 2\frac{dy}{dx} + y = 0$$

given that when $x = 0$, $y = 0$ and when $x = 1$, $y = e$.

Type III — the A.Q.E. with complex roots

15.4   Suppose the roots of the A.Q.E. are $p \pm iq$, where p and $q \in \mathbb{R}$. The sum of the roots is

$$(p + iq) + (p - iq) = 2p$$

and the product of the roots is

$$(p + iq)(p - iq) = p^2 - i^2q^2 = p^2 + q^2$$

so the A.Q.E. can be written

$$m^2 - 2pm + (p^2 + q^2) = 0$$

and the corresponding differential equation is

$$\frac{d^2y}{dx^2} - 2p\frac{dy}{dx} + (p^2 + q^2)y = 0$$

Using the same transformation as that used in the previous section, i.e. putting $y = e^{px}V$, we can write the equation

$$e^{px}\frac{d^2v}{dx^2} + q^2 e^{px}v = 0$$

and dividing through by e^{px}, we have

$$\frac{d^2v}{dx^2} + q^2v = 0$$

This, however, is an equation which the reader should recognise — it is the S.H.M. equation, and we know that its general solution can be written

$$v = A \cos qx + B \sin qx$$

Returning to the original variable y, gives

$$e^{-px}y = A \cos qx + B \sin qx$$
$$y = e^{px}(A \cos qx + B \sin qx)$$

This is the general solution of the Type III differential equation.

Example 3 *Solve the differential equation*

$$\frac{d^2x}{dt^2} + 6\frac{dx}{dt} + 10x = 0$$

given the initial conditions, $x = 0$ and $\dfrac{dx}{dt} = 1$, when $t = 0$.

The A.Q.E. is $m^2 + 6m + 10 = 0$.
Solving by the formula,

$$m = \frac{-6 \pm \sqrt{(36 - 40)}}{2}$$

$$= \frac{-6 \pm \sqrt{(-4)}}{2}$$

$$= \tfrac{1}{2}(-6 \pm 2i)$$

$$= -3 \pm i$$

Since the roots are complex this is a Type III differential equation, and its general solution is

$$x = e^{-3t}(A \cos t + B \sin t)$$

The initial conditions are $x = 0$ and $\dfrac{dx}{dt} = 1$, when $t = 0$, so

$$0 = (A \cos 0 + B \sin 0)$$
$$\therefore A = 0$$

Hence

$$x = e^{-3t} B \sin t$$

and

$$\frac{dx}{dt} = Be^{-3t} \cos t - 3Be^{-3t} \sin t$$

Putting $t = 0$, and $\frac{dx}{dt} = 1$, gives

$$1 = B \cos 0$$
$$\therefore B = 1$$

Hence the solution which fits the initial conditions is

$$x = e^{-3t} \sin t$$

Qu. 7   Find the general solutions of the differential equations:

(a) $\dfrac{d^2 y}{dx^2} - 2\dfrac{dy}{dx} + 50y = 0,$

(b) $\dfrac{d^2 V}{dt^2} + 6\dfrac{dV}{dt} + 34V = 0,$

(c) $36\dfrac{d^2 r}{dt^2} + r = 0.$

Qu. 8   Find the solutions to the equations in Qu. 7, which satisfy the conditions:

(a) when $x = 0$, $y = 0$ and $\dfrac{dy}{dx} = 35,$

(b) when $t = 0$, $V = 1$ and $\dfrac{dV}{dt} = 7,$

(c) when $t = \pi$, $r = 1$, and when $t = 3\pi$, $r = 0$.

Summary

15.5   To solve a differential equation of the form

$$a\frac{d^2 y}{dx^2} + b\frac{dy}{dx} + cy = 0$$

write down its auxiliary quadratic equation

$$am^2 + bm + c = 0$$

and solve it. The general solution of the differential equation then takes one of the three forms shown in the table on the facing page.

type	nature of the roots of the A.Q.E.	general solution of the differential equation
I	real, distinct roots, $m = \alpha$ or β	$y = Ae^{\alpha x} + Be^{\beta x}$
II	real, identical roots, $m = p$	$y = e^{px}(Ax + B)$
III	complex roots, $m = p \pm iq$	$y = e^{px}(A \cos qx + B \sin qx)$

Since solving a second order differential inevitably entails integrating *twice*, the general solution of such an equation must include *two* arbitrary constants.

Example 4   *Solve the differential equations*

(a) $\dfrac{d^2 y}{dx^2} - 4y = 0,$ (b) $\dfrac{d^2 y}{dx^2} + 4y = 0,$ (c) $\dfrac{d^2 y}{dx^2} + 2\dfrac{dy}{dx} + 5y = 0.$

In each case find the solution for which $y = 0$ *and* $\dfrac{dy}{dx} = 2$, *when* $x = 0$ *and sketch its graph.*

(To save space, some of the simpler steps in the working are left to the reader.)

(a) $\dfrac{d^2 y}{dx^2} - 4y = 0.$

The A.Q.E. is

$$m^2 - 4 = 0$$
$$\therefore m = \pm 2$$

Hence the general solution is

$$y = Ae^{2x} + Be^{-2x}$$

Putting in the initial conditions gives $A = \frac{1}{2}$, $B = \frac{1}{2}$, hence the solution which fits the initial conditions is

$$y = \tfrac{1}{2}(e^{2x} - e^{-2x})$$

The sketch of this solution is shown in Fig. 15.1.

(b) $\dfrac{d^2 y}{dx^2} + 4y = 0.$

This is the S.H.M. equation; its general solution is

$$y = A \sin 2x + B \cos 2x$$

and putting in the initial conditions gives

$$y = \sin 2x$$

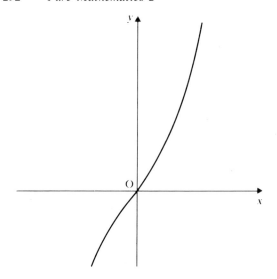

Figure 15.1

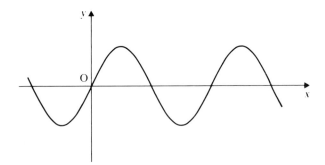

Figure 15.2

The sketch of this solution is shown in Fig. 15.2.

(c) $\dfrac{d^2y}{dx^2} + 2\dfrac{dy}{dx} + 5y = 0.$

The A.Q.E. is $m^2 + 2m + 5 = 0$, and the solution of this is

$$m = -1 \pm 2i$$

consequently the general solution is

$$y = e^{-x}(A \sin 2x + B \cos 2x)$$

Putting $x = 0$, $y = 0$, gives $B = 0$.

$$\therefore \ y = Ae^{-x} \sin 2x$$

and $\dfrac{dy}{dx} = -Ae^{-x} \sin 2x + 2Ae^{-x} \cos 2x$

Putting $x = 0, \dfrac{dy}{dx} = 2$ gives $2 = 2A$, hence $A = 1$; hence the particular solution we require is

$y = e^{-x} \sin 2x$

The solution is shown in Fig. 15.3. (When x is positive, the exponential factor will soon overwhelm $\sin 2x$, so y quickly becomes negligible, for example when $x = 5$, $y = -0.004$. For negative values of x, y still oscillates, but the amplitude of the oscillation increases enormously, due to the exponential factor. For example, when $x = -5$, $y = 80.7$; it is not possible to show this in Fig. 15.3.)

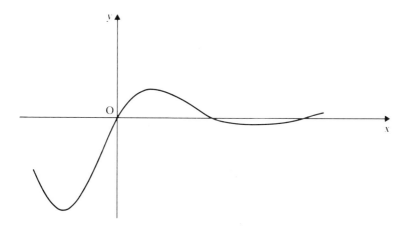

Figure 15.3

Exercise 15a

Find the general solutions of the following differential equations.

1 $\dfrac{d^2 y}{dx^2} - 7\dfrac{dy}{dx} + 10y = 0.$ **2** $\dfrac{d^2 y}{dx^2} - \dfrac{dy}{dx} - 6y = 0.$

3 $\dfrac{d^2 y}{dx^2} - 4\dfrac{dy}{dx} + 29y = 0.$ **4** $25\dfrac{d^2 y}{dx^2} = y.$

5 $\dfrac{d^2 x}{dt^2} - 10\dfrac{dx}{dt} + 25x = 0.$ **6** $16\dfrac{d^2 x}{dt^2} - 8\dfrac{dx}{dt} + x = 0.$

7 $6\dfrac{d^2 u}{dt^2} = 5\dfrac{du}{dt} - u.$ **8** $\dfrac{d^2 y}{dx^2} + 5\dfrac{dy}{dx} = 0.$

9 $2\dfrac{d^2 x}{dt^2} - \dfrac{dx}{dt} - 3x = 0.$ **10** $\dfrac{d^2 r}{d\theta^2} = \dfrac{dr}{d\theta}.$

11 $\dfrac{d^2y}{dx^2} + 4\dfrac{dy}{dx} + 20y = 0.$ **12** $100\dfrac{d^2r}{dt^2} + r = 0.$

13 $\dfrac{d^3y}{dx^3} + 3\dfrac{d^2y}{dx^2} + 2\dfrac{dy}{dx} = 0.$ $\left[\text{Hint: put } \dfrac{dy}{dx} = v.\right]$

14 $\dfrac{d^2y}{dx^2} - 4\dfrac{dy}{dx} + 4y = 8.$ [Hint: put $y = z + 2$.]

15 $\dfrac{d^2x}{dt^2} + 2\dfrac{dx}{dt} + 2x = 1.$ [Hint: put $x = u + \frac{1}{2}$.]

In Nos. 16–20, find the solutions which fit the conditions given.

16 $\dfrac{d^2y}{dx^2} - 5\dfrac{dy}{dx} - 6y = 0.$ When $x = 0$, $y = 5$ and $\dfrac{dy}{dx} = 16$.

17 $\dfrac{d^2u}{dt^2} + 9u = 0.$ When $t = 0$, $u = 4$, and, when $t = \pi/6$, $u = 5$.

18 $\dfrac{d^2r}{dt^2} - 12\dfrac{dr}{dt} + 36r = 0.$ When $t = 0$, $r = 1$ and when $t = 1$, $r = 0$.

19 $\dfrac{d^2z}{dt^2} + \frac{1}{4}z = 0.$ When $t = 0$, $z = 4$ and $\dfrac{dz}{dt} = 0$.

20 $\dfrac{d^2u}{d\theta^2} = u.$ When $\theta = 0$, $u = 1$ and $\dfrac{du}{d\theta} = 1$.

Qu. 9   Solve Nos. 8 and 10 of Exercise 15a by integrating each term and verify that the same solutions are obtained by this method.

The differential equation $a\dfrac{d^2y}{dx^2} + b\dfrac{dy}{dx} + cy = f(x)$

15.6   In the preceding sections we have found methods for solving any differential equation of the form

$$a\dfrac{d^2y}{dx^2} + b\dfrac{dy}{dx} + cy = 0$$

We must now consider the more general second order differential equation, with constant coefficients, in which the R.H.S. of the equation is *not* zero. We shall start by considering the reverse problem, that is, given a function of x containing two arbitrary constants, can we find the second order differential equation which this function satisfies for all values of the two arbitrary constants?

Consider the function

$$y = Ae^{2x} + Be^{3x} + 5x + 3 \tag{1}$$

The first and second derivatives of this are

$$\frac{dy}{dx} = 2Ae^{2x} + 3Be^{3x} + 5 \qquad (2)$$

and

$$\frac{d^2y}{dx^2} = 4Ae^{2x} + 9Be^{3x} \qquad (3)$$

Eliminating B from (1) and (2) gives

$$3y - \frac{dy}{dx} = Ae^{2x} + 15x + 4$$

and eliminating B from (1) and (3) gives

$$9y - \frac{d^2y}{dx^2} = 5Ae^{2x} + 45x + 27$$

Eliminating A from these equations gives

$$5\left(3y - \frac{dy}{dx}\right) - \left(9y - \frac{d^2y}{dx^2}\right) = 30x - 7$$

and hence

$$\frac{d^2y}{dx^2} - 5\frac{dy}{dx} + 6y = 30x - 7 \qquad (4)$$

So (4) is the differential equation which is satisfied by the function (1) for all values of the arbitrary constants. We say that (1) is the **general solution** of the differential equation (4).

Qu. 10 Eliminate the constants A and B from $y = A \sin 4x + B \cos 4x + e^{2x}$ and verify that y satisfies the differential equation

$$\frac{d^2y}{dx^2} + 16y = 20e^{2x}$$

The form of the L.H.S. of the differential equation (4) above, is not very surprising, because the terms in (1) which contain the arbitrary constants A and B, namely $Ae^{2x} + Be^{3x}$, make up the solution of the equation

$$\frac{d^2y}{dx^2} - 5\frac{dy}{dx} + 6y = 0 \qquad (5)$$

This suggests that if we wish to find the general solution of an equation like (4), we should first solve the equation obtained by replacing the R.H.S. by zero, i.e. equation (5) above, and then add to that solution some extra terms to produce the terms of f(x). In the example above, the extra terms needed were $5x + 3$. The art of choosing these extra terms can be developed with practice, but a systematic method for producing them is beyond the scope of this book.

The general solution of an equation such as (4), then, will consist of two parts; the solution of the equation formed by replacing f(x) by zero (this part of the general solution is called **the complementary function**) and the extra terms required to produce f(x) (this part is called **the particular integral**).

In the following sections we shall consider how to find the particular integral when f(x) is (a) a polynomial, (b) an exponential function and (c) a trigonometrical function.

f(x) is a polynomial

15.7 Consider the differential equation

$$\frac{d^2y}{dx^2} - 4\frac{dy}{dx} + 3y = 9x + 6 \tag{1}$$

The first stage in the solution is to find the complementary function by considering

$$\frac{d^2y}{dx^2} - 4\frac{dy}{dx} + 3y = 0$$

The A.Q.E. is

$$m^2 - 4m + 3 = 0$$
$$(m - 3)(m - 1) = 0$$

$$m = 3 \quad \text{or} \quad 1$$

and so the complementary function is

$$y = Ae^{3x} + Be^x$$

The R.H.S. of the equation we are solving is $9x + 6$, and from our experience in §15.6, it would seem reasonable to assume that we shall need some extra terms of the form $px + q$. If we substitute $y = px + q$ into equation (1) we obtain

$$0 - 4p + 3(px + q) = 9x + 6$$

Equating the coefficients of x gives $3p = 9$, and equating the constant terms gives $3q - 4p = 6$. From these equations we obtain

$$p = 3$$

and

$$3q - 12 = 6$$
$$\therefore 3q = 18$$
$$\therefore q = 6$$

Hence the particular integral is $3x + 6$, and so the general solution of the differential equation (1) is

$$y = Ae^{3x} + Be^x + 3x + 6$$

In general, if $f(x)$ is a polynomial of degree n, the 'trial solution' for the particular integral should also be a polynomial of degree n.

Qu. 11 Find the particular integral of the differential equation

$$\frac{d^2y}{dx^2} + 4\frac{dy}{dx} + 5y = 10x^2 + x$$

[Hint: try $y = px^2 + qx + r$.]

$f(x)$ is an exponential function

15.8 Consider the differential equation

$$\frac{d^2y}{dx^2} - 4\frac{dy}{dx} + 3y = 10e^{-2x}$$

The L.H.S. is the same as the L.H.S. of the differential equation in §15.7, so the complementary function is, as before,

$$y = Ae^{3x} + Be^x$$

We must now consider what extra terms are necessary to produce $10e^{-2x}$ on the R.H.S. After a little reflection, the reader will probably agree that the only possible contender is a term of the form pe^{-2x}. So, for the particular integral we shall try

$$y = pe^{-2x}$$

and we substitute this into the L.H.S. of the differential equation. This gives

$$4pe^{-2x} + 8pe^{-2x} + 3pe^{-2x} = 10e^{-2x}$$
$$15pe^{-2x} = 10e^{-2x}$$

and hence the value of p needed is $\frac{2}{3}$. So the particular integral is

$$y = \frac{2}{3}e^{-2x}$$

and consequently the general solution of the differential equation is

$$y = Ae^{3x} + Be^{2x} + \frac{2}{3}e^{-2x}$$

Remember that if the question gives some initial conditions, these must now be used to find the values of A and B. (This must not be tackled *before* the general solution has been found; it must not, for instance, be attempted immediately after finding the complementary function.)

The procedure followed above can be summed up thus — if the R.H.S. of the differential equation has the form ke^{ax}, then a 'trial solution' of the form pe^{ax} should be considered. In most cases this will enable the particular integral to be found. However, if e^{ax} happens to be a term in the complementary function, this method will break down, because when *this* is substituted into the L.H.S. of the differential equation it will produce zero. It is beyond the scope of this book to cover exceptions like this; the reader who wishes to know more should consult a specialised book on differential equations.

f(x) is a trigonometrical function

15.9   Consider the differential equation

$$\frac{d^2y}{dx^2} - 4\frac{dy}{dx} + 3y = 10 \sin 2x + 15 \cos 2x$$

As before, the complementary function is

$$y = Ae^{3x} + Be^x$$

Once again we must ask ourselves what sort of expression is likely to produce a combination of $\sin 2x$ and $\cos 2x$ on the R.H.S. In the light of our experience in the preceding sections, a likely expression would be another linear combination of $\sin 2x$ and $\cos 2x$. So as the 'trial solution' we shall take

$$y = p \sin 2x + q \cos 2x$$

in which case

$$\frac{dy}{dx} = 2p \cos 2x - 2q \sin 2x$$

and

$$\frac{d^2y}{dx^2} = -4p \sin 2x - 4q \cos 2x$$

Substituting these into the differential equation gives

$$(-4p \sin 2x - 4q \cos 2x) - 4(2p \cos 2x - 2q \sin 2x) + 3(p \sin 2x + q \cos 2x)$$
$$= 10 \sin 2x + 15 \cos 2x$$

Hence

$$(-4p + 8q + 3p) \sin 2x + (-4q - 8p + 3q) \cos 2x = 10 \sin 2x + 15 \cos 2x$$

Equating the coefficients of $\sin 2x$ and $\cos 2x$ gives

$$-p + 8q = 10$$

and

$$-8p - q = 15$$

Hence $p = -2$ and $q = 1$.
So the particular integral is

$$y = -2 \sin 2x + \cos 2x$$

and the general solution of the differential equation is

$$y = Ae^{3x} + Be^x - 2 \sin 2x + \cos 2x$$

This procedure will work for most similar differential equations, that is, if the R.H.S. of the differential equation is a combination of $\sin kx$ and $\cos kx$, then a similar combination of these terms should be used as the 'trial solution', but once again there will be difficulties if these terms happen to be part of the complementary function. As before the reader wishing to proceed further will have to turn to a more specialised textbook.

Example 5   *Solve the differential equation*

$$\frac{d^2x}{dt^2} + x = 10e^{2t}$$

given that when $t = 0$, $x = 5$ and $\dfrac{dx}{dt} = 14$.

The A.Q.E. is $m^2 + 1 = 0$,

$$\therefore m = \pm i$$

and hence the complementary function is

$$x = A \cos t + B \sin t$$

Since the R.H.S. is $10e^{2t}$, we take pe^{2t} as a 'trial solution'.

Let

$$x = pe^{2t}$$

$$\therefore \frac{dx}{dt} = 2pe^{2t}$$

and

$$\frac{d^2x}{dt^2} = 4pe^{2t}$$

Substituting these into the differential equation gives

$$4pe^{2t} + pe^{2t} = 10e^{2t}$$

hence

$$5p = 10$$
$$\therefore p = 2$$

So the general solution is

$$x = A \cos t + B \sin t + 2e^{2t}$$

When $t = 0$, $x = 5$,

$$\therefore 5 = A \cos 0 + B \sin 0 + 2$$
$$\therefore 5 = A + 2$$
$$\therefore A = 3$$

Differentiating the general solution gives

$$\frac{dx}{dt} = -A \sin t + B \cos t + 4e^{2t}$$

and, from the initial conditions, we know that $\dfrac{dx}{dt} = 14$, when $t = 0$, so

$$14 = -A \sin 0 + B \cos 0 + 4$$

hence

$$14 = B + 4$$
$$B = 10$$

Hence the solution which fits the initial conditions is

$$x = 3 \cos t + 10 \sin t + 2e^{2t}$$

It is convenient, at this stage, to remind the reader about an alternative notation which can be very useful when dealing with differential equations. If y is a function of x, then $\dfrac{dy}{dx}$ is written y_1, and $\dfrac{d^2y}{dx^2}$ is written y_2; this can save a lot of writing (see §11.7). In this notation the differential equation in §15.7 would be written

$$y_2 - 4y_1 + 3y = 9x + 6$$

This notation is used in the next example.

Example 6 *Solve the following differential equation*

$$y_2 - 7y_1 + 10y = 40x + 2$$

given that when $x = 0$, $y = 6$ and $y_1 = 13$.

First, we find the complementary function by solving

$$y_2 - 7y_1 + 10y = 0$$

The A.Q.E. is

$$m^2 - 7m + 10 = 0$$
$$\therefore (m - 2)(m - 5) = 0$$
$$m = 2 \quad \text{or} \quad 5$$

Hence the complementary function is

$$y = Ae^{2x} + Be^{5x}$$

Since the R.H.S. of the differential equation is a polynomial of degree one, we take a similar expression as the trial particular integral, i.e.

$$y = px + q$$

in this case, $y_1 = p$ and $y_2 = 0$. Substituting these expressions into the L.H.S. of the differential equation gives

$$y_2 - 7y_1 + 10y = -7p + 10(px + q)$$
$$= 10px + (10q - 7p)$$

Equating this to the R.H.S. of the original differential equation produces

$$10px + (10q - 7p) = 40x + 2$$

On equating the coefficients of x and the constant terms, we obtain

$$10p = 40$$
$$\therefore p = 4$$

and

$$10q - 7p = 2$$
$$10q - 28 = 2$$
$$\therefore q = 3$$

So the particular integral is

$$y = 4x + 3$$

The general solution is the sum of the complementary function and the particular integral, i.e.

$$y = Ae^{2x} + Be^{5x} + 4x + 3$$

Finally, we must find the values of A and B which fit the given initial conditions, which were, when $x = 0$, $y = 6$ and $y_1 = 13$. The first derivative of the general solution is

$$y_1 = 2Ae^{2x} + 5Be^{5x} + 4$$

When $x = 0$, $y = 6$, so $6 = A + B + 3$.

$$\therefore A + B = 3 \tag{1}$$

When $x = 0$, $y_1 = 13$, so $13 = 2A + 5B + 4$.

$$\therefore 2A + 5B = 9 \tag{2}$$

Solving equations (1) and (2) gives $A = 2$, $B = 1$. Hence the solution which satisfies the initial conditions is

$$y = 2e^{2x} + e^{5x} + 4x + 3$$

Exercise 15b

Find the general solutions of the following differential equations.

1 $\dfrac{d^2y}{dx^2} - 7\dfrac{dy}{dx} + 12y = 4.$ **2** $\dfrac{d^2y}{dx^2} - 4\dfrac{dy}{dx} - 5y = 17 + 15x.$

3 $\dfrac{d^2y}{dx^2} + 9y = 20e^x.$ **4** $\dfrac{d^2y}{dx^2} + 6\dfrac{dy}{dx} + 9y = 8 \sin x + 6 \cos x.$

5 $y_2 - 6y_1 + 10y = 10x^2 + 18x - 6.$

6 $16\dfrac{d^2r}{dt^2} - 8\dfrac{dr}{dt} + r = 27e^t.$ **7** $\dfrac{d^2z}{dr^2} + 25z = 10.$

8 $5\dfrac{d^2u}{dt^2} + 2\dfrac{du}{dt} + u = 16e^{-t}.$ **9** $\dfrac{d^2V}{d\theta^2} + 3\dfrac{dV}{d\theta} = 12 + \sin\theta.$

10 $10\dfrac{d^2x}{dt^2} + 13\dfrac{dx}{dt} - 3x = 21e^{2t} - 6t + 2.$

In Nos. 11–13, find the solutions of the differential equations, which satisfy the given initial conditions.

11 $\dfrac{d^2y}{dx^2} + 3\dfrac{dy}{dx} + 2y = \cos x.$ When $x = 0$, $y = 0$ and $\dfrac{dy}{dx} = 0$.

12 $\dfrac{d^2y}{dx^2} + 4\dfrac{dy}{dx} + 4y = 8x^2.$ When $x = 0$, $y = 0$ and $\dfrac{dy}{dx} = 0$.

13 $y_2 + 25y = 50e^{5x}.$ When $x = 0$, $y = 3$ and $y_1 = 20$.

14 Show that the general solution of the equation

$$9\dfrac{d^2x}{dt^2} + 6\dfrac{dx}{dt} + x = -50\sin t$$

can be expressed in the form

$$x = (At + B)e^{-(1/3)t} + 5\cos(t - \alpha)$$

where $\tan\alpha = 4/3$.

15 A particle P falls vertically under gravity and the air resistance is taken to be proportional to its speed at any instant. After t seconds, the distance x, which it has fallen, is given by the differential equation

$$\dfrac{d^2x}{dt^2} = g - k\dfrac{dx}{dt}$$

where g and k are positive constants. Express x as a function of t, given that when $t = 0$, $x = 0$ and $\dfrac{dx}{dt} = 0$.

16 The extension x of a spring at time t seconds, is given by the differential equation

$$\dfrac{d^2x}{dt^2} + n^2x = \cos\omega t \qquad \text{where } n \text{ and } \omega \text{ are constants, and } n \neq \omega.$$

Express x as a function of t, given that when $t = 0$, $x = 0$ and $\dfrac{dx}{dt} = V$.

17 The current i in an electrical circuit at time t is given by the differential equation

$$L\dfrac{d^2i}{dt^2} + R\dfrac{di}{dt} + \dfrac{i}{C} = -nE\sin nt$$

where L, R, C, E and n are constants, and $R^2 < 4L/C$.

Find the complementary function and show that as $t \to \infty$, the complementary function tends to zero.

Show that the particular integral can be expressed in the form

$$i = \frac{E}{Z} \sin (nt + \alpha)$$

where Z and α are defined by the triangle in Fig. 15.4, and $Ln - 1/(Cn) > 0$.

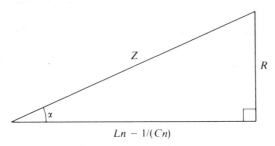

$$Ln - 1/(Cn)$$

Figure 15.4

18 The displacement x of a particle at time t is given by the differential equation

$$\frac{d^2x}{dt^2} + kx + n^2x = P \cos \omega t$$

where P, ω, k and n are positive constants and $k > 2n$. Show that as $t \to \infty$ the motion attains a 'steady state' in which x is given by

$$x = P \left\{ \frac{(n^2 - \omega^2) \cos \omega t + \omega k \sin \omega t}{(n^2 - \omega^2)^2 + k^2 \omega^2} \right\}$$

19 A particle moves on the line Ox so that after time t its displacement from O is x, and

$$\frac{d^2x}{dt^2} = -9x$$

When $t = 0$, $x = 4$ and $\frac{dx}{dt} = 9$. Find:

(a) the position and the velocity of the particle when $t = \pi/6$,
(b) the maximum displacement of the particle from O. (JMB)

20 A bead P moves along a straight wire under certain forces such that its motion is given by

$$\frac{d^2x}{dt^2} + 3 \frac{dx}{dt} + 2x = 10 \cos t$$

where x is the displacement of P at time t from a fixed point O on the wire. Determine x in terms of t, given that P starts from rest at O at time $t = 0$.

Show that for large t, the motion of P is approximately periodic and determine the amplitude of the motion for large t.

21 A particle moves along the x-axis so that its displacement from O at time t satisfies the differential equation

$$\frac{d^2x}{dt^2} + 9x - 18 = 0$$

Its speed when $t = \pi/12$ is $21/\sqrt{2}$ in the direction of the positive x-axis and its speed when $t = \pi/6$ is 15 in the *opposite* direction. Find x at any time t.

What is the greatest distance from O reached by the particle and what is its greatest speed?

If the motion starts at $t = 0$, find the time at which the particle first passes through O.   (O & C)

22 Given that y is a function of x, where $x > 0$, show that, if the substitution $x = \sqrt{t}$ is made, then

(a) $\dfrac{dy}{dx} = 2\sqrt{t}\,\dfrac{dy}{dt}$,   (b) $\dfrac{d^2y}{dx^2} = 4t\,\dfrac{d^2y}{dt^2} + 2\,\dfrac{dy}{dt}$.

Hence or otherwise find the general solution of the differential equation

$$\frac{d^2y}{dx^2} - \frac{1}{x}\left(\frac{dy}{dx}\right) + 4x^2(9y + 6) = 0 \tag{C}$$

23 Show that $y = 2 - \cos x$ is a particular integral of the differential equation

$$\frac{d^2y}{dx^2} + 4y = 8 - 3\cos x$$

and obtain the general solution.

Hence find the general solution such that when $x = 0$, $y = 1.5$ and $\dfrac{dy}{dx} = 0$.

Show that in this case, $5/4 \leqslant y \leqslant 7/2$.   (C)

24 Use the substitution $y = xu$, where $u = f(x)$, in the differential equation

$$x^2\frac{d^2y}{dx^2} - 2x\frac{dy}{dx} + 2(2x^2 + 1)y = 24x^3$$

to obtain a differential equation for u in terms of x.

Hence solve this equation, given that $\dfrac{dy}{dx} = 6$ and $\dfrac{d^2y}{dx^2} = 4$, when $x = 0$.

(O & C)

Chapter 16

Approximations — further expansions in series

Approximation

16.1 To students of elementary mathematics the word *approximation* no doubt conjures up the idea of a 'rough calculation'; but it should also be a reminder that answers may often be relied upon only to a certain degree of accuracy, due to limitations set by data and by the available means of computation.

However, there is a positive aspect of approximation which must be stressed at this stage. In the ever increasing field of application to engineering and complex organisation, mathematics often assumes a character less exact, possibly less aesthetically satisfying, than when it is pursued for its own sake. A problem may arise in which many variables or 'parameters' are involved; only when attention is confined to the more significant of these is the problem susceptible to known mathematical techniques, and even then the functions concerned are often only manageable when reduced to an approximate form. It must therefore be appreciated that approximation, far from always implying a sacrifice of accuracy, can provide the means whereby mathematics may be brought to bear on practical problems which would otherwise be out of reach. In Book 1, Chapter 24, we looked at methods for solving equations by iteration, and, in particular, we met the Newton-Raphson method.

The main object of this chapter is to consider new ways of re-writing functions in an approximate form; already we have used the binomial theorem to this end, and in Chapter 10 we saw how e^x and $\ln(1 + x)$ could be expressed as power series in x. We now start by establishing a basic form of approximation, and applying it to numerical examples.

Linear approximation

16.2 Figure 16.1 represents part of the graph of $y = f(x)$. P is a fixed point $(a, f(a))$, PT is the tangent at P, and Q is a variable point on the curve given by $x = a + h$, where h is small. We shall establish an approximate relationship between $f(a)$ and $f(a + h)$.

$f(a + h) = NQ \approx NT$ since h is small

$NT = MP + RT = f(a) + PR \tan \angle RPT = f(a) + f'(a) h$†

Hence, if h is small,

$f(a + h) \approx f(a) + f'(a) h$

This is called a *linear approximation* since we consider the straight line PT in lieu of the curve PQ. Expressing it another way, when $x \approx a$, the function $f(x)$ may be expressed in an approximate linear form, since

$f(x) \approx f(a) + f'(a)(x - a)$

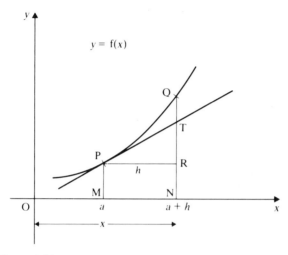

Figure 16.1

Example 1   *Find an approximate value of* $\sin 31°$.

[Here we wish to establish an approximate relationship between $\sin 30°$, which we know, and $\sin 31°$. Since we use the derivative of $\sin x$ we work in radians. Note that $1° = \pi/180$ radians.]

$$31° = \frac{\pi}{6} + \frac{\pi}{180} \text{ radians}$$

Since $f(a + h) \approx f(a) + f'(a) h$,

$$\sin \left(\frac{\pi}{6} + \frac{\pi}{180} \right) \approx \sin \frac{\pi}{6} + \cos \frac{\pi}{6} \times \frac{\pi}{180}$$

$$= 0.5 + \frac{\sqrt{3}}{2} \times \frac{\pi}{180}$$

$$\therefore \sin 31° \approx 0.515$$

†The fact that $RQ \approx RT = f'(a)h$ has been used in Book 1, §7.7; there it was stated in the form $\delta y \approx \dfrac{dy}{dx} \delta x$.

No method of approximation is of much value unless its degree of accuracy is known. However, for the moment we must avoid this issue; the reader should consider the following questions as just a first step in mastering an idea which will be developed more fully later in the chapter.

Qu. 1 Use the method of Example 1 to find an approximate value of tan 45.6°, retaining five significant figures. Compare this with the value you find in four-figure tables or by a calculator.

Qu. 2 Assuming that $\cos^{-1} 0.8 \approx 36° \ 52'$ and $52' \approx 0.0151$ radians, find an approximation for $\cos 36°$.

Qu. 3 If $x \approx \pi/6$, prove that $\cos x \approx \frac{1}{12}(\pi + 6\sqrt{3} - 6x)$, using the second form of linear approximation given in §16.2.

Qu. 4 Use the method of Example 1 to obtain approximations for the following. Retain four decimal places and compare your answers with the values given by four-figure tables or by a calculator.

(a) cosec 61.5°   (take $\sqrt{3}$ as 1.7321),
(b) cot 28.5°   (take $\sqrt{3}$ as 1.7321),
(c) $e^{1.08}$ (take e as 2.7183),
(d) ln 2.001   (take ln 2 as 0.6931).

Qu. 5 Use the fact that if $x \approx a$, $f(x) \approx f(a) + f'(a)(x - a)$ to prove that
(a) if $x \approx 0$, $e^x \approx 1 + x$,   (b) if $x \approx \pi$, $\sin x \approx \pi - x$,

(c) if $x \approx 2$, $\dfrac{1}{(1 + x)^2} \approx \dfrac{1}{27}(7 - 2x)$,   (d) if $x \approx \dfrac{\pi}{4}$, $\tan x \approx 1 - \dfrac{\pi}{2} + 2x$,

(e) if $x \approx 7$, $\sqrt{(2 + x)} \approx \frac{1}{6}(x + 11)$,   (f) if $x \approx 1$, $\ln x \approx x - 1$.

Quadratic approximation

16.3 In Chapter 10 it was established that

$$e^x = 1 + x + \frac{x^2}{2!} + \frac{x^3}{3!} + \frac{x^4}{4!} + \dots$$

If x is small, an approximation for e^x may be found by ignoring high powers of x. Thus the linear approximation obtained in Qu. 5(a) is $1 + x$, the first two terms of the above series. Clearly a better approximation would be obtained by taking more terms; let us see how this fits in with the approach being developed in this chapter.

$y = x + 1$ is the tangent to $y = e^x$ at $(0, 1)$. So when $x = 0$, the graphs of the function and of its linear approximation have equal ordinates and equal gradients. If we take a quadratic approximation, $f(x)$, we can further stipulate that the *rate of change of gradient* of $y = e^x$ and $y = f(x)$ are equal when $x = 0$. $y = f(x)$ is a parabola, and this gives a better approximation to the curve $y = e^x$ over a wider range of values of x.

Suppose

$$f(x) = c_0 + c_1 x + c_2 x^2$$

Then for small values of x,

$$e^x \approx c_0 + c_1 x + c_2 x^2$$

Differentiating twice,

$$e^x \approx \quad c_1 + 2c_2 x$$

and

$$e^x \approx \qquad 2c_2$$

But we have stipulated that, when $x = 0$, these are not approximations but equalities,

$$\therefore c_0 = 1 \qquad c_1 = 1 \qquad c_2 = \tfrac{1}{2}$$

Therefore for small values of x

$$e^x \approx 1 + x + \frac{x^2}{2}$$

and, as expected, we have obtained the first three terms of the series for e^x.

Qu. 6 Sketch with the same axes the graphs of e^x, $1 + x$, $1 + x + x^2/2$ from $x = 0$ to $x = 1$ at intervals of 0.1.

Proceeding on the same lines, we now consider in Qu. 7 the function $\ln x$ when $x \approx 1$, and investigate how we can obtain an improvement on the linear approximation $\ln x \approx x - 1$. (See Qu. 5(f).)

Qu. 7 Given that the graphs of $y = \ln x$ and $y = c_0 + c_1(x - 1) + c_2(x - 1)^2$ have the same ordinate, gradient, and rate of change of gradient when $x = 1$, prove that when $x \approx 1$,

$$\ln x \approx -\frac{3}{2} + 2x - \frac{x^2}{2}$$

The following table gives values of the function $\ln x$, and of the first and second approximations $x - 1$ and $-\frac{3}{2} + 2x - \frac{1}{2}x^2$, in the vicinity of $x = 1$, and the graphs of these functions are shown in Fig. 16.2.

x	0.2	0.3	0.4	0.5	0.6	0.7	0.8	0.9	1
$x - 1$	-0.8	-0.7	-0.6	-0.5	-0.4	-0.3	-0.2	-0.1	0
$-\dfrac{3}{2} + 2x - \dfrac{x^2}{2}$	-1.12	-0.95	-0.78	-0.63	-0.48	-0.345	-0.22	-0.105	0
$\ln x$	-1.61	-1.20	-0.92	-0.69	-0.51	-0.357	-0.223	-0.1054	0

x	1.1	1.2	1.3	1.4	1.5	1.6	1.7	1.8	1.9	2
$x - 1$	0.1	0.2	0.3	0.4	0.5	0.6	0.7	0.8	0.9	1
$-\dfrac{3}{2} + 2x - \dfrac{x^2}{2}$	0.095	0.18	0.255	0.32	0.38	0.42	0.46	0.48	0.495	0.5
$\ln x$	0.0953	0.182	0.262	0.34	0.41	0.47	0.53	0.59	0.64	0.69

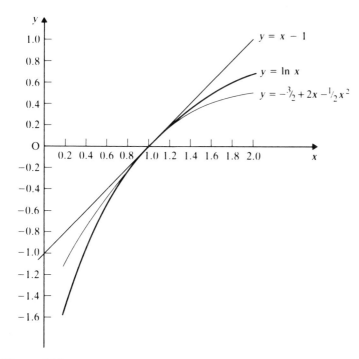

Figure 16.2

Readers who are fortunate enough to have access to a microcomputer with high resolution graphics should use it to plot these graphs. Notice that the approximation is very good for values of x which are near $x = 1$.

Qu. 8   Given that the graphs of $y = f(x)$ and $y = c_0 + c_1(x - a) + c_2(x - a)^2$ have the same ordinate, gradient, and rate of change of gradient when $x = a$, find c_0, c_1, c_2, and hence give an approximation for $f(x)$ when $x \approx a$.

Qu. 9   If a is a constant and h is small, re-write the answer to Qu. 8 so as to give an approximation for $f(a + h)$ in ascending powers of h as far as h^2.

Taylor's theorem

16.4   Pursuing the ideas of Qu. 7 and Qu. 8 we may reasonably suppose that if we add terms of successively higher powers of $(x - a)$ to an approximation for a function $f(x)$ when $x \approx a$, determining the coefficients so that successively higher derivatives of the approximate function are equal to those of f(x) when $x = a$, then we shall obtain ever closer approximations to f(x).

In Chapter 10, e^x and $\ln(1 + x)$ are expressed as infinite series in ascending powers of x. We shall now assume that if f(x) is any function of x, and a is a constant, then provided that f(a) exists and that successive derivatives of f(x) all have finite values when $x = a$, f(x) may be expressed as an infinite series in ascending powers of $(x - a)$.† In what follows, we assume that it is in order to differentiate an infinite series term by term.

Let

$$f(x) = c_0 + c_1(x - a) + c_2(x - a)^2 + c_3(x - a)^3 + \qquad c_4(x - a)^4 + \dots$$

then

$$
\begin{aligned}
f'(x) = &\quad c_1 &&+ 2c_2(x - a) + 3c_3(x - a)^2 + &&4c_4(x - a)^3 + \dots \\
f''(x) = &&&2!c_2 \quad + 3 \times 2c_3(x - a) + &&4 \times 3c_4(x - a)^2 + \dots \\
f'''(x) = &&&&&3!c_3 \qquad + 4 \times 3 \times 2c_4(x - a) + \dots \\
f''''(x) = &&&&&4!c_4 \qquad\qquad + \dots
\end{aligned}
$$

and putting $x = a$ in each line, we find that

$$c_0 = f(a), \quad c_1 = f'(a), \quad c_2 = \frac{f''(a)}{2!}, \quad c_3 = \frac{f'''(a)}{3!}, \quad c_4 = \frac{f''''(a)}{4!}$$

Thus

$$f(x) = f(a) + f'(a)\,(x - a) + \frac{f''(a)}{2!}(x - a)^2 + \frac{f'''(a)}{3!}(x - a)^3 +$$

$$+ \frac{f''''(a)}{4!}(x - a)^4 + \dots$$

or if $x = a + h$,

$$f(x) = f(a + h) = f(a) + f'(a)\,h + \frac{f''(a)}{2!}h^2 + \frac{f'''(a)}{3!}h^3 + \frac{f''''(a)}{4!}h^4 + \dots$$

This result is a form of **Taylor's theorem** (1716).

Example 2 *Use Taylor's theorem to expand* $\sin(\pi/6 + h)$ *in ascending powers of* h *as far as the term in* h^4.

$$\text{Let } f(x) = \sin x = \sin\left(\frac{\pi}{6} + h\right) \qquad f\left(\frac{\pi}{6}\right) = \sin\frac{\pi}{6} = \frac{1}{2}$$

$$f'(x) = \cos x \qquad\qquad\qquad f'\left(\frac{\pi}{6}\right) = \cos\frac{\pi}{6} = \frac{\sqrt{3}}{2}$$

---

†We proved in §10.3 that the expansion of $\ln(1 + x)$ in ascending powers of x is valid only if $-1 < x \leqslant +1$; we may therefore expect some limitations on the value of x in certain cases of the general expansions we are about to discuss. Consideration of this is delayed until §16.8.

$$f''(x) = -\sin x \qquad\qquad f''\left(\frac{\pi}{6}\right) = -\sin\frac{\pi}{6} = -\frac{1}{2}$$

$$f'''(x) = -\cos x \qquad\qquad f'''\left(\frac{\pi}{6}\right) = -\cos\frac{\pi}{6} = -\frac{\sqrt{3}}{2}$$

$$f''''(x) = \sin x \qquad\qquad f''''\left(\frac{\pi}{6}\right) = \sin\frac{\pi}{6} = \frac{1}{2}$$

By Taylor's theorem,

$$f(a+h) = f(a) + f'(a)\,h + \frac{f''(a)}{2!}\,h^2 + \frac{f'''(a)}{3!}\,h^3 + \frac{f''''(a)}{4!}\,h^4 + \dots$$

$$\therefore \sin\left(\frac{\pi}{6}+h\right) = \frac{1}{2} + \frac{\sqrt{3}}{2}h + \frac{(-\frac{1}{2})}{2!}h^2 + \frac{(-\sqrt{3}/2)}{3!}h^3 + \frac{\frac{1}{2}}{4!}h^4 + \dots$$

$$= \frac{1}{2} + \frac{\sqrt{3}}{2}h - \frac{1}{4}h^2 - \frac{\sqrt{3}}{12}h^3 + \frac{1}{48}h^4 + \dots$$

Qu. 10 Using only the first three terms of the expansion obtained in Example 2, obtain a value for sin 31° to five significant figures, taking $\sqrt{3}$ as 1.7321 and 1° as 0.01745 radians. Compare your answer with that of Example 1.

Qu. 11 Use Taylor's theorem to express $\tan(\pi/4 + h)$ as a series in ascending powers of h as far as the term in h^3.

Qu. 12 Use Taylor's theorem to find the first four terms in the expansion of $\cos x$ in ascending powers of $(x - \alpha)$, where $\alpha = \tan^{-1}\frac{4}{3}$.

Maclaurin's theorem

16.5 Bearing in mind the relationship $x = a + h$, where a is a constant, and x and h are variable (see Fig. 16.1), we see that there is a special case given by $a = 0$, when $x = h$, and either form of Taylor's theorem given in §16.4 reduces to

$$f(x) = f(0) + f'(0)\,x + \frac{f''(0)}{2!}x^2 + \frac{f'''(0)}{3!}x^3 + \frac{f''''(0)}{4!}x^4 + \dots$$

This is a form of **Maclaurin's theorem** (1742).

Example 3 *Use Maclaurin's theorem to expand* $\ln(1 + x)$ *in ascending powers of* x *as far as the term in* x^5.

$$f(x) = \ln(1+x) \qquad\qquad f(0) = 0$$
$$f'(x) = (1+x)^{-1} \qquad\qquad f'(0) = 1$$
$$f''(x) = -(1+x)^{-2} \qquad\qquad f''(0) = -1$$
$$f'''(x) = 2(1+x)^{-3} \qquad\qquad f'''(0) = 2!$$
$$f''''(x) = -3 \times 2(1+x)^{-4} \qquad\qquad f''''(0) = -3!$$
$$f'''''(x) = 4 \times 3 \times 2(1+x)^{-5} \qquad\qquad f'''''(0) = 4!$$

By Maclaurin's theorem

$$f(x) = f(0) + f'(0)\,x + \frac{f''(0)}{2!}\,x^2 + \frac{f'''(0)}{3!}\,x^3 + \frac{f''''(0)}{4!}\,x^4 + \frac{f'''''(0)}{5!}\,x^5 + \dots$$

$$\therefore \ln(1 + x) = 0 + 1 \times x + \frac{(-1)}{2!}\,x^2 + \frac{2!}{3!}\,x^3 + \frac{(-3!)}{4!}\,x^4 + \frac{4!}{5!}\,x^5 + \dots$$

$$\therefore \ln(1 + x) = x - \frac{x^2}{2} + \frac{x^3}{3} - \frac{x^4}{4} + \frac{x^5}{5} - \dots$$

Qu. 13   Use Maclaurin's theorem
(a) to expand e^x in ascending powers of x as far as the x^5 term,
(b) to show that when x is small, $\sin x \approx x - \dfrac{x^3}{3!} + \dfrac{x^5}{5!}$,
(c) to find the first three terms of the expansion of $\cos x$ in ascending powers
of x.

Qu. 14   Express $17° \, 11'$ in radians correct to one significant figure. Use the approximation given in Qu. 13(b) to express $\sin 17° \, 11'$ to four significant figures. Check your answer against the value given in four-figure tables, or on a calculator.

Exercise 16a

1 Given that the graphs of $y = \ln x$ and $y = c_0 + c_1(x - 2) + c_2(x - 2)^2$ have the same value of y, gradient and rate of change of gradient when $x = 2$, determine c_0, c_1, c_2 and deduce an approximation for $\ln x$ when $x \approx 2$.
2 Obtain a quadratic approximation for $\sin x$ when $x \approx \alpha$.
3 Apply Taylor's theorem
   (a) to expand $\ln x$ in ascending powers of $(x - e)$ as far as the term in $(x - e)^4$,
   (b) to expand $\operatorname{cosec} x$ in ascending powers of $(x - \pi/2)$ as far as the term in $(x - \pi/2)^4$.
4 Use Taylor's theorem to expand $\cos(\pi/3 + h)$ in ascending powers of h up to the h^3 term. Taking $\sqrt{3}$ as 1.7321 and $5.5°$ as 0.095 99 radians, find the value of $\cos 54.5°$ to three decimal places.
5 Given that the functions $f(x)$ and $c_0 + c_1 x + c_2 x^2 + c_3 x^3 + c_4 x^4 + \dots$ have the same value when $x = 0$, and equal successive derivatives when $x = 0$, deduce the first five terms of the Maclaurin expansion of $f(x)$ in ascending powers of x.
6 We have used Maclaurin's theorem to establish the following expansions, which should be memorised:

$$e^x = 1 + x + \frac{x^2}{2!} + \frac{x^3}{3!} + \dots$$

$$\ln(1 + x) = x - \frac{x^2}{2} + \frac{x^3}{3} - \frac{x^4}{4} + \dots$$

$$\cos x = 1 - \frac{x^2}{2!} + \frac{x^4}{4!} - \frac{x^6}{6!} + \dots$$

$$\sin x = x - \frac{x^3}{3!} + \frac{x^5}{5!} - \frac{x^7}{7!} + \dots$$

Write down the first four terms of the expansions of the following in ascending powers of x:

(a) e^{2x},   (b) $\ln(1-x)$,   (c) $\cos x^2$,   (d) $\sin \frac{x}{2}$.

7 By subtracting the expansion of $\ln(1-x)$ from that of $\ln(1+x)$ deduce that

$$\ln \sqrt{\left(\frac{1+x}{1-x}\right)} = x + \frac{x^3}{3} + \frac{x^5}{5} + \frac{x^7}{7} + \dots$$

8 Find approximations for the following:
   (a) $e^{0.4}$ (correct to five significant figures),
   (b) $\ln 1.2$ (correct to four significant figures),
   (c) $\cos 0.3$ (correct to three significant figures),
   (d) $\sin 0.2$ (correct to three significant figures).

9 Apply Maclaurin's theorem directly (see Example 3) to obtain expansions for the following in ascending powers of x up to the given term:
   (a) $\sin^2 x$, (x^4),   (b) $(1+x)^n$, (x^3),
   (c) 2^x, (x^3),   (d) $\arccos x$, (x^3),
   (e) $e^x \sin x$, (x^5),   (f) $\ln\{x + \sqrt{(x^2+1)}\}$, (x^3).

10 If $f(x) = e^x \sin x$ show that $f^n(x) = (\sqrt{2})^n e^x \sin(x + n\pi/4)$, and use this with Maclaurin's theorem to find an expansion for $f(x)$ in ascending powers of x as far as the x^6 term,

11 Find the expansion of $\ln x$ in ascending powers of $(x-4)$ up to the fourth term

(a) by writing $\ln x$ as $\ln\left\{4\left(1 + \frac{x-4}{4}\right)\right\}$ and applying the expansion for $\ln(1+x)$ given in No. 6,

(b) by applying Taylor's theorem.
Deduce an approximation for $\ln 4.02$ correct to four decimal places.

Expansion by integration

16.6 If we wish to expand $f(x)$ in ascending powers of x, and we find that *an expansion for $f'(x)$ is known* or is easily obtainable, then the required expansion may be obtained from the latter by integration. This is illustrated graphically by saying that if two curves approximate over a certain range of values of x, then the area under these curves will be approximately equal over that range. Since, for example,

$$\frac{d}{dx}\ln(1+x) \quad \text{is} \quad \frac{1}{1+x}$$

which may be expanded by the binomial theorem, this provides an alternative method of obtaining an expansion for $\ln(1+x)$.

Example 4   *Expand* $\ln(1+x)$ *in ascending powers of x as far as the* x^4 *term.*

$$\left[\ln(1+u)\right]_0^x = \int_0^x \frac{1}{1+u}\,du$$

$$= \int_0^x (1 - u + u^2 - u^3 + \ldots)\,du \quad (\text{provided } -1 < u < 1)$$

$$= \left[u - \tfrac{1}{2}u^2 + \tfrac{1}{3}u^3 - \tfrac{1}{4}u^4 + \ldots\right]_0^x$$

$$\therefore \ln(1+x) = x - \tfrac{1}{2}x^2 + \tfrac{1}{3}x^3 - \tfrac{1}{4}x^4 + \ldots$$

Qu. 15   (a) Use Maclaurin's theorem to obtain the coefficients in the expansion of arctan x in ascending powers of x, up to the x^4 term.
(b) Now obtain this expansion up to the x^7 term by using the fact that

$$\left[\arctan u\right]_0^x = \int_0^x \frac{1}{1+u^2}\,du$$

Miscellaneous methods

16.7   Qu. 15 brings out the value of the integration method; this is just one way of avoiding the laborious differentiation sometimes involved in the direct application of Maclaurin's theorem. It is also useful to bear in mind the following possibilities:

(a) the use of a known approximation together with a known expansion (see Example 5),
(b) the use of the product of known, or more easily obtained, expansions (see Qu. 17).

Example 5   *Expand* sec x *in ascending powers of x as far as the* x^6 *term.*

$$\sec x = \frac{1}{\cos x} \approx \left\{1 - \frac{x^2}{2!} + \frac{x^4}{4!} - \frac{x^6}{6!}\right\}^{-1}$$

$$= \left\{1 - \left(\frac{x^2}{2!} - \frac{x^4}{4!} + \frac{x^6}{6!}\right)\right\}^{-1}$$

$$= \left\{1 - \left(\frac{x^2}{2} - \frac{x^4}{24} + \frac{x^6}{720}\right)\right\}^{-1}$$

$$= 1 + \left(\frac{x^2}{2} - \frac{x^4}{24} + \frac{x^6}{720}\right) + \left(\frac{x^2}{2} - \frac{x^4}{24} + \frac{x^6}{720}\right)^2 +$$

$$+ \left(\frac{x^2}{2} - \frac{x^4}{24} + \frac{x^6}{720}\right)^3 + \ldots$$

$$= 1 + \frac{x^2}{2} - \frac{x^4}{24} + \frac{x^6}{720} + \frac{x^4}{4} - \frac{x^6}{24} + \ldots + \frac{x^6}{8} + \ldots$$

$$\therefore \sec x = 1 + \tfrac{1}{2}x^2 + \tfrac{5}{24}x^4 + \tfrac{61}{720}x^6 + \ldots$$

Qu. 16 Use the expansions

$$e^x = 1 + x + \frac{x^2}{2!} + \frac{x^3}{3!} + \frac{x^4}{4!} + \ldots \quad \text{and} \quad \sin x = x - \frac{x^3}{3!} + \ldots$$

to express $e^{\sin x}$ in ascending powers of x as far as the x^4 term.

Qu. 17 Expand $\dfrac{\cos x}{\sqrt{(1-x)}}$ in ascending powers of x as far as the x^4 term, by considering the product of the expansions of $\cos x$ and of $(1-x)^{-1/2}$.

Validity of expansions

16.8 So far we have avoided the issue that some of the expansions obtained in this chapter may be valid only for certain values of x.

For example, the binomial theorem only enables us to expand $(1-x)^{-1}$ as the infinite series $1 + x + x^2 + \ldots$ provided that $-1 < x < 1$. As a reminder of why this is so we may employ an even more elementary method of expansion; by long division

$$\frac{1}{1-x} = 1 + x + \frac{x^2}{1-x} = 1 + x + x^2 + \frac{x^3}{1-x} = 1 + x + x^2 + x^3 + \frac{x^4}{1-x} \text{ etc.}$$

Only if $-1 < x < 1$† is it true that

$$\left|\frac{x^4}{1-x}\right| < \left|\frac{x^3}{1-x}\right| < \left|\frac{x^2}{1-x}\right|$$

and hence that $1 + x$, $1 + x + x^2$, $1 + x + x^2 + x^3$ are progressively better approximations to $1/(1-x)$ since the error involved is progressively decreasing in size; taking the approximation to n terms, the error is $x^n/(1-x)$. If we let $n \to \infty$, we assume that $x^n \to 0$ if $-1 < x < 1$, and so the error $\to 0$; it follows that we may make the approximation

$$1 + x + x^2 + \ldots + x^{n-1}$$

as near as we please to $1/(1-x)$ by taking n sufficiently large. Expressing this in other words, we may say that the infinite series $1 + x + x^2 + \ldots$ *converges* to the sum $1/(1-x)$ provided that $-1 < x < 1$.

†The consequences of taking values of x outside this range are discussed in Book 1, §14.4.

The expansion

$$\ln(1+x) = x - \frac{x^2}{2} + \frac{x^3}{3} - \frac{x^4}{4} + \frac{x^5}{5} - \ldots$$

is valid only if $-1 < x \leqslant +1$. (This is proved in §10.3.) Fig. 16.3 shows the graph of $y = \ln(1+x)$ together with those of the successive approximate functions

(a) $y = x$,

(b) $y = x - \dfrac{x^2}{2}$,

(c) $y = x - \dfrac{x^2}{2} + \dfrac{x^3}{3}$,

(d) $y = x - \dfrac{x^2}{2} + \dfrac{x^3}{3} - \dfrac{x^4}{4}$.

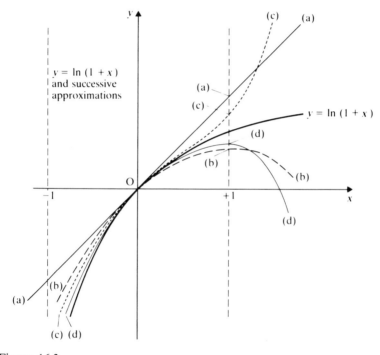

Figure 16.3

$x = -1$ is cut by all the latter, but is an asymptote to $y = \ln(1+x)$, which illustrates the condition $-1 < x$; it can be seen that the values of the successive approximate functions approach that of $\ln(1+x)$ for positive values of x up to *and including* $+1$, but thereafter diverge rapidly from it, hence the condition $x \leqslant 1$.

(Here again, the behaviour of this function and its approximations can be compared very effectively on a microcomputer. Readers are strongly advised to do this if possible.)

An expansion obtained by the use of Maclaurin's theorem is valid only if the series is convergent but general consideration of this matter must be delayed until a later stage. The reader should learn the following conditions for validity, and should always quote them, or derived conditions, whenever they apply.

Expansion	*Condition*
$(1 + x)^n$	$-1 < x < 1$†
$\ln (1 + x)$	$-1 < x \leqslant 1$
$\arctan x$	$-1 \leqslant x \leqslant 1$

Expansions of e^x, $\sin x$, $\cos x$ are valid for all values of x.

Qu. 18 For what ranges of values of x may the following be expanded as infinite series in ascending powers of x?

(a) $\dfrac{1}{2 + x}$, (b) $\ln (1 - x)$, (c) $\arctan (1 + x)$,

(d) e^{3x}, (e) $\ln (1 + 2x)$, (f) $\ln (1 + x^2)$,

(g) $\dfrac{\sqrt{(2 + x)}}{1 + 3x}$, (h) $\ln \dfrac{1 + x}{1 - x}$, (i) $\dfrac{\ln (1 + x)}{1 + x}$.

Qu. 19 What are the conditions that the following may be expanded as an infinite series in ascending powers of $1/x$?

(a) $\sqrt{\left(1 + \dfrac{2}{x}\right)}$, (b) $\ln \left(1 + \dfrac{1}{x}\right)$.

Qu. 20 State the conditions for validity of the following expansions:
(a) $\ln x$ in ascending powers of $(x - 2)$,
(b) $\sin x \sqrt{(1 + x)}$ in ascending powers of $(x - \pi/2)$.

Rate of convergence

16.9 For practical purposes we wish to know not only that a particular expansion is valid, but also that it converges sufficiently rapidly for the value of the variable we are considering; in other words, if we are to be able to obtain a satisfactory approximation by considering reasonably few terms, these must decrease rapidly in size.

Now

$$\arctan x = x - \frac{x^3}{3} + \frac{x^5}{5} - \frac{x^7}{7} + \dots$$

is valid when $-1 \leqslant x \leqslant 1$, and if we put $x = 1$,

$$\arctan 1 = \frac{\pi}{4} = 1 - \tfrac{1}{3} + \tfrac{1}{5} - \tfrac{1}{7} + \dots$$

---

†Also valid for $x = 1$ if $n > -1$, and for $x = -1$ if $n > 0$; but these refinements are ignored elsewhere in this book.

This provides a method of calculating π, but it is not a very good one since the series converges very slowly; a better one is given in Exercise 16b, Nos. 5 and 6.

To take another example,

$$\sin x = x - \frac{x^3}{3!} + \frac{x^5}{5!} - \frac{x^7}{7!} + \dots$$

is valid for all values of x, but only converges rapidly when x is small. If we are concerned with large values of x, where $x \approx a$, it is necessary to use a Taylor expansion in ascending powers of $(x - a)$.

As a third example let us consider the expansion

$$\ln (1 + x) = x - \frac{x^2}{2} + \frac{x^3}{3} - \frac{x^4}{4} + \dots$$

which is valid when $-1 < x \leqslant 1$. Putting $x = 1$, we obtain

$$\ln 2 = 1 - \tfrac{1}{2} + \tfrac{1}{3} - \tfrac{1}{4} + \dots$$

This is an extreme case of a fruitless application of an expansion, since somewhere in the order of 10 000 terms are needed to produce a value of $\ln 2$ correct to four decimal places! A rather more economical method of evaluating $\ln 2$ is given in Exercise 16b. Meanwhile the reader should consider two less contrasting methods of evaluating $\ln 1.5$, which nevertheless stress the practical value of rapid convergence.

Qu. 21 Obtain approximations for $\ln 1.5$ by the following methods, and compare them with the value given in four-figure tables or by a calculator:
(a) use the first five terms of the expansion of $\ln (1 + x)$ in ascending powers of x, putting $x = 0.5$,

(b) find a value of x for which $\dfrac{1 + x}{1 - x} = 1.5$, and substitute this value in the first

three terms of the expansion of $\ln \dfrac{1 + x}{1 - x}$ in ascending powers of x.

Exercise 16b

1 Use the method of integration given in Example 4 to obtain expansions of the following in ascending powers of x up to the given term.
(a) $\arcsin x$ (x^7 term),   (b) $\ln (\sec x + \tan x)$ (x^5 term),
(c) $\arctan (x + 1)$ (x^5 term),   (d) $\arccos x$ (x^5 term).
2 Make use of known expansions to obtain expansions of the following in ascending powers of x up to the given term:
(a) $\tfrac{1}{2}(e^x + e^{-x})$ (x^6 term),   (b) $x \operatorname{cosec} x$ (x^4 term),
(c) $\cos^3 x$ (x^4 term),   (d) $\tan x$ (x^5 term),

(e) $\ln \dfrac{\sin x}{x}$ (x^4 term),   (f) $\ln (1 + \sin x)$ (x^4 term),

(g) $\ln (1 + e^x)$ (x^4 term).

3 Verify the following expansions, and state any limitations on the value of x required:

(a) $e^x \cos x = 1 + x - \frac{1}{3}x^3 - \frac{1}{6}x^4 + \ldots,$

(b) $\dfrac{1}{e^x(1-x)} = 1 + \frac{1}{2}x^2 + \frac{1}{3}x^3 + \ldots,$

(c) $\dfrac{\cos x}{(1-x)^2} = 1 + 2x + \frac{5}{2}x^2 + 3x^3 + \frac{85}{24}x^4 + \ldots,$

(d) $\dfrac{e^{\sin x}}{\sqrt{(4+x^2)}} = \frac{1}{2} + \frac{1}{2}x + \frac{3}{16}x^2 - \frac{1}{16}x^3 - \frac{21}{256}x^4 + \ldots,$

(e) $\dfrac{\ln(1+x)}{\arctan x} = 1 - \frac{1}{2}x + \frac{2}{3}x^2 - \frac{5}{12}x^3 + \frac{2}{9}x^4 + \ldots.$

4 By substituting $x = \dfrac{1}{2m+1}$ in the expansion of $\ln\dfrac{1+x}{1-x}$, show that

$$\ln\frac{m+1}{m} = 2\left\{\frac{1}{2m+1} + \frac{1}{3(2m+1)^3} + \frac{1}{5(2m+1)^5} + \ldots\right\}$$

Use four terms of this expansion to find approximate values for $\ln 1.5$, $\ln 2$, $\ln 3$, and compare them with those given in tables or by a calculator.

5 Show that $\pi/4 = \arctan \frac{1}{2} + \arctan \frac{1}{3}$, and use this relationship to obtain an approximation for π correct to four significant figures.

6 Given that $\tan A = 1/5$ and $\tan B = 1/239$, verify that $\tan(4A - B) = 1$.

Hence show that $\pi/4 = 4 \arctan(1/5) - \arctan(1/239)$, and hence, using Maclaurin's series for $\arctan x$, find π correct to four significant figures.

7 Use Taylor's theorem to obtain an approximation for $\sin 131° \, 28'$ correct to four significant figures. (Assume $11° \, 28' \approx 0.2$ radian.)

8 Obtain the expansion of $\arcsin(\frac{1}{2} - x)$ in ascending powers of x as far as the x^3 term.

Exercise 16c (Miscellaneous)

1 Express $\dfrac{10(x+1)}{(x+3)(x^2+1)}$ in partial fractions. Hence obtain the expansion of the given function in ascending powers of x as far as the term in x^3, stating the necessary restrictions on the value of x. (O & C)

2 Show that

$$\ln(1+x+x^2) = \int \frac{dx}{1-x} - 3\int \frac{x^2 \, dx}{1-x^3} + c$$

where c is an arbitrary constant. By expanding the integrands as far as the terms in x^5, find the first six terms of the series for $\ln(1+x+x^2)$ for small values of x. (JMB)

3 Prove that

$$\frac{x}{1 + x + \sqrt{(2x+1)}} = \frac{1}{x}[1 + x - \sqrt{(2x+1)}]$$

and show that, if x is small, the expression is approximately equal to $\frac{1}{2}x(1-x)$.   (O & C)

4 Use Maclaurin's theorem to show that, if x^5 and higher powers of x are neglected,

$$\ln\{x + \sqrt{(1+x^2)}\} = x - \frac{1}{6}x^3$$ (O & C)

5 State Taylor's theorem for the expansion of $f(a+h)$ in a series of ascending powers of h. Prove that the first four terms in the Taylor expansion of arctan $(1+h)$ are $\frac{1}{4}\pi + \frac{1}{2}h - \frac{1}{4}h^2 + \frac{1}{12}h^3$.   (O & C)

6 Express $E \equiv \dfrac{2x^2 + 7}{(x+2)^2(x-3)}$ in partial fractions. Hence, if x is so large that $1/x^5$ can be neglected, prove that

$$E = \frac{1}{x^4}(2x^3 - 2x^2 + 25x - 17)$$ (O & C)

7 Write down the expansions of e^x and e^{-x}. The limit of

$$\frac{e^{2x} - e^{-2x} - 4x}{f(x)}$$

where $f(x)$ is a polynomial, is 8 as $x \to 0$; show that the term of lowest degree in the polynomial $f(x)$ is $\frac{1}{3}x^3$.   (O & C)

8 Expand $E \equiv \ln\left(\dfrac{2-x}{1-x}\right)$ in ascending powers of x up to x^3, stating the necessary conditions for your expansion. Evaluate E when $x = \frac{1}{3}$ and hence find $\ln 3$ to three places of decimals, given that $\ln 2 \approx 0.6931$.   (O & C)

9 Using Maclaurin's theorem, expand $x \tan(\frac{1}{4}\pi - x)$ in ascending powers of x as far as the term containing x^4.   (L)

10 Prove that, if x is small so that x^6 and higher powers of x may be neglected, then

$$\frac{e^{2x} - e^{-2x}}{e^{2x} + e^{-2x}} = 2x - \frac{8}{3}x^3 + \frac{64}{15}x^5$$ (O & C)

11 Find the coefficient of x^r in the expansion of $(1+3x)e^{-3x}$ as a series of ascending powers of x.   (O & C)

12 Expand the function given by $y = \ln\left\{\dfrac{(2-x)^2}{4-4x}\right\}$ in a series of ascending powers of x as far as x^4, stating the limitations on the value of x, and giving the coefficient of x^n. Prove that, up to x^4,

$$y - \frac{x^2}{4}e^x = \frac{3x^4}{32}$$ (O & C)

13 Using the standard expansion of $\ln (1 + x)$, show that

$$\ln \left(\frac{p}{q}\right) = 2\left\{\frac{p-q}{p+q} + \frac{1}{3}\left(\frac{p-q}{p+q}\right)^3 + \frac{1}{5}\left(\frac{p-q}{p+q}\right)^5 + \ldots\right\}$$

where $p, q \in \mathbb{R}^+$. Hence calculate $\ln (13/12)$ correct to five decimal places.

14 The domain of a function $f(x)$ is $\{x : x \in \mathbb{R}, -1 < x < 1\}$ and

$$f(x) = \frac{p}{x} \ln (1 + x) \qquad \text{when } -1 < x < 0$$

$$f(x) = q \qquad \text{when } x = 0$$

$$f(x) = \frac{x \cos x}{1 - \sqrt{(1 - x)}} \qquad \text{when } 0 < x < 1$$

Use the standard series to find the values of p and q which make $f(x)$ a continuous function throughout its domain.

15 If x is small, obtain a quadratic approximation to the expression

$$f(x) = \left\{\frac{e^x}{1 + x}\right\}^2$$

Draw a rough sketch of the curve $y = f(x)$ in the neighbourhood of the point $(0, 1)$ and its tangent at this point. (O & C)

16 Expand $y = \dfrac{1 - \ln (1 + 2x)}{1 + x}$ as a series in ascending powers of x, giving the first three terms.

What is the equation of the tangent to the graph at the point $(0, 1)$?
(O & C)

17 Using Maclaurin's method, or otherwise, find the expansion of $\tan x$ in ascending powers of x as far as the term in x^3.

Given that $\tan (0.1) = 0.100\,334\,7$ and $\tan (1) = 1.557\,407\,7$ correct to seven decimal places, estimate the percentage error in using the above expansion as an approximation in each case and comment on your results. (L)

18 (a) The functions f and g are defined by

$$f(x) = 1 + \frac{x^2}{2!} + \frac{x^4}{4!} + \ldots + \frac{x^{2n}}{(2n)!} + \ldots$$

$$g(x) = x + \frac{x^3}{3!} + \frac{x^5}{5!} + \ldots + \frac{x^{2n+1}}{(2n+1)!} + \ldots$$

Express $f(x)$ and $g(x)$ in terms of e^x, and hence obtain the expansion of $\{f(x)\}^2 + \{g(x)\}^2$ in a series of powers of x, giving the coefficient of x^{2n}.
(b) Find the first four terms in the expansion of

$$\ln \left(\frac{1 + 2x}{1 - x}\right)$$

in a series of ascending powers of x, and state the set of values of x for which the expansion is valid. If a_n and a_{n+1} are the coefficients of x^n and x^{n+1} respectively in the expansion, show that the value of

$$(n+1)a_{n+1} + 2na_n$$

is independent of n.   (L)

19 Given that $y = \ln(1+x)$, prove by induction that

$$\frac{d^n y}{dx^n} = \frac{(-1)^{n-1}(n-1)!}{(1+x)^n} \qquad \text{for } n \geqslant 1$$

Use this result to find the expansion of $\ln(1+x)$ in ascending powers of x, as far as the term in x^4. Using your result and the expansion

$$\sin x = x - \frac{1}{6}x^3 + \frac{1}{120}x^5 - \dots$$

find the expansion of $y_1 = \ln(1 + a \sin x)$ and $y_2 = \sin\{\ln(1+ax)\}$, where a is a constant $(a \neq 0)$, for each as far as the term in x^4.

Show that the expansions of y_1 and y_2 agree as far as the terms in x^2.

For what values of a do the expansions agree as far as the terms in x^3?

(O & C)

20 Given that $y = e^x \cos x$ and $y_n = \dfrac{d^n y}{dx^n}$, $n > 0$, evaluate y_1 and y_2 at $x = 0$.

Prove that, for all values of x,

$$y_{n+2} = 2(y_{n+1} - y_n)$$

Hence or otherwise, obtain an expression of $e^x \cos x$ as a power series as far as the term in x^5 and use it to estimate the value of y to six decimal places, when $x = 0.1$.   (L)

Chapter 17

Some numerical methods

Introduction

17.1 In Book 1, Chapter 24, we tackled the problem of finding an approximate solution to an equation and developed a method which could be used even when it was impossible to find an exact solution. In this chapter we shall be looking at two methods for estimating the area under a curve which can be used even when integration is out of the question. The first method depends on the formula for the area of a trapezium. For a trapezium like that in Fig. 17.1, in which the lengths of the parallel sides are a and b, and where the distance between them is d, this formula for the area A is

$$A = \left(\frac{a+b}{2}\right)d$$

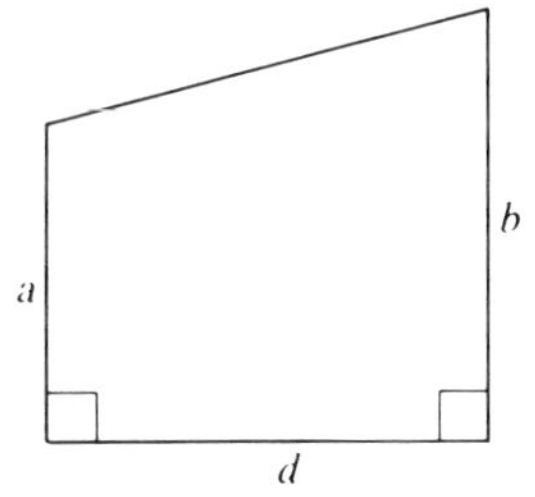

Figure 17.1

Consider the following problem. A cyclist travels along a straight road. He starts from rest and his speed in m/s measured at 2 second intervals is given by the table below.

time in seconds	0	2	4	6	8	10
speed in m/s	0	1.0	2.8	4.9	6.4	7.4

(This information is shown in the graph in Fig. 17.2, which is not drawn to scale.) Find the distance travelled by the cyclist in the 4 seconds from $t = 6$ to $t = 10$.

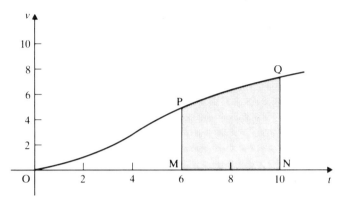

Figure 17.2

When we did problems like this before (see Book 1, §6.2), we used integration, but in this example we do not know the function whose graph is shown in Fig. 17.2. However, we can say that the distance we require is represented by the area bounded by the lines MN, MP, NQ and the curve PQ, and this area is almost the same as the area of the trapezium PQNM; in making this approximation, we have lost the area bounded by the curve PQ and the straight line PQ, but this is only a very small proportion of the total area. We can calculate the area of the trapezium, using the formula above, i.e.

$$A = \left(\frac{4.9 + 7.4}{2}\right) \times 4$$

$$= 24.6$$

So the distance required is approximately 24.6 m.

In the next section we shall see how this method can be applied more generally.

Qu. 1   Use the method above to estimate the distance travelled by the cyclist over the 2 second interval from $t = 4$ to $t = 6$.

Qu. 2   Estimate the distance travelled by the cyclist from $t = 6$ to $t = 10$, by dividing the area into two trapeziums, each two units wide. Would you expect this answer to be better than the one in the text? Justify your answer.

The trapezium rule

17.2   Suppose we wish to find the area under the curve shown in Fig. 17.3. We draw lines parallel to the y-axis at (equal) intervals of d units, and we form an estimate of the area required by calculating the areas of the trapeziums shown. In this diagram there are four trapeziums, but any convenient number may be used; in general the more intervals there are, the better the approximation.

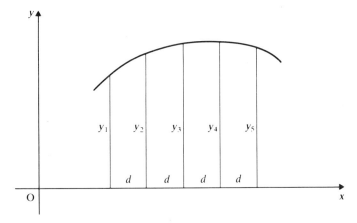

Figure 17.3

The area estimated by this method will be

$$\frac{y_1 + y_2}{2}d + \frac{y_2 + y_3}{2}d + \frac{y_3 + y_4}{2}d + \frac{y_4 + y_5}{2}d$$

$$= \tfrac{1}{2}d(y_1 + 2y_2 + 2y_3 + 2y_4 + y_5) \qquad (1)$$

This is the trapezium rule for five ordinates.†

Qu. 3   Use the trapezium rule to estimate the area, from $x = 0.2$ to $x = 1$, under the curve given by

x	0.20	0.40	0.60	0.80	1.00
y	0.24	0.56	0.96	1.44	2.00

Given that the equation of the curve is $y = x^2 + x$, check your answer by integration.

Qu. 4   Find expressions similar to (1) for (a) eight, (b) nine ordinates. Now express the trapezium rule in words.

Qu. 5   Estimate the area under the curve given by the following table. Beware of the catch!

x	0	10	15	20	25
y	7	9	11	12	10

Another way of looking at the above expression (1) for the area is to take $a = 4d$ so that a is the total interval along the x-axis. In this case the area is estimated to be

$$a\left(\frac{y_1 + 2y_2 + 2y_3 + 2y_4 + y_5}{8}\right) \qquad (2)$$

†Ordinate means y-coordinate, see Book 1, §1.1.

where the expression in brackets appears as the average height of the curve, with a total of eight ordinates (y_2, y_3, y_4 counted twice) divided by 8.

Qu. 6 Obtain the expressions equivalent to (2) for (a) eight ordinates, (b) n ordinates.

The following example has been chosen to illustrate the accuracy of the trapezium rule. We shall compare the answer with that obtained by another rule later.

Example 1 *Use the trapezium rule to estimate the area under the curve* $y = 1/x$ *from* $x = 1$ *to* $x = 2$.

To begin with, let us take six ordinates.

x	1.0	1.2	1.4	1.6	1.8	2.0
y	1	0.8333	0.7143	0.6250	0.5556	0.5

$$
\begin{aligned}
y_1 &= 1.0000 & y_2 &= 0.8333 \\
y_6 &= 0.5000 & y_3 &= 0.7143 \\
\cline{1-2}
 & & y_4 &= 0.6250 \\
 & 1.5000 & y_5 &= 0.5556
\end{aligned}
$$

$$
\begin{aligned}
&2.7282 \\
&\times 2
\end{aligned}
$$

$$5.4564 \longleftarrow 5.4564$$

$$6.9564$$

$$\tfrac{1}{2}d = 0.1$$

$\therefore$ estimated area $= 0.696$, correct to three significant figures.

Now by integration the area is

$$\int_1^2 \frac{1}{x}\,dx = \left[\ln x\right]_1^2$$

$$= \ln 2$$

$$= 0.693, \quad \text{correct to three significant figures.}$$

Qu. 7 Repeat the calculation of Example 1 but with eleven ordinates instead of six.

Qu. 8 Use the trapezium rule to find the distance travelled by the cyclist in §17.1 in the first ten seconds.

Readers who have access to a microcomputer should certainly write a program for evaluating definite integrals by the trapezium method. Since the computer will be doing the arithmetic, a large number of strips can be used and hence a high degree of accuracy can be achieved.

Simpson's rule

17.3   It will have been clear from Fig. 17.3 that the trapezium rule with a small number of strips will not be very accurate for curves like the one illustrated. If, on the other hand, we were to join the tops of the ordinates by a smooth curve, we might expect to get a better estimate. The question then arises as to what curve to use — and there are a number of possibilities. But if we take three ordinates we can find a parabola in the form

$$y = ax^2 + bx + c$$

to pass through the three corresponding points.

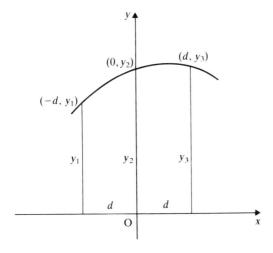

Figure 17.4

Given a curve with three ordinates y_1, y_2, y_3 at *equal* intervals of d apart, take the y-axis along the middle ordinate and the x-axis through its foot as in Fig. 17.4.
    Let

$$y = ax^2 + bx + c$$

be the *parabola* through the points $(-d, y_1)$, $(0, y_2)$, (d, y_3); its equation is therefore satisfied by their coordinates.

$$\therefore y_1 = ad^2 - bd + c,$$
$$y_2 = \qquad\quad c,$$
$$y_3 = ad^2 + bd + c.$$

The area under the parabola is

$$\int_{-d}^{d} (ax^2 + bx + c)\, dx = \left[\frac{ax^3}{3} + \frac{bx^2}{2} + cx\right]_{-d}^{d}$$

$$= \tfrac{2}{3}ad^3 + 2cd$$

(Note that we do not need to find the equation of the parabola because we can express this area in terms of the data y_1, y_2, y_3, d.)

Now

$$y_1 + y_3 - 2y_2 = 2ad^2$$
$$\therefore\ y_1 + 4y_2 + y_3 = 2ad^2 + 6c$$
$$\therefore\ \tfrac{1}{3}d(y_1 + 4y_2 + y_3) = \tfrac{2}{3}ad^3 + 2cd$$

So an approximation for the area under the given curve is

$$\tfrac{1}{3}d(y_1 + 4y_2 + y_3)$$

This result is known as Simpson's rule and was published by Thomas Simpson in 1743.

Note: it makes very little difference to the proof exactly what points we are given originally. If, for instance, we are told that the curve passes through (x_1, y_1), (x_2, y_2), (x_3, y_3), where $x_2 = \tfrac{1}{2}(x_1 + x_3)$, we can at once take new axes, parallel to the given ones, with the new origin at $(x_2, 0)$. Let $d = x_3 - x_2 = x_2 - x_1$; the rest of the proof is as above.

In practice we usually require the area under a curve with more than three ordinates and so, provided there is an *odd* number of ordinates, we may apply Simpson's rule a number of times. Thus with seven ordinates (see Fig. 17.5) the area is

$$\tfrac{1}{3}d(y_1 + 4y_2 + y_3) + \tfrac{1}{3}d(y_3 + 4y_4 + y_5) + \tfrac{1}{3}d(y_5 + 4y_6 + y_7)$$

$$= \tfrac{1}{3}d(y_1 + 4y_2 + 2y_3 + 4y_4 + 2y_5 + 4y_6 + y_7)$$

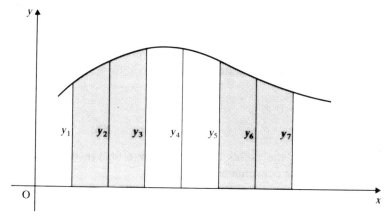

Figure 17.5

Qu. 9 Find similar expressions for the area with (a) five, (b) nine ordinates. Now express Simpson's rule for an odd number of ordinates in words.

The next example is the same as Example 1. This is so that the reader may compare the accuracy of Simpson's rule and the trapezium rule for this case.

Example 2 *Use Simpson's rule to find an approximation for the area under the curve* $y = 1/x$ *between* $x = 1$ *and* $x = 2$.

Five ordinates have been used.

x	1	1.25	1.5	1.75	2
y	1	0.8000	0.6667	0.5714	0.5

$$y_1 = 1.0000 \qquad y_3 = 0.6667 \qquad y_2 = 0.8000$$
$$y_5 = 0.5000 \qquad\qquad \times 2 \qquad\quad y_4 = 0.5714$$

$$\begin{array}{ccc} \underline{} & \underline{} & \underline{} \\ 1.5000 & 1.3334 & 1.3714 \\ 1.3334 & & \times 4 \\ 5.4856 & & \underline{} \\ \underline{} & & 5.4856 \\ 8.3190 & & \end{array}$$

$$\tfrac{1}{3}d = \tfrac{1}{12}$$

$\therefore$ the area $= \dfrac{8.3190}{12} = 0.693,$ correct to three significant figures.

This is a nearer value for ln 2 than the result obtained with the trapezium rule using eleven ordinates (see Qu. 7).

(If the arithmetic in Example 2 is done on a calculator, it will be seen that the result is 0.693 253 97, correct to eight decimal places, whereas the exact answer, ln 2, equals 0.693 147 18, correct to eight decimal places, so in this case the approximate method has yielded the first three significant figures correctly. The reader should notice that the accuracy of the result depends on the method selected, the number of strips used and the shape of the graph, so there is no virtue in presenting an answer which includes all the figures shown by the calculator; indeed this could give a totally false impression.)

Qu. 10 Evaluate approximately $\displaystyle\int_1^2 \frac{1}{x}\,dx$ using Simpson's rule with eleven ordinates.

Qu. 11 Repeat Qu. 3, using Simpson's rule.

Readers are advised to do some of the questions on the trapezium rule and Simpson's rule from Exercise 17, before continuing with the next section.

Numerical solution of differential equations

17.4   Although it is always pleasing to find an exact solution of a mathematical problem, there are many mathematical problems for which no such solution exists. Sometimes it is possible to turn to an approximate method, and we have already done this to solve certain equations (see Book 1, Chapter 24) and to evaluate definite integrals. We shall now look at three methods which can be used to find an approximate solution to a first order differential equation of the form

$$\frac{\mathrm{d}y}{\mathrm{d}x} = F(x, y)$$

A differential equation like this gives the gradient of a curve, at any point $P(x, y)$, as a function of x and y. (Throughout the rest of this chapter $F(x, y)$ will be used to represent the general form of this function.) An *exact* solution will be the equation of the curve. If the equation cannot be found, we may either establish, step by step, the coordinates of a number of points which lie as close as possible to that part of the curve with which we are concerned (see methods 1 and 2 below); or we may find a polynomial function whose graph approximates to that of the exact solution (method 3).

Normally one only turns to approximate methods when an exact solution is unobtainable. However, at this stage in studying the subject, it is instructive to use an example for which the exact solution is known, so that we can compare the results. We shall consider the equation in which $F(x, y) = x + y$, in other words

$$\frac{\mathrm{d}y}{\mathrm{d}x} = x + y$$

with initial conditions $x = 0$, $y = 1$.

Qu. 12   Show, by using the integrating factor method, that the exact solution of the differential equation above is $y = 2e^x - x - 1$.

Method 1

The first approximate method we shall consider is called Euler's method; it depends on the linear approximation

$$\delta y \approx \frac{\mathrm{d}y}{\mathrm{d}x} \delta x$$

(See §16.2 and Book 1, §7.7.) Since we are considering problems for which $\frac{\mathrm{d}y}{\mathrm{d}x} = F(x, y)$, this becomes

$$\delta y \approx F(x, y)\delta x \tag{1}$$

and for the particular example we are considering this is

$$\delta y \approx (x + y)\delta x$$

In Fig. 17.6, the curve represents the (unknown) solution, $y = f(x)$, of the differential equation, and the point $P_0(x_0, y_0)$ is the initial point; in our case this is (0, 1). We shall consider a sequence of values of x:

$$x_1 = h, \quad x_2 = 2h, \quad x_3 = 3h, \quad \ldots$$

making h small in order to make the approximation as good as possible. Ideally we would like to find the point P_1, *on* the curve, for which $x = x_1$, but we shall settle for the point $Q_1(x_1, y_1)$, which is on the tangent to the solution curve at P_0. Consequently, using the approximation (1), with $\delta x = h$,

$$y_1 = y_0 + h\, F(x_0, y_0)$$

For our example, this becomes

$$y_1 = y_0 + h(x_0 + y_0)$$
$$= 1 + h$$

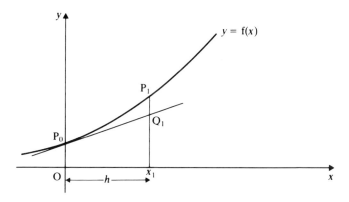

Figure 17.6

Then, using x_1, the value of y_1 just calculated and the value of the gradient function at the point (x_1, y_1), we repeat the process to find y_2, i.e.

$$y_2 = y_1 + h\, F(x_1, y_1)$$

We can then repeat this step as many times as we please, using

$$y_{n+1} = y_n + h\, F(x_n, y_n)$$

In doing so we shall produce the sequence of points $Q_1, Q_2, Q_3, Q_4, \ldots$ at the vertices of the polygon shown in Fig. 17.7. Each step is likely to take the polygon further and further away from the curve, so we should not expect too much from this method. Nevertheless it will produce reasonable results if h is small.

For the example we have been considering, the successive values of y are given by

$$y_{n+1} = y_n + h(x_n + y_n)$$

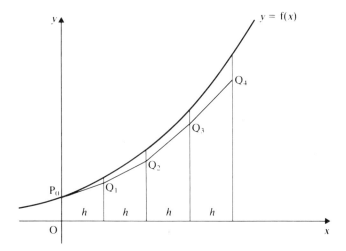

Figure 17.7

The two tables below show the results obtained for $h = 0.1$ (Table (a)) and $h = 0.01$ (Table (b)). In each case the last column, headed $f(x_n)$, shows the results produced by the exact solution (see Qu. 12) for comparison. (For convenience, only four decimal places are shown.)

Table (a) $h = 0.1$

n	x_n	y_n	$0.1\,(x_n + y_n)$	$f(x_n)$
0	0	1	0.1	1
1	0.1	1.1	0.12	1.1103
2	0.2	1.22	0.142	1.2428
3	0.3	1.362	0.1662	1.3997
4	0.4	1.5282	0.1928	1.5836
5	0.5	1.7210	0.2221	1.7974

Table (b) $h = 0.01$

n	x_n	y_n	$0.01(x_n + y_n)$	$f(x_n)$
0	0	1	0.01	1
1	0.01	1.01	0.0102	1.0101
2	0.02	1.0202	0.0104	1.0204
3	0.03	1.0306	0.0106	1.0309
4	0.04	1.0412	0.0108	1.0416
5	0.05	1.0520	0.0110	1.0525

In Table (b), we can see that after five cycles the accuracy is still quite good.
     Improving the accuracy by reducing the magnitude of h by a factor of ten inevitably means ten times as much work. (This may not matter too much if a

computer is used.) If we do not wish to do this then we have to look for a more efficient method. Many such methods can be found in books that specialise in numerical solution of differential equations. We shall consider now one such method.

Method 2

One of the reasons for Euler's method being unsatisfactory is that it uses the value of the gradient *at one end* of each interval to estimate the value of y at the other end. In the method below, the value of the gradient *in the middle* of the interval is used to relate the values of y at the ends.

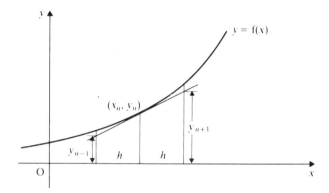

Figure 17.8

In Fig. 17.8, (x_n, y_n) is a point on the solution curve $y = f(x)$ and the approximate values of y on either side of it are given, as before, by

$$y_{n+1} = y_n + h\,F(x_n, y_n)$$

and

$$y_{n-1} = y_n - h\,F(x_n, y_n)$$

(Remember that $F(x, y)$ gives the gradient at the point (x, y).)
   Subtracting gives

$$y_{n+1} - y_{n-1} = 2h\,F(x_n, y_n)$$
$$\therefore\ y_{n+1} = y_{n-1} + 2h\,F(x_n, y_n) \tag{1}$$

This can be used to produce a sequence of values $y_1, y_2, y_3, y_4, \ldots$ starting from the given value of y_0. However, there is a snag; in order to find y_2, we need both y_0 and y_1. So we shall still have to use the old method in order to get started. In other words, we use

$$y_1 = y_0 + h\,F(x_0, y_0)$$

and thereafter

$$y_{n+2} = y_n + 2h\,F(x_{n+1}, y_{n+1})$$

(This is equation (1), moved up one step.)

[The only point which is definitely *on* the curve is (x_0, y_0); all the others are approximations. We assume that, provided h is small, the subsequent points are sufficiently close to the curve to make the polygon formed by them a reasonable approximation.]

Table (c) shows the results produced by this method for the differential equation we used before.

Table (c) $h = 0.1$

n	x_n	y_n	$0.2(x_n + y_n)$
0	0	1	
1	0.1	1.1	0.24
2	0.2	1.24	0.288
3	0.3	1.388	0.3376
4	0.4	1.5776	0.3955
5	0.5	1.7835	

(The dotted lines indicate the connection between each new value of y_n and the preceding lines.)

Comparing these values of y_n with the values of $f(x_n)$ shown in the last column of Table (a), we can see that these results are better. (However we are relying on our intuition to claim that this method is better; we have not *proved* that this is so.)

Method 3

The third method we shall consider is fairly easy to understand, as it depends on a simple application of Maclaurin's expansion (see §16.5).

$$f(x) = f(0) + f'(0)x + \frac{f''(0)}{2!}x^2 + \frac{f'''(0)}{3!}x^3 + \dots$$

Once again we shall use the differential equation

$$\frac{dy}{dx} = x + y$$

with initial conditions $x = 0$, $y = 1$, to explain the method.

The differential equation may be written

$$f'(x) = x + f(x)$$

By repeated differentiation

$$f''(x) = 1 + f'(x)$$

and $f'''(x) = f''(x)$, $f''''(x) = f'''(x)$, etc.

We know that $f(0) = 1$, and putting $x = 0$ in the higher derivatives we obtain
$$f'(0) = 1, \quad f''(0) = 2, \quad f'''(0) = 2, \quad f''''(0) = 2, \dots$$

Substituting these values into Maclaurin's series gives

$$f(x) = 1 + x + \frac{2}{2!}x^2 + \frac{2}{3!}x^3 + \frac{2}{4!}x^4 + \ldots$$

Thus an approximate solution of the differential equation, for small values of x, is

$$y = 1 + x + x^2 + \frac{1}{3}x^3 + \frac{1}{12}x^4 + \ldots$$

Table (d) shows the results given by this method.

Table (d)

x_n	y_n
0	1
0.1	1.1103
0.2	1.2428
0.3	1.3997
0.4	1.5835
0.5	1.7969

The results obtained by this method can be seen to compare very favourably with those obtained by methods 1 and 2 and they could be improved by using more terms of Maclaurin's series.

Exercise 17

1 Evaluate $\displaystyle\int_0^{\pi/4} \tan x \, dx$,

(a) by integration, (b) by using the trapezium rule with four strips,
(c) by Simpson's rule with four strips.
Comment on the accuracy of your answers.

2 Repeat No. 1, parts (b) and (c), for $\displaystyle\int_0^1 e^{x^2} \, dx$.

3 (x_1, y_1), (x_2, y_2), (x_3, y_3), where $x_2 = \frac{1}{2}(x_1 + x_3)$, are three points on the parabola $y = ax^2 + bx + c$. Prove that the area under the curve between the lines $x - x_1 = 0$, $x - x_3 = 0$ is equal to $\frac{1}{3}(x_2 - x_1)(y_1 + 4y_2 + y_3)$.
Use this formula to find the area between the parabola $y = x(10 - x)$ and the x-axis. Check your answer by integration.

4 Evaluate $\displaystyle\int_0^1 e^{-x^2} \, dx$ by Simpson's rule taking ten intervals.

5 Estimate the area of a quadrant of a circle of radius 8 cm by dividing it into eight intervals and using (a) the trapezium rule and (b) Simpson's rule. Use the better of these results to find an approximate value of π.

6 The area in square centimetres of the cross-section of a model boat 28 cm

long at intervals of 3.5 cm is as follows:

0	11.5	15.3	16.3	16.2	13.4	9.3	4.9	0

Find the volume of the boat.

7 A jug of circular cross-section is 16 cm high inside and its internal diameter is measured at equal intervals from the bottom:

height (cm)	0	4	8	12	16
diameter (cm)	10.2	13.8	15.3	9.3	9.9

What volume of liquid will the jug hold if filled to the brim?

8 Using tables, or a calculator, where necessary, calculate the value of

$$\int_{0.1}^{0.5} e^{-x}\, dx$$

(a) by direct integration,
(b) by Simpson's rule, using five ordinates spaced at intervals of 1/10 unit. (Give your answers to four places of decimals.)   (JMB)

9 By means of Simpson's rule and taking unit intervals of x from $x = 8$ to $x = 12$, find approximately the area enclosed by the curve $y = \log_{10} x$, the lines $x = 8$ and $x = 12$, and the x-axis. Deduce the average value of $\log_{10} x$ between $x = 8$ and $x = 12$.   (JMB)

10 The coordinates of three points on the curve $y = A + Bx + Cx^2$ are (x_1, y_1), (x_2, y_2) and (x_3, y_3), where $x_2 = \frac{1}{2}(x_3 + x_1)$. Prove that the area under the curve between the lines $x = x_1$ and $x = x_3$ is equal to

$$\tfrac{1}{6}(x_3 - x_1)(y_1 + 4y_2 + y_3)$$

Deduce Simpson's rule for five ordinates.

Using five ordinates, apply Simpson's rule to evaluate the integral

$$4\int_0^1 \frac{dx}{1 + x^2}$$ and thus to find a value for π correct to three places of decimals.

(O & C)

11 The coordinates of three points on the curve $y = ax^3 + bx^2 + cx + d$ are (x_1, y_1), (x_2, y_2), (x_3, y_3). Prove that, if $x_2 - x_1 = x_3 - x_2 = h$, the area under the curve between the lines $x = x_1$, $x = x_3$ is $\frac{1}{3}h(y_1 + 4y_2 + y_3)$.

Find the area between the curve $y = x(x - 2)^2$ and the x-axis by means of Simpson's rule with three ordinates. Use integration to check that your answer is exact.

12 Show that the area under the curve $y = 1/x$, from $x = n - 1$ to $x = n + 1$, is $\ln\{(n + 1)/(n - 1)\}$, provided $n > 1$.

By applying Simpson's rule to this area, deduce that, approximately,

$$\ln \frac{n + 1}{n - 1} = \frac{1}{3}\left(\frac{1}{n - 1} + \frac{4}{n} + \frac{1}{n + 1}\right)$$

and show that the error in this approximation is $4/(15n^5)$, when higher powers of $1/n$ are neglected. (JMB)

13 Use the binomial theorem to expand $(1 + x^3)^{10}$ in ascending powers of x, up to and including the term in x^9. Hence estimate I, where

$$I = \int_0^{0.2} (1 + x^3)^{10} \, dx$$

to three decimal places.

Make another estimate of I, again to three places, by using Simpson's rule with three ordinates, showing all your working. (L)

14 Tabulate, to three decimal places, the values of the function

$$f(x) = \sqrt{(1 + x^2)}$$

for values of x from 0 to 0.8 at intervals of 0.1. Use these values to estimate

$$\int_0^{0.8} f(x) \, dx$$

(a) by the trapezium rule, using all the ordinates,
(b) by Simpson's rule, using only the ordinates at intervals of 0.2. (L)

15 By considering suitable areas, or otherwise, show that, for any $n > 0$,

$$\tfrac{1}{2} \leqslant \int_0^1 (1 + x^n)^{-1} \, dx \leqslant 1$$

When $n = 4$, find a value (to three significant figures) for the integral, using Simpson's rule with five ordinates. (O & C)

16 The integral $\int_0^{1/2} \sqrt{(1 - x^2)} \, dx$ is denoted by I. The value of I is estimated by using the trapezium rule, and T_1, T_2 denote the estimates obtained when one and two strips respectively are used. Calculate T_1 and T_2, giving your answers correct to three decimal places.

Assuming the error when using the trapezium rule is proportional to h^3, where h denotes the width of a strip, show that an improved estimate of I is given by $(8T_2 - T_1)/7$, and evaluate this expression correct to three decimal places.

Given that $y^2 = 1 - x^2$ is the equation of a circle whose centre is the origin and whose radius is 1 unit, show that $I = \dfrac{\pi}{12} + \dfrac{\sqrt{3}}{8}$. Hence calculate an estimate for the value of π. (C)

17 Using Euler's method, find a numerical solution for the differential equation

$$\frac{dy}{dx} = xy$$

with initial conditions $x = 0$, $y = 1$, giving the values of y which correspond to
(a) $x = 0, 0.1, 0.2, 0.3, 0.4, 0.5$,   (b) $x = 0, 0.01, 0.02, 0.03, 0.04, 0.05$.
(You are recommended to set out your results in the form of a table, as in the text.)

18 Repeat No. 17 for $\dfrac{dy}{dx} = x + 2y$.

19 Repeat No. 17 for $\dfrac{dy}{dx} = x^2 + y^2$, with initial conditions $x = 0.5$, $y = 0$, and using steps of (a) 0.1, and (b) 0.01.

20 Repeat No. 17 for $\dfrac{dy}{dx} = e^{-x^2}$.

21 Use the modified version of Euler's method, described in the text, using $h = 0.1$, to solve the differential equation in No. 17.

22 Repeat No. 21, for the differential equation in No. 18.

23 Find the first four terms of the Maclaurin's series for the solution to the differential equation in No. 17. Use this to estimate y when $x = 0.1$ and $x = 0.2$.

24 Repeat No. 23 for the differential equation $\dfrac{dy}{dx} = x + 2y$, with initial conditions $x = 0$, $y = 1$.

25 Repeat No. 23 for the differential equation $\dfrac{dy}{dx} = x^2 + y^2$, with initial conditions $x = 0$, $y = 1$.

Chapter 18

Hyperbolic functions

Hyperbolic cosine and sine

18.1 We shall begin by defining two new functions, the hyperbolic cosine and the hyperbolic sine. No attempt to explain the reason for adopting these definitions will be given at present, as more knowledge of complex numbers is needed if the reason is to be fully appreciated. (See §20.6.) The reader will, however, very soon find some strong similarities between the hyperbolic functions and the familiar trigonometrical functions which, to save confusion, are often referred to as the *circular functions*. These similarities would not, by themselves, justify the inclusion of a study of the hyperbolic functions in this book: they are being introduced because they will very quickly extend the reader's powers of integration, and the reader may begin to need them in mechanics. But first we shall study the functions themselves. They were introduced by J. H. Lambert in a paper read in 1768.

Definitions

The hyperbolic cosine of x

$$\cosh x = \tfrac{1}{2}(e^x + e^{-x})$$

and the hyperbolic sine of x

$$\sinh x = \tfrac{1}{2}(e^x - e^{-x})$$

cosh x is pronounced as it is spelled; sinh x may be pronounced 'sinch x' (or 'shine x').

First we sketch their graphs. Starting with the graph of e^x in Fig. 18.1(i), that of e^{-x} has been shown dotted in the same figure. cosh x is half the sum of these two functions (see Fig. 18.1(ii)) and sinh x is half the difference (see Fig. 18.1(ii)). The two graphs are distinct in the first quadrant but they approach so close that, when they are sketched to this scale, the lines run together.

In general, the properties of hyperbolic functions are easily proved and this will be left to the reader to do in Exercise 18a. We shall first prove one important identity.

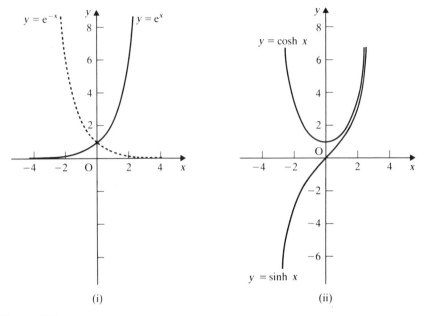

Figure 18.1

Example 1   *Prove the identity* $\cosh^2 x - \sinh^2 x = 1.$

From the definitions of $\cosh x$ and $\sinh x$,

$$\cosh^2 x - \sinh^2 x = \{\tfrac{1}{2}(e^x + e^{-x})\}^2 - \{\tfrac{1}{2}(e^x - e^{-x})\}^2$$
$$= \tfrac{1}{4}(e^{2x} + 2 + e^{-2x}) - \tfrac{1}{4}(e^{2x} - 2 + e^{-2x})$$
$$= \tfrac{1}{4}e^{2x} + \tfrac{1}{2} + \tfrac{1}{4}e^{-2x} - \tfrac{1}{4}e^{2x} + \tfrac{1}{2} - \tfrac{1}{4}e^{-2x}$$

$\therefore \cosh^2 x - \sinh^2 x = 1$

Definitions

The hyperbolic tangent, cotangent, secant, cosecant are defined as follows:

$$\tanh x = \frac{\sinh x}{\cosh x} \qquad\qquad \coth x = \frac{1}{\tanh x}$$

$$\operatorname{sech} x = \frac{1}{\cosh x} \qquad\qquad \operatorname{cosech} x = \frac{1}{\sinh x}$$

Exercise 18a

Most of the following properties of the hyperbolic functions should be deduced from the definitions. Work all of Nos. 1–14.

1 From a sketch of $\cosh x$ and $\sinh x$ referred to the same axes, sketch the graph of $\tanh x$.

2 Prove that (a) cosh $(-x)$ = cosh x,   (b) sinh $(-x)$ = $-$ sinh x.

3 Prove that cosh x > sinh x. [Show that cosh x $-$ sinh x > 0.] Prove also that, when x < 0, cosh x > |sinh x|. Deduce the values between which tanh x lies.

4 From sketches of cosh x and sinh x, sketch the graphs of sech x and cosech x.

5 Prove that cosh $x \geqslant 1$. [See hint in No. 3.]

6 Prove that

$$\text{cosh } x + \text{sinh } x = e^x$$
$$\text{cosh } x - \text{sinh } x = e^{-x}$$

Deduce the identity cosh^2 x $-$ sinh^2 x = 1.

7 Prove that the point (a cosh t, b sinh t) lies on one branch of the hyperbola

$$\frac{x^2}{a^2} - \frac{y^2}{b^2} = 1$$

[Hence the name hyperbolic functions. Use the result of No. 6.]

8 Prove that sinh $2x$ = 2 sinh x cosh x.

9 Prove that cosh $2x$ = cosh^2 x + sinh^2 x
$$= 2 \text{ cosh}^2 \ x - 1$$
$$= 1 + 2 \text{ sinh}^2 \ x.$$

10 Use the results of Nos. 8 and 9 to show that

$$\text{tanh } 2x = 2 \text{ tanh } x/(1 + \text{tanh}^2 \ x).$$

11 Prove that

$$\text{sech}^2 \ x = 1 - \text{tanh}^2 \ x$$
$$\text{cosech}^2 \ x = 1 - \text{coth}^2 \ x$$

[Use the identity connecting cosh x and sinh x.]

12 Prove that

$$\text{cosh } (A + B) = \text{cosh } A \text{ cosh } B + \text{sinh } A \text{ sinh } B$$

Deduce a similar expression for cosh $(A - B)$.

13 Prove that

$$\text{sinh } (A + B) = \text{sinh } A \text{ cosh } B + \text{cosh } A \text{ sinh } B$$

Deduce a similar expression for sinh $(A - B)$.

14 Use the results of Nos. 12 and 13 to find expressions for tanh $(A + B)$, tanh $(A - B)$ in terms of tanh A, tanh B.

15 Solve the equation 8 cosh x + 17 sinh x = 20.

16 Find the condition that the equation a cosh x + b sinh x = c should have equal roots.

17 If $a > b > 0$, prove that $b < \dfrac{a \ e^x + b \ e^{-x}}{e^x + e^{-x}} < a$.

18 If $|a| < |b|$, prove that the equation a cosh x + b sinh x = 0 has one and only one root.

19 Prove that sinh 3θ = 3 sinh θ + 4 sinh^3 θ.

20 Prove that cosh^2 x sin^2 x $-$ sinh^2 x cos^2 x = $\frac{1}{2}$(1 $-$ cosh $2x$ cos $2x$).

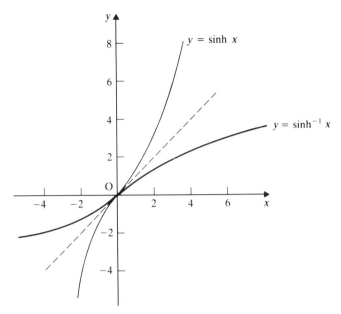

Figure 18.2

$f^{-1}(x) = \sqrt{x}$.) However this technical difficulty can be overcome by restricting the domain of cosh x to *non-negative* real numbers. The graphs of $y = \cosh x$ and $y = \cosh^{-1} x$ are illustrated by the unbroken curves in Fig. 18.3.

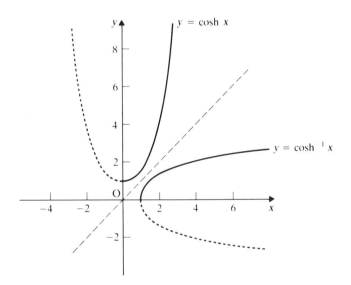

Figure 18.3

The domain of $\cosh^{-1} x$ is $\{x : x \in \mathbb{R},\ x \geqslant 1\}$ and the corresponding range is $\{y : y \in \mathbb{R},\ y \geqslant 0\}$.

Exercise 18b

1 Prove the following results by first expressing the functions concerned in terms of cosh x and sinh x:

(a) $\dfrac{d}{dx}(\tanh x) = \text{sech}^2 x$,

(b) $\dfrac{d}{dx}(\coth x) = -\text{cosech}^2 x$,

(c) $\dfrac{d}{dx}(\text{sech } x) = -\text{sech } x \tanh x$,

(d) $\dfrac{d}{dx}(\text{cosech } x) = -\text{cosech } x \coth x$.

2 Write down $\int \cosh x \, dx$, $\int \sinh x \, dx$.

3 Write down the derivatives of
(a) $\cosh 3x$,   (b) $\sinh 2x$,   (c) $\cosh^2 x$,
(d) $2 \sinh^3 x$,   (e) $3 \tanh 2x$,   (f) $\frac{1}{2} \text{sech}^2 x$,
(g) $\sinh^2 3x$,   (h) $\sqrt{(\coth x)}$,   (i) $2 \tanh^2 \frac{1}{2}x$.

*4 Sketch, on the same pair of axes, the graphs of $y = \tanh x$ and $y = \tanh^{-1} x$. State the domain and range of the inverse function, $y = \tanh^{-1} x$. State, giving a reason, whether $\tanh x$ is an odd or even function.

5 Differentiate the following functions with respect to x, simplifying your answers:

(a) $\ln \tanh x$,   (b) $e^x \sinh x$,   (c) $\dfrac{e^x - 1}{e^x + 1}$.

6 Find:

(a) $\int \text{sech}^2 2x \, dx$,   (b) $\displaystyle\int \dfrac{\sinh x}{\cosh^2 x} \, dx$.

7 Find the minimum value of $5 \cosh x + 3 \sinh x$.

8 Prove that

(a) $\dfrac{d}{dx}(\cosh^{-1} x) = \dfrac{1}{\sqrt{(x^2 - 1)}}$,   (b) $\dfrac{d}{dx}(\tanh^{-1} x) = \dfrac{1}{1 - x^2}$,

and find an expression for $\dfrac{d}{dx}(\sinh^{-1} x)$.

[Hint: see the method of §11.3.]

9 Prove that $\dfrac{d}{dx}\{\tan^{-1}(e^x)\} = \frac{1}{2} \text{sech } x$.

10 Find $\dfrac{d}{dx}[\ln\{x + \sqrt{(1 + x^2)}\} - \sinh^{-1} x]$.

11 Find: (a) $\int \cosh 2x \sinh 3x \, dx$, (b) $\int \cosh x \cosh 3x \, dx$.

12 Find the distance from the y-axis of the centroid of the area formed by $y = \sinh x$, $x - 1 = 0$ and the x-axis.

13 Find the equations of the tangent and normal to the hyperbola
$$b^2 x^2 - a^2 y^2 = a^2 b^2$$

*Qu. 9 (A repeat of Exercise 18b, No. 17.) If $\tanh^{-1} x = y$, show that $x = (e^{2y} - 1)/(e^{2y} + 1)$. Hence prove that

$$\tanh^{-1} x = \frac{1}{2} \ln \frac{1+x}{1-x}$$

The inverse hyperbolic functions expressed in terms of logarithms

18.7 The result of Qu. 9 suggests that it may be possible to express $\cosh^{-1} x$, $\sinh^{-1} x$ in terms of logarithms.

Let $y = \sinh^{-1} x$, then

$$\sinh y = x \tag{1}$$

Also $\cosh^2 y = 1 + \sinh^2 y$
$$\therefore \cosh y = \sqrt{(1 + x^2)} \tag{2}$$

[$\cosh y > 0$, so the negative square root does not give a real value of y.]
   Now $\cosh y + \sinh y = \frac{1}{2}(e^y + e^{-y}) + \frac{1}{2}(e^y - e^{-y})$
$$= e^y$$

But from (1), (2),

$$\cosh y + \sinh y = x + \sqrt{(1 + x^2)}$$
$$\therefore e^y = x + \sqrt{(1 + x^2)}$$
$$\therefore y = \ln \{x + \sqrt{(1 + x^2)}\}$$

That is,

$$\sinh^{-1} x = \ln \{x + \sqrt{(1 + x^2)}\}$$

An expression for $\cosh^{-1} x$ may be obtained in a similar manner. Remember that $x \geqslant 1$ and $\cosh^{-1} x \geqslant 0$.
Let $y = \cosh^{-1} x$, then

$$\cosh y = x$$

Now, $\sinh^2 y = \cosh^2 y - 1$
$$= x^2 - 1$$
$$\therefore \sinh y = \sqrt{(x^2 - 1)}$$

(The positive square root is used because we know that y, and hence $\sinh y$, is positive.)
Now

$$e^y = \cosh y + \sinh y$$
$$= x + \sqrt{(x^2 - 1)}$$
$$\therefore y = \ln \{x + \sqrt{(x^2 - 1)}\}$$

That is

$$\cosh^{-1} x = \ln \{x + \sqrt{(x^2 - 1)}\}$$

Qu. 10 Use the formulae in §18.7 to find the values of (a) $\sinh^{-1} 1$, (b) $\cosh^{-1} 2$, (c) $\sinh^{-1} 0.58$, giving your answers correct to four decimal places.

Once the reader has grasped the forms which require the substitution of a hyperbolic function, the integrations in Exercise 18c should present no new difficulty. Only in exceptional cases as, for instance, in Example 5, is the treatment of hyperbolic functions completely different from the treatment of circular functions. If the reader is unable to integrate any particular function in Exercise 18c, he or she should refer back to Chapters 1 and 13 for help. The following examples illustrate how a knowledge of integrating with circular functions helps with the present work.

Example 2 *Find* $\displaystyle\int \frac{1}{\sqrt{(x^2 + 2x + 10)}}\,dx.$

First complete the square:

$$x^2 + 2x + 10 = (x + 1)^2 + 9$$

[The substitution $x + 1 = 3 \sinh\theta$ makes $(x+1)^2 + 9 = 9\cosh^2\theta$.]

$$\int \frac{1}{\sqrt{(x^2 + 2x + 10)}}\,dx = \int \frac{1}{\sqrt{\{(x+1)^2 + 9\}}}\,\frac{dx}{d\theta}\,d\theta$$

$$= \int \frac{1}{3\cosh\theta}\,3\cosh\theta\,d\theta$$

$$= \int 1\,d\theta = \theta + c$$

Let $x + 1 = 3\sinh\theta$.

$$\therefore\ \frac{dx}{d\theta} = 3\cosh\theta.$$

$$\therefore\ \int \frac{1}{\sqrt{(x^2 + 2x + 10)}}\,dx = \sinh^{-1}\frac{x+1}{3} + c$$

Example 3 *Evaluate* $\displaystyle\int_2^3 \cosh^{-1} x\,dx.$

[$\int \cos^{-1} x\,dx$ we integrate by parts as $\int 1 \times \cos^{-1} x\,dx.$]

$$\int_2^3 1 \times \cosh^{-1} x\,dx = \left[x\cosh^{-1} x \right]_2^3 - \int_2^3 x\,\frac{1}{\sqrt{(x^2 - 1)}}\,dx$$

$$= 3\cosh^{-1} 3 - 2\cosh^{-1} 2 - \left[\sqrt{(x^2 - 1)} \right]_2^3$$

$$= 3\ln(3 + \sqrt{8}) - 2\ln(2 + \sqrt{3}) - (\sqrt{8} - \sqrt{3})$$

$$= 1.56,\quad \text{correct to three significant figures}$$

Example 4 *Find* $\int \sinh^3\theta\,d\theta.$

$$\int \sinh^3\theta\,d\theta = \int (\cosh^2\theta - 1)\sinh\theta\,d\theta$$

$$= \tfrac{1}{3}\cosh^3\theta - \cosh\theta + c$$

12 Show that $y = A \cosh nx + B \sinh nx$ is a solution of the differential equation

$$\frac{d^2 y}{dx^2} - n^2 y = 0, \text{ for all values of } A \text{ and } B.$$

Find the solution when $n = 2$, given that when $x = 0$, $y = 5$ and $\dfrac{dy}{dx} = 8$.

13 Find the possible values of $\sinh x$ if

$$\begin{vmatrix} \cosh x & -\sinh x \\ \sinh x & \cosh x \end{vmatrix} = 2 \tag{L}$$

14 (a) Show that when x is small, $\ln (\cosh x) \approx x^2/2 - x^4/12$, and that when x is large, $\ln (\cosh x) \approx x - \ln 2$.

Sketch the graph of $\ln (\cosh x)$.

(b) Find the condition that must be satisfied by the constant k if $(\cosh x + k \sinh x)$ is to have a minimum value. Find the minimum value of $3 \cosh x + 2 \sinh x$. (L)

15 If $\sec \theta = \cosh u$, with $u > 0$ and $0 < \theta < \frac{1}{2}\pi$, express (a) $\tan \theta$, (b) $\dfrac{d\theta}{du}$ in terms of u. Hence, or otherwise, evaluate $\displaystyle\int_0^\infty \frac{1}{\cosh u}\, du$. (O & C: SMP)

16 Prove that $\sinh^{-1} x = \ln \{x + \sqrt{(x^2 + 1)}\}$.

Show that $\dfrac{d}{dx} (\sinh^{-1} x) = \dfrac{1}{\sqrt{(x^2 + 1)}}$.

Evaluate $\displaystyle\int_1^8 \frac{1}{\sqrt{(x^2 - 2x + 2)}}\, dx$, expressing your answer as a natural logarithm.

Show that $\displaystyle\int_1^8 \frac{1}{\sqrt{(x^2 - 2x + 2)}}\, dx = \int_1^2 \frac{3}{\sqrt{(x^2 - 2x + 2)}}\, dx$. (JMB)

17 Prove that $(\cosh x + \sinh x)^n = \cosh nx + \sinh nx$, for all positive integral values of n.

Show that $\dfrac{1 + \tanh x}{1 - \tanh x} = \cosh 2x + \sinh 2x$. (JMB)

18 (a) Find $\displaystyle\int \frac{1}{\sqrt{(x^2 + 4x - 5)}}\, dx$.

(b) The parametric equations of a curve are $x = a \cosh t$, $y = a \sinh t$, where a is a positive constant. Prove that the area of the region in the first quadrant bounded by the curve, the x-axis between $x = a$ and $x = 2a$, and the line $x = 2a$, may be expressed as

$$\tfrac{1}{2}a^2 \int_0^k (\cosh 2t - 1)\, dt$$

where $\cosh k = 2$, and hence show that this area is approximately $1.07a^2$. (C)

19 Prove that $\cosh^{-1} x = \ln \{x + \sqrt{(x^2 - 1)}\}$ and that $\tanh^{-1} x = \frac{1}{2} \ln \left(\dfrac{1 + x}{1 - x} \right)$.

Prove that the equation in x, $a \cosh x + b \sinh x = 1$, has no real solution if $a^2 > b^2 + 1$. If however, $0 < b < a < \sqrt{(1 + b^2)}$, prove that the equation has two distinct real solutions. Obtain all the real solutions (if any) when $a = 1/5$ and $b = 1$.

The parametric coordinates of a point on a curve are given by

$$x = a \cosh t, \quad y = b \sinh t \qquad (a > b > 0)$$

Obtain the condition for the line $lx + my = n$ to touch this curve. (O & C)

20 Define $\sinh x$, $\cosh x$ and prove from these definitions that

$$\tanh (x + y) = \frac{\tanh x + \tanh y}{1 + \tanh x \tanh y}$$

(a) If $\tanh^{-1} u + \tanh^{-1} v = \frac{1}{2} \ln 5$, prove that $v = (2 - 3u)/(3 - 2u)$.

(b) Prove that $\cosh x > \sinh x > x$, for all $x > 0$ and deduce that, for all $x > 0$,

$$1 + \sinh x > \cosh x > 1 + \tfrac{1}{2}x^2$$

and

$$\sinh x - x > \cosh x - 1 - \tfrac{1}{2}x^2 > \tfrac{1}{24} x^4. \tag{O & C}$$

Chapter 19

Some geometrical applications of calculus[†]

Area of a sector

19.1   A good slogan for a reader who finds the formulae in this chapter difficult to remember, or prefers to work them out when needed, is, 'When in doubt, differentiate'. By differentiation we mean, in this instance, take a small increment — which is the fundamental step in differentiating a new function. With this approach we now work out an expression for areas of sectors and closed curves in polar coordinates.

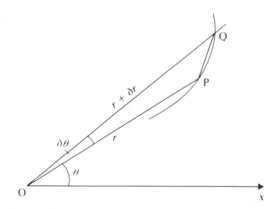

Figure 19.1

In Fig. 19.1, the radius vectors OP, OQ are $r, r + \delta r$; the angles between them and the fixed line OX are $\theta, \theta + \delta\theta$. If $\delta\theta$ is small, the area of sector OPQ is approximately equal to the area of triangle OPQ.

∴ sector OPQ $\approx \frac{1}{2}r(r + \delta r) \sin \delta\theta$

[†]The reader should work all the questions in the text.

but

$$\sin \delta\theta = \delta\theta - \frac{(\delta\theta)^3}{3!} + \dots$$

and $\delta r \sin \delta\theta$ is small compared with $\delta\theta$ so, correct to the term in $\delta\theta$,

$$\text{sector OPQ} = \tfrac{1}{2}r^2 \; \delta\theta$$

(Here it is assumed that the difference between the sector OPQ and triangle OPQ is small compared to $\delta\theta$.)

Summing for all the elements in the sector concerned and proceeding to the limit,

$$\textbf{area of sector} = \int_\alpha^\beta \tfrac{1}{2}r^2 \; \textbf{d}\theta$$

where α, β are the values of θ corresponding to the bounding radius vectors of the sector.

Qu. 1 Sketch the curve given by $r = a$. What does the integral $\int_\alpha^\beta \tfrac{1}{2}r^2 \; d\theta$ represent in this case?

Qu. 2 Sketch the cardioid $r = a(1 + \cos \theta)$ and find the area enclosed by it.

Qu. 3 Find the area swept out by the radius vector of the equiangular spiral $r = a \, e^{k\theta}$ as θ increases from $-\pi$ to π. Show this area on a sketch.

Qu. 4 Sketch the trefoil $r = a \sin 3\theta$ and find the area of one of its loops.

Qu. 5 Regarding the limaçon $r = 1 + 2 \cos \theta$ as having a small loop contained within a larger one, find the area of the larger loop.

*Qu. 6 If x and y are functions of a parameter t, show that

$$\int \tfrac{1}{2}r^2 \; d\theta = \int \tfrac{1}{2}\left(x\frac{dy}{dt} - y\frac{dx}{dt}\right)dt$$

*Qu. 7 The vertices of a triangle are $O(0, 0)$, $P(x, y)$, $Q(x + \delta x, y + \delta y)$. Show that the area of the triangle is $\tfrac{1}{2}(x \, \delta y - y \, \delta x)$. Hence show that the area of a sector may be found from the expression

$$\int_{t_1}^{t_2} \tfrac{1}{2}\left(x\frac{dy}{dt} - y\frac{dx}{dt}\right)dt$$

where t_1, t_2 are the values of t corresponding to the bounding radius vectors of the sector.

Qu. 8 Find the area enclosed by the ellipse $x = a \cos \theta$, $y = b \sin \theta$.

Qu. 9 Find the area enclosed by the loop of the curve given by $x = t^2 - 4$, $y = t^3 - 4t$.

Qu. 10 Find the area of one loop of the curve given by $x = \sin \theta$, $y = \sin 2\theta$. Why does the formula of Qu. 7 give a negative answer?

Qu. 11 Find the area between the cycloid $x = a(\theta - \sin \theta)$, $y = a(1 - \cos \theta)$ and the portion of the x-axis between the points determined by $\theta = 0$ and $\theta = 2\pi$.

Arc length: parametric equations

19.3   Now suppose that we were given a curve in the form $x = f(t)$, $y = g(t)$ as, for example, $x = at^2$, $y = 2at$. It would then be more convenient in finding the length of an arc to have an integral with respect to t.

From (1) (p. 356),

$$P_{r-1}P_r{}^2 = \left\{\left(\frac{\delta x}{\delta t}\right)^2 + \left(\frac{\delta y}{\delta t}\right)^2\right\}(\delta t)^2$$

$$\therefore P_{r-1}P_r = \sqrt{\left\{\left(\frac{\delta x}{\delta t}\right)^2 + \left(\frac{\delta y}{\delta t}\right)^2\right\}}\,\delta t$$

Hence

$$s = \int_{t_1}^{t_2}\sqrt{\left\{\left(\frac{dx}{dt}\right)^2 + \left(\frac{dy}{dt}\right)^2\right\}}\,dt$$

where t_1, t_2 are the values of t corresponding to the ends of the arc. Care should be taken to ensure that the integrand is positive throughout the range of integration. This applies particularly to Qu. 18.

Qu. 16   Find the length of the arc from $\theta = 0$ to $\theta = \alpha$ of the curve given by $x = a\cos\theta$, $y = a\sin\theta$. What is this curve?

Qu. 17   Find an expression for the distance measured along the curve from the origin to any point on the locus $x = at^2$, $y = at^3$.

Qu. 18   Sketch the astroid given by $x = a\cos^3 t$, $y = a\sin^3 t$ and find the length of its circumference.

Qu. 19   Sketch the arc of the cycloid $x = a(\theta - \sin\theta)$, $y = a(1 - \cos\theta)$ from $\theta = 0$ to $\theta = 2\pi$. Find its length.

Arc length: polar equations

19.4   In Qu. 20 the formula for arc length in polar coordinates is introduced.

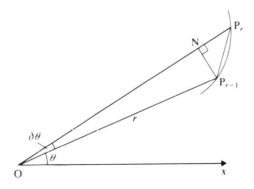

Figure 19.4

*Qu. 20 In Fig. 19.4, $P_{r-1}N$ is the perpendicular from P_{r-1} on to a neighbouring radius OP_r of a curve given in polar coordinates.
(a) Find approximations for NP_r, NP_{r-1} in terms of r, θ.
(b) Obtain the expression for arc length

$$s = \int_{\alpha}^{\beta} \sqrt{\left\{ r^2 + \left(\frac{dr}{d\theta}\right)^2 \right\}} \, d\theta$$

where α, β are the values of θ corresponding to the ends of the arc.
Qu. 21 Find the length of the equiangular spiral $r = a\,e^{k\theta}$ from $\theta = 0$ to $\theta = 2\pi$.
Qu. 22 Find the length of the spiral of Archimedes $r = a\theta$ from $\theta = 0$ to $\theta = \pi$.
Qu. 23 What is the length of the circumference of the cardioid $r = a(1 + \cos\theta)$?
[Make sure the integrand is positive.]

Area of surface of revolution

19.5 In considering areas of surfaces, we come up against a difficulty straight away. How can we measure the area of a curved surface? With a cylinder or cone, the surface can be 'developed', i.e. laid out on a plane surface, but with other figures, such as a sphere, this cannot be done. It is beyond the scope of this book to define, in mathematical terms, what is meant by the area of a curved surface and we shall assume that the following method of finding an expression for the area of a surface of revolution is valid.

First let us consider a fairly simple problem, namely to find the area of the curved surface of the frustum of a cone bounded by circles of radius r and R; see Fig. 19.5.

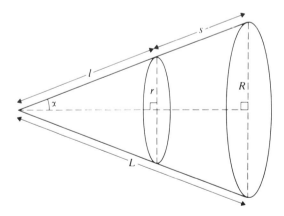

Figure 19.5

From elementary work, we know that the area of the curved surface of a cone is $\pi r l$, where r is the base radius and l the slant height; if we work in terms of the semi-vertical angle α, this becomes $\pi l^2 \sin\alpha$. In the notation of Fig. 19.5, the area

the x-axis is

$$\rho(y_2 - y_1)\,\delta x \times \tfrac{1}{2}(y_2 + y_1) = \tfrac{1}{2}\rho(y_2{}^2 - y_1{}^2)\,\delta x$$

If $\bar{y}$ is the centre of mass of the lamina, taking moments about the x-axis,

$$\rho A \times \bar{y} = \int_a^b \tfrac{1}{2}\rho(y_2{}^2 - y_1{}^2)\,dx$$

where a, b are the extreme values of x.

If we multiply each side by $2\pi/\rho$, we obtain

$$A \times 2\pi\bar{y} = \int_a^b \pi(y_2{}^2 - y_1{}^2)\,dx \tag{1}$$

where the R.H.S. is the volume of the solid generated when the lamina is rotated through 2π radians about the x-axis. Hence we may write

$$area \times distance\ moved\ by\ centroid = volume\ of\ revolution \tag{2}$$

Note that, although we have considered the special case where the area is rotated through an angle 2π, we might have written (1) as

$$A \times \alpha\bar{y} = \int_a^b \tfrac{1}{2}\alpha(y_2{}^2 - y_1{}^2)\,dx$$

for a general rotation α, so that (2) remains valid.

Example 1 *Use Pappus' theorem to obtain the volume of a right circular cylinder, base radius r, height h.*

Rotate a rectangle of height h, base r about one of the sides of length h through four right angles. This generates the cylinder. The area of the rectangle is rh. The centre of mass of the rectangle moves a distance $2\pi \times \tfrac{1}{2}r = \pi r$. So, by Pappus' theorem, the volume of the cylinder is $rh \times \pi r = \pi r^2 h$.

Example 2 *Find the centre of mass of a semicircular lamina of radius a.*

Rotate the semicircle through one revolution about its diameter. The area of the semicircle is $\tfrac{1}{2}\pi a^2$. The volume swept out is $4\pi a^3/3$. Let $\bar{y}$ be the distance of the centre of mass from the diameter then, by Pappus' theorem,

$$\tfrac{1}{2}\pi a^2 \times 2\pi\bar{y} = \tfrac{4}{3}\pi a^3$$

$$\therefore \bar{y} = \frac{4a}{3\pi}$$

Therefore the centre of mass of the semicircular lamina is $4a/(3\pi)$ from the bounding diameter.

Now consider the area of the surface of revolution swept out when an arc is rotated about the x-axis. We consider only arcs which do not cut the axis (Fig. 19.8).

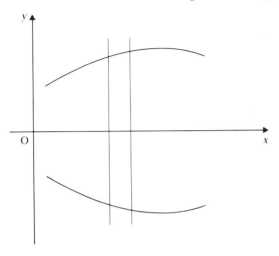

Figure 19.8

Think of the arc as having a density σ per unit length. An element of length of arc has a mass $\sigma\delta s$ with moment $y \times \sigma\delta s$ about the x-axis. If the total length of the arc is s and if its centre of mass is at a distance $\bar{y}$ from the x-axis, taking moments about the x-axis,

$$\sigma s \times \bar{y} = \int y\sigma \ \mathrm{d}s$$

the integral being evaluated between the appropriate limits.

$$\therefore \ s \times 2\pi\bar{y} = \int 2\pi y \ \mathrm{d}s$$

But this integral represents the area of the surface of revolution so that we may write

> *length of arc × distance moved by centre of mass*
> $\qquad\qquad\qquad\qquad$ *= area of surface of revolution*

If the reader can remember that both of Pappus' theorems involve the distance moved by a centre of mass, the idea of dimensions will help in working out what the formulae must be, e.g. to find a volume $[L^3]$, the distance $[L]$ moved by the centre of mass must be multiplied by an area $[L^2]$.

Example 3   *An inflated inner tube of a bicycle tyre has a section (through a plane of symmetry) as shown in Fig. 19.9. Find the surface area of the tube.*

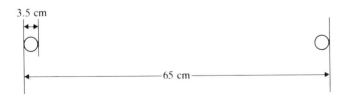

Figure 19.9   (Not to scale)

So it only remains to find $\dfrac{d\psi}{dx}$ in terms of x, y. The gradient of a curve is

$$\tan \psi = \frac{dy}{dx}$$

Differentiating with respect to x,

$$\sec^2 \psi \, \frac{d\psi}{dx} = \frac{d^2y}{dx^2}$$

But $\sec^2 \psi = 1 + \tan^2 \psi$,

$$\therefore \left\{1 + \left(\frac{dy}{dx}\right)^2\right\} \frac{d\psi}{dx} = \frac{d^2y}{dx^2}$$

$$\therefore \frac{d\psi}{dx} = \frac{d^2y}{dx^2} \bigg/ \left\{1 + \left(\frac{dy}{dx}\right)^2\right\}$$

So from (1),

$$\kappa = \frac{d^2y}{dx^2} \bigg/ \sqrt{\left\{1 + \left(\frac{dy}{dx}\right)^2\right\}^3}$$

Qu. 39 Find the curvature of the parabola $y = x^2$ (a) at $(1, 1)$, (b) at the origin.

*Qu. 40 If the coordinates x, y of a point on a curve are given in terms of a parameter t, use the equation $\kappa = \dfrac{d\psi}{dt} \bigg/ \dfrac{ds}{dt}$ to show that

$$\kappa = \left(\frac{dx}{dt}\frac{d^2y}{dt^2} - \frac{dy}{dt}\frac{d^2x}{dt^2}\right) \bigg/ \sqrt{\left\{\left(\frac{dx}{dt}\right)^2 + \left(\frac{dy}{dt}\right)^2\right\}^3}$$

Qu. 41 Find the curvature of the parabola $y^2 = 4ax$ at $(at^2, 2at)$, (a) by treating y as a function of x, (b) by the formula of Qu. 40.

Qu. 42 Find the least curvature of the cycloid

$$x = a(\theta - \sin \theta), \quad y = a(1 - \cos \theta)$$

Qu. 43 If tangents to a curve make angles ψ and $\psi + \delta\psi$ with the x-axis (see Fig. 19.11), find the angle between the corresponding normals. Let the normals intersect at C. Find the limiting distance of C from the curve (measured along one normal) as $\delta\psi \to 0$. Take s as the arc length.

The point C is called the **centre of curvature**. Note that: (a) it is on a normal at a distance ρ from the curve on the concave side, (b) it is the limiting position of the point of intersection of neighbouring normals. The circle with centre C and radius ρ is called the **circle of curvature**.

Qu. 44 Find the equation of the circle of curvature at the point (c, c) on the rectangular hyperbola $xy = c^2$.

Qu. 45 The equation of the normal at $(at^2, 2at)$ to the parabola $y^2 = 4ax$ is

$$xt + y - at^3 - 2at = 0$$

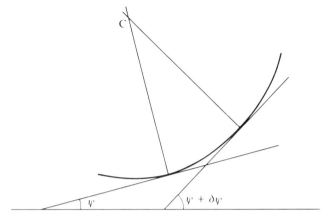

Figure 19.11

Treating this as a cubic in t, show that the condition that it should have two equal roots is

$$27ay^2 = 4(x - 2a)^3$$

What is this the locus of?

*Qu. 46   A curve touches the x-axis at the origin. Write down the first non-zero term in the expansion of y in ascending powers of x. Show that the radius of curvature at the origin is the limiting value of $x^2/2y$ as $x, y \to 0$. This is known as Newton's formula.

Qu. 47   Use Newton's formula to find the radius of curvature at the origin for
(a) $y = x^2$,
(b) $y = x^2/(1 - x^2)$,
(c) $x = 2y^2/(1 + y^2)$ (which touches the y-axis),
(d) $x = a \sin t, \quad y = b \tan^2 t$.

Exercise 19

1 Sketch the limaçon $r = 2 + \cos \theta$ and find its area.

2 Express the equation $(x^2 + y^2)^2 = a^2(x^2 - y^2)$ (Bernoulli's lemniscate) in polar coordinates and find the sum of the areas enclosed by the loops.

3 P, Q are the points $(ca, c/a), (c/a, ca)$ on the rectangular hyperbola $xy = c^2$. Find the area bounded by OP, OQ and the arc PQ.

4 An arc AB of a circle, radius a, subtends an angle 2α at the centre O. The sector OAB is rotated through four right angles about the diameter parallel to AB. Write down the area of the surface of revolution generated by the perimeter of the sector and hence find the distance from O of the centroid of the perimeter.

5 $P(at^2, 2at)$ is a point on the parabola $y^2 = 4ax$ and S is the focus. Show that the area bounded by the parabola, its axis and the line PS is $\frac{1}{3}a^2(3t + t^3)$.

6 Find the area enclosed by the astroid $x = a \cos^3 t, y = a \sin^3 t$.

7 A cylindrical hole of radius r is made in a sphere of radius a so that the axis of

Chapter 20

Further complex numbers

Revision

20.1 In Book 1, Chapter 10, the reader was introduced to the complex numbers, i.e. the set of numbers

$$\mathbb{C} = \{x + iy : x, y \in \mathbb{R}\} \qquad \text{where } i^2 = -1$$

and the rules for adding, subtracting, multiplying and dividing complex numbers were explained. These were as follows: if $z_1 = x_1 + iy_1$ and $z_2 = x_2 + iy_2$, then

$$z_1 + z_2 = (x_1 + x_2) + i(y_1 + y_2)$$

$$z_1 - z_2 = (x_1 - x_2) + i(y_1 - y_2)$$

$$z_1 \times z_2 = (x_1 x_2 - y_1 y_2) + i(x_1 y_2 + x_2 y_1)$$

$$\frac{z_1}{z_2} = \frac{x_1 + iy_1}{x_2 + iy_2}$$

$$= \frac{x_1 + iy_1}{x_2 + iy_2} \times \frac{x_2 - iy_2}{x_2 - iy_2} \qquad \text{(i.e. multiplying the top and the bottom by } z_2^*, \text{ the complex conjugate of } z_2\text{)}$$

$$= \frac{(x_1 x_2 + y_1 y_2) + i(x_2 y_1 - x_1 y_2)}{x_2^2 + y_2^2}$$

These four rules can be summarised by saying that complex numbers obey the same rules as real numbers, but wherever i^2 appears, it is replaced by -1. There is absolutely no need to commit these rules to memory.

The reader will also remember (or should see Book 1, §10.9 to revise it) that there is a very useful method for illustrating complex numbers, namely the Argand diagram (see Fig. 20.1) in which the vector $\overrightarrow{OP}$ represents the complex number $z = x + iy$. (Sometimes it is convenient to say the *point* P represents z.) The modulus of z, written $|z|$, and the argument, written $\arg(z)$, are given by

$$|z| = OP = \sqrt{(x^2 + y^2)}$$

$$\arg(z) = \theta = \arctan\left(\frac{y}{x}\right)$$

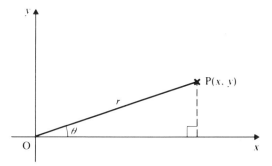

Figure 20.1

(In this chapter, arg(z) will normally be expressed in radians.)

This gave us some useful alternatives to the rules above for multiplying and dividing; if we write $|z| = r$ and $\arg(z) = \theta$, then

$$|z_1 z_2| = r_1 r_2 = |z_1| \times |z_2|$$

and

$$\arg(z_1 z_2) = \theta_1 + \theta_2 = \arg(z_1) + \arg(z_2)$$

Similarly,

$$\left|\frac{z_1}{z_2}\right| = \frac{r_1}{r_2} = \frac{|z_1|}{|z_2|}$$

and

$$\arg\left(\frac{z_1}{z_2}\right) = \theta_1 - \theta_2 = \arg(z_1) - \arg(z_2)$$

The following special cases are important:

(a) $|z^2| = |z|^2$ and $\arg(z^2) = 2\,\arg(z)$

(b) $\left|\dfrac{1}{z}\right| = \dfrac{1}{|z|}$ and $\arg\left(\dfrac{1}{z}\right) = -\arg(z)$

The reader's attention is drawn to the very close analogy between adding vectors and adding complex numbers. This analogy is very useful when interpreting complex numbers geometrically. In particular, diagrams like that in Fig. 20.2 are very common. We could regard this diagram either as an Argand diagram illustrating the complex numbers z, a, and $z - a$, or as a vector triangle representing vectors **z**, **a**, and **z** − **a**. Consequently, a statement in complex numbers, such as

$$|z - a| = r$$

in which z represents a (variable) complex number $x + iy$, a is a (constant) complex number and r is a given real constant, tells us that the length AP is equal to r; in other words the variable point P lies on a circle, centre A, radius r.

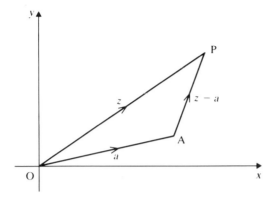

Figure 20.2

Similarly the equation

$$|z - a| = |z - b|$$

where a and b are complex numbers represented by fixed points A and B, tells us that the variable point P which represents z, is equidistant from A and B.

It is essential that the reader should be thoroughly conversant with these elementary concepts. The purpose of Exercise 20a is to revise this work before we proceed further.

Exercise 20a

1 Given that $z_1 = 3 + 4i$ and $z_2 = 1 + i$, find
    (a) $z_1 + z_2$,   (b) $z_1 - z_2$,   (c) $z_1 z_2$,   (d) z_1/z_2,
    (e) $z_1{}^2$,   (f) $z_1{}^3$,   (g) $1/z_1$,   (h) $1/z_2$.
2 Write down the modulus and argument of z_1 and z_2 in No. 1. Find also the modulus and argument of the complex number in each part of No. 1. (Leave surds in the moduli; express the arguments in radians, correct to three decimal places.)
3 Solve the equation $z^2 - 4z + 53 = 0$, expressing the roots in the form $a + ib$, where $a, b \in \mathbb{R}$. Verify that the sum of the roots is 4 and their product is 53.
4 Draw Argand diagrams to illustrate the loci:
    (a) $|z - 10| = 5$,   (b) $\arg(z) = \pi/6$,
    (c) $|z - 1| = |z - i|$,   (d) $\arg(z - a) = \pi/4$, where $a = 1 + i$.
5 Writing $z = x + iy$, find the equations of the following loci in terms of x and y:
    (a) $|z - 10| = 5$,   (b) $|z - 1| = |z - i|$.
6 Draw Argand diagrams to illustrate the regions
    (a) $\text{Re}(z) > 0$,   (b) $\text{Im}(z) > 0$,   (c) $|z| < 3$,   (d) $3 < |z| < 5$,
    (e) $|z - 3| > |z - 5|$,   (f) $0 < \arg(z) < \pi/4$,
    (g) $0 < \arg(z - a) < \pi/2$, where $a = 1 + i$,
    (h) $4 < |z + a| < 5$, where $a = 1 + i$.

7 Given that $z = r$ and $\arg(z) = \theta$, show that z can be expressed in the form

$r(\cos \theta + i \sin \theta)$

Verify that

$$z^2 = r^2(\cos 2\theta + i \sin 2\theta) \quad \text{and that} \quad z^3 = r^3(\cos 3\theta + i \sin 3\theta)$$

*8 Given that $z = \cos \theta + i \sin \theta$, prove by induction that

$$z^n = \cos n\theta + i \sin n\theta$$

where n is a positive integer.

9 The points A, B and C represent the complex numbers z_1, z_2 and z_3, respectively, and ABC is an isosceles triangle, with a right angle at A. Prove that

$$2z_1{}^2 + z_2{}^2 + z_3{}^2 = 2(z_3 z_1 + z_1 z_2)$$

10 The coordinates of the points A and B are (x_1, y_1) and (x_2, y_2), respectively, and O is the origin. Given that OAB is an equilateral triangle, prove that

$$x_2 + iy_2 = \pm\left(\cos\frac{\pi}{3} + i\sin\frac{\pi}{3}\right)(x_1 + iy_1)$$

Hence, or otherwise, show that it is not possible to draw an equilateral triangle with its three vertices at lattice points. (A lattice point is a point whose coordinates are integers.)

Functions of a complex variable

20.2 In real numbers we have already studied many functions, e.g.

$x \mapsto x^2$

and we have either plotted or sketched their graphs ($y = x^2$ in this case). We shall now consider the corresponding problem in complex numbers; in order to keep the discussion fairly simple, we shall take the function

$z \mapsto z^2$

as an illustration. For convenience, in this book, we shall use $z = x + iy$ as a typical member of the domain (i.e. the set of complex numbers to which the rule is applied) and we shall use $w = u + iv$ as the corresponding point in the co-domain. We shall illustrate the effect of the function by drawing a *pair* of Argand diagrams, the left-hand one will always be the z-plane, i.e. the domain, and the right-hand one the w-plane, i.e. the co-domain. First let us consider the effect of the function $z \mapsto z^2$ on a particular point, say $z = 2 + i$, then

$w = (2 + i)^2$
$\quad = 3 + 4i$

Figure 20.3 shows P(2, 1) and its image P'(3, 4).

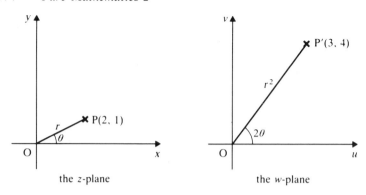

Figure 20.3

More generally if $|z| = r$ and $\arg(z) = \theta$ then, under this function, $|w| = r^2$ and $\arg(w) = 2\theta$.

Qu. 1   Verify that this last statement is true for the points P and P' above.

We shall now consider the effect of this function on a particular set of points in the z-plane: let us take, for example, the circle, centre O, radius 2 units, in which case, $|z| = 2$, and let $\arg(z) = \theta$. Then $|w| = 4$ and $\arg(w) = 2\theta$ (see Fig. 20.4).

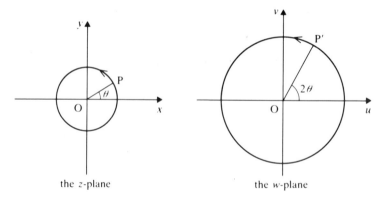

Figure 20.4

The image, in the w-plane, is also a circle, centre O, but its radius is 4 units and OP' is inclined at an angle 2θ to the u-axis. If we imagine the radius OP rotating anti-clockwise at a constant speed, then OP' would be rotating anti-clockwise at twice the speed; when P has completed one full circle P' will have been around its path twice.

Example 1   *In the z-plane, the point* P(x, y) *moves around the semi-circle* $|z| = 2$, $0 \leqslant \arg(z) \leqslant \pi$. *Describe the locus of its image* P'(u, v) *in the w-plane, where* $w = 1/z$.

Under the function $w = 1/z$,

$$|w| = \frac{1}{|z|} \quad \text{and} \quad \arg(w) = -\arg(z)$$

so P' lies on the semi-circle, centre O, radius $\frac{1}{2}$. As P moves anti-clockwise, from $\theta = 0$ to $\theta = \pi$, the point P' moves clockwise from $\theta = 0$ to $\theta = -\pi$ (see Fig. 20.5).

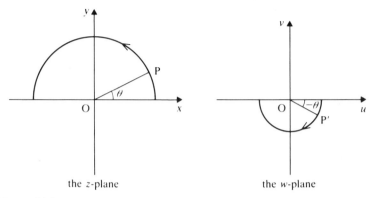

the z-plane   the w-plane

Figure 20.5

Example 2 *Given that $w = 1/z$, find the image of the straight line $x = 1$. Find the images of $(1, -1), (1, 0)$ and $(1, 1)$. Describe the locus of P', the image of P(1, y) as y increases from $-\infty$ to $+\infty$.*

In this example $|z|$ is not constant and so the modulus–argument method of Example 1 is not the best approach; instead we tackle the problem algebraically.

The image P'(u, v) of the point P(1, y) is given by

$$u + iv = \frac{1}{1 + iy}$$

$$= \frac{1}{1 + iy} \times \frac{1 - iy}{1 - iy}$$

$$= \frac{1 - iy}{1 + y^2}$$

Hence, equating the real and imaginary parts,

$$u = \frac{1}{1 + y^2} \quad \text{and} \quad v = \frac{-y}{1 + y^2}$$

From these equations we can see that the images of $(1, -1), (1, 0)$, and $(1, 1)$ are $(\frac{1}{2}, \frac{1}{2}), (1, 0)$ and $(\frac{1}{2}, -\frac{1}{2})$ respectively. The equations for u and v, in terms of y, can be regarded as parametric equations, with y as the parameter. To find the equation relating u and v we must eliminate y. Dividing v by u, we obtain

$$\frac{v}{u} = -y$$

and using this to eliminate y, gives

$$u = \frac{1}{1 + (-v/u)^2}$$

$$= \frac{u^2}{u^2 + v^2}$$

$$\therefore\ u(u^2 + v^2) = u^2$$

Dividing by u (which is never zero),

$$u^2 + v^2 = u$$

This is the equation of a circle in the w-plane; we can find its centre and radius by completing the square,

$$(u - \tfrac{1}{2})^2 + v^2 = \tfrac{1}{4}$$

From this we see that the locus is a circle, centre $(\tfrac{1}{2}, 0)$, radius $\tfrac{1}{2}$ but excluding the point $(0, 0)$, since $u > 0$. (See Fig. 20.6; note in particular that as y increases, P′ moves clockwise.)

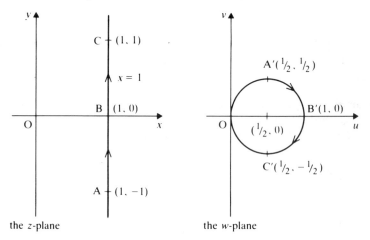

the z-plane   the w-plane

Figure 20.6

Qu. 2   Find the image of $y = 1$ when $w = 1/z$.

Exercise 20b

In Nos. 1 and 2, $a, b \in \mathbb{R}$ and $c = a + ib$. (If you find this difficult, try $a = 3$ and $b = 4$, and then return to the general case.)

1 Find the image of the unit square OIRJ, whose vertices are $(0, 0)$, $(1, 0)$, $(1, 1)$ and $(0, 1)$ respectively, under each of the following transformations:
   (a) $w = az$,   (b) $w - a + z$,   (c) $w = c \mid z$,   (d) $w = cz$,   (e) $w - z^*$.
   Describe each of these transformations in words.

2 Show that the matrix transformation

$$\begin{pmatrix} x \\ y \end{pmatrix} \mapsto \begin{pmatrix} a & -b \\ b & a \end{pmatrix} \begin{pmatrix} x \\ y \end{pmatrix}$$

has the same effect as the complex number transformation $z \mapsto cz$.

In Nos. 3–6, draw Argand diagrams to show the effect of the given function on the region indicated.

3 $w = z + 1 + i$; $0 \leqslant \arg(z) \leqslant \pi/4$.
4 $w = (\cos \pi/4 + i \sin \pi/4)z$; $0 \leqslant \arg(z) \leqslant \pi/4$.
5 $w = z^2$; $1 \leqslant |z| \leqslant 2$.
6 $w = z^2$; $0 \leqslant \arg(z) \leqslant \pi/4$.

7 Find the range of the function $w = z^2$, given that the domain is

$$\{z: \quad |z| = 5, \quad 0 \leqslant \arg(z) \leqslant \pi/6\}$$

8 Given that $w = z^2$, find the images of
  (a) the hyperbola $xy = 10$,   (b) the line $y = x$,   (c) the line $y = -x$.
9 Find the image of the region $1 \leqslant x \leqslant 2$ under the transformation $z \mapsto 1/z$.
10 Given that $w = z^2$, find the images of the lines $x = k$ and $y = k$, where $k \in \mathbb{R}$.

de Moivre's theorem

20.3 In Exercise 20a, No. 8, the reader was introduced to what is known as de Moivre's theorem. This theorem was published by de Moivre in 1730, but the substance of it had appeared in the posthumous publication of Cotes in 1722. The theorem states that, for any rational value of n, one value of $(\cos \theta + i \sin \theta)^n$ is given by

$$(\cos \theta + i \sin \theta)^n = \cos n\theta + i \sin n\theta$$

The reason for saying *one* value of $(\cos \theta + i \sin \theta)^n$ is that there is more than one value for expressions such as $(\cos \theta + i \sin \theta)^{3/2}$: we return to this in §20.4. A proof of the theorem may proceed as follows:

Stage 1.   Prove that

$$(\cos \theta + i \sin \theta)(\cos \phi + i \sin \phi) = \cos (\theta + \phi) + i \sin (\theta + \phi).$$

[Expand the left-hand side.]

Stage 2.   Use induction to prove the theorem for positive integral values of n.

Stage 3.   Use the identity $(\cos \theta + i \sin \theta)(\cos \theta - i \sin \theta) = 1$ to show that

(a) $(\cos \theta + i \sin \theta)^{-1} = \cos \theta - i \sin \theta = \cos (-\theta) + i \sin (-\theta)$,
(b) if $n = -m$, where m is a positive integer,

$$(\cos \theta + i \sin \theta)^n = \cos n\theta + i \sin n\theta$$

In accordance with the usual laws of algebra we take

$$(\cos \theta + i \sin \theta)^0 = 1$$

Stage 4.   If $n = 1/q$, where q is an integer (positive or negative, but not zero), show that *one* value of $(\cos \theta + i \sin \theta)^n$ is $\cos \dfrac{1}{q} \theta + i \sin \dfrac{1}{q} \theta$ by finding the value of

$$\left(\cos \frac{1}{q} \theta + i \sin \frac{1}{q} \theta \right)^q .$$

Stage 5.   If n is a rational number say $\dfrac{p}{q}$, where p, q are integers, we have by Stage 4 one value of

$$(\cos \theta + i \sin \theta)^{1/q} = \cos \frac{1}{q} \theta + i \sin \frac{1}{q} \theta$$

$$(\cos \theta + i \sin \theta)^{p/q} = \left(\cos \frac{1}{q} \theta + i \sin \frac{1}{q} \theta \right)^p$$

$$= \cos \frac{p}{q} \theta + i \sin \frac{p}{q} \theta$$

Qu. 3   Express in the form $x + iy$:
(a) $(\cos \theta + i \sin \theta)^5$,   (b) $1/(\cos \theta + i \sin \theta)^2$,
(c) $(\cos \theta - i \sin \theta)^{-3}$,   (d) $(\cos \theta + i \sin \theta)^2 (\cos \theta + i \sin \theta)^3$,

(e) $\dfrac{\cos \theta + i \sin \theta}{\cos \phi + i \sin \phi}$,   (f) $\dfrac{\cos \theta + i \sin \theta}{\cos \phi - i \sin \phi}$.

Qu. 4   Find one value of:
(a) $\sqrt{(\cos 2\theta + i \sin 2\theta)}$,   (b) $\sqrt[3]{(\cos 2\pi + i \sin 2\pi)}$,
(c) $\sqrt{(\cos \theta + i \sin \theta)^3}$.

Complex roots of unity

20.4   In Book 1, §10.6, we referred to the theorem that an equation of the nth degree has n roots. This means that the equation

$$z^3 - 1 = 0$$

has three roots: one of them is 1 but what are the others? Qu. 5 gives one method of finding out.

Qu. 5   Use the identity $z^3 - 1 = (z - 1)(z^2 + z + 1)$ to find the three cube roots of unity.

Still, the method of Qu. 5 would not work for, say, the roots of $z^7 - 1 = 0$, and the reader may well have been wondering what this has got to do with de Moivre's theorem. Now, we can express 1 as a complex number in infinitely many ways:

...cos $(-2\pi) + i \sin (-2\pi)$, cos $0 + i \sin 0$,

$$\cos 2\pi + i \sin 2\pi, \cos 4\pi + i \sin 4\pi,...$$

or, in general,

cos $2k\pi + i \sin 2k\pi$ where k is an integer

By de Moivre's theorem, one value of

$$\sqrt[3]{(\cos \theta + i \sin \theta)} = (\cos \theta + i \sin \theta)^{1/3} = \cos \tfrac{1}{3}\theta + i \sin \tfrac{1}{3}\theta$$

Therefore values of $\sqrt[3]{1}$ are given by

$$...\cos\left(-\frac{2\pi}{3}\right) + i \sin\left(\frac{2\pi}{3}\right), \cos\frac{0}{3} + i \sin\frac{0}{3},$$

$$\cos\frac{2\pi}{3} + i \sin\frac{2\pi}{3}, \cos\frac{4\pi}{3} + i \sin\frac{4\pi}{3},...$$

or, in general,

$$\cos\frac{2k\pi}{3} + i \sin\frac{2k\pi}{3} \qquad \text{(see Fig. 20.7)}$$

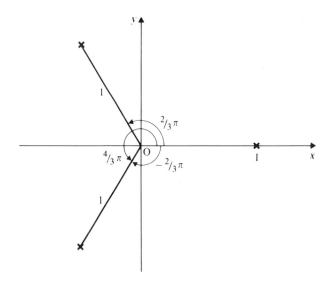

Figure 20.7

Qu. 6 Show that the expression cos $\tfrac{2}{3}k\pi + i \sin \tfrac{2}{3}k\pi$ represents the same complex number when k is replaced by (a) $k + 3$, (b) $k + 3m$, ($k, m \in \mathbb{Z}$).

Qu. 7 By writing $-1 = \cos \pi + i \sin \pi$, and in two other ways, find the cube roots of -1.

Qu. 8 Writing $\omega = \cos \tfrac{2}{3}\pi + i \sin \tfrac{2}{3}\pi$, show that the cube roots of 1 are 1, ω, ω^2. Prove that $1 + \omega + \omega^2 = 0$ in two different ways. [For a hint, see Qu. 5.] Show also that ω is the square of ω^2.

Real and imaginary parts

20.5   One advantage of using complex numbers is that sometimes it is possible to do two pieces of working simultaneously! The possibility rests on the following result: *if $a + ib = c + id$, where a, b, c, d are real, then $a = c$ and $b = d$.* [This is referred to as 'equating real and imaginary parts'. See Book 1, §10.6.]

Example 3   *Prove that* $\tan 3\theta = \dfrac{3 \tan \theta - \tan^3 \theta}{1 - 3 \tan^2 \theta}$.

By de Moivre's theorem,

$$\cos 3\theta + i \sin 3\theta = (\cos \theta + i \sin \theta)^3$$

The R.H.S. may be written

$$\cos^3 \theta + 3i \cos^2 \theta \sin \theta - 3 \cos \theta \sin^2 \theta - i \sin^3 \theta$$

$$\therefore \cos 3\theta + i \sin 3\theta = \cos^3 \theta - 3 \cos \theta \sin^2 \theta + i(3 \cos^2 \theta \sin \theta - \sin^3 \theta)$$

Equating real and imaginary parts,

$$\cos 3\theta = \cos^3 \theta - 3 \cos \theta \sin^2 \theta$$
$$\sin 3\theta = 3 \cos^2 \theta \sin \theta - \sin^3 \theta$$

By division,

$$\frac{\sin 3\theta}{\cos 3\theta} = \frac{3 \cos^2 \theta \sin \theta - \sin^3 \theta}{\cos^3 \theta - 3 \cos \theta \sin^2 \theta}$$

Dividing numerator and denominator of the R.H.S. by $\cos^3 \theta$,

$$\tan 3\theta = \frac{\sin 3\theta}{\cos 3\theta} = \frac{3 \tan \theta - \tan^3 \theta}{1 - 3 \tan^2 \theta}$$

Qu. 9   Show that $\tan 4\theta = (4 \tan \theta - 4 \tan^3 \theta)/(1 - 6 \tan^2 \theta + \tan^4 \theta)$.

Exercise 20c

Simplify Nos. 1–8:

1 $(\cos \theta + i \sin \theta)^3(\cos 2\theta + i \sin 2\theta)$.

2 $(\cos \theta + i \sin \theta)^2(\cos \theta - i \sin \theta)^{-2}$.

3 $\dfrac{\cos 4\theta + i \sin 4\theta}{(\cos \theta + i \sin \theta)^3}$.

4 $\dfrac{(\cos 2\theta + i \sin 2\theta)^3}{(\cos \theta + i \sin \theta)^5}$.

5 $\dfrac{\cos 5\theta + i \sin 5\theta}{(\cos \theta - i \sin \theta)^3}$.

6 $\dfrac{\cos \theta + i \sin \theta}{\cos 2\theta - i \sin 2\theta}$.

7 $(\cos \phi + i \sin \phi)^2(\cos \theta + i \sin \theta)^3$.

8 $(\cos 2\theta + i \sin 2\theta)^3(\cos 3\phi - i \sin 3\phi)^2$.

9 By writing 1 in the form $\cos \theta + i \sin \theta$, where $\theta = -2\pi, 0, 2\pi, 4\pi$, find the fourth roots of unity.

What are the fourth roots of -1?

*10 By writing 1 in the form $\cos 2k\pi + i \sin 2k\pi$ $(k = -2, -1, 0, 1, 2)$, find the fifth roots of unity.

Show that, if z, z^* are conjugate complex numbers, the expansion of $(x - z^*)(x - z)$ is a quadratic in x with real coefficients. Hence write $x^5 - 1$ in the form $(x - 1)q_1 q_2$, where q_1, q_2 are quadratic expressions with real coefficients.

11 Find the sixth roots of 1 and show the corresponding vectors on the Argand diagram. Deduce the real quadratic factors of $x^4 + x^2 + 1$.

12 Show that the nth roots of -1, together with the nth roots of 1, form a complete set of $2n$th roots of 1. How can the $2n$th roots of -1 be deduced from these? Illustrate your answer by showing the corresponding vectors on the Argand diagram for the case $n = 5$.

13 If α is a seventh root of unity, other than 1, show that the other roots are $\alpha^2, \alpha^3, \alpha^4, \alpha^5, \alpha^6, 1$. Show further that

$$1 + \alpha + \alpha^2 + \alpha^3 + \alpha^4 + \alpha^5 + \alpha^6 = 0$$

Do analogous properties hold for all other nth roots of unity?

14 Find the real factors of

(a) $x^7 - 1$, (b) $x^5 + 1$, (c) $x^{2n} - 1$.

*15 If $z = \cos \theta + i \sin \theta$, show that

$$\frac{1}{z} = \cos \theta - i \sin \theta, \qquad z^3 = \cos 3\theta + i \sin 3\theta, \qquad \frac{1}{z^3} = \cos 3\theta - i \sin 3\theta$$

Show further that $(z + 1/z)^3 = 8 \cos^3 \theta$, and, by expanding $(z + 1/z)^3$, prove that

$$2 \cos 3\theta + 6 \cos \theta = 8 \cos^3 \theta$$

Hence express $\cos 3\theta$ in terms of powers of $\cos \theta$.

16 With the notation of No. 15, show that $(z - 1/z)^3 = -8i \sin^3 \theta$ and hence express $\sin 3\theta$ in terms of $\sin \theta$.

17 Use the method of No. 15 to prove that

$$\cos^4 \theta = \tfrac{1}{8}(\cos 4\theta + 4 \cos 2\theta + 3)$$

and express $\cos 4\theta$ in terms of $\cos \theta$. [Expand $(z + 1/z)^4$.]

18 Prove that $\sin^5 \theta = \tfrac{1}{16}(\sin 5\theta - 5 \sin 3\theta + 10 \sin \theta)$.

19 Prove that $\cos^6 \theta = \tfrac{1}{32}(\cos 6\theta + 6 \cos 4\theta + 15 \cos 2\theta + 10)$.

20 Show that $\tan 5\theta = \dfrac{5 \tan \theta - 10 \tan^3 \theta + \tan^5 \theta}{1 - 10 \tan^2 \theta + 5 \tan^4 \theta}$.

[Use the method of Example 3.]

21 Find expressions in terms of $\tan \theta$ for

(a) $\tan 6\theta$, (b) $\tan 2n\theta$, (c) $\tan (2n + 1)\theta$.

22 Show that $\tan 4\theta = \dfrac{4t - 4t^3}{1 - 6t^2 + t^4}$, where $t = \tan \theta$.

Hence find the roots of the equation $t^4 + 4t^3 - 6t^2 - 4t + 1 = 0$ correct to three significant figures.

23 Solve the equation $t^5 - 10t^4 - 10t^3 + 20t^2 + 5t - 2 = 0$ correct to three significant figures.

24 If $w = u + iv$, $z = x + iy$, and $w = z^3$, express u, v in terms of x, y.

25 Find $(a + ib) \div (c + id)$ by equating real and imaginary parts in the equation $(c + id)(p + iq) = a + ib$.

26 (a) Find the square roots of $-5 + 12i$ by equating real and imaginary parts of $(a + ib)^2 = -5 + 12i$.

    (b) Find the square roots of i
      (i) by the method above
      (ii) by using de Moivre's theorem.

*27 Let $\quad C = 1 + \cos \theta + \cos 2\theta + \dots + \cos(n - 1)\theta$
$\qquad\qquad S = \sin \theta + \sin 2\theta + \dots + \sin (n - 1)\theta$

$\qquad Z = C + iS, \qquad z = \cos \theta + i \sin \theta$

    (a) Show that $Z = (1 - z^n)/(1 - z)$.
    (b) Express Z with a real denominator.
    (c) Deduce expressions for C and S by equating real and imaginary parts.

28 Sum the series

$$C = 1 + a \cos \theta + a^2 \cos 2\theta + \dots + a^n \cos n\theta$$
$$S = a \sin \theta + a^2 \sin 2\theta + \dots + a^n \sin n\theta$$

29 Examine the answers to Nos. 27 (c), and 28. Show how the series in these two questions could be summed by multiplying both sides of the equations by some expression.

30 Find the sum to infinity of the series

$$\cos \theta \cos \theta + \cos^2 \theta \cos 2\theta + \cos^3 \theta \cos 3\theta + \dots$$

e^z, $\cos z$, $\sin z$, where $z \in \mathbb{C}$

20.6   In this section it is intended to give the reader a glimpse of the way in which it is possible to extend the above functions to cover complex variables.

First consider the function e^x, where x is real. We have to find some property of the function which we can use to define what we mean by e^z where z is complex. The most obvious way to start is to go back to our definition of e^x but unfortunately this does not lend itself to an extension to complex numbers; so we have to find some other property of e^x. Now, e^x can be expanded as a power series:

$$e^x = 1 + x + \frac{x^2}{2!} + \dots + \frac{x^n}{n!} + \dots$$

and we can readily give e^z a meaning by *defining* it by the series

$$e^z = 1 + z + \frac{z^2}{2!} + \dots + \frac{z^n}{n!} + \dots$$

We know that the series for e^x is convergent for all real values of x, and we shall assume that the series for e^z is convergent for complex values of z, but the reader should be aware that this does not follow automatically.

The definition of e^z by a series poses another problem: will the usual laws of indices still hold? In particular, we must satisfy ourselves that

$$e^w \times e^z = e^{w+z}$$

This is something which has to be deduced from our definition of e^z: that is, it is necessary to show that

$$\left(1 + w + \frac{w^2}{2!} + \dots + \frac{w^n}{n!} + \dots\right)\left(1 + z + \frac{z^2}{2!} + \dots + \frac{z^n}{n!} + \dots\right)$$

$$= 1 + (w + z) + \frac{(w + z)^2}{2!} + \dots + \frac{(w + z)^n}{n!} + \dots$$

The conditions under which infinite series may be multiplied are outside the scope of this book and so a proof will not be given here, but the reader should work Qu. 10.

Qu. 10 Show that term-by-term multiplication of the first few terms of the series for e^w and e^z gives the first few terms of the series for e^{w+z}.

Also find the terms of degree n in the product and show that they reduce to $(w + z)^n/n!$

Now let us see what happens if we write $z = x + iy$ in the expression e^z.

$$e^z = e^{x+iy} = e^x \times e^{iy}$$

e^x is real (and familiar), so we shall examine the function e^{iy}.

$$e^{iy} = 1 + iy + \frac{i^2 y^2}{2!} + \frac{i^3 y^3}{3!} + \frac{i^4 y^4}{4!} + \dots + \frac{i^n y^n}{n!} + \dots$$

$$= 1 \qquad - \frac{y^2}{2!} \qquad + \frac{y^4}{4!} - \dots$$

$$+ i\left(y \qquad - \frac{y^3}{3!} + \dots\right)$$

But

$$1 - \frac{y^2}{2!} + \frac{y^4}{4!} - \dots = \cos y \quad \text{and} \quad y - \frac{y^3}{3!} + \frac{y^5}{5!} - \dots = \sin y$$

so that

$$e^{iy} = \cos y + i \sin y$$

A result equivalent to this, discovered by Cotes, was published in 1722, after his death.

Qu. 11 Use the results

$$(\cos\theta + i\sin\theta)(\cos\theta - i\sin\theta) = 1 \quad \text{and} \quad e^w \times e^z = e^{w+z}$$

to show that

$$e^{-i\theta} = \cos\theta - i\sin\theta$$

Note. In most texts the reader will find the series

$$1 + z + \frac{z^2}{2!} + \ldots + \frac{z^n}{n!} + \ldots$$

denoted by exp z.

We are now in a position to turn to the problem of assigning meanings to the functions sin z, cos z, where z is complex.

From above, $e^{i\theta} = \cos\theta + i\sin\theta$, and from Qu. 11, $e^{-i\theta} = \cos\theta - i\sin\theta$. Hence

$$\cos\theta = \frac{e^{i\theta} + e^{-i\theta}}{2}$$

$$\sin\theta = \frac{e^{i\theta} - e^{-i\theta}}{2i}$$

Since e^z has been defined for complex z, we may use these last two equations to define cos z and sin z:

$$\cos z = \frac{e^{iz} + e^{-iz}}{2} \tag{1}$$

$$\sin z = \frac{e^{iz} - e^{-iz}}{2i} \tag{2}$$

These results were given by Euler in 1748, twenty years before Lambert introduced the hyperbolic functions.

The hyperbolic functions cosh x, sinh x were defined for real values of x in Chapter 18:

$$\cosh x = \tfrac{1}{2}(e^x + e^{-x}), \qquad \sinh x = \tfrac{1}{2}(e^x - e^{-x})$$

These definitions can be used to define the functions cosh z, sinh z of a complex variable z:

$$\cosh z = \tfrac{1}{2}(e^z + e^{-z}), \qquad \sinh z = \tfrac{1}{2}(e^z - e^{-z})$$

Replacing z by iz,

$$\cosh iz = \tfrac{1}{2}(e^{iz} + e^{-iz}), \qquad \sinh iz = \tfrac{1}{2}(e^{iz} - e^{-iz}) \tag{3}$$

and so from (1), (2), (3) above we obtain the following relations connecting circular and hyperbolic functions:

$$\cosh iz = \cos z \qquad \sinh iz = i \sin z$$

Qu. 12 Confirm these last relationships for real x by replacing z by x in equations (3) and expressing e^{ix}, e^{-ix} in terms of $\cos x$, $\sin x$.

Qu. 13 Express $\cosh z$, $\sinh z$ in terms of the corresponding circular functions.

Qu. 14 Use the series $e^w = 1 + w + \dfrac{w^2}{2!} + \ldots + \dfrac{w^n}{n!} + \ldots$ to show that

$$\cos z = 1 - \frac{z^2}{2!} + \frac{z^4}{4!} - \ldots$$

$$\sin z = z - \frac{z^3}{3!} + \frac{z^5}{5!} - \ldots$$

Much of the earlier part of this book was concerned with identities connecting sines and cosines: will they still hold for $\cos z$ and $\sin z$ when z is complex?

Qu. 15 Deduce from equations (1) and (2) above that
(a) $\cos^2 z + \sin^2 z = 1$,   (b) $\cos(-z) = \cos z$,
(c) $\sin(-z) = -\sin z$,   (d) $\cos(w + z) = \cos w \cos z - \sin w \sin z$,
(e) $\sin(w + z) = \sin w \cos z + \cos w \sin z$.

The other trigonometrical identities follow from these (if you cannot satisfy yourself about this, turn to Book 1, chapter 17, where they were proved for real numbers); and so trigonometrical functions of a complex variable may be manipulated by the same identities as those for a real variable.

Qu. 16 A function of a complex variable is said to be periodic with period p if $f(z + p) = f(z)$ for all z. Show that $\cos z$, $\sin z$ have period 2π.

Exercise 20d (Miscellaneous)

1 Express in the form $a + ib$:

(a) $(x + iy)^6$,   (b) $\dfrac{1}{3\cos\theta + 2i\sin\theta - 1}$,   (c) $\sqrt{\dfrac{\cos\theta + i\sin\theta}{\cos 2\theta - i\sin 2\theta}}$,

(d) $\dfrac{z + 1}{z - 1}$ where $z = x + iy$.

2 Express in the form $a + ib$:

(a) $\dfrac{(\cos\theta + i\sin\theta)^3}{(\cos 2\theta + i\sin 2\theta)^2}$,   (b) $\dfrac{(\cos 2\theta + i\sin 2\theta)^2}{(\cos\theta - i\sin\theta)^4}$,

(c) $1 - \text{cis}\,\theta + \text{cis}\,2\theta - \ldots + (-1)^{n-1}\,\text{cis}\,(n-1)\theta$,
  where $\text{cis}\,r\theta = \cos r\theta + i\sin r\theta$.

3 P_1, P_2 are points on the Argand diagram corresponding to the complex numbers z_1, z_2. Show that the mid-point of $P_1 P_2$ corresponds to $\frac{1}{2}(z_1 + z_2)$;

hence proves that the mid-points of the sides of a plane quadrilateral are the vertices of a parallelogram.

4 What is the locus given by $zz^* + 2(z + z^*) = 0$, where $z = x + iy$ and z^* is its conjugate?

Express as an equation connecting z, z^* the condition that (x, y) should lie on the circle centre $(2, -1)$ radius 3.

5 (a) Prove that $\cos^5 \theta = \frac{1}{16}(\cos 5\theta + 5 \cos 3\theta + 10 \cos \theta)$, and find an expression for $\sin^4 \theta$ in terms of $\cos 2\theta$, $\cos 4\theta$.

(b) Express $\tan (\theta_1 + \theta_2 + \theta_3 + \theta_4)$ in terms of $\tan \theta_1$, $\tan \theta_2$, $\tan \theta_3$, $\tan \theta_4$.

6 (a) Express -1 in the form $\cos \theta + i \sin \theta$. Hence obtain the three linear factors of $z^3 + 1$.

(b) Find the real factors of $x^5 - a^5$.

7 Show that, if a quadratic equation is satisfied by a complex number, then the conjugate complex number is also a root of the equation.

Hence find the quadratic equation satisfied by the complex number $2 - 3i$. Find the four roots of the equation $z^4 - 3z^3 + 4z^2 - 3z + 1 = 0$.

8 (a) Solve the equation $z^4 - 6z^2 + 25 = 0$.

(b) Given that one root of the equation $z^4 - 6z^3 + 23z^2 - 34z + 26 = 0$ is $1 + i$, find the others.

9 If $z + \dfrac{1}{z} = -1$, prove that $z^5 + \dfrac{1}{z^5} = -1$ and find the value of $z^{11} + \dfrac{1}{z^{11}}$.

10 Prove that, if $a + ib = c + id$ where a, b, c, d are real, then $a = c$, $b = d$.

Find the sum to infinity of $1 - \frac{1}{2} \cos \theta + \frac{1}{4} \cos 2\theta - \frac{1}{8} \cos 3\theta + \dots$.

11 Find the sum of the series $1 + 2 \cos 2\theta + 4 \cos 4\theta + \dots + 2^n \cos 2n\theta$.

12 Prove by induction that, for positive integral values of n,

$$(\cos \theta + i \sin \theta)^n = \cos n\theta + i \sin n\theta$$

Hence show that the identity also holds for negative integral values of n. Find the sum to infinity of the series

$$\sin \theta \sin \theta + \sin^2 \theta \sin 2\theta + \sin^3 \theta \sin 3\theta + \dots$$

In Nos. 13–15 x, y, u, v are real numbers, z, w are the complex numbers $x + iy$, $u + iv$.

13 If w, z are connected by the equation $w = z + 1$, find the locus of (u, v) when
(a) $x + 1 = 0$, (b) $x + y - 1 = 0$, (c) $|z + 1| = 1$, (d) $|z - 3| = 2$.

14 If $w = 3z$, find the locus of (u, v) when
(a) $x - 2y = 0$, (b) $y - 1 = 0$, (c) $|z| = 1$, (d) $|z - 2| = 2$.

15 The point (x, y) moves once round each of the circles (a) $|z| = 1$, (b) $|z| = 2$ in a counter-clockwise sense. Describe the corresponding motions of (u, v) if $w = z^2$.

16 The point $P(x, y)$ represents a complex number $z = x + iy$. Given that $|z + 1| = 2|z - 1|$, find, in terms of x and y, the equation of the locus of P and describe the locus in words.

17 Repeat No. 16 for $|z - i| = 3|z + i|$.

18 (a) Given that $(2 + 3i)z = 4 - i$, find the complex number z, giving your answer in the form $a + ib$, where $a, b \in \mathbb{R}$.

(b) Find the modulus and argument of the complex number $5 - 3i$.

The complex number w is represented in an Argand diagram by the point W. Describe geometrically the locus of W in each of the following cases:

(i) $|w| = |5 - 3i|$,   (ii) $\arg(w - 5 + 3i) = \arg(5 - 3i) + \frac{1}{2}\pi$.   (C)

19 Express z_1, where $z_1 = \dfrac{10 - i\,2\sqrt{3}}{1 - i\,3\sqrt{3}}$, in the form $p + iq$, where p and q are real.

Given that $z_1 = r(\cos \theta + i \sin \theta)$, where $r > 0$ and $-\pi < \theta \leqslant \pi$, obtain values for r and θ. Hence determine $z_1{}^9$.

Sketch on an Argand diagram the locus of the points representing the complex number z such that $|z - z_1| = \sqrt{3}$.   (L)

20 The point P represents the complex number $z = x + iy$ on an Argand diagram. Describe the locus of P if

(a) $|2z + 1 - 2i| = 3$,   (b) $\arg\left(\dfrac{z + i}{z - i}\right) = \frac{1}{4}\pi$,   (c) $|z - i| + |z + i| = 4$.

If $|z| = 1$, find the locus of the point representing the complex number $z + \dfrac{1}{z}$.

     (L)

21 (a) Find the modulus and argument of each of the following complex numbers:

(i) $\dfrac{5 + i}{2 + 3i}$,   (ii) $\dfrac{(5 + i)^4}{(2 + 3i)^4}$,   (iii) $\dfrac{(\cos \pi/6 - i \sin \pi/6)^4}{(\sin \pi/6 + i \cos \pi/6)^3}$.

(b) Find, in the form $re^{i\theta}$, all the complex numbers z, such that

$$z^3 = \frac{5 + i}{2 + 3i}$$ (C)

22 The transformation $w = (z + 1)^2 + 3$ maps the complex number $z = x + iy$ to the complex number $w = u + iv$. Show that as z moves along the y-axis from the origin to the point $(0, 2)$ in the z-plane, w moves from the point $(4, 0)$ to the point $(0, 4)$ along a curve in the w-plane. Write down the equation of this curve.   (JMB)

23 (a) Given that the real part of $(z - 2i)/(z + 4)$ is zero, prove that, in the Argand diagram, the locus of z is a circle. Find the centre of the circle and show that the radius is $\sqrt{5}$.

(b) Find the image of the circle $|z| = 1$ under each of the transformations given in (i) and (ii) below. If the image is a circle, give its centre and radius. If the image is a straight line, give its equation (in any form).

(i) $w = \dfrac{2}{i + 2z}$,   (ii) $w = \dfrac{2 + z}{i - z}$.   (C)

24 Two complex numbers w and z are connected by the relation

$$w = 2\left(\frac{1 + z}{1 - z}\right)$$

Prove that, as the point P representing z in the Argand diagram follows the locus $i\lambda$ with λ decreasing from ∞ to 0, the locus of Q, representing w is a semi-circle C and determine the centre and radius of this semi-circle.

Determine the locus C' of the point Q' corresponding to the point P', as P' describes the real axis in the positive direction from the origin.

Indicate on a diagram the loci C and C' and the directions in which the loci are described.   (O & C)

25 The point P represents the complex number z and the point Q represents the complex number w, where $w = 1/(z-1)$. Prove that if $|z| = 1$, then $|w| = |w+1|$.

The point P moves anti-clockwise once round the circle C with centre the origin and radius 1,

(a) describe the locus of Q;

(b) given that P starts from $z = 1$, describe carefully the locus of a point R which represents the complex number $i/(z+i)$.   (O & C)

26 Express $\cos z$ and $\sin z$ in terms of the hyperbolic cosine and sine, and deduce the following identities:

(a) $\sinh 2z = 2 \sinh z \cosh z$,   (b) $\cosh 2z = \cosh^2 z + \sinh^2 z$,

(c) $\cosh(w - z) = \cosh w \cosh z - \sinh w \sinh z$,

(d) $\cosh w + \cosh z = 2 \cosh \frac{1}{2}(w + z) \cosh \frac{1}{2}(w - z)$.

27 Define $\cosh z$, $\sinh z$ when z is complex. Deduce from your definitions the identities:

(a) $\cosh^2 z - \sinh^2 z = 1$,

(b) $\sinh(w + z) = \sinh w \cosh z + \cosh w \sinh z$.

Express $\cosh z$, $\sinh z$ in terms of the circular cosine and sine, and deduce identities corresponding to the two above.

28 (a) Write down the values of $|z|$ and $\arg(z)$, where $z = x + iy$. Illustrate by means of an Argand diagram.

The numbers c and p are given, c being real and p being complex, with $p = a + ib$; z^* and p^* denote the conjugates of z and p respectively. Prove that, if

$$zz^* - p^*z - pz^* + c = 0$$

then the point on the Argand diagram which represents z lies on a certain circle whose centre and radius should be determined.

(b) Prove that, if

$$x + iy = \frac{1}{\lambda + i\mu}$$

then the points on the Argand diagram defined by making λ constant lie on a circle, and the points defined by making μ a constant also lie on a circle.

Prove also that, whatever be the values of the constants, the centres of the two systems of circles obtained lie on two fixed perpendicular lines.

(O & C)

29 The coordinates (x, y) of a point P are expressible in terms of real variables u and v by the formula

$$x + iy = (u + iv)^2$$

Prove that the locus of P is a parabola (a) when u varies and v is constant, and also (b) when v varies and u is constant. Prove also that all the parabolas have a common focus and a common axis.

Prove that through a given point (x_0, y_0) there pass two parabolas, one of each system $u = $ constant, $v = $ constant, which cut at right angles. (O & C)

30 (a) Complex numbers z_1 and z_2 are given by the formulae

$$z_1 = R_1 + i\omega L \qquad z_2 = R_2 - \frac{i}{\omega C}$$

and z is given by the formula

$$\frac{1}{z} = \frac{1}{z_1} + \frac{1}{z_2}$$

Find the value of ω for which z is a real number.

(b) Use de Moivre's theorem to prove that, if

$$2 \cos \theta = x + \frac{1}{x}$$

then $\quad 2 \cos n\theta = x^n + \frac{1}{x^n}$

Hence, or otherwise, solve the equation

$$5x^4 - 11x^3 + 16x^2 - 11x + 5 = 0 \qquad \text{(O \& C)}$$

Chapter 21

Further vector methods

Some further examples on scalar products and planes

21.1 In the previous chapter on vectors (see Book 1, §15.13), we found the equation of the plane through three given points. Another way of describing a plane is to specify a vector which is perpendicular to the plane, and give a point through which the plane passes.

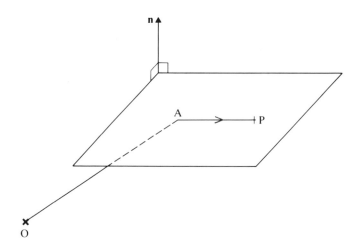

Figure 21.1

Suppose the plane is perpendicular to a given vector $\mathbf{n}$, where $\mathbf{n} = \begin{pmatrix} a \\ b \\ c \end{pmatrix}$; that is, every line in the plane is perpendicular to $\mathbf{n}$. Let the given point through which the plane passes be $A(x_1, y_1, z_1)$ and let $P(x, y, z)$ be any point in the plane (Fig. 21.1), then AP is perpendicular to $\mathbf{n}$. This fact can be expressed in terms of a scalar product, namely

$$\begin{pmatrix} a \\ b \\ c \end{pmatrix} \cdot \begin{pmatrix} x - x_1 \\ y - y_1 \\ z - z_1 \end{pmatrix} = 0$$

Hence $a(x - x_1) + b(y - y_1) + c(z - z_1) = 0$.

$$\therefore \ ax + by + cz = d$$

where $d = ax_1 + by_1 + cz_1$. Notice that the coefficients a, b, c of x, y and z form the vector $\mathbf{n}$; consequently if we are given the equation of a plane, we can immediately write down the vector which is perpendicular to it.

Example 1 *Write down the unit vector which is perpendicular to the plane* $2x + 3y + 6z = 10$.

The vector $\begin{pmatrix} 2 \\ 3 \\ 6 \end{pmatrix}$ is normal to the plane and the magnitude of this is

$\sqrt{(4 + 9 + 36)} = \sqrt{49} = 7$, so the unit vector required is $\frac{1}{7} \begin{pmatrix} 2 \\ 3 \\ 6 \end{pmatrix}$.

Example 2 *Find the equation of the plane through the point* (1, 2, 3) *and perpendicular to the vector* $4\mathbf{i} + 5\mathbf{j} + 6\mathbf{k}$.

Any plane which is perpendicular to the given vector will have an equation of the form

$$4x + 5y + 6z = d$$

where d is a constant. To find the equation of the plane which passes through (1, 2, 3), we must choose the value of d, so that the equation is satisfied when $x = 1$, $y = 2$ and $z = 3$, i.e.

$$4 \times 1 + 5 \times 2 + 6 \times 3 = d$$

i.e. $\qquad\qquad\qquad\qquad d = 32$

Hence the equation we require is $4x + 5y + 6z = 32$.

Example 3 *Find the angle between the planes* $4x + 3y + 12z = 10$ *and* $8x - 6y = 14$.

The angle required is the angle between the normal vectors and these are $\mathbf{m} = 4\mathbf{i} + 3\mathbf{j} + 12\mathbf{k}$ and $\mathbf{n} = 8\mathbf{i} - 6\mathbf{j}$. We shall find the angle between $\mathbf{m}$ and $\mathbf{n}$ by finding the scalar product $\mathbf{m}.\mathbf{n}$ in two forms and equating them. Firstly,

$$\begin{aligned} \mathbf{m}.\mathbf{n} &= (4\mathbf{i} + 3\mathbf{j} + 12\mathbf{k}).(8\mathbf{i} - 6\mathbf{j}) \\ &= 32 - 18 \\ &= 14 \end{aligned}$$

Alternatively, using **m.n** = *mn* cos θ, where *m* and *n* are the magnitudes of the vectors **m** and **n**, and θ is the angle between them, we obtain

$$\begin{aligned}\textbf{m.n} &= \sqrt{(16+9+144)} \times \sqrt{(64+36)} \times \cos\theta\\ &= \sqrt{169} \times \sqrt{100} \times \cos\theta\\ &= 130\cos\theta\end{aligned}$$

Equating these two expressions for **m.n** gives

$$130\cos\theta = 14$$

$$\therefore \cos\theta = \frac{14}{130}$$

$$\therefore \theta = 83.8°$$

The angle between the planes is $83.8°$, correct to one decimal place.

Example 4   *Find the distance of the point* A(25, 5, 7) *from the plane* $12x + 4y + 3z = 3$.

Let P be the point in the plane such that $\overrightarrow{AP}$ is perpendicular to the plane. Then the distance required is the length of the vector $\overrightarrow{AP}$.

We know that $12\mathbf{i} + 4\mathbf{j} + 3\mathbf{k}$ is perpendicular to the plane, so let $\overrightarrow{AP} = t(12\mathbf{i} + 4\mathbf{j} + 3\mathbf{k})$. Then, since $\overrightarrow{OP} = \overrightarrow{OA} + \overrightarrow{AP}$,

$$\begin{aligned}\overrightarrow{OP} &= (25\mathbf{i} + 5\mathbf{j} + 7\mathbf{k}) + t(12\mathbf{i} + 4\mathbf{j} + 3\mathbf{k})\\ &= (25 + 12t)\mathbf{i} + (5 + 4t)\mathbf{j} + (7 + 3t)\mathbf{k}\end{aligned}$$

Hence P is the point $(25 + 12t, 5 + 4t, 7 + 3t)$ and, since this point lies in the plane, its coordinates satisfy the equation of the plane; consequently,

$$12(25 + 12t) + 4(5 + 4t) + 3(7 + 3t) = 3$$
$$\therefore 169t + 341 = 3$$
$$\therefore t = -2$$

Hence $\overrightarrow{AP} = -2(12\mathbf{i} + 4\mathbf{j} + 3\mathbf{k}) = -24\mathbf{i} - 8\mathbf{j} - 6\mathbf{k}$, and so

$$\begin{aligned}\text{AP}^2 &= 24^2 + 8^2 + 6^2\\ &= 576 + 64 + 36\\ &= 676\end{aligned}$$
$$\therefore \text{AP} = 26$$

The distance from the point A to the plane is 26 units.

Exercise 21a

1 Find **a.b**, **a.c** and **a.(b + c)**, given that
   (a) $\mathbf{a} = 2\mathbf{i} + 2\mathbf{j} + \mathbf{k}$,   (b) $\mathbf{a} = 3\mathbf{i} + 7\mathbf{j} - 5\mathbf{k}$,
       $\mathbf{b} = 3\mathbf{i} + 4\mathbf{j} + 5\mathbf{k}$, $\mathbf{b} = 2\mathbf{i} + 6\mathbf{j} + 3\mathbf{k}$,
       $\mathbf{c} = 4\mathbf{i} + \mathbf{j} - 8\mathbf{k}$, $\mathbf{c} = 4\mathbf{i} - 8\mathbf{j} - 2\mathbf{k}$.

2 Find the angle between the vectors $\mathbf{a} = 2\mathbf{i} + \mathbf{k}$ and $\mathbf{b} = 3\mathbf{i} + 4\mathbf{j} + 5\mathbf{k}$.

3 Find the coordinates of the point N, where the perpendicular from (37, 9, 10) meets the plane $12x + 4y + 3z = 3$.

4 Find the reflection of the point (5, 7, 11) in the plane $2x + 3y + 5z = 10$.

5 Given that the vectors $\mathbf{a}$ and $\mathbf{b}$ have equal magnitudes, prove that $(\mathbf{a} + \mathbf{b}).(\mathbf{a} - \mathbf{b}) = 0$ and interpret this result in terms of the parallelogram which has $\mathbf{a}$ and $\mathbf{b}$ as a pair of adjacent sides.

*6 In Fig. 21.2, $\overrightarrow{OA} = \mathbf{a}$, $\overrightarrow{OB} = \mathbf{b}$ and the angle AOB is θ.

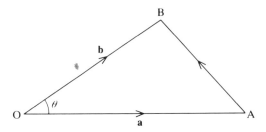

Figure 21.2

By considering the scalar product $\overrightarrow{AB}.\overrightarrow{AB}$, prove that

$$AB^2 = OA^2 + OB^2 - 2 \times OA \times OB \times \cos \theta$$

7 The points $A(x_1, y_1, z_1)$ and $B(x_2, y_2, z_2)$ lie in the plane

$$ax + by + cz = d$$

Show that $a(x_2 - x_1) + b(y_2 - y_1) + c(z_2 - z_1) = 0$. Hence prove that the vector $\overrightarrow{AB}$ is perpendicular to the vector $a\mathbf{i} + b\mathbf{j} + c\mathbf{k}$.

*8 The point O is the centre of the circumcircle of triangle ABC (the circumcircle is the circle which passes through the vertices of a triangle) and G is its centroid. H is a point on OG such that $\overrightarrow{OH} = 3\overrightarrow{OG}$. Prove that $\overrightarrow{AH}$ is perpendicular to $\overrightarrow{BC}$. Prove also that $\overrightarrow{BH}$ is perpendicular to $\overrightarrow{AC}$ and $\overrightarrow{CH}$ is perpendicular to $\overrightarrow{AB}$. (The point H is called the *orthocentre* of the triangle.)

9 In the triangle OPQ the angle POQ is a right angle. The point R lies on PQ and PR:RQ = 1:3. Express the position vector of R in terms of $\mathbf{p}$ and $\mathbf{q}$, the position vectors of P and Q. Given that $\overrightarrow{OR}$ is perpendicular to $\overrightarrow{PQ}$, prove that OP:OQ = $1:\sqrt{3}$.

*10 In Fig. 21.3, $\overrightarrow{OA}$ and $\overrightarrow{OB}$ are unit vectors making angles α and β respectively with the x-axis. Express in terms of α and β:
(a) $\overrightarrow{OA}$, (b) $\overrightarrow{OB}$, (c) the angle AOB.

Write down the scalar product $\overrightarrow{OA}.\overrightarrow{OB}$ in two ways. Hence prove that

$$\cos (\alpha - \beta) = \cos \alpha \cos \beta + \sin \alpha \sin \beta$$

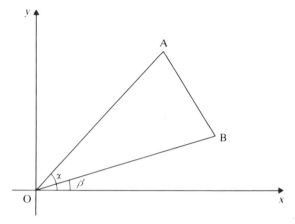

Figure 21.3

The vector product

21.2   The *scalar* product, which we met in Book 1, Chapter 15, combines two vectors to produce a scalar result. The *vector* product, which we shall now examine, combines two vectors and produces a vector result, i.e. vector multiplication is closed (see Book 1, §25.8). The vector product is a useful concept in mechanics, where it is used, amongst other things, when finding the moment of a force about a point (for further applications the reader should consult a suitable book on mechanics).

Definition

The vector product of two vectors **a** and **b**, which is written **a** ∧ **b**, is given by

$$\mathbf{a} \wedge \mathbf{b} = ab \sin \theta \, \hat{\mathbf{n}}$$

where a and b are the magnitudes of the vectors **a** and **b**,
$\quad\quad\quad\theta$ is the angle between **a** and **b**,
and $\quad\hat{\mathbf{n}}$ is the unit vector, perpendicular to both **a** and **b**, and such that **a**, **b** and $\hat{\mathbf{n}}$, *taken in that order*, form a right-handed set (see Fig. 21.4).

The notation **a** × **b** is frequently used instead of **a** ∧ **b**, and when this is done it is usual to call it 'the cross product' of **a** and **b**. (**a** ∧ **b** is usually read as '**a** *vec* **b**'.)

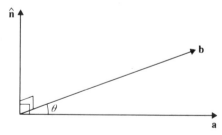

Figure 21.4

Several features of the vector product $\mathbf{a} \wedge \mathbf{b}$ should be noted immediately:

(a) $\mathbf{b} \wedge \mathbf{a} = -\mathbf{a} \wedge \mathbf{b}$. This follows from the fact that reversing the order of $\mathbf{a}$ and $\mathbf{b}$ in the right-handed set will have the effect of reversing the direction of the unit vector. (We say that vector multiplication is not commutative. See Book 1, §25.8.)

(b) Given that $\mathbf{i}$, $\mathbf{j}$ and $\mathbf{k}$ represent the usual base vectors,

$$\mathbf{j} \wedge \mathbf{k} = \mathbf{i}, \qquad \mathbf{k} \wedge \mathbf{i} = \mathbf{j}, \qquad \mathbf{i} \wedge \mathbf{j} = \mathbf{k},$$

but

$$\mathbf{k} \wedge \mathbf{j} = -\mathbf{i}, \quad \mathbf{i} \wedge \mathbf{k} = -\mathbf{j}, \quad \mathbf{j} \wedge \mathbf{i} = -\mathbf{k}.$$

(c) For any vector $\mathbf{a}$, $\mathbf{a} \wedge \mathbf{a} = \mathbf{0}$ (and thus $\mathbf{i} \wedge \mathbf{i} = \mathbf{j} \wedge \mathbf{j} = \mathbf{k} \wedge \mathbf{k} = \mathbf{0}$).

These features of the vector product should be contrasted with those of the scalar product $\mathbf{a}.\mathbf{b} = \mathbf{b}.\mathbf{a} = ab \cos \theta$, set out in Book 1, §15.15.

Qu. 1 Given that $\mathbf{a} = 5\mathbf{i}$ and $\mathbf{b} = \sqrt{3}\mathbf{i} + \mathbf{j}$, verify that $b = 2$ and that the angle between $\mathbf{a}$ and $\mathbf{b}$ is $\pi/6$. Hence, from the definition above, show that $\mathbf{a} \wedge \mathbf{b} = 5\mathbf{k}$. Verify that the same result is obtained by assuming that vector multiplication obeys the distributive law (i.e. that for any vectors $\mathbf{p}$, $\mathbf{q}$, $\mathbf{r}$, $\mathbf{p} \wedge (\mathbf{q} + \mathbf{r}) = \mathbf{p} \wedge \mathbf{q} + \mathbf{p} \wedge \mathbf{r}$; this will be proved in §21.4.)

The scalar triple product

21.3 For scalar multiplication, the triple product $\mathbf{a}.(\mathbf{b}.\mathbf{c})$ is meaningless, because $\mathbf{b}.\mathbf{c}$ is not a vector. However, because $\mathbf{b} \wedge \mathbf{c}$ *is* a vector, the triple product $\mathbf{a}.(\mathbf{b} \wedge \mathbf{c})$ can be found. We shall show that it is equal to the volume of the *parallelepiped*, whose six faces are the parallelograms formed by taking the vectors $\mathbf{a}$, $\mathbf{b}$ and $\mathbf{c}$, two at a time (see Fig. 21.5).

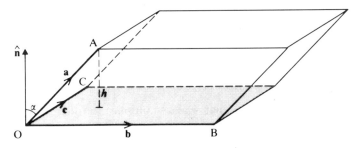

Figure 21.5

First notice that $\mathbf{b} \wedge \mathbf{c}$ is equal in magnitude to the area of the parallelogram formed by $\mathbf{b}$ and $\mathbf{c}$. It will be convenient to call this area S. The direction of $\mathbf{b} \wedge \mathbf{c}$ is perpendicular to the plane of $\mathbf{b}$ and $\mathbf{c}$; see the unit vector $\hat{\mathbf{n}}$ in Fig. 21.5.

When we tackle $\mathbf{a}.(\mathbf{b} \wedge \mathbf{c})$, we obtain a scalar whose magnitude is

$$S a \cos \alpha$$

where α is the angle between $\hat{\mathbf{n}}$ and $\mathbf{a}$ (see Fig. 21.5). However, notice that $a \cos \alpha$ is h, the perpendicular height of the point A above the plane of $\mathbf{b}$ and $\mathbf{c}$. Thus we can write

$$\mathbf{a}.(\mathbf{b} \wedge \mathbf{c}) = Sh$$

but the product of the area of the base and the perpendicular height of the parallelepiped is equal to its volume. So we have shown that

$$\mathbf{a}.(\mathbf{b} \wedge \mathbf{c}) = \text{the volume of the parallelepiped}$$

Notice that $\mathbf{a}.(\mathbf{b} \wedge \mathbf{c}) = \mathbf{b}.(\mathbf{c} \wedge \mathbf{a}) = \mathbf{c}.(\mathbf{a} \wedge \mathbf{b})$, because they all represent the volume of the same parallelepiped.

Qu. 2   Verify that the result in the preceding section is true for the cuboid formed by the vectors

$$\mathbf{a} = 4\mathbf{i}, \quad \mathbf{b} = 3\mathbf{j}, \quad \mathbf{c} = 2\mathbf{k}$$

Qu. 3   Describe the circumstances under which $\mathbf{a}.(\mathbf{b} \wedge \mathbf{c})$ will be (a) zero, (b) negative.

The distributive law

21.4   Before we can go far in our study of vector products, we shall need the distributive law. This is the law which enables us to remove brackets to obtain results such as

$$\mathbf{a} \wedge (\mathbf{b} + \mathbf{c}) = \mathbf{a} \wedge \mathbf{b} + \mathbf{a} \wedge \mathbf{c}$$

Working in the reverse direction enables us to factorise the R.H.S. of such an expression. The distributive law will also enable us to expand products of vectors expressed in terms of the base vectors $\mathbf{i}, \mathbf{j}, \mathbf{k}$.

We shall prove the distributive law by considering the vector

$$\mathbf{v} = \mathbf{a} \wedge (\mathbf{b} + \mathbf{c}) - \mathbf{a} \wedge \mathbf{b} - \mathbf{a} \wedge \mathbf{c}$$

and showing that it is zero.

Scalar multiplying both sides by $\mathbf{v}$ gives

$$\mathbf{v}.\mathbf{v} = \mathbf{v}.\mathbf{a} \wedge (\mathbf{b} + \mathbf{c}) - \mathbf{v}.\mathbf{a} \wedge \mathbf{b} - \mathbf{v}.\mathbf{a} \wedge \mathbf{c}$$

(we can do this because we know that scalar multiplication *does* obey the distributive law; see Book 1, §15.16.)

From the last section we know that we can permute the factors in a scalar triple product, so this expression can be written

$$\mathbf{v}.\mathbf{v} = (\mathbf{b} + \mathbf{c}).\mathbf{v} \wedge \mathbf{a} - \mathbf{b}.\mathbf{v} \wedge \mathbf{a} - \mathbf{c}.\mathbf{v} \wedge \mathbf{a}$$

Now we use the distributive law for scalar multiplication to factorise the R.H.S.

$$\begin{aligned} \mathbf{v}.\mathbf{v} &= [(\mathbf{b} + \mathbf{c}) - \mathbf{b} - \mathbf{c}].(\mathbf{v} \wedge \mathbf{a}) \\ &= [\mathbf{b} + \mathbf{c} - \mathbf{b} - \mathbf{c}].(\mathbf{v} \wedge \mathbf{a}) \\ &= 0 \end{aligned}$$

Hence

$$v^2 = 0$$
$$\therefore \mathbf{v} = \mathbf{0}$$

Thus we have shown that

$$\mathbf{a} \wedge (\mathbf{b} + \mathbf{c}) - \mathbf{a} \wedge \mathbf{b} - \mathbf{a} \wedge \mathbf{c} = 0$$

$$\therefore \mathbf{a} \wedge (\mathbf{b} + \mathbf{c}) = \mathbf{a} \wedge \mathbf{b} + \mathbf{a} \wedge \mathbf{c}$$

which completes the proof of the distributive law.

Exercise 21b

1 Find $\mathbf{p} \wedge \mathbf{q}$, when
(a) $\mathbf{p} = \mathbf{i} + \mathbf{j}, \quad \mathbf{q} = \mathbf{k},$ (b) $\mathbf{p} = \mathbf{i} + \mathbf{j}, \quad \mathbf{q} = \mathbf{i} - \mathbf{j},$
(c) $\mathbf{p} = -2\mathbf{i} + 3\mathbf{j} + \mathbf{k}, \quad \mathbf{q} = 5\mathbf{i},$ (d) $\mathbf{p} = \mathbf{i} + 2\mathbf{j} + 3\mathbf{k}, \quad \mathbf{q} = 4\mathbf{i} - \mathbf{j} + 7\mathbf{k}.$

2 Find $\mathbf{p}.(\mathbf{q} \wedge \mathbf{r})$, when
(a) $\mathbf{p} = \mathbf{i}, \quad \mathbf{q} = \mathbf{i} + \mathbf{j}, \quad \mathbf{r} = \mathbf{k},$
(b) $\mathbf{p} = \mathbf{i} + \mathbf{j}, \quad \mathbf{q} = \mathbf{i} + 2\mathbf{j} + 3\mathbf{k}, \quad \mathbf{r} = 5\mathbf{j} - \mathbf{k},$
(c) $\mathbf{p} = \mathbf{i} + \mathbf{j} + \mathbf{k}, \quad \mathbf{q} = 2\mathbf{i} + 3\mathbf{j}, \quad \mathbf{r} = 4\mathbf{i} + 5\mathbf{j} + 2\mathbf{k},$
(d) $\mathbf{p} = \mathbf{i} + \mathbf{j}, \quad \mathbf{q} = -\mathbf{i} + \mathbf{j}, \quad \mathbf{r} = -\mathbf{k}.$

3 Using the vectors in No. 2, find $(\mathbf{p} \wedge \mathbf{q}).\mathbf{r}$ in each part.

*4 Show that if $\mathbf{a} = a_1\mathbf{i} + a_2\mathbf{j} + a_3\mathbf{k}$ and $\mathbf{b} = b_1\mathbf{i} + b_2\mathbf{j} + b_3\mathbf{k}$, then

$$\mathbf{a} \wedge \mathbf{b} = \begin{vmatrix} \mathbf{i} & \mathbf{j} & \mathbf{k} \\ a_1 & a_2 & a_3 \\ b_1 & b_2 & b_3 \end{vmatrix}$$

Comment on the properties of the determinant and their relationship to the geometrical properties of $\mathbf{a}$ and $\mathbf{b}$, when $\mathbf{a} \wedge \mathbf{b} = \mathbf{0}$.

5 Use the determinant form of $\mathbf{a} \wedge \mathbf{b}$ in No. 4 to find $\mathbf{a} \wedge \mathbf{b}$, when
(a) $\mathbf{a} = 2\mathbf{i} + 3\mathbf{j} + \mathbf{k}, \mathbf{b} = 5\mathbf{i} + 4\mathbf{j} - \mathbf{k},$
(b) $\mathbf{a} = 7\mathbf{i} + 4\mathbf{j} - \mathbf{k}, \mathbf{b} = 2\mathbf{i} - \mathbf{j} + \mathbf{k}.$
(Many people find this a very convenient method for evaluating vector products.)

6 Using the vectors in No. 2, find $\mathbf{p} \wedge (\mathbf{q} \wedge \mathbf{r})$ and in each part show that $\mathbf{p} \wedge (\mathbf{q} \wedge \mathbf{r}) = (\mathbf{p}.\mathbf{r})\mathbf{q} - (\mathbf{p}.\mathbf{q})\mathbf{r}$.

7 Using notation similar to that in No. 4, show that

$$\mathbf{a}.(\mathbf{b} \wedge \mathbf{c}) = \begin{vmatrix} a_1 & a_2 & a_3 \\ b_1 & b_2 & b_3 \\ c_1 & c_2 & c_3 \end{vmatrix}$$

Comment on the case $\mathbf{a}.(\mathbf{b} \wedge \mathbf{c}) = 0$.
(This form of the triple product can be used to prove that

$$\mathbf{a}.(\mathbf{b} \wedge \mathbf{c}) = \mathbf{b}.(\mathbf{c} \wedge \mathbf{a}) = \mathbf{c}.(\mathbf{a} \wedge \mathbf{b}))$$

8 Find a unit vector which is perpendicular to the vectors

$$\mathbf{i} + \mathbf{j} + \mathbf{k} \quad \text{and} \quad \mathbf{i} + 2\mathbf{j} + 3\mathbf{k}$$

9 Prove that if $\mathbf{p} + \mathbf{q} + \mathbf{r} = \mathbf{0}$, then $\mathbf{p} \wedge \mathbf{q} = \mathbf{q} \wedge \mathbf{r} = \mathbf{r} \wedge \mathbf{p}$.

10 If $\mathbf{a}, \mathbf{b}, \mathbf{c}$ and $\mathbf{d}$ are four given vectors and $\lambda\mathbf{a} + \mu\mathbf{b} + \nu\mathbf{c} = \mathbf{d}$, where $\lambda, \mu, \nu \in \mathbb{R}$, prove that

$$\lambda = \frac{\mathbf{d}.(\mathbf{b} \wedge \mathbf{c})}{\mathbf{a}.(\mathbf{b} \wedge \mathbf{c})}$$

Using this, and corresponding expressions for μ and ν, solve the equations

$$\lambda + 4\mu + 2\nu = 0$$
$$2\lambda - \mu + \nu = 0$$
$$8\lambda + 5\mu + 6\nu = 6$$

The vector triple product

21.5   As $\mathbf{b} \wedge \mathbf{c}$ is a vector, the possibility of forming a triple product $\mathbf{a} \wedge (\mathbf{b} \wedge \mathbf{c})$ is a viable proposition. (The reader should now do Qu. 4, below.)

Qu. 4   Given that $\mathbf{a} = \mathbf{i}$, $\mathbf{b} = \mathbf{j}$ and $\mathbf{c} = 3\mathbf{i} + 4\mathbf{j}$, show that

$$\mathbf{a} \wedge (\mathbf{b} \wedge \mathbf{c}) = 3\mathbf{j}$$
$$(\mathbf{a} \wedge \mathbf{b}) \wedge \mathbf{c} = 3\mathbf{j} - 4\mathbf{i}$$

This brings us face to face with a rather disturbing fact: the triple products $\mathbf{a} \wedge (\mathbf{b} \wedge \mathbf{c})$ and $(\mathbf{a} \wedge \mathbf{b}) \wedge \mathbf{c}$ are not the same. (We say that vector multiplication is not associative. See Book 1, §25.8.) However this is not so surprising if we consider the direction of the vector $\mathbf{a} \wedge (\mathbf{b} \wedge \mathbf{c})$. We know that $\mathbf{b} \wedge \mathbf{c}$ is perpendicular to the plane containing $\mathbf{b}$ and $\mathbf{c}$, and that $\mathbf{a} \wedge (\mathbf{b} \wedge \mathbf{c})$ is perpendicular to the plane containing $\mathbf{a}$ and $\mathbf{b} \wedge \mathbf{c}$. Consequently $\mathbf{a} \wedge (\mathbf{b} \wedge \mathbf{c})$ is parallel to the plane of $\mathbf{b}$ and $\mathbf{c}$. By similar reasoning $(\mathbf{a} \wedge \mathbf{b}) \wedge \mathbf{c}$ is parallel to the plane of $\mathbf{a}$ and $\mathbf{b}$. Consequently we should expect $\mathbf{a} \wedge (\mathbf{b} \wedge \mathbf{c})$ and $(\mathbf{a} \wedge \mathbf{b}) \wedge \mathbf{c}$, in general, to be different.

From the last paragraph we can deduce that $\mathbf{a} \wedge (\mathbf{b} \wedge \mathbf{c})$ is a linear combination of $\mathbf{b}$ and $\mathbf{c}$. In other words we can write

$$\mathbf{a} \wedge (\mathbf{b} \wedge \mathbf{c}) = \lambda\mathbf{b} + \mu\mathbf{c}$$

where λ and μ are scalars.

Clearly the scalars λ and μ will depend on the vectors $\mathbf{a}, \mathbf{b}$ and $\mathbf{c}$. We shall now show that $\lambda = \mathbf{a}.\mathbf{c}$ and $\mu = -\mathbf{a}.\mathbf{b}$, i.e. that

$$\mathbf{a} \wedge (\mathbf{b} \wedge \mathbf{c}) = (\mathbf{a}.\mathbf{c})\mathbf{b} - (\mathbf{a}.\mathbf{b})\mathbf{c} \tag{1}$$

There is no loss of generality if we fix the axes to suit our own convenience (provided they form a right-handed set of mutually perpendicular lines). We shall choose the direction of $\mathbf{c}$ as one of the axes, and it will be especially convenient to make this the z-axis. Nor is there any loss of generality if we choose the scale to suit ourselves, so we shall choose a scale such that $\mathbf{c}$ is a unit vector. In other words we choose

$$\mathbf{c} = \mathbf{k} \tag{2}$$

We are still free to choose *one* of the other axes (the remaining axis must complete the right-handed set). We shall choose for the *y*-axis a line perpendicular to the axis which we have already fixed, and in the plane containing **b** and **c**. This enables us to write **b** as a linear combination of **j** and **k**. We shall write

$$\mathbf{b} = \beta\mathbf{j} + \gamma\mathbf{k} \tag{3}$$

We cannot have any choice over how we write **a** and so this will have to be expressed in terms of all three base vectors, i.e.

$$\mathbf{a} = a_1\mathbf{i} + a_2\mathbf{j} + a_3\mathbf{k} \tag{4}$$

Now we are ready to start! First we shall find $\mathbf{b} \wedge \mathbf{c}$:

$$\begin{aligned}\mathbf{b} \wedge \mathbf{c} &= (\beta\mathbf{j} + \gamma\mathbf{k}) \wedge \mathbf{k} \\ &= \beta\mathbf{i}\end{aligned}$$

Now consider $\mathbf{a} \wedge (\mathbf{b} \wedge \mathbf{c})$:

$$\begin{aligned}\mathbf{a} \wedge (\mathbf{b} \wedge \mathbf{c}) &= (a_1\mathbf{i} + a_2\mathbf{j} + a_3\mathbf{k}) \wedge \beta\mathbf{i} \\ &= -\beta a_2\mathbf{k} + \beta a_3\mathbf{j}\end{aligned}$$

This is a simple looking result, but we have to transform it into a linear combination of **b** and **c**. This can be done by writing

$$\begin{aligned}-\beta a_2\mathbf{k} + \beta a_3\mathbf{j} &= a_3(\beta\mathbf{j} + \gamma\mathbf{k}) + (-\beta a_2 - \gamma a_3)\mathbf{k} \\ &= a_3\mathbf{b} - (\beta a_2 + \gamma a_3)\mathbf{c}\end{aligned}$$

We have now arrived at the required form $\lambda\mathbf{b} + \mu\mathbf{c}$, but we still have to express λ and μ in terms of *scalar* products, as in (1).
From (2) and (4)

$$\mathbf{a.c} = (a_1\mathbf{i} + a_2\mathbf{j} + a_3\mathbf{k}).\mathbf{k} = a_3 = \lambda$$

From (3) and (4)

$$\mathbf{a.b} = (a_1\mathbf{i} + a_2\mathbf{j} + a_3\mathbf{k}).(\beta\mathbf{j} + \gamma\mathbf{k}) = \beta a_2 + \gamma a_3 = -\mu$$
$$\therefore \mathbf{a} \wedge (\mathbf{b} \wedge \mathbf{c}) = (\mathbf{a.c})\mathbf{b} - (\mathbf{a.b})\mathbf{c}$$

Qu. 5 Given that

$$\mathbf{a} = \mathbf{i} + \mathbf{j} + \mathbf{k}$$
$$\mathbf{b} = \mathbf{i} - \mathbf{j} - \mathbf{k}$$
$$\mathbf{c} = \mathbf{i} + 2\mathbf{j} + 3\mathbf{k}$$

find $\mathbf{a} \wedge (\mathbf{b} \wedge \mathbf{c})$ and verify that it is equal to $(\mathbf{a.c})\mathbf{b} - (\mathbf{a.b})\mathbf{c}$.

There is no need for a separate proof of the formula for the triple product $(\mathbf{a} \wedge \mathbf{b}) \wedge \mathbf{c}$; it can be deduced from the one above, as follows:

$$(\mathbf{a} \wedge \mathbf{b}) \wedge \mathbf{c} = -\mathbf{c} \wedge (\mathbf{a} \wedge \mathbf{b})$$

and now we only have to permute the letters in the formula we have already proved, i.e.

$$(\mathbf{a} \wedge \mathbf{b}) \wedge \mathbf{c} = -[(\mathbf{c}.\mathbf{b})\mathbf{a} - (\mathbf{c}.\mathbf{a})\mathbf{b}]$$
$$= (\mathbf{c}.\mathbf{a})\mathbf{b} - (\mathbf{c}.\mathbf{b})\mathbf{a}$$

Qu. 6   Prove that $(\mathbf{a} \wedge \mathbf{b}) \wedge (\mathbf{c} \wedge \mathbf{d}) = \mathbf{b}(\mathbf{a}.\mathbf{c} \wedge \mathbf{d}) - \mathbf{a}(\mathbf{b}.\mathbf{c} \wedge \mathbf{d})$.

The perpendicular distance of a point from a line

21.6   In Fig. 21.6 suppose that we are given the coordinates of the point P and the equation of the line *l*.

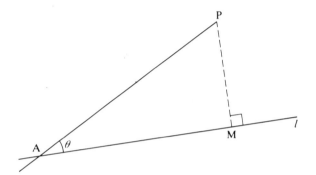

Figure 21.6

In order to calculate the distance of P from *l*, we shall need to calculate the length of PM, where M is the foot of the perpendicular from P to *l*.
Now

$$PM = AP \sin \theta$$
$$= |\overrightarrow{AP} \wedge \hat{\mathbf{u}}|$$

where $\hat{\mathbf{u}}$ is a unit vector along the line *l*. So the perpendicular distance of P from *l* is given by

$$|(\mathbf{p} - \mathbf{a}) \wedge \hat{\mathbf{u}}| \tag{1}$$

When we are given the equation of a line, it is always possible to find a point A on it, and a unit vector parallel to it; the next example illustrates how to use formula (1).

Example 5   *Find the perpendicular distance of the point* P(0, 14, 10) *from the line whose equation is* $\mathbf{r} = (\mathbf{i} + 2\mathbf{j} + 3\mathbf{k}) + \lambda(3\mathbf{i} + 4\mathbf{k})$.

By putting $\lambda = 0$, we can see that the point (1, 2, 3) lies on the line, so this will be the point A in (1) above. Also we know that the line is parallel to (3**i** + 4**k**), but this is not a *unit* vector, so we must divide by its magnitude, i.e. 5, thus taking $\hat{\mathbf{u}}$ to be $\frac{1}{5}(3\mathbf{i} + 4\mathbf{k})$. Using (1), the perpendicular distance required is the magnitude of

$(\mathbf{p} - \mathbf{a}) \wedge \mathbf{u}$. In this example,

$$(\mathbf{p} - \mathbf{a}) = (14\mathbf{j} + 10\mathbf{k}) - (\mathbf{i} + 2\mathbf{j} + 3\mathbf{k}) = -\mathbf{i} + 12\mathbf{j} + 7\mathbf{k}$$

$$\therefore (\mathbf{p} - \mathbf{a}) \wedge \mathbf{u} = \tfrac{1}{5}(-\mathbf{i} + 12\mathbf{j} + 7\mathbf{k}) \wedge (3\mathbf{i} + 4\mathbf{k})$$

$$= \tfrac{1}{5} \begin{vmatrix} \mathbf{i} & \mathbf{j} & \mathbf{k} \\ -1 & 12 & 7 \\ 3 & 0 & 4 \end{vmatrix} \qquad \text{(See Exercise 21(b), No. 4)}$$

$$= \tfrac{1}{5}(48\mathbf{i} + 25\mathbf{j} - 36\mathbf{k})$$

The magnitude of this (and hence the required perpendicular distance of P from the line) is equal to

$$\tfrac{1}{5} \times \sqrt{(48^2 + 25^2 + 36^2)}$$
$$= \tfrac{1}{5} \times 65$$
$$= 13$$

(For another method, see Exercise 21c, Nos. 12 and 13.)

Qu. 7 Find the perpendicular distance of the point P(2, 3, 4) from the line

$$\mathbf{r} = (\mathbf{i} + 15\mathbf{j} + 11\mathbf{k}) + \lambda(4\mathbf{i} - 12\mathbf{j} - 3\mathbf{k})$$

Exercise 21c (Miscellaneous)

1 Given that $\mathbf{a} = 2\mathbf{i} - 3\mathbf{j} + \mathbf{k}$, $\mathbf{b} = \mathbf{i} - 2\mathbf{j} + \mathbf{k}$, $\mathbf{c} = 3\mathbf{i} + 2\mathbf{j} - \mathbf{k}$, find:
   (a) $\mathbf{a}.(\mathbf{b} \wedge \mathbf{c})$,   (b) $\mathbf{a} \wedge (\mathbf{b} \wedge \mathbf{c})$,   (c) $(\mathbf{a} \wedge \mathbf{b}) \wedge \mathbf{c}$.
2 Find a vector which is perpendicular to both $\mathbf{p}$ and $\mathbf{q}$, where

$$\mathbf{p} = \mathbf{i} + 3\mathbf{j} + 5\mathbf{k} \quad \text{and} \quad \mathbf{q} = 4\mathbf{i} - \mathbf{j} + 2\mathbf{k}$$

Hence write down the equation of the plane through the origin, parallel to $\mathbf{p}$ and $\mathbf{q}$.
3 Prove that
   (a) $\mathbf{p} \wedge (\mathbf{q} \wedge \mathbf{r}) + \mathbf{q} \wedge (\mathbf{r} \wedge \mathbf{p}) + \mathbf{r} \wedge (\mathbf{p} \wedge \mathbf{q}) = \mathbf{0}$.
   (b) $(\mathbf{p} \wedge \mathbf{q}).(\mathbf{r} \wedge \mathbf{s}) = \begin{vmatrix} \mathbf{p}.\mathbf{r} & \mathbf{p}.\mathbf{s} \\ \mathbf{q}.\mathbf{r} & \mathbf{q}.\mathbf{s} \end{vmatrix}$.
4 Prove that if the points A, B, C, whose position vectors are $\mathbf{a}$, $\mathbf{b}$, $\mathbf{c}$, respectively, are coplanar, then $\mathbf{a}.(\mathbf{b} \wedge \mathbf{c}) = 0$. Hence or otherwise, find the equation of the plane through the points (6, 3, 0), (2, 2, 2) and (3, 3, 1). [Hint: if P(x, y, z) is a general point in the plane, then the vectors $\overrightarrow{AP}$, $\overrightarrow{AB}$ and $\overrightarrow{AC}$ are coplanar.]
5 Let $\mathbf{a}$ be the unit vector along a line l, and let $\mathbf{m} = \mathbf{r} \wedge \mathbf{a}$, where $\mathbf{r}$ is the position vector of any point P on l with respect to a fixed origin O. Show that $\mathbf{m}$ is independent of the choice of P.
   If another line l' has corresponding vectors $\mathbf{a}'$ and $\mathbf{m}'$, prove that, if l and l' intersect, then $\mathbf{a}.\mathbf{m}' + \mathbf{a}'.\mathbf{m} = 0$.   (O)
6 The non-collinear points A, B and C have position vectors $\mathbf{a}$, $\mathbf{b}$ and $\mathbf{c}$ respectively with respect to an origin O, and $\mathbf{b} \wedge \mathbf{c} = \mathbf{p}$, $\mathbf{c} \wedge \mathbf{a} = \mathbf{q}$, $\mathbf{a} \wedge \mathbf{b} = \mathbf{r}$.

Show, from the definitions of the scalar and vector products, that

(a) $\mathbf{a.q} = \mathbf{a.r} = 0$,

(b) the normal to the plane ABC is parallel to $\mathbf{p} + \mathbf{q} + \mathbf{r}$.

Given that A, B and C lie on a sphere of unit radius with centre O, and that the normal through O to the plane ABC meets the sphere at a point D, whose position vector with respect to O is $\mathbf{d}$, show that

$$(\mathbf{a.p})\mathbf{d} = (\mathbf{p} + \mathbf{q} + \mathbf{r})\cos \theta$$

where θ is the angle AOD.   (JMB)

7 The points A and B have coordinates (2, 1, 1) and (0, 5, 3) respectively. Find the equation of the line AB in terms of a parameter. If C is the point (5, −4, 2) find the coordinates of the point D on AB such that CD is perpendicular to AB.

Find the equation of the plane containing AB and perpendicular to the line CD.   (O & C: MEI)

8 The parametric equations of two planes are

$$\mathbf{r} = \begin{pmatrix} 1 \\ 0 \\ 1 \end{pmatrix} + t \begin{pmatrix} 1 \\ -2 \\ 1 \end{pmatrix} + s \begin{pmatrix} 0 \\ 3 \\ 2 \end{pmatrix}$$

and

$$\mathbf{r} = \begin{pmatrix} 2 \\ 1 \\ -1 \end{pmatrix} + u \begin{pmatrix} 0 \\ 0 \\ 5 \end{pmatrix} + v \begin{pmatrix} -2 \\ 4 \\ 3 \end{pmatrix}$$

(a) Find the cosine of the acute angle between the planes.

(b) The line of intersection is l. Find, in the form $\mathbf{r} = \mathbf{a} + \lambda\mathbf{b}$, the equation of l.

(c) Show that the length of the perpendicular from the point (1, 5, 1) to the line l is $\sqrt{2}$.   (C)

9 Show that if it is possible to find a vector $\mathbf{r}$ such that $\mathbf{r} \wedge \mathbf{a} = \mathbf{b}$, where $\mathbf{a}$ and $\mathbf{b}$ are given vectors, then $\mathbf{a.b} = 0$.

Find the set of vectors $\mathbf{r}$ which satisfy $\mathbf{r} \wedge \mathbf{a} = \mathbf{b}$ in the following cases:

(a) $\mathbf{a} = \begin{pmatrix} 2 \\ 1 \\ -3 \end{pmatrix}, \mathbf{b} = \begin{pmatrix} 3 \\ -3 \\ 1 \end{pmatrix}$;   (b) $\mathbf{a} = \begin{pmatrix} 4 \\ -1 \\ 5 \end{pmatrix}, \mathbf{b} = \begin{pmatrix} 8 \\ 0 \\ 7 \end{pmatrix}$.   (C)

10 (a) Find the image of the origin by reflection in the plane

$$x + 2y + 3z = 14$$

(b) Find the coordinates of the foot of the perpendicular from the origin to the line

$$\frac{x-1}{1} = \frac{y-1}{2} = \frac{z-1}{3}$$

(c) Find the equations of the spheres which touch the plane containing the y- and z-axes at O and also touch the plane $x + 2y + 2z = 1$.   (O & C)

*11 Prove that the line through the point (x_1, y_1, z_1) perpendicular to the plane $ax + by + cz = d$ meets the plane at a point whose coordinates are $(x_1 + ta, \ y_1 + tb, \ z_1 + tc)$, where $t = \dfrac{d - (ax_1 + by_1 + cz_1)}{a^2 + b^2 + c^2}$. Hence show that the perpendicular distance from the point to the plane is

$$\left| \frac{d - (ax_1 + by_1 + cz_1)}{\sqrt{(a^2 + b^2 + c^2)}} \right|$$

12 Given that $\mathbf{a} = \begin{pmatrix} 4 \\ 2 \\ -3 \end{pmatrix}$, $\mathbf{n} = \begin{pmatrix} 3 \\ 0 \\ 4 \end{pmatrix}$ and $\mathbf{p} = \begin{pmatrix} 1 \\ 3 \\ 3 \end{pmatrix}$ and that R is a point on the line $\mathbf{r} = \mathbf{a} + t\mathbf{n}$, express PR^2 in terms of t. Show that, as t varies, the least value of PR^2 is 37 and verify that, in this case, PR is perpendicular to the line.

13 Given that $\mathbf{a}$ is a constant vector which is perpendicular to a unit vector $\hat{\mathbf{u}}$, and that R is any point on the line $\mathbf{r} = \mathbf{a} + t\hat{\mathbf{u}}$, show that the distance to R from a fixed point P, whose position vector is $\mathbf{p}$, is given by

$$PR^2 = (\mathbf{a} - \mathbf{p}).(\mathbf{a} - \mathbf{p}) - 2t(\hat{\mathbf{u}}.\mathbf{p}) + t^2$$

Hence show that the least value of PR^2, as t varies, is $(\mathbf{a} - \mathbf{p}).(\mathbf{a} - \mathbf{p}) - (\hat{\mathbf{u}}.\mathbf{p})^2$. Prove that $\overrightarrow{PR}$ is then perpendicular to the given line.

14 Use the formula of §21.6 to show that the least value of PR in No. 12 is $\sqrt{37}$.

15 Use the formula in §21.6 to show that the perpendicular distance of the point P from the line in No. 13 is

$$\sqrt{\{(\mathbf{a} - \mathbf{p}).(\mathbf{a} - \mathbf{p}) - (\hat{\mathbf{u}}.\mathbf{p})^2\}}$$

Answers

Although either $\sin^{-1} x$, etc. or $\arcsin x$, etc. can be used for the inverse trigonometrical functions, for convenience, the former notation is used throughout the answers.

Chapter 1

Qu. 1 (a) $16x(2x^2 + 3)^3$, (b) $\dfrac{x-1}{\sqrt{(x^2 - 2x + 1)}}$, (c) $-4(2x-1)^{-3}$,

(d) $4\cos(4x - 7)$, (e) $3\tan^2 x \sec^2 x$, (f) $-6\cos 3x \sin 3x$.

Qu. 2 (a) $\frac{1}{6}(x^2 + 1)^3 + c$, (b) $\frac{1}{10}(2x + 1)^5 + c$, (c) $\frac{1}{7}x^7 + \frac{3}{5}x^5 + x^3 + x + c$,

(d) $-\frac{1}{6}\cos 3x + c$, (e) $\frac{2}{9}(x^3 + 1)^{3/2} + c$, (f) $\frac{1}{2}\tan^2 x + c$,

Qu. 3 (a) $\frac{1}{3}\cos^3 x - \cos x + c$, (b) $\sin x - \frac{2}{3}\sin^3 x + \frac{1}{5}\sin^5 x + c$.

Qu. 4 (a) $\frac{1}{3}\sin^3 x - \frac{1}{5}\sin^5 x + c$, (b) $\frac{1}{5}\cos^5 x - \frac{1}{3}\cos^3 x + c$.

Qu. 5 $\frac{1}{3}\sec^3 x - \sec x + c$.

Qu. 6 $\frac{1}{12}\sin^3 4x + c$.

Qu. 7 $-\cos x + \frac{2}{3}\cos^3 x - \frac{1}{5}\cos^5 x + c$.

Qu. 8 (a) $\frac{1}{15}(2x + 1)^{3/2}(3x - 1) + c$, (b) $\frac{1}{15}(2x + 1)^{3/2}(3x - 1) + c$,

(c) $\frac{1}{504}(3x - 2)^7(21x + 2) + c$.

Qu. 9 (a) $45°$, $\pi/4$ rad, (b) $30°$, $\pi/6$ rad, (c) $45°$, $\pi/4$ rad, (d) $60°$, $\pi/3$ rad,

(e) $30°$, $\pi/6$ rad, (f) $0°$, 0 rad, (g) $60°$, $\pi/3$ rad, (h) $30°$, $\pi/6$ rad,

(i) $30°$, $\pi/6$ rad, (j) $60°$, $\pi/3$ rad, (k) $90°$, $\pi/2$ rad.

Qu. 10 (a) $\pi/9$, (b) $7\pi/18$, (c) $5\pi/6$, (d) $5\pi/3$, (e) $9\pi/4$.

Qu. 11 (a) $57.3°$, (b) $1.7°$, (c) $71.6°$, (d) $41.0°$, (e) $36°$.

Qu. 12 (a) 1.29, (b) 0.927, (c) 0.784.

Qu. 13 (a) $3\cos u$, $\sin^{-1}\dfrac{x}{3}$, (b) $\cos u$, $\sin^{-1} 5x$, (c) $2\cos u$, $\sin^{-1}\dfrac{3x}{2}$,

(d) $\sqrt{7}\cos u$, $\sin^{-1}\dfrac{x}{\sqrt{7}}$, (e) $\cos u$, $\sin^{-1}\sqrt{3}x$, (f) $\sqrt{3}\cos u$, $\sin^{-1}\sqrt{\frac{2}{3}}x$.

Qu. 14 (a) $\sin^{-1}\dfrac{x}{2} + c$, (b) $\dfrac{1}{\sqrt{3}}\sin^{-1}\sqrt{3}x + c$, (c) $\dfrac{1}{3}\sin^{-1}\dfrac{3x}{4} + c$.

Qu. 16 $\tan^{-1} x + c.$

Qu. 17 (a) $9 \sec^2 u$, $\tan^{-1} \dfrac{x}{3}$, (b) $\sec^2 u$, $\tan^{-1} 2x$, (c) $25 \sec^2 u$, $\tan^{-1} \dfrac{3x}{5}$,

(d) $3 \sec^2 u$, $\tan^{-1} \dfrac{x}{\sqrt{3}}$, (e) $\sec^2 u$, $\tan^{-1} \sqrt{5}x$, (f) $7 \sec^2 u$, $\tan^{-1} \sqrt{\tfrac{3}{7}}x$.

Qu. 18 (a) $\tfrac{1}{2} \tan^{-1} \dfrac{x}{2} + c$, (b) $\tfrac{1}{4} \tan^{-1} 4x + c$, (c) $\dfrac{1}{2\sqrt{3}} \tan^{-1} \dfrac{2x}{\sqrt{3}} + c.$

Qu. 19 (a) 0.866, (b) 0.479, (c) 2.57, (d) -0.990, (e) 1.04.

Qu. 20 (a) $\tfrac{1}{2}\pi$, (b) $-\tfrac{1}{4}\pi$, (c) $\tfrac{1}{6}\pi$, (d) 1.11.

Exercise 1a, page 3

1 (a) $30x(5x^2 - 1)^2$, (b) $\dfrac{2 - 8x}{(2x^2 - x + 3)^3}$, (c) $\tfrac{2}{3}x(x^2 + 4)^{-2/3}$, (d) $-5 \csc^2 5x$,

(e) $-5 \sin(5x - 1)$, (f) $\tfrac{1}{3} \sin \tfrac{2}{3}x$, (g) $\dfrac{1}{2\sqrt{x}} \sec^2 \sqrt{x}$, (h) $4 \sec^2 2x \tan 2x$,

(i) $-\tfrac{1}{2} \cot x \sqrt{\csc x}.$

2 (a) $\tfrac{1}{12}(x^2 - 3)^6 + c$, (b) $\tfrac{1}{18}(3x - 1)^6 + c$, (c) $\tfrac{1}{4}x^4 + \tfrac{4}{3}x^3 + 2x^2 + c$,

(d) $-\tfrac{1}{2}(x^2 + 1)^{-1} + c$, (e) $-\tfrac{1}{4}(x^2 + 2x - 5)^{-2} + c$, (f) $\tfrac{1}{3}(x^2 - 3x + 7)^3 + c$,

(g) $-\tfrac{1}{4}(4x^2 - 7)^{-1} + c$, (h) $\tfrac{2}{9}(3x^2 - 5)^{3/2} + c$, (i) $\tfrac{1}{7}x^7 + \tfrac{1}{2}x^4 + x + c$,

(j) $\tfrac{2}{3}\sqrt{(x^3 - 3x)} + c$, (k) $-\tfrac{1}{2}(2x^2 - 4x + 1)^{-1/2} + c$, (l) $\tfrac{8}{7}x^7 - \tfrac{12}{5}x^5 + 2x^3 - x + c.$

3 (a) $\sin 3x + c$, (b) $-\tfrac{1}{2}\cos(2x + 3) + c$, (c) $\tfrac{1}{2}\sin^2 x + c$, (d) $\tfrac{1}{6}\sin 2x + c$,

(e) $-\tfrac{1}{9}\cos^3 3x + c$, (f) $\tfrac{1}{3}\tan^3 x + c$, (g) $\tfrac{1}{5}\sec^5 x + c$, (h) $\tfrac{2}{3}\sin^{3/2} x + c$,

(i) $-\tfrac{1}{2}\cot x^2 + c$, (j) $2 \sin \sqrt{x} + c$, (k) $-\tfrac{1}{3}\csc^3 x + c.$

4 (a) $\sin x - \tfrac{1}{3}\sin^3 x + c$, (b) $2 \sin \tfrac{1}{2}x - \tfrac{4}{3}\sin^3 \tfrac{1}{2}x + \tfrac{2}{5}\sin^5 \tfrac{1}{2}x + c$,

(c) $\tfrac{1}{6}\cos^3 2x - \tfrac{1}{2}\cos 2x + c$, (d) $\tfrac{1}{2}\sin(2x + 1) - \tfrac{1}{6}\sin^3(2x + 1) + c$,

(e) $-\tfrac{1}{3}\cos^3 x + \tfrac{2}{5}\cos^5 x - \tfrac{1}{7}\cos^7 x + c$, (f) $\tfrac{1}{4}\sin^4 x - \tfrac{1}{6}\sin^6 x + c$,

(g) $\tan x + \tfrac{1}{3}\tan^3 x + c$, (h) $\csc x - \tfrac{1}{3}\csc^3 x + c$,

(i) $\sec x - \tfrac{2}{3}\sec^3 x + \tfrac{1}{5}\sec^5 x + c.$

5 (a) $\tfrac{1}{4}\sec^4 x + c$, (b) $\tfrac{1}{2}\tan^2 x + \tfrac{1}{4}\tan^4 x + k.$

6 $A + \tfrac{1}{2}\sin^2 x$, $B - \tfrac{1}{2}\cos^2 x$, $C - \tfrac{1}{4}\cos 2x.$

7 (a) $\tfrac{1}{2}(1 - \cos x)$, (b) $\tfrac{1}{2}(1 + \cos 6x).$

8 (a) $\tfrac{1}{2}x + \tfrac{1}{4}\sin 2x + c$, (b) $\tfrac{1}{2}x - \tfrac{1}{2}\sin x + c$, (c) $\tfrac{1}{2}x + \tfrac{1}{12}\sin 6x + c.$

9 $\tfrac{1}{4} - \tfrac{1}{2}\cos 2x + \tfrac{1}{4}\cos^2 2x$, $\tfrac{1}{2}(1 + \cos 4x).$

10 $\tfrac{3}{8}x + \tfrac{1}{4}\sin 2x + \tfrac{1}{32}\sin 4x + c.$

11 (a) $\tfrac{1}{2}x - \tfrac{1}{4}\sin 2x + c$, (b) $\tfrac{1}{2}x + \tfrac{3}{4}\sin \tfrac{2}{3}x + c$, (c) $\tfrac{3}{8}x - \tfrac{1}{8}\sin 4x + \tfrac{1}{64}\sin 8x + c$,

(d) $\tfrac{3}{8}x + \tfrac{1}{2}\sin x + \tfrac{1}{16}\sin 2x + c.$

13 (a) $2\sqrt{2} \sin \dfrac{x}{2} + c$, (b) $-\dfrac{1}{\sqrt{2}}\csc x + c$, (c) $\tfrac{1}{2}\sin^4 x + c$, (d) $-\dfrac{8}{3}\cos^3 \dfrac{x}{2} + c.$

14 (a) $2 \sin 2x \cos x$, (b) $\sin 5x + \sin x$, (c) $-\tfrac{1}{10}\cos 5x - \tfrac{1}{2}\cos x + c.$

15 (a) $\tfrac{1}{4}\cos 2x - \tfrac{1}{8}\cos 4x + c$, (b) $\tfrac{1}{2}\sin 2x + \sin x + c$,

(c) $\tfrac{1}{6}\sin 3x - \tfrac{1}{10}\sin 5x + c.$

Exercise 1b, page 6

1 (a) $\tfrac{1}{20}(4x - 1)^{3/2}(6x + 1) + c$, (b) $\tfrac{2}{375}(5x + 2)^{3/2}(15x - 4) + c$,

(c) $\tfrac{1}{224}(2x - 1)^7(14x + 1) + c$, (d) $\tfrac{2}{3}(x + 4)\sqrt{(x - 2)} + c$,

Page 7

(e) $\frac{1}{30}(x-1)^5(5x+13)+c$, (f) $\frac{1}{168}(x-2)^6(21x^2+156x+304)+c$,

(g) $\dfrac{x^2-4x+8}{x-2}+c$, (h) $\frac{1}{3}(x-6)\sqrt{(2x+3)}+c$.

3 (a) $\frac{2}{135}(3x-4)^{3/2}(9x+8)+c$, $\frac{1}{9}(3x^2-4)^{3/2}+c$,
   (b) $\frac{1}{14}(x^2+5)^7+c$, $\frac{1}{56}(x+5)^7(7x-5)+c$,
   (c) $\frac{2}{3}(x+2)\sqrt{(x-1)}+c$, $\sqrt{(x^2-1)}+c$.

4 (a) $\frac{1}{6}(2x^2+1)^{3/2}+c$, (b) $-\frac{1}{2}(x^3-x+4)^{-2}+c$, (c) $\frac{2}{15}(2x-1)^{3/2}(3x+1)+c$,
   (d) $\frac{1}{2}\sin 2x-\frac{1}{6}\sin^3 2x+c$, (e) $-\frac{2}{3}(\cos x)^{3/2}+c$, (f) $-\frac{1}{3}\cot^3 x+c$,
   (g) $\frac{1}{16}(4x^2-1)^4+c$, (h) $\frac{1}{2}\sqrt{(2x^2-5)}+c$, (i) $-2(8+x)\sqrt{(4-x)}+c$,
   (j) $-2\cos\sqrt{x}+c$.

Exercise 1c, page 8

1 (a) $\frac{26}{15}$, (b) $\frac{1}{30}$, (c) $\sqrt{3}-\frac{2}{3}$, (d) $-\frac{7}{20}$, (e) $\frac{67}{48}$.
2 (a) $\frac{7}{36}$, (b) $\frac{8}{15}$, (c) $\frac{1}{6}(4-\sqrt{2})$.
3 (a) $1-\frac{1}{2}\sqrt{3}$, (b) $\frac{256}{15}$, (c) $-\frac{1}{10}$, (d) $\frac{4}{3}$, (e) $\frac{23}{108900}$, (f) 24.3, (g) $\frac{4}{3}$, (h) $\frac{74}{27}$, (i) $\frac{2}{3}$.
4 $2\sqrt{2}-\sqrt{3}$.
5 $\frac{9}{8}$.
6 $\frac{1}{4}\pi^2$.
7 $\frac{1}{8}\pi$.

Exercise 1d, page 12

1 (a) $45°$, $\pi/4$ rad, (b) $45°$, $\pi/4$ rad, (c) $17.3°$, $\pi\sqrt{3}/18$ rad, (d) $60°$, $\pi/3$ rad,
   (e) $52.0°$, $\pi\sqrt{3}/6$ rad, (f) $15°$, $\pi/12$ rad, (g) $67\frac{1}{2}°$, $3\pi/8$ rad, (h) $15°$, $\pi/12$ rad.
2 (a) 0.559, (b) 1.05, (c) 0.0992, (d) 4.11.
3 (a) $114.6°$, (b) $4.6°$, (c) $78.0°$, (d) $30°$.
4 (a) 0.927, (b) 0.588, (c) 1.12.

5 (a) $4\cos u$, $\sin^{-1}\dfrac{x}{4}$, (b) $\cos u$, $\sin^{-1}3x$, (c) $3\cos u$, $\sin^{-1}\frac{2}{3}x$,

   (d) $\sqrt{10}\cos u$, $\sin^{-1}\dfrac{x}{\sqrt{10}}$, (e) $\cos u$, $\sin^{-1}\sqrt{6x}$, (f) $\sqrt{5}\cos u$, $\sin^{-1}\sqrt{\frac{3}{5}}x$.

6 (a) $\sin^{-1}\dfrac{x}{5}+c$, (b) $\frac{1}{2}\sin^{-1}2x+c$, (c) $\dfrac{1}{3}\sin^{-1}\dfrac{3x}{2}+c$, (d) $\sin^{-1}\dfrac{x}{\sqrt{3}}+c$,

   (e) $\dfrac{1}{\sqrt{7}}\sin^{-1}\sqrt{7x}+c$, (f) $\dfrac{1}{\sqrt{3}}\sin^{-1}\sqrt{\frac{3}{2}}x+c$.

7 (a) $16\sec^2 u$, $\tan^{-1}\dfrac{x}{4}$, (b) $\sec^2 u$, $\tan^{-1}3x$, (c) $4\sec^2 u$, $\tan^{-1}\dfrac{\sqrt{3}}{2}x$,

   (d) $2\sec^2 u$, $\tan^{-1}\dfrac{x}{\sqrt{2}}$, (e) $\sec^2 u$, $\tan^{-1}\sqrt{3x}$, (f) $5\sec^2 u$, $\tan^{-1}\sqrt{\frac{2}{5}}x$.

8 (a) $\dfrac{1}{5}\tan^{-1}\dfrac{x}{5}+c$, (b) $\frac{1}{6}\tan^{-1}6x+c$, (c) $\dfrac{1}{4\sqrt{3}}\tan^{-1}\dfrac{\sqrt{3}}{4}x+c$,

   (d) $\dfrac{1}{\sqrt{5}}\tan^{-1}\dfrac{x}{\sqrt{5}}+c$, (e) $\dfrac{1}{\sqrt{6}}\tan^{-1}\sqrt{6x}+c$, (f) $\dfrac{1}{\sqrt{30}}\tan^{-1}\sqrt{\frac{10}{3}}x+c$.

Page 13

9 (a) $\dfrac{1}{3\sqrt{2}} \tan^{-1} \dfrac{\sqrt{2}}{3} x + c$, (b) $\dfrac{3}{\sqrt{5}} \sin^{-1} \dfrac{\sqrt{5}}{2} x + c$, (c) $\dfrac{1}{\sqrt{2}} \sin^{-1} \sqrt{\dfrac{2}{3}} x + c$,

(d) $\dfrac{2}{\sqrt{15}} \tan^{-1} \sqrt{\dfrac{5}{3}} x + c$.

10 (a) $\frac{1}{6}\pi$, (b) $\frac{1}{4}\pi$, (c) π, (d) $\frac{1}{12}\pi$, (e) $\frac{1}{18}\pi$, (f) $\frac{1}{6}\pi$.

11 (a) (i) $\sin^{-1} \dfrac{x}{3} + c$, (ii) $-\cos^{-1} \dfrac{x}{3} + c = -\cos^{-1} \dfrac{x}{3} + \dfrac{\pi}{2} + k = \sin^{-1} \dfrac{x}{3} + k$.

(b) (i) $\frac{1}{3}\pi$, (ii) $\frac{1}{3}\pi$.

12 (a) $\sin^{-1} \dfrac{x+1}{2} + c$, (b) $\frac{1}{3} \tan^{-1} \dfrac{x-3}{3} + c$.

13 (a) $\dfrac{1}{5} \tan^{-1} \dfrac{x+3}{5} + c$, (b) $\sin^{-1} \dfrac{x-1}{2} + c$, (c) $\dfrac{1}{\sqrt{15}} \tan^{-1} \dfrac{(x-2)\sqrt{3}}{\sqrt{5}} + c$,

(d) $\dfrac{1}{\sqrt{3}} \sin^{-1} \dfrac{x+1}{\sqrt{3}} + c$.

14 (a) (i) $(x-3)^2 + 7$, (ii) $3(x-2)^2 + 2$, (iii) $2(x-1)^2 + 3$.

(b) (i) $\dfrac{1}{2} \tan^{-1} \dfrac{x-1}{2} + c$, (ii) $\dfrac{1}{3\sqrt{2}} \tan^{-1} \dfrac{(x+1)\sqrt{2}}{3} + c$,

(iii) $\dfrac{1}{3} \tan^{-1} \dfrac{x-2}{3} + c$, (iv) $\dfrac{1}{2\sqrt{3}} \tan^{-1} \dfrac{2(x-1)}{\sqrt{3}} + c$.

15 (a) (i) $4 - (x+1)^2$, (ii) $9 - (2-x)^2$, (iii) $7\frac{1}{2} - 2(x-\frac{1}{2})^2$.

(b) (i) $\sin^{-1} \dfrac{x+1}{2} + c$, (ii) $\frac{1}{2} \sin^{-1} \dfrac{2}{\sqrt{5}}(x-1) + c$, (iii) $\sin^{-1} \dfrac{x-2}{4} + c$,

(iv) $\dfrac{1}{\sqrt{2}} \sin^{-1} \dfrac{(x-3)\sqrt{2}}{3} + c$.

16 (a) $\frac{1}{4}\pi$, (b) $\frac{1}{2}\pi$.

17 (a) $3 \sin^{-1} x + \sqrt{(1-x^2)} + c$, (b) $3 \sin^{-1} \dfrac{x}{2} - 2\sqrt{(4-x^2)} + c$.

18 (a) $\frac{1}{2} \sin^{-1} x - \frac{1}{2}x\sqrt{(1-x^2)} + c$, (b) $\dfrac{1}{54} \tan^{-1} \dfrac{x}{3} + \dfrac{x}{18(9+x^2)} + c$,

(c) $\dfrac{1}{2} \sin^{-1} \dfrac{x^2}{2} + c$.

19 (a) $\dfrac{x}{\sqrt{(1-9x^2)}} + c$, (b) $-\dfrac{1}{x}\sqrt{(1-x^2)} + c$, (c) $\sec^{-1} x + c$.

Exercise 1e, page 16

1 (a) 0.966, (b) 0.997, (c) 0.0808, (d) 1.05

3 (a) 1.25, (b) $(\sqrt{3}/3)\pi$, (c) $\pi/6$, (d) 1.96, (e) 1.25.

4 (a) $x = \dfrac{1+y}{1-y}$, (b) $x = \frac{1}{2}\sqrt{3}y + \frac{1}{2}\sqrt{(1-y^2)}$.

5 (a) $\pi/6$, (b) 0.325.

6 (a) $\pi/4$, (b) 0.322.

7 (a) 10, 0, -10, 0, 10 m, (b) After $\pi/3$ s, (c) After $\frac{2}{3}\pi$ s.

Page 17
8 (a) 0, 5, 0, -5, 0 m, (b) $2 \sin^{-1} \frac{3}{5} \approx 1.29$ s.
9 (a) 0.464, (b) 0.927, (c) 0.0183, (d) 0.0947, (e) 0.168, (f) 0.300.

Exercise 1f, page 17

1 $\frac{2}{9}(x^3 - 1)^{3/2} + c.$ **2** $-\frac{1}{4}(x^2 - 1)^{-2} + c.$ **3** $-\frac{1}{6}\cos^3 2x + c.$

4 $2\sqrt{\tan x} + c.$ **5** $\frac{1}{4}\sin 4x - \frac{1}{12}\sin^3 4x + c.$ **6** $\frac{1}{2}x - \frac{3}{4}\sin \frac{2}{3}x + c.$

7 $\frac{3}{8}x + \frac{1}{8}\sin 4x + \frac{1}{64}\sin 8x + c.$ **8** $-2\sqrt{2}\cos \dfrac{x}{2} + c.$

9 $-3\cos^4 \dfrac{x}{6} + c.$ **10** $\frac{1}{10}\sin 5x + \frac{1}{2}\sin x + c.$

11 $\frac{1}{270}(15x + 7)(3x - 7)^5 + c.$ **12** $\frac{2}{3}(x - 10)\sqrt{(5 + x)} + c.$

13 $\dfrac{1}{\sqrt{5}}\sin^{-1}\sqrt{\dfrac{5}{6}}x + c.$ **14** $\dfrac{1}{\sqrt{8}}\tan^{-1}\sqrt{8x} + c.$

15 $\sin^{-1}\dfrac{x + 2}{3} + c.$ **16** $\dfrac{1}{\sqrt{6}}\tan^{-1}\sqrt{\dfrac{3}{2}}(x + 1) + c.$

17 $\sin^{-1}\dfrac{x}{\sqrt{5}} - \sqrt{(5 - x^2)} + c.$ **18** $\frac{1}{2}x\sqrt{(9 - x^2)} + \dfrac{9}{2}\sin^{-1}\dfrac{x}{3} + c.$

19 $\dfrac{x}{4\sqrt{(4 - x^2)}} + c.$ **20** $-\dfrac{\sqrt{(16 - x^2)}}{16x} + c.$

21 (a) $-\sqrt{(1 - x^2)} + c,$ (b) $2\tan^{-1} x + c,$ (c) $-\frac{1}{3}(1 - x^2)^{3/2} + c,$
(d) $2\sin^{-1} x + c,$ (e) $2\sin^{-1} x + \sqrt{(1 - x^2)} + c.$

22 $\frac{2}{27}(3x + 20)\sqrt{(3x - 1)} + c.$ **23** $\frac{1}{6}(x^2 + 2)^{3/2} + c.$ **24** $-\dfrac{\sqrt{(1 - x^2)}}{x} + c.$

25 $\dfrac{2}{5}\cos^5 \dfrac{x}{2} - \dfrac{2}{3}\cos^3 \dfrac{x}{2} + c.$ **26** $-\frac{1}{3}\sqrt{(4 - x^2)} + c.$ **27** $3\sin^{-1}\dfrac{x}{6} + c.$

28 $\frac{1}{168}(x + 3)^6(21x^2 - 66x + 61) + c.$ **29** $2\sin^{-1}\dfrac{x}{2} - \dfrac{1}{2}x\sqrt{(4 - x^2)} + c.$

30 $2\sin^{-1}\dfrac{x}{3} - \sqrt{(9 - x^2)} + c.$ **31** $\dfrac{1}{2\sqrt{3}}\tan^{-1}\dfrac{\sqrt{3}}{2}(x - 2) + c.$

32 $\frac{1}{2}x - \frac{5}{4}\sin \frac{2}{5}x + c.$ **33** $\frac{2}{3}\sqrt{2}\sin \frac{3}{2}x + c.$
34 $2\sin^{-1}\frac{1}{2}x + \frac{1}{2}x\sqrt{(4 - x^2)} + c.$ **35** $\frac{1}{5}\sec^5 x + c.$ **36** $\frac{1}{9}\sin^{-1}\frac{3}{2}x^3 + c.$

37 $-\frac{1}{2}\cos 2x + \frac{1}{3}\cos^3 2x - \frac{1}{10}\cos^5 2x + c.$ **38** $\dfrac{1}{2}\tan^{-1} x + \dfrac{x}{2(1 + x^2)} + c.$

39 $\frac{1}{2}\cos x - \frac{1}{8}\cos 4x + c.$ **40** $\dfrac{1}{\sqrt{2}}\sec x + c.$ **41** $\frac{1}{2}\tan^{-1} x^2 + c.$

42 $\dfrac{1}{2}\sin x\sqrt{(1 - 2\sin^2 x)} + \dfrac{1}{2\sqrt{2}}\sin^{-1}(\sqrt{2}\sin x) + c.$

43 $\frac{1}{2}\tan^{-1}(2\tan x) + c.$

44 $\dfrac{1}{3}\sec^{-1}\dfrac{x}{3} + c.$ **45** $\sqrt{\dfrac{(1 + x)}{(1 - x)}} + c.$ **46** $\frac{3672}{125}.$ **47** $\frac{1}{2}.$ **48** 1.

49 $\frac{3}{4}\pi.$ **50** $\sqrt{3}/4.$

Chapter 2

Qu. 1 The larger a, the larger the gradient.

Qu. 2 The reflection of $y = 2^x$ in the y-axis.

Qu. 3 (a) 0.7×2^x, 1.1×3^x, (b) 1.4, 1.9.

Qu. 4 0.7×2^x.

Qu. 5 1.08.

Qu. 6 (a) $30x^2(2x^3 + 1)^4$, (b) $6x^2 \cos(2x^3)$, (c) $6x^2 e^{2x^3}$, (d) $2y e^{y^2} \dfrac{dy}{dx}$,

(e) $-2x e^{-x^2}$, (f) $\sec^2 x \times e^{\tan x}$, (g) $\dfrac{1}{2\sqrt{x}} e^{\sqrt{x}}$, (h) $\cos y \times e^{\sin y} \dfrac{dy}{dx}$.

Qu. 7 (a) $-\frac{1}{2}(x^2 + 1)^{-1} + c$, (b) $-\frac{1}{2} \cos x^2 + c$, (c) $\frac{1}{2} e^{x^2} + c$, (d) $-e^{\cos x} + c$,

(e) $6 e^{x/3} + c$, (f) $\frac{3}{2} e^{2x} + c$, (g) $\frac{1}{12} e^{3x^2} + c$, (h) $-\frac{1}{2} e^{\cot 2x} + c$.

Qu. 8 (a) $\log_{10}(100a^2 b^{-1/3})$, (b) $\log_c \dfrac{B(1+x)}{1-x}$.

Qu. 9 (a) $\log_c 2 + \log_c a$, (b) $2 \log_c a$, (c) $-\log_c a$, (d) $\log_c 2 - \log_c a$,

(e) $\frac{1}{2} \log_c a$, (f) $\log_c a - \log_c 2$, (g) $-2 \log_c a$, (h) $-\log_c 2 - \log_c a$.

Qu. 10 (a) $x = \frac{3}{2}$, (b) $x = 6.02$.

Qu. 11 (b) 4.61.

Qu. 12 0.693.

Qu. 13 (a) $\ln(xy/e)$, (b) 2.

Qu. 14 (a) x^2, (b) $1/x$, (c) $\sqrt{x}$, (d) $\sin x$, (e) $\frac{1}{2}x^2$, (f) $\frac{1}{2}x$.

Qu. 15 (a) $10x(x^2 - 2)^4$, (b) $-2x \operatorname{cosec} x^2 \cot x^2$, (c) $2x e^{x^2}$, (d) $\dfrac{2x}{x^2 - 2}$,

(e) $2 \cot x$, (f) $2x \cot x^2$.

Qu. 16 $\dfrac{3x + 2}{2x(x + 1)}$.

Qu. 17 (a) $\dfrac{1}{x}$, (b) $\dfrac{1}{x}$, (c) $\dfrac{3}{3x + 1}$, (d) $\dfrac{1}{y}\dfrac{dy}{dx}$, (e) $\dfrac{3}{x}$, (f) $\dfrac{3x^2}{x^3 - 2}$, (g) $\dfrac{3}{x - 1}$, (h) $\dfrac{1}{t}\dfrac{dt}{dx}$,

(i) $\cot x$, (j) $-3 \tan 3x$, (k) $-3 \tan x$, (l) $6 \cot 3x$, (m) $\dfrac{x}{x^2 - 1}$,

(n) $\dfrac{1+x}{x(1-x)}$.

Qu. 18 $4^x \ln 4$, $16 \ln 4$.

Qu. 19 (a) 0.6931, (b) 1.0986.

Qu. 20 (a) $10^x \ln 10$, (b) $2^{3x} \ln 64$.

Qu. 21 $5^x \ln 5$, $\dfrac{5^x}{\ln 5} + c$.

Qu. 22 $x 2^{x^2} \ln 4$, $\dfrac{2^{x^2}}{\ln 4} + c$.

Qu. 23 (a) $\dfrac{3^{2x}}{\ln 9} + c$, (b) $\frac{1}{3} e^{x^3} + c$, (c) $\dfrac{2^{\tan x}}{\ln 2} + c$.

Qu. 24 (a) $2 \ln x + c$, (b) $\frac{1}{3} \ln x + c$, (c) $\frac{1}{3} \ln(3x - 2) + c$, (d) $\frac{1}{3} \ln(x - 2) + c$.

Qu. 25 (a) $\frac{1}{2} \ln(2x + 3) + c$, (b) $\ln \dfrac{1}{1-x} + c$.

Qu. 26 ln 8.

Qu. 27 (a) $c = \ln A$, (b) $c = \ln (k\sqrt{2})$.

Qu. 28 (a) $\dfrac{1}{3(2 - x^3)} + c$, (b) $\frac{1}{3} \sin x^3 + c$, (c) $\frac{1}{3} e^{x^3} + c$, (d) $\ln \{k\sqrt[3]{(x^3 - 2)}\}$,

(e) $\ln \{k\sqrt{(x^2 - 2x)}\}$, (f) $\ln \dfrac{k}{3 - x^2}$, (g) $\ln (k \sin x)$.

Qu. 29 $x + \ln (x - 1) + c$.

Qu. 30 (a) $\ln \frac{3}{4}$, (b) $-\ln 2$.

Qu. 31 $-\ln 2$.

Qu. 32 No.

Qu. 33 $\dfrac{1}{x - 3}, \dfrac{1}{x - 3}$.

Qu. 34 (a) $\ln \frac{3}{2}$, (b) $-\ln 5$.

Qu. 35 $-\ln 3$.

Exercise 2a, page 24

2 Yes. (a) Neither, (b) even.

3 Even. $0 < y \leqslant 1$.

4 1.1×3^x.

5 (a) $4 e^x$, (b) $3 e^{3x}$, (c) $2 e^{2x+1}$, (d) $4x e^{2x^2}$, (e) $-2 e^{-2x}$, (f) $3 e^{3y} \dfrac{dy}{dx}$,

(g) $2x e^{x^2 + 3}$, (h) $-2x^{-3} e^{x^{-2}}$, (i) $-5x^{-2} e^{5/x}$, (j) $\frac{1}{3} x^{-2/3} e^{x^{1/3}}$, (k) $2ax e^{ax^2 + b}$,

(l) $\dfrac{1}{2\sqrt{t}} e^{\sqrt{t}} \dfrac{dt}{dx}$.

6 (a) $-e^{\cos x} \sin x$, (b) $e^{\sec x} \sec x \tan x$, (c) $e^{3 \tan y} 3 \sec^2 y \dfrac{dy}{dx}$,

(d) $2 e^{\sin 2x} \cos 2x$, (e) $e^{-\cot x} \csc^2 x$, (f) $-2 e^{\csc^2 x} \csc^2 x \cot x$,

(g) $-\dfrac{\sin x}{2\sqrt{\cos x}} e^{\sqrt{\cos x}}$, (h) $ab \, e^{a \sin bx} \cos bx$, (i) $3 e^{\sin 3t} \cos 3t \dfrac{dt}{dx}$,

(j) $2x e^{\tan x^2} \sec^2 x^2$.

7 (a) $\dfrac{x}{\sqrt{(x^2 + 1)}} e^{\sqrt{(x^2 + 1)}}$, (b) $\dfrac{2x}{(1 - x^2)^2} e^{(1 - x^2)^{-1}}$, (c) $4 e^{\sin^2 4x} \sin 8x$,

(d) $2x e^{\tan (x^2 + 1)} \sec^2 (x^2 + 1)$, (e) $6 e^{\sec^2 3x} \sec^2 3x \tan 3x$,

(f) $e^{-\csc x} \csc x \cot x$, (g) $2x^{-3} e^{-x^{-2}}$, (h) $(\sin x + x \cos x) e^{x \sin x}$,

(i) $e^{xy} \left(y + x \dfrac{dy}{dx} \right)$, (j) $e^{x + e^x}$.

8 (a) $e^x(x^2 + 2x)$, (b) $(x - 1) e^x/x^2$, (c) $\frac{1}{2} e^{\sin x} (1 + x \cos x)$,

(d) $e^x \csc x (2x - \cot x)$, (e) $e^x \csc x (1 - \cot x)$,

(f) $-e^{-x} x^{-2}(x \sin x + x \cos x + \cos x)$, (g) $(1 + x) e^{x + xe^x}$,

(h) $e^{ax} \sec bx (a + b \tan bx)$, (i) $e^{ax} \csc bx (a - b \cot bx)$,

(j) $n e^x \tan^{n-1} e^x \times \sec^2 e^x$, (k) $2 e^x \cos x$.

9 (a) $6 e^{x/2} + c$, (b) $-e^{-x} + c$, (c) $3 e^{x/3} + c$, (d) $\frac{2}{3} e^{3x-1} + c$, (e) $\frac{1}{4} e^{x^2} + c$,

(f) $-\frac{1}{3} e^{-x^3} + c$, (g) $-e^{\cos x} + c$, (h) $e^{\tan x} + c$, (i) $-e^{\cot x} + c$, (j) $-e^{1/x} + c$.

10 $y = e^a(x - a + 1)$, $y - e^2x + e^2 = 0$.

11 $\frac{1}{2}\pi(e^2 - 1)$.

Page 24

12 $e^x(1 + x)$, $e^x(x - 1) + c$.

13 Minimum $-\dfrac{1}{e}$ when $x = -1$, $e^2 y + x + 4 = 0$.

18 $25\,e^{4x}\cos(3x + 2\alpha)$, $125\,e^{4x}\cos(3x + 3\alpha)$.

19 $13^n\,e^{5x}\sin(12x + n\beta)$.

Exercise 2b, page 28

1 (a) $\log_{10}\dfrac{a^3}{50}$, (b) $\log_e\dfrac{x^2 e^3}{3}$, (c) $\log_e\dfrac{(x - 3)^4}{(x - 2)^3}$, (d) $\log_e\{k\sqrt{(1 - y^2)}\}$.

2 (a) $\log_e a + \log_e 3$, (b) $3\log_e a$, (c) $\log_e a - \log_e 3$, (d) $-3\log_e a$,
(e) $\log_e 3 - \log_e a$, (f) $-\log_e 3 - \log_e a$, (g) $\tfrac{1}{3}\log_e a$.

3 (a) $\log_e \cos x - \log_e \sin x$, (b) $2\log_e \sin x - 2\log_e \cos x$,
(c) $\log_e(x - 2) + \log_e(x + 2)$, (d) $\tfrac{1}{2}\log_e(x + 1) - \tfrac{1}{2}\log_e(x - 1)$,
(e) $\log_e 3 + 2\log_e \sin x$.

4 (a) $a = 100$, (b) $y = 100$ or $y = \tfrac{1}{100}$.

5 (a) $x = \tfrac{2}{5}$, (b) $x = 1.26$.

6 1.10.

7 $x = 1$ or $x = -2$.

8 $a = 10$ or $a = 100$.

9 $x = +\tfrac{2}{9}$.

10 $x = 1$, $y = 10$ or $x = 10$, $y = 1$.

Exercise 2c, page 32

1 (a) $\dfrac{1}{x}$, (b) $\dfrac{4}{x}$, (c) $\dfrac{2}{2x - 3}$, (d) $\dfrac{1}{y}\dfrac{dy}{dx}$, (e) $\dfrac{1}{x - 1}$, (f) $\dfrac{4}{x}$, (g) $\dfrac{2x}{x^2 - 1}$, (h) $\dfrac{2}{x}$, (i) $\dfrac{6}{x}$,

(j) $\dfrac{2}{x + 1}$, (k) $\dfrac{3}{t}\dfrac{dt}{dx}$, (l) $-\dfrac{1}{x}$, (m) $\dfrac{1}{x}$, (n) $\dfrac{1}{2x}$, (o) $-\dfrac{1}{x}$, (p) $-\dfrac{1}{x}$, (q) $-\dfrac{2}{x}$,

(r) $\dfrac{1}{x \ln 10}$, (s) $-\dfrac{3}{t}\dfrac{dt}{dx}$, (t) $\dfrac{1}{3x}$.

2 (a) $-\tan x$, (b) $2\cot x$, (c) $6\operatorname{cosec} 6x$, (d) $-6\tan 2x$, (e) $-4\operatorname{cosec} 2x$,
(f) $-4\tan 2x$, (g) $\operatorname{cosec} x$, (h) $\tan x$, (i) $\sec x$, (j) $-2x\cot x^2$, (k) $2\sec 2x$.

3 (a) $\dfrac{1}{x^2 - 1}$, (b) $\dfrac{2x^2 - 1}{x(x^2 - 1)}$, (c) $\dfrac{3x - 5}{2(x^2 - 1)}$, (d) $\dfrac{1}{\sqrt{(x^2 - 1)}}$.

4 (a) $\dfrac{1}{t}\dfrac{dt}{dx}$, (b) $1 + \ln x$, (c) $x + 2x\ln x$, (d) $\dfrac{1}{x^2}(1 - \ln x)$, (e) $\ln y + \dfrac{x}{y}\dfrac{dy}{dx}$,

(f) $\dfrac{y}{x} + \dfrac{dy}{dx}\ln x$, (g) $\dfrac{1 - 2\ln x}{x^3}$, (h) $\dfrac{\ln x - 1}{(\ln x)^2}$, (i) $\dfrac{2}{x}\ln x$, (j) $\dfrac{1}{x \ln x}$, (k) $\cos x$.

5 (a) $5^x \ln 5$, (b) $x\,2^{x^2}\ln 4$, (c) $\tfrac{2}{3}\,3^{2x}\ln 3$, (d) 1.

6 (a) $3^x \ln 3$, $\dfrac{3^x}{\ln 3} + c$, (b) $x\,2^{x^2}\ln 4$, $\dfrac{2^{x^2}}{\ln 4} + c$.

7 (a) $\dfrac{1}{\ln 10}\,10^x + c$, (b) $\dfrac{2^{3x}}{\ln 8} + c$, (c) $\dfrac{3^{x^2}}{\ln 9} + c$, (d) $-\dfrac{2^{\cos x}}{\ln 2} + c$.

Page 33

8 $1 + \ln x$, $x(\ln x - 1) + c$.

9 $2^x(1 + x \ln 2)$, $\dfrac{x \, 2^x}{\ln 2} - \dfrac{2^x}{(\ln 2)^2} + c$.

10 (a) $1/(x - 2)$, (b) $1/(x - 2)$.

12 (a) $\{ y : y \geqslant 0 \}$, (b) $\mathbb{R}$.

Exercise 2d, page 39

1 (a) $\ln (kx^{1/4})$, (b) $\ln (kx^5)$, (c) $\ln \{ k\sqrt{(2x - 3)} \}$, (d) $\ln \{ k\sqrt{(x + 4)} \}$,
(e) $\ln \{ k(3 - 2x)^{-1/2} \}$, (f) $\ln \{ k(1 - x^2)^{-1/2} \}$, (g) $\ln \{ k(x^2 - 1)^{3/2} \}$,
(h) $\ln \{ k(x^2 + x - 2) \}$, (i) $\ln \{ k\sqrt[3]{(3x^2 - 9x + 4)} \}$, (j) $x - \ln \{ k(x + 2)^2 \}$,
(k) $\frac{3}{2}x - \ln \{ k(2x + 3)^{9/4} \}$, (l) $-2x - \ln \{ k(3 - x)^6 \}$, (m) $-x - \ln \{ k(2 - x) \}$,
(n) $-2x - \ln \{ k(x - 4)^5 \}$, (o) $\ln (k \sec x)$, (p) $\ln \left(k \sin^2 \dfrac{x}{2} \right)$,
(q) $\ln \{ k\sqrt{\sin (2x + 1)} \}$, (r) $\ln \left(k \cos^3 \dfrac{x}{3} \right)$, (s) $\ln \{ k(x - \sin^2 x) \}$,
(t) $\ln \{ k(\sin x + \cos x) \}$, (u) $\ln \{ k(x + \tan x) \}$.

2 (b) $\dfrac{2}{2x - 1}$, $\dfrac{2}{2x - 1}$, (c) $\frac{1}{2} \ln 3$, $-\frac{1}{2} \ln 5$.

3 (a) $\ln \frac{2}{3}$, (b) $\ln 2$.

4 (a) $\dfrac{1}{3 - x}$, $\dfrac{1}{3 - x}$, (b) $-\ln 1.5$, (c) $\ln 1.5$.

5 (a) $\ln 2$, (b) $\frac{1}{3} \ln 2$, (c) $-\ln 2$, (d) $\frac{1}{2} \ln 5 - \ln 3$, (e) $\frac{1}{2} \ln 3$, (f) $-\frac{1}{2} \ln 2$,
(g) $\ln 7$, (h) $2 + \ln 4$, (i) $2 + \ln 4$, (j) $-\frac{1}{2} - \ln \frac{3}{2}$, (k) $\frac{1}{2} \ln \frac{4}{3}$, (l) $\frac{1}{2} \ln 2$, (m) $\frac{1}{2} \ln 3$.

Exercise 2e, page 40

1 $x = 4$ or $x = \sqrt{2}$.

2 $x = 0.178$.

3 $x = 3$ or $x = 1$.

4 (a) $x = 7.13$, (b) $x = 0.304$, (c) $x = 0.1$.

5 $y = 100x^{-2/3}$.

6 -1.20.

7 $(3, 9)$, $(9, 3)$.

8 (a) $\mathbb{R}$, $\{ y : y \in \mathbb{R}, \; y \geqslant 1 \}$, $\mathbb{R}^+$, (b) $g_1 g_2(x) = \ln (2 + 2x + x^2)$, neither;
$g_2 g_1(x) = 1 + \ln (1 + x^2)$, even.

9 (a) $x = 1.7$, (b) 2.0.

10 (a) $a = 0.693$, $b = 0.366$, (b) $x = 2.32$.

11 $a = 2.4$, $b = 1.6$, $n = -0.631$.

14 $a = 3$, $b = 4$. (a) $\pi - \tan^{-1} \frac{4}{3} \approx 2.21$, (b) $\tan^{-1} \frac{4}{3} \approx 0.93$.

15 Minimum 1 when $x = 0$, maximum $2/\sqrt{e}$ when $x = \pm 1/\sqrt{2}$.

16 $\frac{3}{4}\pi$, $\frac{7}{4}\pi$, $\frac{11}{4}\pi$.

17 (a) $\dfrac{-7 \sin x}{(3 + 4 \cos x)(4 + 3 \cos x)}$.

Page 42

18 $\{y: y \in \mathbb{R}, 0 < y < 1\}$, $\mathbb{R}^+$, $\{y: y \in \mathbb{R}, y > 1\}$; $f^{-1}: x \mapsto \ln (1/x)$, $(0 < x < 1)$;

$g^{-1}: x \mapsto 1 - 1/x$, $(x > 0)$; $(g \circ f)^{-1}: x \mapsto \ln \left(\dfrac{x}{x-1} \right)$, $(x > 1)$.

19 $\ln (2 + \sqrt{5}) \approx 1.44$.

20 Minimum 0, maximum $4e^{-2}$, $ey = x$.

21 0.53.

22 (a) 2π, (b) even, (d) no inverse, (e) $D = \{x: 0 \leqslant x \leqslant \pi\}$.

Chapter 3

Qu. 1 (a) $\dfrac{3-x}{1-x^2}$, (b) $\dfrac{(x+2)(x-1)}{(x^2+1)(x+1)}$, (c) $\dfrac{3x^2-x+4}{(x-1)^2(x+1)}$.

Qu. 2 (a) $\dfrac{1}{x-2} - \dfrac{1}{x+2}$, (b) $\dfrac{1}{2(1-x)} + \dfrac{1}{2(1+x)}$, (c) $\frac{1}{2} - \frac{1}{3}$, (d) $\dfrac{1}{n} - \dfrac{1}{n+1}$.

Qu. 3 (a) $12A - 3B + 4C = 17$, $6A - 4B + 3C = 5$, $10A - 15B + 6C = -1$,

(b) $A = 2$, $B = 1$, $C = -1$.

Qu. 4 (a) $A = 1$, $B = -3$, $C = 4$, (b) $A = 2$, $B = 1$, $C = -3$,

(c) $A = 5$, $B = -1$.

Qu. 5 $A = 3$, $B = -2$, $C = -1$.

Qu. 6 (a) $A = -\frac{2}{3}$, $B = \frac{2}{3}$, $C = 1$, (b) No.

Qu. 7 (a) $\dfrac{1}{x-3} - \dfrac{1}{x+3}$, (b) $\dfrac{1}{2(2-x)} - \dfrac{1}{2(2+x)}$, (c) $\dfrac{1}{x-2} - \dfrac{2}{3x-5}$,

(d) $\dfrac{1}{2(x+1)} - \dfrac{1}{x+2} + \dfrac{1}{2(x-3)}$, (e) $\dfrac{2}{1+2x} - \dfrac{1}{2-x}$.

Qu. 8 (a) $\dfrac{1}{1-x} + \dfrac{2+x}{4+x^2}$, (b) $\dfrac{1}{x+1} - \dfrac{2x-1}{2x^2+x+3}$, (c) $\dfrac{1}{x+1} + \dfrac{1}{x-2} - \dfrac{2}{x+2}$,

(d) $\dfrac{1}{2-x} + \dfrac{x}{3+x^2}$.

Qu. 9 (a) $\dfrac{1}{x+3} - \dfrac{2}{(x+3)^2}$, (b) $\dfrac{5}{x-2} - \dfrac{3}{x-1} - \dfrac{4}{(x-1)^2}$.

Qu. 10 $A = 3$, $B = -2$, $C = 1$, $D = 5$.

Qu. 11 (a) $x + \dfrac{x+2}{(x-1)(x+3)}$, (b) $3 + \dfrac{x-1}{(x-2)(x+1)}$.

Qu. 12 (a) $1 + \dfrac{2}{x+1} - \dfrac{1}{x-2}$, (b) $x - 1 - \dfrac{3}{4(x-2)} + \dfrac{3}{4(x+2)}$.

Qu. 14 (a) $\frac{1}{6} \ln (x-3) - \frac{1}{6} \ln (x+3) + c$, (b) $\frac{1}{2} \ln (2x-3) - \frac{5}{2}(2x-3)^{-1} + c$.

Qu. 15 $-\dfrac{1}{2} \ln (2-x) - \dfrac{1}{2} \ln (2+x) + c = \ln \dfrac{k}{\sqrt{(4-x^2)}}$.

Qu. 16 No.

Qu. 17 (a) $\ln \frac{25}{32}$, (b) $\ln 3$.

Exercise 3a, page 47

1 (a) $\dfrac{x-12}{(x+3)(x-2)}$, (b) $\dfrac{7-3x-5x^2}{(x+2)^2(3x-1)}$, (c) $\dfrac{2-4x-3x^2}{(2+3x^2)(1-x)}$,

(d) $\dfrac{-x^3+6x^2-7x+6}{(x^2+1)(x-1)^2}$.

2 (a) $\dfrac{1}{3-x}-\dfrac{1}{3+x}$, (b) $\dfrac{1}{2(a-b)}+\dfrac{1}{2(a+b)}$, (c) $\frac{1}{5}-\frac{1}{6}$, (d) $\dfrac{1}{1-p}+\dfrac{1}{p}$.

3 (a) $A=3$, $B=7$, (b) $A=1$, $B=-1$, $C=2$, (c) $A=2$, $B=-1$, $C=-3$,
(d) $A=1$, $B=-2$, $C=3$.

4 (a) $A=1$, $B=-1$, (b) $A=2$, $B=1$, (c) $A=2$, $B=-1$, $C=3$,
(d) $A=1$, $B=-2$, $C=3$.

5 (a) $A=3$, $B=-\frac{1}{2}$, $C=-\frac{1}{2}$, (b) No, (c) $A=2$, $B=3$, $C=1$, (d) No,
(e) $A=1$, $B=3$, $C=2$, $D=-1$.

6 $A=1$, $B=1$, $C=1$. (a) $(x+1)(x^2-x+1)$, (b) $(x-2)(x^2+2x+4)$,
(c) $(x+3)(x^2-3x+9)$, (d) $(2x-3)(4x^2+6x+9)$,
(e) $(3x+5)(9x^2-15x+25)$.

7 $x(x-1)(x-2)+3x(x-1)+x+1$.

8 $a=60$, $b=25$.

9 $\alpha+\beta=-b/a$, $\alpha\beta=c/a$.

10 $\alpha+\beta+\gamma=-q/p$, $\beta\gamma+\gamma\alpha+\alpha\beta=r/p$, $\alpha\beta\gamma=-s/p$.

Exercise 3b, page 52

1 (a) $\dfrac{2}{x+3}-\dfrac{1}{x-4}$, (b) $\dfrac{1}{2(5-x)}-\dfrac{1}{2(5+x)}$, (c) $\dfrac{4}{x+1}+\dfrac{2}{x-2}-\dfrac{3}{x-3}$,

(d) $\dfrac{3}{x-1}-\dfrac{1}{x}+\dfrac{2}{x+1}$, (e) $\dfrac{1}{x+2}+\dfrac{2}{2x+1}-\dfrac{2}{3x+2}$,

(f) $2x-1+\dfrac{1}{x+3}-\dfrac{3}{x-2}$.

2 (a) $\dfrac{2}{x-3}+\dfrac{3x-1}{x^2+4}$, (b) $\dfrac{2}{x+1}-\dfrac{1}{x^2+2}$, (c) $\dfrac{1}{x-1}+\dfrac{2x}{x^2+5}$,

(d) $\dfrac{3}{2x-3}+\dfrac{1-3x}{2x^2+1}$, (e) $\dfrac{3}{x-3}-\dfrac{2}{x+3}-\dfrac{1}{x+5}$, (f) $2+\dfrac{5}{x-3}+\dfrac{x}{x^2+1}$.

3 (a) $\dfrac{1}{x-2}-\dfrac{3}{(x-2)^2}$, (b) $\dfrac{1}{x-1}-\dfrac{1}{x+2}+\dfrac{2}{(x+2)^2}$,

(c) $\dfrac{23}{4(3x+1)}-\dfrac{1}{4(x+1)}-\dfrac{7}{2(x+1)^2}$, (d) $x+\dfrac{1}{x+2}+\dfrac{2}{x-1}+\dfrac{1}{(x-1)^2}$.

4 (a) $\dfrac{1}{x-2}+\dfrac{2}{x+1}-\dfrac{3}{(x+1)^2}+\dfrac{1}{(x+1)^3}$,

(b) $\dfrac{3}{x-1}-\dfrac{1}{(x-1)^2}-\dfrac{3}{x+1}-\dfrac{2}{(x+1)^2}$.

5 (a) $x+2+\dfrac{1}{x-3}-\dfrac{2}{x+3}$, (b) $3-\dfrac{2}{x-1}-\dfrac{1}{x+2}$, (c) $2x-\dfrac{6}{x-2}+\dfrac{12}{x^2+3}$,

Page 52

(d) $x - 2 - \dfrac{6}{x+1} + \dfrac{2}{(x+1)^2} + \dfrac{3}{x}$.

6 $\dfrac{1}{6(x+2)} - \dfrac{7}{2x} + \dfrac{10}{3(x-1)}$.   **7** $\dfrac{3}{2x^2} - \dfrac{3}{4x} + \dfrac{3}{4(x+2)}$.

8 $2x + 4 - \dfrac{1}{3(x-2)} - \dfrac{5x+61}{3(x^2+5)}$.   **9** $\dfrac{5}{3+x} + \dfrac{2}{4-x} - \dfrac{3}{4+x}$.

10 $\dfrac{1}{x-1} - \dfrac{x}{x^2+x+1}$.   **11** $\dfrac{2}{(2x+1)^2} - \dfrac{5}{2x+1} + \dfrac{3}{x-3}$.

12 $\dfrac{13}{29(2x-5)} - \dfrac{5}{29(3x+7)}$.   **13** $\dfrac{3}{x-1} - \dfrac{3}{x} - \dfrac{2}{x^2} - \dfrac{1}{(x-1)^2}$.

14 $\dfrac{2}{5(5x-2)} - \dfrac{10x+1}{5(25x^2+10x+4)}$.   **15** $\dfrac{2}{x} - \dfrac{2x}{x^2+3} + \dfrac{2-5x}{(x^2+3)^2}$.

16 $\dfrac{1}{x^2+2} - \dfrac{1}{x^2+3}$.   **17** $\dfrac{\sqrt{3}}{36(x-\sqrt{3})} - \dfrac{\sqrt{3}}{36(x+\sqrt{3})} - \dfrac{1}{6(x^2+3)}$.

Exercise 3c, page 56

1 $\dfrac{1}{n} - \dfrac{1}{n+2}$.

2 $\dfrac{2}{n-1} - \dfrac{3}{n} + \dfrac{1}{n+1}$.

3 (a) $\dfrac{n+4}{n(n+1)(n+2)}$, (b) $\dfrac{3}{2} - \dfrac{n+3}{(n+1)(n+2)}$, (c) $1\frac{1}{2}$.

4 2.

5 (a) $\dfrac{11}{18} - \dfrac{3n^2+12n+11}{3(n+1)(n+2)(n+3)}$, (b) $\dfrac{n}{4(n+1)}$, (c) $\dfrac{n}{9(n+1)}$,

(d) $\dfrac{3}{16} - \dfrac{2n+3}{8(n+1)(n+2)}$, (e) $\dfrac{11}{96} - \dfrac{1}{8(n+1)} - \dfrac{1}{8(n+2)} + \dfrac{1}{8(n+3)} + \dfrac{1}{8(n+4)}$,

(f) $\dfrac{1}{6} - \dfrac{n+2}{(n+3)(n+4)}$.

6 (a) $\dfrac{2n}{2n+1}$, (b) $\dfrac{1}{12} - \dfrac{1}{4(2n+1)(2n+3)}$, (c) $\dfrac{5}{24} - \dfrac{4n+5}{8(2n+1)(2n+3)}$.

7 (a) $\dfrac{1}{2} \ln \dfrac{k(x-2)}{x}$, (b) $\dfrac{1}{17} \ln \dfrac{k(5x-2)}{x+3}$, (c) $\ln \dfrac{kx}{3x+1} - \dfrac{2}{x}$, (d) $\ln \dfrac{k}{\sqrt{(16-x^2)}}$,

(e) $\dfrac{1}{6} \ln \dfrac{k(x-5)}{x+1}$, (f) $\ln \{k(x^2 - 4x - 5)^{1/2}\}$, (g) $\ln \{k(x+2)\sqrt{(x^2+3)}\}$,

(h) $\ln \dfrac{k(3+x)^2(2-x)}{(4-x)^3}$, (i) $2 \ln \{k(2x+1)\} - \frac{1}{2} \ln \{(x-3)(x+3)^3\}$,

(j) $\dfrac{1}{3} \ln \dfrac{k(x-2)}{x+1} - \dfrac{4}{x-2}$, (k) $\ln \{k(2x+1)^{1/2}\} - \dfrac{4}{3} \tan^{-1} \dfrac{x}{3}$,

(l) $\ln \{k(3x+2)^{1/3}\} - \frac{1}{6} \ln (9x^2 - 6x + 4)$, (m) $\dfrac{1}{2}x^2 + 3x + \ln \dfrac{k(x-5)^2}{x+2}$,

(n) $\frac{1}{4} \ln \{k(x-3)\} - \frac{1}{8} \ln (1 + 4x^2) - \frac{3}{2} \tan^{-1} 2x$.

8 (a) $\tan^{-1} x + c$, (b) $\ln \{k(1 + x^2)^{1/2}\}$, (c) $\tan^{-1} x + \ln \{k(1 + x^2)^{1/2}\}$,

(d) $\ln \left(k \sqrt{\dfrac{1 + x}{1 - x}} \right)$, (e) $\ln \dfrac{k}{\sqrt{(1 - x^2)}}$, (f) $c - \sqrt{(1 - x^2)}$, (g) $\sin^{-1} x + c$,

(h) $\sin^{-1} x - \sqrt{(1 - x^2)} + c$, (i) $-\ln \{k(1 - x)\}$, (j) $-\ln \{k(x - 1)\}$,

(k) $x - \ln (1 + x) + c$, (l) $\dfrac{1}{1 - x} + c$, (m) $\dfrac{1}{1 - x} + \ln (1 - x) + c$.

9 (a) $\ln \frac{4}{3} \approx 0.288$, (b) $\frac{1}{2} \ln 2 + \frac{1}{4}\pi \approx 1.13$, (c) $\ln \frac{45}{64} \approx -0.352$,

(d) $-3 \ln 2 - \frac{1}{2} \ln 3 \approx -2.63$.

10 $\pi/2$, $(2 + 2 \ln 2, 0)$.

Chapter 4

Qu. 1 (a) $(-1)^r(r + 1)x^r$, (b) $3^r x^r$, (c) $(\frac{1}{2})^{r+1}(r + 1)(r + 2)x^r$,

(d) $\frac{1}{6}(-1)^r(r + 1)(r + 2)(r + 3)x^r$.

Qu. 2 (a) $\dfrac{20!(21 - 2r)}{(21 - r)!r!} x^r$, (b) $\dfrac{10!(33 - 5r)(-1)^r x^r}{(11 - r)! \, r!}$.

Qu. 3 (a) $(-1)^{r-1}(3r + 1)x^r$.

Qu. 4 (a) $\dfrac{1}{x} - \dfrac{1}{x^2} + \dfrac{1}{x^3} - \dots + \dfrac{(-1)^{r+1}}{x^r} + \dots$, $|x| > 1$.

(b) $\dfrac{1}{x^2} - \dfrac{4}{x^3} + \dfrac{12}{x^4} - \dots + \dfrac{(r - 1)(-1)^r 2^{r-2}}{x^r} + \dots$, $|x| > 2$,

(c) $\dfrac{1}{9x^2} - \dfrac{2}{27x^3} + \dfrac{1}{27x^4} - \dots + \dfrac{(-1)^r(r - 1)}{3^r x^r} + \dots$, $|x| > \dfrac{1}{3}$,

(d) $\dfrac{3}{x^2} + \dfrac{9}{x^3} + \dfrac{21}{x^4} + \dots + \dfrac{3(2^{r-1} - 1)}{x^r} + \dots$, $|x| > 2$,

(e) $\dfrac{1}{x} - \dfrac{1}{x^3} + \dfrac{1}{x^5} - \dots + \dfrac{(-1)^r}{x^{2r+1}} + \dots$, $|x| > 1$.

Qu. 5 $792 \times 4^7 \times 3^5$.

Qu. 6 (a) $^{n+2}C_{r+1} = c_{r+1} + 2c_r + c_{r-1}$, $(1 \leqslant r \leqslant n - 1)$;

(b) $^{n+2}C_{r+2} = c_{r+2} + 2c_{r+1} + c_r$, $(0 \leqslant r \leqslant n - 2)$.

Exercise 4a, page 64

1 (a) $1 - 3x + 9x^2 - \dots + (-1)^r 3^r x^r + \dots$, $|x| < \frac{1}{3}$.

(b) $1 + 2x + 4x^2 + \dots + 2^r x^r + \dots$, $|x| < \frac{1}{2}$.

(c) $1 - 2x + 3x^2 - \dots + (-1)^r(r + 1)x^r + \dots$, $|x| < 1$.

(d) $1 + x + \dfrac{3}{4}x^2 + \dots + \dfrac{(r + 1)}{2^r} x^r + \dots$, $|x| < 2$.

(e) $1 - 3x + 6x^2 - \dots + \frac{1}{2}(r + 1)(r + 2)(-1)^r x^r + \dots$, $|x| < 1$.

(f) $\frac{1}{2} - \frac{1}{4}x + \frac{1}{8}x^2 - \dots + (-1)^r x^r/2^{r+1} + \dots$, $|x| < 2$.

(g) $\frac{1}{9} + \frac{2}{27}x + \frac{1}{27}x^2 + \dots + (r + 1)x^r/3^{r+2} + \dots$, $|x| < 3$.

(h) $\frac{1}{8} + \frac{9}{16}x + \frac{27}{16}x^2 + \dots + (r + 1)(r + 2)3^r x^r/2^{r+4} + \dots$, $|x| < \frac{2}{3}$.

(i) $1 + \dfrac{1}{2}x - \dfrac{1}{8}x^2 + \dots + (-1)^{r-1} \dfrac{1 \times 3 \dots (2r - 3)}{2^r r!} x^r + \dots$, $|x| < 1$.

Page 64

2 (a) $\dfrac{1}{1-x}+\dfrac{2}{1+2x}$, $3-3x+9x^2+\ldots+\{1-(-2)^{r+1}\}x^r+\ldots$, $|x|<\dfrac{1}{2}$.

(b) $\dfrac{1}{1+x}-\dfrac{1}{2+x}$, $\dfrac{1}{2}-\dfrac{3}{4}x+\dfrac{7}{8}x^2+\ldots+(-1)^r\{1-(\tfrac{1}{2})^{r+1}\}x^r+\ldots$, $|x|<1$.

(c) $\dfrac{1}{x+1}-\dfrac{2}{(x+1)^2}$, $-1+3x-5x^2+\ldots+(-1)^{r+1}(2r+1)x^r+\ldots$, $|x|<1$.

(d) $\dfrac{3}{1-3x}+\dfrac{2}{1+2x}$, $5+5x+35x^2+\ldots+\{3^{r+1}+(-1)^r2^{r+1}\}x^r+\ldots$, $|x|<\tfrac{1}{3}$.

(e) $\dfrac{1}{x-2}+\dfrac{5}{(x-2)^2}$, $\dfrac{3}{4}+x+\dfrac{13}{16}x^2+\ldots+(5r+3)x^r/2^{r+2}+\ldots$, $|x|<2$.

(f) $\dfrac{3}{2(x-1)}-\dfrac{1}{2(x+1)}$, $-2-x-2x^2+\ldots-\{3+(-1)^r\}x^r/2+\ldots$, $|x|<1$.

3 (a) $1-x^2+x^4-\ldots+(-1)^rx^{2r}+\ldots$, $|x|<1$.

(b) $x+x^3+x^5+\ldots+x^{2r+1}+\ldots$, $|x|<1$.

(c) $1-2x+2x^2-\ldots+2(-1)^rx^r+\ldots$, $|x|<1$.

(d) $1-9x+35x^2-\ldots+\dfrac{10!(-1)^r(11-2r)}{(11-r)!r!}x^r+\ldots$, all x.

(e) $\dfrac{4}{3}-\dfrac{16}{9}x+\dfrac{52}{27}x^2-\ldots+(-1)^r2\left(1-\dfrac{1}{3^{r+1}}\right)x^r+\ldots$, $|x|<1$.

(f) $-\tfrac{5}{3}-\tfrac{28}{9}x-\tfrac{110}{27}x^2-\ldots-(6-13\times 2^r/3^{r+1})x^r+\ldots$, $|x|<1$.

(g) $-\tfrac{7}{2}+\tfrac{19}{4}x-\tfrac{57}{8}x^2+\ldots-\{(\tfrac{1}{2})^{r+1}+(2r+3)(-1)^r\}x^r+\ldots$, $|x|<1$.

4 (a) $\dfrac{1}{x}-\dfrac{2}{x^2}+\dfrac{4}{x^3}-\ldots+\dfrac{(-1)^{r-1}2^{r-1}}{x^r}+\ldots$, $|x|>2$.

(b) $-\dfrac{1}{x^3}-\dfrac{9}{x^4}-\dfrac{54}{x^5}-\ldots-\dfrac{1}{2}(r-2)(r-1)\dfrac{3^{r-3}}{x^r}+\ldots$, $|x|>3$.

(c) $\dfrac{1}{4x^2}+\dfrac{1}{4x^3}+\dfrac{3}{16x^4}+\ldots+\dfrac{r-1}{2^rx^r}+\ldots$, $|x|>\tfrac{1}{2}$.

(d) $1+\dfrac{1}{x}-\dfrac{1}{x^2}+\ldots+\dfrac{(-1)^{r+1}}{x^r}+\ldots$, $|x|>1$.

(e) $\dfrac{1}{x}-\dfrac{5}{x^2}+\dfrac{16}{x^3}-\ldots+(-1)^{r-1}(3r-1)\dfrac{2^{r-2}}{x^r}+\ldots$, $|x|>2$.

(f) $\dfrac{1}{x^2}+\dfrac{5}{x^3}+\dfrac{19}{x^4}+\ldots+\dfrac{3^{r-1}-2^{r-1}}{x^r}+\ldots$, $|x|>3$.

(g) $\dfrac{2}{x}+\dfrac{6}{x^3}-\dfrac{12}{x^4}+\ldots+\dfrac{3\{1+(-1)^{r-1}3^{r-2}\}}{2x^r}+\ldots$, $|x|>3$

(h) $-\left\{\dfrac{2}{x}+\dfrac{2}{x^3}+\dfrac{2}{x^5}+\ldots+\dfrac{2}{x^{2r+1}}+\ldots\right\}$, $|x|>1$.

(i) $-\dfrac{1}{x^3}-\dfrac{1}{x^4}-\dfrac{1}{x^7}-\dfrac{1}{x^8}-\ldots-\dfrac{1}{x^{4r-1}}-\dfrac{1}{x^{4r}}+\ldots$, $|x|>1$.

5 $x^{1/2}-x^{-1/2}-\tfrac{1}{2}x^{-3/2}$; 1.4142.   **6** 1.25992.   **7** 2.00993.

8 3.014963.   **9** 0.009920.   **10** 0.2425.   **11** $1+2x+\tfrac{5}{2}x^2$.

12 $1-x-x^2+3x^3$.   **13** $1-4x+6x^2+4x^3$.   **14** $1+2x-\tfrac{1}{2}x^2$.

Page 66

15 $1 - \frac{5}{3}x + \frac{14}{9}x^2 - \frac{130}{81}x^3$.   **16** $-1 - \frac{1}{2}x + \frac{1}{4}x^2$.   **17** $x - x^2 - \frac{2}{3}x^3$.

18 $\pm 5\%$.   **19** $\dfrac{86400}{86400 - x}$ s, 17 s.   **20** 21.6 s.   **21** $2x + y - z$.

22 $\frac{1}{2}(p - q - r)$.   **23** $\frac{2}{3}$.   **24** $1/\sqrt{(1 + 2x)}$, $|x| < \frac{1}{2}$.

25 $1/(1 - 2x)^2$, $|x| < \frac{1}{2}$.   **26** $\frac{9}{16}$.   **27** $\sqrt[3]{\frac{3}{2}}$.   **28** $1/(1 + 2x)^3$, $|x| < \frac{1}{2}$.

29 $(1 - 2x)^{1/2}$, $|x| < \frac{1}{2}$.   **30** $\sqrt[3]{4}$.

Exercise 4b, page 70

1 (a) 15360, (b) 20412, (c) $792 \times 4^7 \times 3^5 = 2^{17} \times 3^7 \times 11$,
(d) $15504 \times 2^5 \times 5^{15}$ or $38760 \times 2^6 \times 5^{14} = 2^9 \times 3 \times 5^{15} \times 17 \times 19$,
(e) $330 \times 2^4/3^4 = 2^5 \times 5 \times 11/3^3$, (f) $126 \times 3^5 \times 2^4 = 2^5 \times 3^7 \times 7$,
(g) $11^{11}/12^{12}$, (h) $3 \times 5^5/7^7$.

2 (a) $66 \times 2^2 \times 9^{10} = 2^3 \times 3^{21} \times 11$, (b) $210(\frac{1}{2})^4(\frac{2}{3})^6 = \frac{280}{243}$,
(c) $56(\frac{4}{3})^3(\frac{5}{2})^5 = 2^4 \times 5^5 \times 7/3^3$, (d) $3^{18} \times 2^{-8} \times 5^{-10}$.

Exercise 4c, page 70

1 $\dfrac{3}{1 - x} - \dfrac{2}{1 + x^2}$, $1 + 3x + 5x^2 + 3x^3$.

2 $A = 1$, $B = -1$, $C = 1$, $2x + 2x^2 + 2x^5$; $|x| < 1$.

3 (a) 323, (b) $1 - x - x^2 - \frac{5}{3}x^3$; 1.71.

4 $\dfrac{1}{x} - \dfrac{1}{x^2} + \dfrac{4}{x^3} - \dfrac{1}{x + 2}$, $-7 - 14y - 29y^2$; $|y| < 1$.

5 (a) $\dfrac{3}{3x - 1} - \dfrac{1}{x + 1} + \dfrac{1}{(x + 1)^2}$, $n(-1)^n - 3^{n+1}$;
(b) $1 - \frac{1}{3000} - \frac{1}{9} \times 10^{-6}$, 3.332 222.

6 $1 + \frac{1}{2}x - \frac{1}{8}x^2$, $1 - \frac{1}{2}x + \frac{3}{8}x^2$.

7 0.000 103 2.

9 (a) $2p + 4q - r$,
(b) (i) $1 + 4x + 9x^2 + 16x^3$, $|x| < 1$; (ii) $-\dfrac{1}{x^2} - \dfrac{4}{x^3} - \dfrac{9}{x^4} - \dfrac{16}{x^5}$, $|x| > 1$.

10 (a) 10 201 810 000, (b) $1 + 2x + 3x^2 + \dots + (n + 1)x^n + \dots$.

11 $1 - \frac{1}{2}y + \frac{3}{8}y^2 - \frac{5}{16}y^3$, $c_1 = \cos\theta$, $c_2 = \frac{1}{2}(3\cos^2\theta - 1)$,
$c_3 = \frac{1}{2}\cos\theta(5\cos^2\theta - 3)$.

12 (b) $1 + \frac{1}{2}x - \frac{1}{8}x^2 + \frac{1}{16}x^3$, (c) $8n$, $-(8n + 4)$.

13 $1 + \frac{1}{2}x - \frac{1}{8}x^2 + \frac{1}{16}x^3 - \frac{5}{128}x^4$.

15 $\dfrac{4}{9(2x + 1)} - \dfrac{2}{9(x + 2)} - \dfrac{1}{3(x + 2)^2}$,
$\dfrac{1}{4} - \dfrac{3}{4}x + \dfrac{27}{16}x^2 - \dfrac{7}{2}x^3$, $\dfrac{4}{9}(-2)^n - \dfrac{3n + 7}{36}\left(-\dfrac{1}{2}\right)^n$, $-\dfrac{1}{8}$.

16 $1 - x - \frac{1}{2}x^2 - \dots - 1 \times 3 \times 5 \dots (2n - 3)x^n/n! - \dots$, 3.316 66...; correct to 4 places of decimals.

18 (a) $\frac{21}{16}$, (b) $1 + x - x^2 + \frac{5}{3}x^3$, 1.009 90.

19 $A = 4$, $B = 1$, $C = 2$; $5 + 9x + \frac{57}{2}x^2 + \frac{287}{4}x^3$.

Page 73
22 Tenth.
23 (a) $r > \frac{1}{2}(n + 1)$, (b) 1.221 3.
24 55/53.
25 (a) $q = \frac{1}{2}p(p - 1)$, $p = -3$, $q = 6$, $a_6 = 16$.

Chapter 5

Qu. 3 (a) 46.4°, (b) 87.3°.
Qu. 4 44.9°.
Qu. 7 67.4°.

Exercise 5a, page 80

1 (a) 50.1°, (b) 51.5°, (c) 36.9°. **2** (a) 67.4°, (b) 71.6°, (c) 28.1°.
3 (a) 22.0 cm, (b) 65.2°. **4** 29.4°. **5** 97.9°. **6** 54.7°.
7 (a) $1/\sqrt{3}$, (b) $1/\sqrt{5}$. **8** 6.53 cm, 54.7°, 70.5°. **9** $\tan^{-1}(1/\sqrt{2})$.
10 22.2°. **11** 75.2°. **12** $\sqrt{\frac{2}{3}}$. **13** 28.1°. **14** 80.4°.
15 $\tan^{-1}\dfrac{c\sqrt{(a^2 + b^2)}}{ab}$. **16** $\dfrac{2\sqrt{3}}{\sqrt{133}}$.

Exercise 5b, page 84

12 $S \cos^{-1}(\frac{5}{4}\tan\alpha)$ E or W.

Chapter 6

Qu. 1 (a) Yes, (b) No.
Qu. 2 (a) No, (b) Yes.
Qu. 3 No.
Qu. 4 Not necessarily.
Qu. 6 (a) i, ii, iii, (b) ii, (c) iii, (d) iii, (e) i, ii, iii, (f) iii, (g) i, ii, iii, (h) iii.
Qu. 7 (a) $x - y = 0$, (b) $x + y = 0$.
Qu. 9 (a) As $x \to 0, 2, 4$, $\dfrac{dy}{dx} \to \infty$,

   (b) When $x = 0$, $\dfrac{dy}{dx} = \pm\sqrt{2}$; as $x \to -2$, $\dfrac{dy}{dx} \to \infty$.

Exercise 6a, page 90

1 (a) $x < \frac{1}{2}, x > 2$, (b) $\frac{1}{2} < x < 2$.
2 (a) $x < -2, x > -\frac{1}{2}$, (b) $-2 < x < -\frac{1}{2}$.
3 $-2 < x < 3$. **4** $x > 2\frac{1}{2}, x < -2\frac{1}{3}$. **5** $-3 < x < 2\frac{1}{2}$.
6 $x < -2, x > 2\frac{1}{2}$. **7** $x < -1\frac{1}{4}, x > 1$. **8** $\frac{3}{5} < x < 7$.
9 $1 < x < 2, 3\frac{1}{4} < x < 3\frac{2}{3}$. **10** $x < -1, 1 < x < 2, x > 3$.
11 $\frac{3}{4}$. **12** -1. **13** $\frac{7}{8}$. **14** $-\frac{11}{12}$.
15 $x < 0, x > 1; |y| > \frac{1}{2}$. **16** $|x| > 2, |y| > \sqrt{3}$.

Page 91

17 $x < -1, 0 < x < 1$.   **18** Discontinuities when $x = 2, 3$.
19 $y < -\frac{1}{2}, y > 4\frac{1}{2}$.   **20** $x \geqslant 0$.   **21** $x \leqslant 0$.   **22** $|x| \leqslant \sqrt{2}$.
25 $(2x - 3y)^2 + (x - 2)^2 > 0$, unless $x = 2, y = 1\frac{1}{3}$.

Exercise 6b, page 96

In Nos. 10–15, y *cannot* lie in the following intervals:
10 $-4 < y < 0$.
11 $\frac{1}{6} < y < 1\frac{1}{2}$.
12 $y < 0, y > 1$.
13 $y < -\frac{1}{2}, y > 4\frac{1}{2}$.
14 $y > 1$.
15 $\frac{2}{9}(1 - \sqrt{10}) < y < \frac{2}{9}(1 + \sqrt{10})$.
16 $(2, 4)$ min.
17 $(-6 - 2\sqrt{10}, (11 + 2\sqrt{10})/18)$ min., $(-6 + 2\sqrt{10}, (11 - 2\sqrt{10})/18)$ max.
18 $((-1 - \sqrt{7})/2, (-23 - 8\sqrt{7})/9)$ max., $((-1 + \sqrt{7})/2, (-23 + 8\sqrt{7})/9)$ min.
19 $(-\frac{1}{5}, 4\frac{1}{12})$ max.

Exercise 6d, page 103

1 (a) $x < 1$ or $x > 3$, (b) $x < 1$ or $x > 3$, (c) $1 < x < 2$ or $x > 3$.
2 $p - 1 \leqslant x \leqslant p + 1$.
3 $b^2 \geqslant 4ac, a, -b, c$ all of the same sign.
4 $2 < \lambda < 2\frac{2}{3}$.
5 $-2 < k < 6, 0 < k < 6$.
6 $(a + b + c)^2 = 4(bc + ca + ab)$.
7 $\pi/12 \leqslant \alpha \leqslant 7\pi/12, 13\pi/12 \leqslant \alpha \leqslant 19\pi/12$.
9 $k = 3, a = -2, b = -2; k = -7, a = 8, b = \frac{1}{2}$.
12 (a) $(x + 1)(y + 1) > 0$, (b) $-1 \leqslant a \leqslant 1\frac{2}{3}$.
13 (a) (i) $\frac{1}{4} < x < 1\frac{1}{2}$, (ii) $x < -1, 1 < x < 2$.
14 -1.22.
16 $x = 0, x = -1; x > 0, x < -1$.
17 (a) $x > 2, -3 < x < -1$, (b) $|x| > 2$.
18 (a) $x \geqslant 4, x \leqslant 1$, (b) $x \geqslant 4, 2 < x < 3, x \leqslant 1$.
19 $-\frac{3}{4} < x < 3$.
20 $x < -1, 0 < x < 1$.
21 $y \geqslant -\frac{1}{9}, y \leqslant -1$.
22 $x = 1, y = x + 1, (-1, -2), (3, 6)$.
23 $(0, 0), (-2, -4)$.
25 $(0, 0), y = 1, x = 1, x = -1$.

Chapter 7

Qu. 7   (a) 3, (b) 2, (d) -2, (f) 4.
Qu. 8   (a) $a^3 + b^3 + c^3$, (b) $a(b + c) + b(c + a) + c(a + b)$,
            (c) $1/a + 1/b + 1/c$, (d) $ab^2c^2 + bc^2a^2 + ca^2b^2$.

Qu. 12 (a) $\frac{4}{3}, \frac{2}{3}, -\frac{5}{3}$, (b) 0, 0, 1, (c) 0, $\frac{6}{7}, \frac{5}{7}$, (d) $-2, -1, 3$, (e) 5, 0, -2,
(f) $-1, 1, -1$.

Qu. 13 (a) $x^3 - 6x^2 + 11x - 6 = 0$, (b) $x^3 - 13x + 12 = 0$,
(c) $x^3 - 14x^2 + 288 = 0$.

Qu. 14 (a) $x = \sqrt{y}$, (b) $x = y + 2$, (c) $x = \frac{1}{2}(y - 1)$, (d) $x = -7y/3$.

Qu. 17 $3ac = b^2$, $27a^2d = b^3$. (Other relations are possible. Those given
determine the ratios $c:a$, $d:a$ in terms of $b:a$.)

Exercise 7a, page 110

1 $-2, -1, 3, 4$.   **2** $-7, -1, -1, 5$.   **3** $\pm \ln 2$.   **4** 0, 2.
5 $\pm 1, \pm 3$.   **6** 1, 8.   **7** 3.   **8** 9.   **9** 4, *not* 1.   **10** 5, $1\frac{4}{9}$.
11 5.   **12** 5, *not* $22\frac{7}{9}$.   **13** 14, *not* 6.   **14** $(4, 1), (-2, -2)$.
15 $(4, 5), (-3, 4)$.   **16** $(3, 1), (-5, -5)$.   **17** $(0, 0), (8, -6)$.
18 $(-3, 4), (-6, 3)$.   **19** $\pm(\sqrt{5}, \frac{4}{5}\sqrt{5}), \pm(2\sqrt{5}, \frac{2}{5}\sqrt{5})$.   **20** $2(a + b)/b$.
21 (a) $-(c + d)/c$, (b) $-(a + b)/(a - b)$.
22 $t^2 x - y - ct^3 + c/t = 0; (-c/t^3, -ct^3)$.
23 $(a(1 - p)^2, 2a(1 - p))$.
24 $(-c/t^2, -ct^2)$.
25 $(a(a^4 - b^4)/(a^4 + b^4), -2a^2 b^3/(a^4 + b^4))$.
26 $(3 - a)^2 = (6 - ab)(b - 2)$.
28 $x = \dfrac{Ca - cA}{Ab - aB}, y = \dfrac{Bc - bC}{Ab - aB}; (aC - cA)^2 = (bC - cB)(aB - bA)$.
29 $\frac{1}{3}, \frac{1}{2}, 2, 3$.
30 $-\frac{1}{4}, -4$.

Exercise 7b, page 114

1 $\dfrac{2a - c}{2b - d}$.   **2** $\dfrac{3a - 4c}{3b - 4d}$.   **3** $\dfrac{2a + 3c}{2b + 3d}$.   **4** $\dfrac{3a - c}{3b - d}$.   **16** 1, 3, 2.
17 $\frac{1}{2}, \frac{1}{4}, 1$.   **18** $-3, 2, -1$.   **19** $3\frac{25}{71}, \frac{1}{71}, \frac{20}{71}$.   **20** $\frac{6}{7}, 8\frac{2}{7}, 6\frac{4}{7}$.

Exercise 7c, page 119

1 $3:2, -4:3$.   **2** $-5:2, 6:1$.   **3** $-1:1, 2:1, 2:1$.
4 $1:1, -2:3, -3:2$.   **5** $\pm 3:1, \pm 1:2$.   **6** $10: -11: -13$.
7 $1:14:11$.   **8** $-c:c:a - b$.   **9** $a: -(a^2 + 1):a$.
10 $(\sin \theta - \cos \theta):(\sin \theta + \cos \theta):1$.
11 $x^4 + y^4 + z^4$.
12 $1/(yz) + 1/(zx) + 1/(xy)$.
13 $x^2(y + z) + y^2(z + x) + z^2(x + y)$.
14 $x^2 y + y^2 z + z^2 x$.
15 $xy^2 + yz^2 + zx^2$.
18 $(1 - t)(1 + t + t^2)$.
19 $(4x + y)(16x^2 - 4xy + y^2)$.
20 $(2 + 3z)(4 - 6z + 9z^2)$.
21 $(5y - z^2)(25y^2 + 5yz^2 + z^4)$.

Page 119

22 $(a - b)(a + b)(a^2 + ab + b^2)(a^2 - ab + b^2)$.

24 $(x^n - a^n)/(x - a)$.

Exercise 7d, page 124

1 (a) $\frac{4}{3}$, $-\frac{1}{3}$, $-\frac{2}{3}$, (b) 0, $\frac{5}{4}$, $\frac{3}{2}$, (c) -5, 4, -4, (d) $-\frac{7}{3}$, $\frac{13}{3}$, $-\frac{1}{3}$.

2 (a) $x^3 - 2x^2 + 5 = 0$, (b) $x^3 + 3x^2 + 2x - 6 = 0$, (c) $x^3 - x - 5 = 0$.

3 (a) $2y^3 - 3y^2 - 13y + 7 = 0$, (b) $7y^3 + 13y^2 - 3y - 2 = 0$,

(c) $2y^3 + 15y^2 + 23y - 5 = 0$, (d) $4y^3 + 26y^2 - 21y - 49 = 0$.

4 (a) $y^3 + 3y^2 - 21y + 17 = 0$, (b) $y^3 + 3y^2 - 21y - 63 = 0$,

(c) $100y^3 + 30y^2 - 3y - 1 = 0$.

5 (a) $y^3 + 6hy^2 + 9h^2y - g^2 = 0$, (b) $g^2y^3 - 9h^2y^2 - 6hy - 1 = 0$,

(c) $y^3 + 3gy^2 + 3(g^2 + 9h^3)y + g^3 = 0$.

6 (a) $a^2y^3 + (2ac - b^2)y^2 + (c^2 - 2bd)y - d^2 = 0$,

(b) $d^2y^3 + (2bd - c^2)y^2 + (b^2 - 2ac)y - a^2 = 0$,

(c) $a^3y^3 + (3a^2d + b^3 - 3abc)y^2 + (3ad^2 - 3bcd + c^3)y + d^3 = 0$.

7 (a) $-6h$, (b) $9h^2$, (c) $18h^2$.

8 $-6h$, $-3g$, $18h^2$.

9 (a) $9a^2 - 6b$, (b) $-27a^3 + 27ab - 3c$, (c) $81a^4 - 108a^2b + 12ac + 18b^2$.

10 (a) $2b^3 - 9abc + 27a^2d = 0$,

(b) $b^3d - ac^3 = 0$, $dy^3 + cy^2 + by + a = 0$, $2c^3 - 9bcd + 27ad^2 = 0$.

11 $b^3 - 4abc + 8a^2d = 0$.

12 $\frac{1}{2}$, $\frac{2}{3}$, $\frac{8}{9}$.

13 $\frac{3}{4}$, $\frac{5}{4}$, $\frac{7}{4}$.

14 $\frac{5}{2}$, $\frac{5}{2}$, $-\frac{2}{3}$.

15 $\frac{2}{3}$, $\frac{2}{3}$, $-\frac{5}{2}$.

16 $\frac{3}{2}$, $\frac{3}{2}$, $-\frac{4}{3}$.

17 $g^2 = 4h^3$, $g = h = 0$.

18 $(\frac{1}{4}t^2, -\frac{1}{8}t^3)$.

19 $tx + y - at^3 - 2at = 0$; $27ay^2 = 4(x - 2a)^3$.

20 $2x + t^3y - 3t = 0$; $-1:4$, $2:1$.

21 -1, 2, -3 and permutations.

22 2, -3, 5 and permutations.

23 1, 4, -5 and permutations.

24 1, 3, -2 and permutations.

25 $-b/a$, c/a, $-d/a$, e/a; $ey^4 + dy^3 + cy^2 + by + a = 0$,

$a^2y^4 + (2ac - b^2)y^3 + (c^2 + 2ae - 2bd)y^2 + (2ce - d^2)y + e^2 = 0$.

26 -0.81, 0.39, 6.4.

27 -0.9397, 0.1736, 0.7660.

28 -1.83, 0.226, 1.61.

29 Between -4 and -3; -3.

30 $f = b - a^2$, $g = 2a^3 - 3ab + c$. Reduce equation to the form

$x^3 + 3fx + g = 0$ and draw across $y = x^3$ the line $y + 3fx + g = 0$.

Exercise 7e, page 126

1 (a) $\pm 1/27$, ± 8, (b) $(-1, 2)$, $(-\frac{1}{2}, 1\frac{3}{4})$.

Page 126

2 (a) k, -3, $-k$, (b) 4.
3 (a) -2, 0, (b) $(-3, 2)$, $(3, -2)$, $(-4, 1\frac{1}{2})$, $(4, -1\frac{1}{2})$.
4 (a) $A = -1$, $B = -2$.
5 (b) $\frac{1}{2}(3 \pm \sqrt{5})$.
6 (b) $t^2 + qt - p^3 = 0$, $2^{1/3} - 2^{2/3}$.
7 $(b - c)(c - a)(a - b)(bc + ca + ab)$.
8 -32, 48.
9 18.
10 (a) $\frac{2}{7}$, $\frac{4}{7}$, $\frac{6}{7}$; $-1\frac{1}{2}$, -3, $-4\frac{1}{2}$; (b) $a = 3$, $b = 2$, $x^2 + 2x + 3$, $x^2 - 2x + 3$.
11 (a) -2, 1, (b) 3, -1, 2.
12 $(b_1 c_2 - b_2 c_1):(c_1 a_2 - c_2 a_1):(a_1 b_2 - a_2 b_1)$, (a) 4, 2, -6,
    (b) $(c_1 a_2 - c_2 a_1)^2 = (b_1 c_2 - b_2 c_1)(a_1 b_2 - a_2 b_1)$.
13 (a) -2, $\frac{1}{2}(3 \pm \sqrt{5})$, (b) $1 - \frac{1}{7}\sqrt{7}$.
14 (a) 3, (b) $-\frac{2}{3}$, $-2 \pm \sqrt{6}$.
15 (a) 3, $-\frac{2}{7}$, $-\frac{1}{3}$.
16 (b) $x^3 - 2x^2 + 5x - 11 = 0$.
17 $-\frac{1}{2}$.
18 (a) $(x - 2)(x^2 + 4x + 9)$, (b) 3, (c) $k = -8$, $l = 4$.
19 $\pm\sqrt{3}$, $(-1 \pm \sqrt{3}i)/2$, $\pm\sqrt{3}i$, $(1 \pm \sqrt{5})/2$.

Chapter 8

Qu. 1   6.

Qu. 4 -20, $\begin{matrix} +1 & -3 & -5 \\ -5 & -5 & +5 \\ -3 & +9 & -5 \end{matrix}$

Qu. 7 $\begin{pmatrix} 4 & -1 & -1 \\ -8 & 2 & 2 \\ 4 & -1 & -1 \end{pmatrix}$, $\begin{pmatrix} 0 & 0 & 0 \\ 0 & 0 & 0 \\ 0 & 0 & 0 \end{pmatrix}$.

Qu. 8 $\begin{pmatrix} -1 & 2 & -3 \\ -2 & 1 & 0 \\ 2 & -1 & 3 \end{pmatrix}$, $\begin{pmatrix} 3 & 0 & 0 \\ 0 & 3 & 0 \\ 0 & 0 & 3 \end{pmatrix}$.

Exercise 8a, page 132

1 (a) 14, (b) -9, (c) 0, (d) 0.
2 (a) $x^2 + y^2$, (b) 0, (c) 1, (d) x^2.
4 (a) 0, 5/3, (b) 1, 4.
7 (a) 21, (b) 24, (c) 0, (d) 0.
8 2, $-3 \pm \sqrt{6}$.
10 $\mathbf{AB} = \begin{pmatrix} 10 & -4 & 2 \\ 3 & 3 & 0 \\ 11 & -5 & 7 \end{pmatrix}$, $\det(AB) = 198$.

Exercise 8b, page 136

1 (a) 7, (b) 3.
2 (a) 0, (b) 1.
3 (a) 0, (b) 1/480.
5 1, 2, -3.
6 $(p + q + r)(p^2 + q^2 + r^2 - qr - rp - pq)$.
7 $-(x - y)(y - z)(z - x)(yz + zx + xy)$.
8 $(x - 1)^2(x^2 + x + 1)^2$.

Exercise 8c, page 142

1 (1, 2, 3).   2 (1, -1, 2).   3 (1, -3, -5).   4 (7, 5, 0).
5 (-0.1, -0.1, $+0.1$).   6 $a = -1$, $k \neq 5$.   7 (b), $(t, 3t/7, -11t/28)$.
8 (a), (0, 0, 0).   9 (b), $(t, (t + 26)/7, -(8 + 3t)/7)$.   10 (c).

Exercise 8d, page 146

1 $\dfrac{1}{18}\begin{pmatrix} -2 & -1 & 7 \\ -8 & 5 & 1 \\ 6 & 3 & -3 \end{pmatrix}$.   2 No inverse.   3 $\dfrac{1}{2}\begin{pmatrix} 1 & 2 & 1 \\ 2 & 3 & 5 \\ 3 & 4 & 7 \end{pmatrix}$.

4 No inverse.   5 $\dfrac{1}{8}\begin{pmatrix} 1 & -1 & 1 \\ 2 & 1 & 0 \\ 1 & 0 & 3 \end{pmatrix}$.   6 $\begin{pmatrix} 1 & -p & pr - q \\ 0 & 1 & -r \\ 0 & 0 & 1 \end{pmatrix}$.

7 $\begin{pmatrix} 1 & 0 & 0 \\ 0 & \cos \alpha & \sin \alpha \\ 0 & \sin \alpha & -\cos \alpha \end{pmatrix}$.   8 1, -1, 1.   9 2, 1, 3.   10 1, 3, -1.

11 (a) $\begin{pmatrix} -1 & -2 & 2 \\ 2 & 5 & -4 \\ 1 & 1 & -1 \end{pmatrix}$, (b) $\dfrac{1}{6}\begin{pmatrix} 3 & -3 & 3 \\ -1 & 1 & 1 \\ 2 & 4 & -2 \end{pmatrix}$,

(c) $\begin{pmatrix} 3 & 5 & 1 \\ 2 & 0 & 3 \\ 4 & 2 & 4 \end{pmatrix}$, (d) $\dfrac{1}{6}\begin{pmatrix} -6 & -18 & 15 \\ 4 & 8 & -7 \\ 4 & 14 & -10 \end{pmatrix}$.

12 (a) $\dfrac{1}{17}\begin{pmatrix} 7 & -1 & 2 \\ 3 & 2 & -4 \\ -11 & 4 & 9 \end{pmatrix}$ (b) $\dfrac{1}{3}\begin{pmatrix} 3 & -2 & 4 \\ 3 & -1 & 2 \\ -6 & 4 & -5 \end{pmatrix}$,

(c) $\begin{pmatrix} -1 & 7 & 2 \\ 8 & 17 & 12 \\ -1 & 1 & -1 \end{pmatrix}$, (d) $\dfrac{1}{51}\begin{pmatrix} -29 & 9 & 50 \\ -4 & 3 & 28 \\ 25 & -6 & -73 \end{pmatrix}$.

Exercise 8e, page 147

1 (a) 121, (b) 35, (c) -14.
2 (a) 0, (b) 1, (c) $-2 \times 7 \times 7 \times 7 \times 13 \times 17 = -151\,606$.
3 (a) 1, (b) 1.
4 (a) $(y - z)(z - x)(x - y)$, (b) $x(y - x)(z - y)$.

Page 147

5 1, $\begin{pmatrix} \cos\alpha\cos\beta & -\sin\beta & \sin\alpha\cos\beta \\ \cos\alpha\sin\beta & \cos\beta & \sin\alpha\sin\beta \\ -\sin\alpha & 0 & \cos\alpha \end{pmatrix}$.

9 $t, 2t, 3t$.

12 $\begin{pmatrix} a_1 \\ a_2 \\ a_3 \end{pmatrix}, \begin{pmatrix} b_1 \\ b_2 \\ b_3 \end{pmatrix}, \begin{pmatrix} c_1 \\ c_2 \\ c_3 \end{pmatrix}$; (a) $\begin{pmatrix} 1 & 0 & 0 \\ 0 & 1 & 0 \\ 0 & 0 & -1 \end{pmatrix}$,

(b) $\begin{pmatrix} 0 & -1 & 0 \\ 1 & 0 & 0 \\ 0 & 0 & 1 \end{pmatrix}$, (c) $\begin{pmatrix} 0 & 1 & 0 \\ 1 & 0 & 0 \\ 0 & 0 & 1 \end{pmatrix}$.

13 (1) is (2) rotated through 45°.

14 (1) is (2) rotated through $-60°$ and enlarged $\times 2$.

15 $k = 1,\ \begin{pmatrix} x \\ y \end{pmatrix} = \begin{pmatrix} 3 \\ -2 \end{pmatrix}$, or any multiple; $k = 2,\ \begin{pmatrix} x \\ y \end{pmatrix} = \begin{pmatrix} 2 \\ -1 \end{pmatrix}$, or any multiple.

16 $k = 15,\ \begin{pmatrix} x \\ y \end{pmatrix} = \begin{pmatrix} 1 \\ -1 \end{pmatrix}$, or any multiple; $k = 20,\ \begin{pmatrix} x \\ y \end{pmatrix} = \begin{pmatrix} 4 \\ 1 \end{pmatrix}$, or any multiple.

18 $\dfrac{1}{5}\begin{pmatrix} -1 & 9 & -3 \\ -2 & 3 & -1 \\ 2 & -13 & 6 \end{pmatrix}$;

$(-a + 9b - 3c)/5,\ (-2a + 3b - c)/5,\ (2a - 13b + 6c)/5$.

19 $\mathbf{M} = \begin{pmatrix} 3 & 2 & 1 \\ 2 & 1 & 0 \\ 1 & 0 & 0 \end{pmatrix},\ \mathbf{M}^{-1} = \begin{pmatrix} 0 & 0 & 1 \\ 0 & 1 & -2 \\ 1 & -2 & 1 \end{pmatrix}$.

21 $\begin{pmatrix} 6 & 2 & -2 \\ -21 & -7 & 7 \\ 3 & 1 & -1 \end{pmatrix}$.

22 $\mathbf{A}^{-1} = \begin{pmatrix} 1 & 0 & 0 \\ 1 & 1 & 0 \\ -5 & -2 & 1 \end{pmatrix},\ \mathbf{B}^{-1} = \begin{pmatrix} 1 & -4 & 14 \\ 0 & 1 & -3 \\ 0 & 0 & 1 \end{pmatrix}$.

$(\mathbf{AB})^{-1} = \begin{pmatrix} -73 & -32 & 14 \\ 16 & 7 & -3 \\ -5 & -2 & 1 \end{pmatrix},\ x = 5,\ y = -1,\ z = 0.$

23 (a) 3, -2 , 4, (b) $t,\ (9 - 5t)/3,\ (11t - 21)/3$.

24 (a) 3/2, $-4/3$, (b) 0, 1, $1\frac{1}{2}$; $a = 3/2$, possible, $a = -4/3$, not possible.

25 $-(a + b + c)\{(a - b)^2 + (b - c)^2 + (c - a)^2\}$.

Chapter 9

Qu. 1 (a) $y^2 = -4ax$, (b) $x^2 = 4by$.

Qu. 4 $y_1 x + 2ay - y_1(x_1 + 2a) = 0$.

Qu. 5 $x - ty + at^2 = 0$.
Qu. 6 $(\frac{5}{4}, 3)$, $y - \frac{13}{4} = 0$.
Qu. 7   7.2 cm, 12 cm.
Qu. 9   4, 3.
Qu. 10 $\frac{3}{5}$.
Qu. 11 $(\pm 3\sqrt{3}/2, 0)$.

Exercise 9a, page 156

1 $(at_1t_2, a(t_1 + t_2))$.
2 $t_1t_2 = -1$.
4 $(0, \frac{1}{2})$.
5 $(x - 2)^2 = 4(y + 1)$.
6 $(x + y)^2 + 4(x - y + 1) = 0$.
7 $-t$, $2/(t + t_1)$, $(a(t + 2/t)^2, -2a(t + 2/t))$.
9   (a) $(\frac{1}{2}, -2)$, (b) $(8, 8)$.
11 $(a(t_1{}^2 + t_1t_2 + t_2{}^2 + 2), -at_1t_2(t_1 + t_2))$.
14 $4a$.
15 $x - y + a = 0$, $x - 16y + 256a = 0$.

Exercise 9b, page 161

1 $y + 1 = 0$.   **2** $(1, \frac{1}{2})$, $\frac{3}{4}$.   **3** $(-\frac{3}{4}, -6)$.   **5** $a^2 = bc$.
6 $y = 4 - x^2$, $2x + y - 5 = 0$.   **8** $y = 2a/k$.   **9** $x = a$.
10 $y^2 - 4ax = \frac{1}{4}k^2$.   **11** $y^2 = 2a(x - a)$.   **12** $x(x - a)^2 = ay^2$.
13 $\left(-\dfrac{b}{2a}, -\dfrac{b^2 - 4ac}{4a} + \dfrac{1}{4a}\right)$, $y = -\dfrac{b^2 - 4ac}{4a} - \dfrac{1}{4a}$.
14 $yy_1 = 2a(x + x_1)$.
15 $(3a, \pm 2\sqrt{3}a)$.
16 $2a$.
18 $2x(h - x - 2a) + ky = 0$.
19 $3y^2 = 16ax$.
20 $y^2 - 2ax - 2ay + 2a^2 = 0$.

Exercise 9c, page 167

1 (a) $(\pm\sqrt{5}, 0)$, $x = \pm 9\sqrt{5}/5$; (b) $(\pm 5\sqrt{15}/4, 0)$, $x = \pm 4\sqrt{15}/3$.
2 (a) $2x \cos\theta + 3y \sin\theta - 6 = 0$, (b) $9x + 16y - 25 = 0$.
3 (a) $16x - 9y - 7 = 0$, (b) $4x + y - 2 = 0$.   **4** $3x - 2y - 5 = 0$.
6 $\dfrac{\cos\frac{1}{2}(\theta - \phi)}{\cos\frac{1}{2}(\theta + \phi)} = \pm e$.   **8** $ex + y - a = 0$.
10 $\left(a\dfrac{\cos\frac{1}{2}(\theta + \phi)}{\cos\frac{1}{2}(\theta - \phi)}, b\dfrac{\sin\frac{1}{2}(\theta + \phi)}{\cos\frac{1}{2}(\theta - \phi)}\right)$.   **12** $\dfrac{(2x - ae)^2}{a^2} + \dfrac{4y^2}{b^2} = 1$.
13 $\dfrac{x^2}{a^2} + \dfrac{y^2}{b^2} = \cos^2\frac{1}{2}k$.   **14** $4a^2x^2 + 4b^2y^2 = (a^2 - b^2)^2$.
15 $a^2y^2 + b^2x^2 = 4x^2y^2$.   **16** $b^2x^2 + a^2y^2 = 2a^2b^2$.

Exercise 9d, page 170

1   (a) $2x - y \pm 5 = 0$, (b) $x + y \pm 2 = 0$, (c) $x - 2y \pm 10 = 0$.

2   (a) $(\frac{9}{13}, -\frac{4}{13})$, (b) $(-\frac{1}{2}, \frac{1}{5})$, (c) $(11, -6)$.

4   $3x + 2y \pm 2\sqrt{10} = 0$.

5   ± 5, $(16/5, -9/5)$, $(-16/5, 9/5)$.

6   $c^2 < a^2m^2 + b^2$.

8   $(x^2 + y^2)^2 = a^2x^2 + b^2y^2$.

9   $a^2y_1x - b^2x_1y - x_1y_1(a^2 - b^2) = 0$.

10   (a) $\left(\dfrac{-a^2mc}{a^2m^2 + b^2}, \dfrac{b^2c}{a^2m^2 + b^2}\right)$, (b) $\left(\dfrac{2am^2 - nl}{l^2}, -\dfrac{2am}{l}\right)$.

11   $8x - 27y = 0$.

12   $mx - y - mae = 0$, $\left(\dfrac{a^3em^2}{b^2 + a^2m^2}, -\dfrac{ab^2em}{b^2 + a^2m^2}\right)$, $a^2y^2 + b^2x^2 - ab^2ex = 0$.

13   $a^2y^2 + b^2x(x - a) = 0$.

14   $(a^2 + b^2 - x^2 - y^2)^2 = 4(b^2x^2 + a^2y^2 - a^2b^2)$.

15   $b^2hx + a^2ky - (b^2h^2 + a^2k^2) = 0$.

16   $(a^2y^2 + b^2x^2)^2 = a^2(a^4y^2 + b^4x^2)$.

17   $(y^2 - 2ax)^2 = 4a^4 + b^2y^2$.

18   $a^4y^2 + b^4x^2 = 4x^2y^2$.

Exercise 9e, page 179

2   $9b^2x^2 - 9a^2y^2 - 12ab^2x + 12a^2by - a^2b^2 = 0$.

4   $(a^2 + b^2)^2 = 4(a^2x^2 - b^2y^2)$.

5   $(-c/t^3, -ct^3)$.

6   $xy = c^2$.

7   $2xyc^2 = c^4 - y^4$.

8   $y_1x + x_1y - 2x_1y_1 = 0$.

10   $(ct, -ct^3)$.

11   $(x^2 + y^2)^2 = 4c^2xy$.

12   $n^2 = 4lmc^2$.

14   $x^2 + y^2 = a^2 - b^2$.

15   $(x^2 + y^2)^2 = a^2x^2 - b^2y^2$.

18   $bx \cos \frac{1}{2}(\theta - \phi) - ay \sin \frac{1}{2}(\theta + \phi) - ab \cos \frac{1}{2}(\theta + \phi) = 0$.

Exercise 9f, page 180

1   $x^2/36 + y^2/20 = 1$.

2   $x^2/36 + 4y^2/119 = 1$.

4   $3x \pm 2y = 0$; $3x \pm 2y + 1 = 0$.

5   $25x + 20y + 64 = 0$, $4/5$.

6   $x \pm y \pm \frac{1}{6}\sqrt{3} = 0$.

9   $(4, -3)$.

10   $(0, 3)$, $3\sqrt{2}$.

11   $y + 1 = 0$.

13   1, $(\frac{1}{2}, 2)$; $2x - 2y + 1 = 0$, $2x + 2y - 3 = 0$.

17 (a) $x = -4a$, (b) $(4a, 0)$.

18 $\frac{1}{2}ab \cos\mathrm{ec}\ 2\theta$.

19 $h = \dfrac{2cpq}{p+q}$, $k = \dfrac{2c}{p+q}$, (b) $xy = 2c^2$, least distance $= 2c$.

20 $(V^2 \sin 2\alpha/(2g), -V^2 \cos 2\alpha/(2g))$, $y = V^2/(2g)$.

Chapter 10

Qu. 1 (a) $1 - x + \dfrac{x^2}{2!} - \dfrac{x^3}{3!} + \ldots + (-1)^n\dfrac{x^n}{n!} + \ldots,$

(b) $1 + x^2 + \dfrac{x^4}{2!} + \dfrac{x^6}{3!} + \ldots + \dfrac{x^{2n}}{n!} + \ldots,$

(c) $1 + 3x + \dfrac{9x^2}{2} + \dfrac{9x^3}{2} + \ldots + \dfrac{3^n x^n}{n!} + \ldots,$

(d) $1 + \dfrac{1}{x} + \dfrac{1}{2x^2} + \dfrac{1}{3!x^3} + \ldots + \dfrac{1}{n!x^n} + \ldots,$

(e) $1 - \dfrac{1}{x^2} + \dfrac{1}{2x^4} - \dfrac{1}{3!x^6} + \ldots + \dfrac{(-1)^n}{n!x^{2n}} + \ldots.$

Qu. 3 (a) $\dfrac{1}{4}x - \dfrac{1}{32}x^2 + \dfrac{1}{192}x^3 - \ldots + (-1)^{n-1}\dfrac{x^n}{4^n n} + \ldots,\ -4 < x \leqslant 4,$

(b) $\ln 3 - \dfrac{1}{3}x - \dfrac{1}{18}x^2 - \ldots - \dfrac{x^n}{3^n n} - \ldots,\ -3 \leqslant x < 3,$

(c) $-2x - x^2 - \dfrac{2}{3}x^3 - \ldots - \dfrac{2x^n}{n} - \ldots,\ -1 \leqslant x < 1.$

Qu. 4 $-1 + x - \frac{1}{2}x^2.$

Exercise 10a, page 187

1 (a) 1.1052, (b) 0.3679, (c) 1.6487.

2 $1 + x^3 + \frac{1}{2}x^6 + \frac{1}{6}x^9 + \ldots + x^{3n}/n! + \ldots$

3 $1 + \frac{1}{3}x + \frac{1}{18}x^2 + \frac{1}{162}x^3 + \ldots + x^n/(n!3^n) + \ldots.$

4 $1 - 2x + 2x^2 - \frac{4}{3}x^3 + \ldots + (-1)^n 2^n x^n/n! + \ldots.$

5 $e^2\{1 + x + \frac{1}{2}x^2 + \frac{1}{6}x^3 + \ldots + x^n/n! + \ldots\}.$

6 $1 - \frac{1}{2}x + \frac{1}{8}x^2 - \frac{1}{48}x^3 + \ldots + (-1)^n x^n/(n!2^n) + \ldots.$

7 $1 + 2x + \frac{3}{2}x^2 + \frac{2}{3}x^3 + \ldots + (n+1)x^n/n! + \ldots.$

8 $1 - 2x^2 + \frac{8}{3}x^3 - 2x^4 + \ldots + (-1)^{n-1}(n-1)2^n x^n/n! + \ldots.$

9 $1 + 4x + 8x^2 + \frac{32}{3}x^3 + \ldots + 4^n x^n/n! + \ldots.$

10 $2 + 3x + \frac{5}{2}x^2 + \frac{3}{2}x^3 + \ldots + (1 + 2^n)x^n/n! + \ldots.$

11 $10^9/9!,\ 10^{10}/10!$

12 $1 + 2x + 3x^2 + \frac{10}{3}x^3.$

13 $e(1 - 3x + \frac{11}{2}x^2 - \frac{15}{2}x^3).$

14 $1 + \frac{1}{2}x^2 - \frac{1}{3}x^3.$

15 $1 - x + \frac{1}{2}x^2 - \frac{1}{6}x^3.$

16 (a) $\frac{1}{4}$, (b) $2\frac{2}{3}$, (c) 1.

Page 187

17 $(1 + x)e^x$.

18 $(e^{3x} - 1)/(3x)$.

19 $\frac{1}{2}(e^x + e^{-x})$.

20 $\frac{1}{2}(e^x - e^{-x})$.

Exercise 10b, page 193

1 (a) $\ln 3 + \dfrac{1}{3}x - \dfrac{1}{18}x^2 + \dfrac{1}{81}x^3 - \dots + (-1)^{n-1}\dfrac{x^n}{3^n \times n} + \dots,\ -3 < x \leqslant 3.$

  (b) $-\dfrac{1}{2}x - \dfrac{1}{8}x^2 - \dfrac{1}{24}x^3 - \dfrac{1}{64}x^4 - \dots - \dfrac{x^n}{2^n \times n} \dots,\ -2 \leqslant x < 2.$

  (c) $\ln 2 - \dfrac{5}{2}x - \dfrac{25}{8}x^2 - \dfrac{125}{24}x^3 - \dots - \dfrac{5^n x^n}{2^n \times n} - \dots,\ -\dfrac{2}{5} \leqslant x < \dfrac{2}{5}.$

  (d) $-x^2 - \dfrac{1}{2}x^4 - \dfrac{1}{3}x^6 - \dfrac{1}{4}x^8 \dots - \dfrac{x^{2n}}{n} - \dots,\ -1 < x < 1.$

  (e) $\dfrac{2}{3}x + \dfrac{2}{81}x^3 + \dfrac{2}{1215}x^5 + \dots + \dfrac{2x^{2n-1}}{(2n-1)\,3^{2n-1}} + \dots,\ -3 < x < 3.$

  (f) $-\dfrac{3x}{2} - \dfrac{9}{32}x^3 - \dfrac{243}{2560}x^5 - \dots - 2 \times \dfrac{3^{2n-1}x^{2n-1}}{(2n-1)4^{2n-1}} - \dots,\ -\dfrac{4}{3} < x < \dfrac{4}{3}.$

2 $\ln\dfrac{2}{3} - \dfrac{1}{6}x - \dfrac{5}{72}x^2 - \dots - \left(\dfrac{1}{2^n} - \dfrac{1}{3^n}\right)\dfrac{x^n}{n} - \dots,\ -2 \leqslant x < 2.$

3 $-\ln 3 + \frac{4}{3}x + \frac{20}{9}x^2 + \dots + 2^n\{1 + (-1)^n(\frac{1}{3})^n\}x^n/n + \dots,\ -\frac{1}{2} \leqslant x < \frac{1}{2}.$

4 $-6x^2 + 28x^3 - 111x^4 + \dots + (-1)^{n-1}(4^{n-1} - 3^{n-1})12x^n/n + \dots,\ -\frac{1}{4} < x \leqslant \frac{1}{4}.$

5 $\frac{1}{2}\ln 2 + \frac{3}{4}x - \frac{5}{16}x^2 + \dots + (-1)^{n-1}\{1 + (\frac{1}{2})^n\}x^n/(2n) + \dots,\ -1 < x \leqslant 1.$

6 $x + \dfrac{1}{2}x^2 - \dfrac{2}{3}x^3 + \dots - \dfrac{2}{3n}x^{3n} + \dfrac{x^{3n+1}}{3n+1} + \dfrac{x^{3n+2}}{3n+2} - \dots,\ -1 \leqslant x < 1.$

7 $1 - \frac{1}{2}x + \frac{1}{3}x^2 - \dots + (-1)^n x^n/(n+1) + \dots,\ -1 < x \leqslant 1.$

8 $-x + \frac{1}{2}x^2 + \frac{2}{3}x^3 + \dots - (-1)^{3n}\dfrac{2x^{3n}}{3n} + (-1)^{3n+1}\dfrac{x^{3n+1}}{3n+1} +$

  $\qquad\qquad\qquad + (-1)^{3n+2}\dfrac{x^{3n+2}}{3n+2} + \dots,\ -1 < x \leqslant 1.$

9 $x + \frac{1}{2}x^2 + \frac{5}{6}x^3,\ |x| < 1.$

10 $x + \frac{1}{2}x^2 + \frac{1}{3}x^3,\ -1 < x \leqslant 1.$

11 $1 + \frac{3}{2}x + \frac{5}{12}x^2,\ -1 < x \leqslant 1.$

12 $x^2 + x^3 + \frac{11}{12}x^4,\ -1 \leqslant x < 1.$

13 0.693 1, 1.099.

14 2.302 6, 0.434 3.

15 1.945 9.

16 1.041 4.

17 (a) -1, (b) 1, (c) $-\frac{2}{3}$, (d) 0.

18 $\ln\frac{4}{3}$.

19 $\ln 2$.

20 $\frac{1}{2}\ln\frac{5}{3}$.

21 $\frac{5}{2}\ln\frac{7}{5}$.

Page 194

22 ln 3.

25 s_n tends to a limit between 0 and 1.

Exercise 10c, page 195

1 For standard logarithmic series, see pages 188, 191.
$$2\left\{\frac{m-n}{m+n}+\frac{1}{3}\left(\frac{m-n}{m+n}\right)^3+\frac{1}{5}\left(\frac{m-n}{m+n}\right)^5+\ldots+\frac{1}{2n-1}\left(\frac{m-n}{m+n}\right)^{2n-1}+\ldots\right\},$$
2.079 44.

2 $1/(1-x^2)$.

4 $e^y = 1 + y + \frac{1}{2}y^2 + \ldots;\ \frac{1}{2}(\ln a)^2 - 1/a.$

5 (a) 2.3979, (b) $(-1)^{n-1}\dfrac{(n-1)3^n}{n!}.$

6 $a(1-r^n)/(1-r),\ 0.095\,310.$

7 (b) $-2/(3n),\ 1/(3n+1),\ 1/(3n+2),$ (c) $-\frac{2}{3}x^3 + \frac{1}{8}x^4.$

8 (c) $2e - 5.$

9 (a) $2x - \frac{1}{3}x^3 + \frac{2}{5}x^5 - \frac{1}{2}x^6,\ |x| < 1,$ (b) $e^x = 1 + x + \dfrac{x^2}{2!} + \dfrac{x^3}{3!} + \ldots + \dfrac{x^n}{n!} + \ldots$

$$e^{-x} = 1 - x + \frac{x^2}{2!} - \frac{x^3}{3!} + \ldots + (-1)^n\frac{x^n}{n!} + \ldots.$$

10 $p = q = \frac{1}{2};\ a = \frac{1}{48}.$

11 $a = \frac{2}{3},\ b = \frac{1}{6}.$

13 $\ln 2 + \dfrac{3}{2}x + \dfrac{3}{8}x^2 + \dfrac{3}{8}x^3 + \ldots + \dfrac{x^n}{n}\left[1 - \left(-\dfrac{1}{2}\right)^n\right] + \ldots,\ -1 \leqslant x < 1.$

15 $\dfrac{1}{2} + \dfrac{1}{4}\left(\dfrac{x^2-1}{x}\right)\ln\left(\dfrac{1+x}{1-x}\right);\ \dfrac{N}{2N+1}.$

16 $(1+x)^n = 1 + nx + \dfrac{n(n-1)}{2!}x^2 + \dfrac{n(n-1)(n-2)}{3!}x^2 + \ldots;\ \ln 2.$

17 (b) $10e - 4.$

18 (a) $e^x(1+x),$ (b) $1 + (1/x - 1)\ln(1-x),$ (c) $9\sqrt{3}.$

Chapter 11

Qu. 1 (a) $2\ln a + \ln b,$ (b) $3\ln a - 3\ln b,$ (c) $\frac{1}{2}\ln a + \frac{1}{2}\ln b + \frac{1}{2}\ln c,$
(d) $\ln a + \frac{1}{3}\ln b - 3\ln c,$ (e) $-4\ln c,$ (f) $b\ln a.$

Qu. 2 (a) 3, (b) $-2,$ (c) 4, (d) 2, (e) $2x,$ (f) $3x^2.$

Qu. 3 (a) $1/x,$ (b) $2/(1+2x),$ (c) $-1/(1-x),$ (d) $3/x,$ (e) $\cot x,$
(f) $\sec x \csc x = 2\csc 2x.$

Qu. 4 (a) $6y\dfrac{dy}{dx},$ (b) $3y^2\dfrac{dy}{dx},$ (c) $-\sin y\dfrac{dy}{dx},$ (d) $\dfrac{1}{y}\dfrac{dy}{dx},$ (e) $20y^3\dfrac{dy}{dx},$

(f) $-\dfrac{6}{y^3}\dfrac{dy}{dx},$ (g) $\dfrac{1}{2\sqrt{y}}\dfrac{dy}{dx},$ (h) $\sec^2 y\dfrac{dy}{dx}.$

Qu. 5   (a) $-\frac{2}{3}(x+1)^{-2/3}(x-1)^{-4/3}$, (b) $-\dfrac{2x^2+x+4}{(2x-1)^3\sqrt{(x^2+1)}}$,

(c) $\dfrac{x\,e^x(x^2-2x-2)}{(x-1)^4}$.

Qu. 6 $\dfrac{10^x}{\ln 10}$, $\dfrac{10^x}{\ln 10}+c$.

Qu. 7   (a) $2^x \ln 2$, (b) $3^x \ln 3$, (c) $-(\tfrac{1}{2})^x \ln 2$, (d) $5(\ln 10)10^{5x}$, (e) $2x\,10^{x^2} \ln 10$.

Qu. 8   (a) $\dfrac{2^x}{\ln 2}+c$, (b) $\dfrac{3^x}{\ln 3}+c$, (c) $-\dfrac{(\tfrac{1}{2})^x}{\ln 2}+c$, (d) $\dfrac{10^{5x}}{5\ln 10}+c$.

Qu. 9   (a) $\frac{2}{9}(3x+1)^{3/2}+c$, (b) $-\frac{1}{6}\cos^6 x+c$, (c) $\frac{1}{2}(1+\cos x)^{-2}+c$,
(d) $5^x/\ln 5+c$, (e) $2^{2x}/(2\ln 2)+c$, (f) $x\ln x-x+c$.

Qu. 10   (a) $\tan y=x$, (b) $x=\sec y$, (c) $p=\cos q$.

Qu. 11   (a) $2y\dfrac{dy}{dx}$, (b) $\cos y\dfrac{dy}{dx}$, (c) $\sec^2 y\dfrac{dy}{dx}$, (d) $\sec y\tan y\dfrac{dy}{dx}$.

Qu. 12   (a) $-1/\sqrt{(1-x^2)}$, (b) $-1/(1+x^2)$, (c) $1/\sqrt{(-x-x^2)}$.

Qu. 13   Maximum at $(1, 1/e)$.

Qu. 15 $(3, 8)$.

Exercise 11a, page 205

1   (a) $3\ln a+4\ln b$, (b) $\ln a-\ln b$, (c) $\frac{3}{2}\ln a-\frac{1}{2}\ln b$,
(d) $2\ln a+\ln b-\frac{1}{2}\ln c$, (e) $\frac{1}{2}\ln a+\frac{1}{2}\ln b-\frac{1}{2}\ln c$,
(f) $-\frac{1}{2}\ln a-\frac{1}{2}\ln b-\frac{1}{2}\ln c$.

2   (a) 5, (b) 3, (c) 4, (d) $\frac{1}{2}$, (e) x^3, (f) $-2x$.

3 $2(3-2x)\sqrt{\dfrac{2x+3}{(1-2x)^3}}$.

4 $\dfrac{e^{x/2}}{2x^5}(x\sin x-8\sin x+2x\cos x)$.

5 $\dfrac{x-5x^3}{3\sqrt{(x^2+1)^3}\,\sqrt[3]{(x^2-1)^4}}$.   **6** $\dfrac{x\tan x-x-1}{x^2\,e^x\cos x}$.   **7** $7^x \ln 7$.

8 $10^{3x}\,3\ln 10$.   **9** $-\frac{1}{2}\,10^{-x/2}\ln 10$.   **10** $-\dfrac{\ln 10}{10^x}$.   **11** $\dfrac{5^x}{\ln 5}+c$.

12 $\dfrac{8^x}{\ln 8}+c$.   **13** $-\dfrac{(\frac{1}{3})^x}{\ln 3}+c$.   **14** $\dfrac{3^{2x}}{2\ln 3}+c$.   **15** $a^x \ln a$.

16 $\dfrac{a^x}{\ln a}+c$.   **17** $\dfrac{1}{1+x^2}$.   **18** $\dfrac{1}{x\sqrt{(x^2-1)}}$.   **19** $\dfrac{1}{\sqrt{(-x^2-2x)}}$.

20 $-\dfrac{1}{\sqrt{(x-x^2)}}$.   **21** $-\dfrac{2x}{x^4+1}$.   **22** $\dfrac{-10}{\sqrt{(1-25x^2)}}$.

23   (a) 0, (b) 0. The angles are complementary.

24   (a) $\dfrac{2}{\sqrt{(1-4x^2)}}$, (b) $\dfrac{2x}{\sqrt{(1-x^4)}}$, (c) $\dfrac{1}{\sqrt{(1-x^2)}}$.

25 $\frac{1}{6}(4x+3)^{3/2}+c$.

26 $\frac{1}{16}(2x^2+1)^4+c$.

27 $x\ln x-x+c$.

Page 206

28 $x \sin^{-1} x + \sqrt{(1 - x^2)} + c.$

29 $-(1 + \ln x)x^{-x}.$

30 $\left(\cos x \ln x + \dfrac{1}{x} \sin x \right) x^{\sin x}.$

Exercise 11b, page 211

1 (1, 4) max., (3, 0) min.

2 (2, 3) min.

3 (2, −1) max., (4, 3) min.

4 (1, 1) min.

5 (±3, −405).

6 None.

8 $(\frac{1}{3}\pi, \frac{3}{4}\sqrt{3})$ max., $(\pi, 0)$ infl., $(\frac{5}{3}\pi, -\frac{3}{4}\sqrt{3})$ min.

9 $(\frac{1}{2}\pi, \frac{1}{2})$ max.

10 (0, −1) max., (2π, −1) max.

11 $(0, \frac{2}{3})$ min., $(\frac{1}{4}\pi, \frac{2}{3}\sqrt{2})$ max., $(\frac{3}{4}\pi, -\frac{2}{3}\sqrt{2})$ min., $(\pi, -\frac{2}{3})$ max., $(\frac{5}{4}\pi, -\frac{2}{3}\sqrt{2})$ min., $(\frac{7}{4}\pi, \frac{2}{3}\sqrt{2})$ max., $(2\pi, \frac{2}{3})$ min.

12 (1, 1/e) max.

13 (2, 10 tan^{-1} 2 − 2) max.

15 $\left(2n\pi + \dfrac{1}{4}\pi, \dfrac{1}{\sqrt{2}} e^{2n\pi + \pi/4} \right)$ max., $\left(2n\pi + \dfrac{5}{4}\pi, -\dfrac{1}{\sqrt{2}} e^{2n\pi + 5\pi/4} \right)$ min.

16 $\sqrt{2}:1.$

18 $\sqrt{2}:1.$

20 1:2.

Exercise 11c, page 216

1 (a) $\dfrac{2x}{(1 - x^2)^2}$, (b) $\dfrac{-1}{x\sqrt{(x^2 - 1)}}$, (c) $e^{2x} (2 \cos 3x - 3 \sin 3x).$

2 (a) $\dfrac{x - 3}{2(x - 1)^{3/2}}$, (b) $3 \sin^2 x \cos^3 x - 2 \sin^4 x \cos x$, (c) ln $x.$

3 (a) $-\dfrac{12x - 1}{(2x + 1)^2(3x - 2)^2}$, (b) $-2 \sin 2x \, e^{\cos 2x}$, (c) $2 \cot 2x.$

4 (a) $\dfrac{-2}{\sqrt{(2x + 1)}\sqrt{(2x - 1)^3}}$, (b) $\dfrac{1}{1 + x^2}$, (c) $-4 \, e^{-4x}.$

5 (a) $x(x^2 + 1)(x^3 + 1)^2(13x^3 + 9x + 4),$

(b) $-\dfrac{\sec^2 x}{\sqrt{(1 - \tan^2 x)}} = - \sec x\sqrt{(\sec 2x)}$, (c) $\dfrac{1 - 2 \ln x}{x^3}.$

6 (a) $\dfrac{\sin^3 x(4 + \sin^2 x)}{\cos^6 x}$, (b) $\dfrac{1}{1 + x^2}.$

7 (a) $2^{x^2 + 1}x \ln 2$, (b) $2(\ln x + 1)x^{2x}.$

10 (a) $-b/(a^2 \sin^3 \theta).$

11 (a) $\dfrac{3}{4t} + \dfrac{1}{t^3}$, (b) $\frac{1}{3} \sec^4 t \, \text{cosec } t$, (c) $-\dfrac{1}{4a} \text{cosec}^4 \frac{1}{2}\theta.$

Page 217

14 $2\sqrt{(1-x^2)}$, $\frac{1}{2}x\sqrt{(1-x^2)} + \frac{1}{2}\sin^{-1}x + c$.

15 $\tan^{-1}x$, $x\tan^{-1}x - \frac{1}{2}\ln(1+x^2) + c$.

16 1 max., -1 min.

17 $(2^{1/3}, 2^{2/3})$ max., $(0, 0)$ min.

18 Point of inflexion at $(0, 0)$.

20 $\cos(x + \frac{1}{2}\pi)$, $\cos(x + \frac{1}{2}n\pi)$.

23 $\dfrac{-16}{21(x+5)^2} + \dfrac{4}{21(4x-1)^2}$, $\dfrac{32}{21(x+5)^3} - \dfrac{32}{21(4x-1)^3}$.

$(1, \frac{1}{9})$ max., $(-\frac{1}{3}, \frac{9}{49})$ min., $(2, \frac{5}{49})$ infl.

24 $\left(\dfrac{\pi}{4}, \dfrac{2\sqrt{2}-1}{14}\right)$ max., $\left(\dfrac{5\pi}{4}, -\dfrac{2\sqrt{2}+1}{14}\right)$ min.

25 $4a(\sqrt{5}-1)$.

26 $5c$.

30 (a) $e^x(x^3 + 9x^2 + 18x + 6)$, (b) $x\cos x + 4\sin x$,

(c) $2^{n-2}e^{2x}\{4x^2 + 4nx + n(n-1)\}$, (d) $(1-x^2)y_n - 2nxy_{n-1} - n(n-1)y_{n-2}$.

Chapter 12

Qu. 1 (a) $n\pi$, (b) $(2n+1)\pi$, (c) $n\pi + \frac{1}{4}\pi$, (d) $2n\pi + \frac{1}{2}\pi$, (e) $n\pi + \frac{1}{2}\pi$,

(f) $n\pi + (-1)^n\pi/6$, (g) $n\pi - \frac{1}{4}\pi$, (h) $n\pi + (-1)^n\frac{1}{4}\pi$, (i) $2n\pi \pm \frac{3}{4}\pi$.

Qu. 2 (a) $-\pi/4$, (b) $-\pi/6$, (c) π, (d) $\pi/3$, (e) $\pi/2$, (f) 0.

Qu. 3 (a) 0.322, (b) 1.824, (c) 0.010, (d) -0.201, (e) -1.249, (f) 0.927.

Qu. 4 $n\pi$, $(2n+1)\pi/4$.

Qu. 5 $(4n+1)\pi/18$, $(4n-1)\pi/2$.

Exercise 12a, page 223

1 $n\pi \pm \frac{1}{6}\pi$.   **2** $180n° - 32.7°$.   **3** $360n° \pm 47.6°$.   **4** $n\pi$, $2n\pi \pm \pi/3$.

5 $\frac{1}{6}\pi + 2n\pi/3$.   **6** $(2n+1)\pi/2$, $n\pi + (-1)^n\pi/6$.   **7** $180n° + (-1)^n17.6°$.

8 $n\pi$.   **9** $n\pi/2$.   **10** $n\pi/2$.   **11** $n\pi/3$.   **12** $n\pi/2$, $n\pi - (-1)^n\pi/6$.

13 $n\pi/3$.   **14** $\frac{1}{2}\pi + 2n\pi/3$.   **15** $2n\pi - \frac{1}{2}\pi$, $n\pi + (-1)^n\pi/6$.

16 $(4n+1)\pi/8$, i.e. $180n° + 22.5°$, $180n° - 67.5°$.   **17** $n\pi$, $n\pi + (-1)^n\pi/6$.

18 $2n\pi$, $2n\pi \pm \pi/3$.   **19** $n\pi + (-1)^n\pi/6$.   **20** $(2n+1)\pi/2$, $(2n+1)\pi/4$.

21 $180n° + 8.1°$, $180n° - 12.5°$.   **22** $n\pi - (-1)^n\pi/6$.   **23** $2n\pi$, $2n\pi + 2\pi/3$.

24 $360n° - 53.1°$, $360n° + 36.9°$.   **25** $360n° + 163.7°$.

Exercise 12b, page 225

1 (a) $\pi/3$, (b) $\pi/4$, (c) $\pi/6$, (d) $\pi/2$, (e) $-\pi/3$, (f) $-\pi/2$.

2 (a) 1.107, (b) 0.643, (c) 1.159, (d) -0.340, (e) -0.464, (f) 1.318.

3 $n\pi$, $2n\pi \pm \cos^{-1}\frac{1}{6}$.

4 $(2n+1)\pi$, $2n\pi \pm \cos^{-1}\frac{2}{3}$.

5 $n\pi$, $n\pi \pm \sin^{-1}\frac{5}{6}$.

6 $n\pi$, $n\pi \pm \tan^{-1}(\frac{1}{2}\sqrt{2})$.

7 $2n\pi + \pi/4$.

8 $2n\pi$, $2n\pi - 2\pi/3$.

Page 225

9 $n\pi + \tan^{-1} (3 \tan \alpha)$.

10 $n\pi + \alpha - \tan^{-1} \frac{3}{4}$.

11 $2n\pi \pm \alpha, (2n + 1)\pi \pm \cos^{-1} (\frac{1}{2} \cos \alpha)$.

12 $(2n + 1)\pi - \tan^{-1} \frac{4}{3}, (2n + 1)\pi - \tan^{-1} \frac{12}{5}$.

13 $(4n + 1)\pi/2, (2n + 1)\pi - \tan^{-1} \frac{20}{21}$.

14 $n\pi/2$.

15 $2n\pi, (2n + 1)\pi/5$.

16 $n\pi/3$.

17 $(4n + 1)\pi/10$.

18 $(4n + 1)\pi/14$.

19 $n\pi/6$.

20 $(4n - 1)\pi/10, (4n + 1)\pi/2$.

21 $2n\pi/5; 4 \cos^3 \theta - 2 \cos^2 \theta - 3 \cos \theta + 1 = 0; -\frac{1}{4}(1 + \sqrt{5})$.

Exercise 12c, page 225

1 (a) $217°, 323°$, (b) $60°, 90°, 270°, 300°$, (c) $76.7°, 209.6°$.

2 (a) $0°, 45°, 135°, 180°$, (b) $35.3°, 144.7°$, (c) $90°$.

3 (a) $18°, 72°$, (b) $0°, 30°, 90°, 150°, 180°, 210°, 270°, 330°, 360°$.

4 (a) $(2n + 1)\pi/4, 2n\pi \pm \frac{1}{3}\pi$, (b) $360n° + 29.6°, 360n° + 256.7°$.

5 (a) $(2k - \frac{1}{2})\pi/5, (2k + \frac{1}{2})\pi/3$, (b) $(2k \pm \frac{1}{3})\pi$, (c) $k\pi/6$.

6 (a) $n\pi + (-1)^n \pi/6, n\pi - (-1)^n \sin^{-1} \frac{3}{4}$, (b) $(4n - 1)\pi/2, (4n + 1)\pi/10$.

7 (a) $16c^5 - 20c^3 + 5c, \frac{1}{4}\sqrt{(10 - 2\sqrt{5})}$, (b) $k\pi/3$.

8 $2x^2 - x - 1 = 0; 1, -\frac{1}{2}; \theta = 0, \phi = 2\pi/3; \theta = 2\pi/3, \phi = 0$.

9 (a) $120°, 240°, 300°$.

10 $n\pi/5$.

11 (a) $45°, 105°, 165°, 225°, 285°, 345°$, (b) $210°, 330°$.

12 $26.6°, 90°, 206.6°, 270°$.

13 (a) $\pi/2, 11\pi/6$; (b) $5\pi/4, 3\pi/2$; $\dfrac{1 - \sqrt{2}}{2} \leqslant k \leqslant \dfrac{1 + \sqrt{2}}{2}$.

14 (a) $\alpha + \pi/6 + 2n\pi$ or $\pi/2 - \alpha + 2n\pi$, (b) $11\pi/12$ or $5\pi/4$.

15 $\frac{1}{3}, \frac{1}{2}$.

16 (a) No, (b) $51°, 111°, 171°, 231°, 291°, 351°$.

17 (a) $a = 2, b = 1$, (b) $2/\cos 2x$.

18 $0 < \theta < \pi/3; \cos 3\theta = 1 - 1/(2p^2); 0.37$.

19 $\tan 3A = \dfrac{3 \tan A - \tan^3 A}{1 - 3 \tan^2 A}; 0°, 40.2°, 139.8°, 180°$.

20 (a) $-\sqrt{3}/2 < x < 1/2; x > \sqrt{3}/2$, (b) $-\pi/3 + 2n\pi < x < \pi/6 + 2n\pi$; $5\pi/6 + 2n\pi < x < 4\pi/3 + 2n\pi; \pi/3 + 2n\pi < x < 2\pi/3 + 2n\pi$, (c) $\mathbb{R}$.

Chapter 13

Qu. 3   (a) $\sin x - x \cos x + c$, (b) $\frac{1}{2}x \sin 2x + \frac{1}{4} \cos 2x + c$,
(c) $\frac{1}{2}x^2 \ln x - \frac{1}{4}x^2 + c$, (d) $xe^x - e^x + c$.

Qu. 4 $2xe^{x^2}, \frac{1}{2}e^{x^2} (x^2 - 1) + c$.

Qu. 5 $x \ln x - x + c$.

Qu. 6 (a) $(1 - x^2)^{-1/2}$, (b) $x \sin^{-1} x + \sqrt{(1 - x^2)} + c$.

Qu. 8 (a) $(x^2 - 2) \sin x + 2x \cos x + c$, (b) $e^x(x^2 - 2x + 2) + c$.

Qu. 9 (a) $3(1 - 9x^2)^{-1/2}$, (b) $2(1 + 4x^2)^{-1}$, (c) $(9 - x^2)^{-1/2}$,
     (d) $-2(1 - 4x^2)^{-1/2}$, (e) $\frac{3}{2}(1 + 9x^2)^{-1}$, (f) $6(4 + x^2)^{-1}$,
     (g) $\frac{1}{2}(2x - x^2)^{-1/2}$, (h) $4(5 + 2x + x^2)^{-1}$.

Qu. 10 (a) $\tan^{-1} \dfrac{x}{2} + c$, (b) $\frac{3}{2} \tan^{-1} 2x + c$, (c) $4 \sin^{-1} \dfrac{x}{3} + c$, (d) $\frac{1}{3} \sin^{-1} 3x + c$,

   (e) $\dfrac{1}{5\sqrt{2}} \tan^{-1} \dfrac{5x}{\sqrt{2}} + c$, (f) $\sin^{-1} \dfrac{2x}{\sqrt{3}} + c$, (g) $\dfrac{1}{\sqrt{2}} \tan^{-1} \dfrac{x - 1}{\sqrt{2}} + c$,

   (h) $5 \sin^{-1} \dfrac{x + 2}{3} + c$.

Qu. 11 $\ln \tan \frac{1}{2}x + c$.

Qu. 12 $-2 \ln \cos \frac{1}{2}\theta + c$.

Qu. 14 (a) $(1 + t^2)^{-1}$, (b) $(4 + 4t^2)^{-1}$, (c) $2(3 + 3t^2)^{-1}$.

Qu. 15 (a) $\frac{1}{2} \ln \tan x + c$, (b) $-\frac{2}{3}(1 + \tan \frac{3}{2}\theta)^{-1} + c$, (c) $\ln \{x + \sqrt{(x^2 - 1)}\} + c$.

Qu. 16 (a) $\dfrac{1}{\sqrt{2}} \tan^{-1} \left(\dfrac{1}{\sqrt{2}} \tan x \right) + c$, (b) $-\ln (1 - \tan^2 x) + c$.

Qu. 17 (a) $\ln (x^2 + 2x + 10) + \frac{1}{3} \tan^{-1} \dfrac{x + 1}{3} + c$,

   (b) $5 \sin^{-1} \dfrac{x + 2}{3} + 2\sqrt{(5 - 4x - x^2)} + c$, (c) $\frac{1}{2}x - \frac{1}{2} \ln (\sin x + \cos x) + c$,

   (d) $\frac{3}{2}x - \frac{5}{2} \ln (3 \cos x + \sin x) + c$.

Qu. 18 (a) 1, (b) $\frac{1}{2}\pi$.

Qu. 19 (b) $\frac{1}{2}\pi$, (e) 1, (f) $\pi/3$.

Qu. 20 $\frac{1}{13}e^{2x}(2 \sin 3x - 3 \cos 3x) + c$.

Qu. 21 (a) and (b) $\frac{1}{5}e^x(\cos 2x + 2 \sin 2x) + c$, no.

Qu. 22 $\frac{1}{4} \sin x \cos^3 x + \frac{3}{8} \sin x \cos x + \frac{3}{8}x + c$.

Qu. 24 (a) $\frac{1}{3} \sin x \cos^2 x + \frac{2}{3} \sin x + c$, (b) $\sin x - \frac{1}{3} \sin^3 x + c$.

Qu. 25 $\frac{8}{15}$.

Qu. 26 (a) $35\pi/256$, (b) $128/315$, (c) $63\pi/512$.

Qu. 29 (a) 7, (b) $2/\pi$, (c) $4/3$, (d) $1/2$.

Exercise 13a, page 230

1 (a) $2 \sin x - 2x \cos x + c$, (b) $\frac{1}{2}(x - 1)e^x + c$, (c) $\frac{1}{4} \sin 2x - \frac{1}{2}x \cos 2x + c$,
   (d) $\frac{1}{9}x^3(3 \ln x - 1) + c$, (e) $x \sin (x + 2) + \cos (x + 2) + c$,
   (f) $\frac{1}{72}(1 + x)^8(8x - 1) + c$, (g) $\frac{1}{2}xe^{2x} - \frac{1}{4}e^{2x} + c$, (h) $\frac{1}{2}e^{x^2} + c$,
   (i) $-\dfrac{1}{x} (\ln x + 1) + c$, (j) $x \tan x + \ln \cos x + c$,
   (k) $(n + 1)^{-2}x^{n+1}\{(n + 1) \ln x - 1\} + c$, (l) $(\ln 3)^{-2} \times 3^x(x \ln 3 - 1) + c$.

2 (a) $x \ln 2x - x + c$, (b) $x \sin^{-1} 3x + \frac{1}{3}\sqrt{(1 - 9x^2)} + c$, (c) $2y(\ln y - 1) + c$,
   (d) $\theta \tan^{-1} \dfrac{\theta}{2} - \ln (4 + \theta^2) + c$, (e) $t \cos^{-1} t - \sqrt{(1 - t^2)} + c$,
   (f) $2e^{\sqrt{x}}(\sqrt{x} - 1) + c$.

Page 230

3 (a) $\frac{1}{3}e^{x^3}(x^3 - 1) + c$, (b) $-\frac{1}{2}e^{-x^2} + c$, (c) $-\frac{1}{2}e^{-x^2}(1 + x^2) + c$,
   (d) $\frac{1}{2}x^2 \sin x^2 + \frac{1}{2}\cos x^2 + c$, (e) $\frac{1}{2}x^2 \tan x^2 + \frac{1}{2}\ln \cos x^2 + c$.
4 (a) $\frac{1}{3}x^2 \sin 3x + \frac{2}{9}x \cos 3x - \frac{2}{27}\sin 3x + c$, (b) $e^x(x^3 - 3x^2 + 6x - 6) + c$,
   (c) $\frac{1}{8}\cos 2x(1 - 2x^2) + \frac{1}{4}x \sin 2x + c$, (d) $-e^{-x}(x^2 + 2x + 2) + c$,
   (e) $\frac{1}{6}x^3 + \frac{1}{8}(2x^2 - 1) \sin 2x + \frac{1}{4}x \cos 2x + c$, (f) $\frac{1}{4}x^2\{1 - 2 \ln x + 2(\ln x)^2\} + c$.
5 (a) $\frac{1}{8}\sin 2x - \frac{1}{4}x \cos 2x + c$, (b) $-e^{-x}(1 + x) + c$, (c) $\frac{1}{168}(1 + 2x)^6(12x - 1) + c$,
   (d) $\frac{1}{2}(\ln y)^2 + c$, (e) $\frac{1}{2}(1 + u^2) \tan^{-1} u - \frac{1}{2}u + c$, (f) $-\frac{1}{2}e^{-x^2} + c$,
   (g) $-e^{-x}(x^3 + 3x^2 + 6x + 6) + c$, (h) $-\frac{1}{14}(1 - x^2)^7 + c$,
   (i) $\frac{1}{4}t^2 - \frac{1}{4}t \sin 2t - \frac{1}{8}\cos 2t + c$, (j) $\frac{1}{9}e^{3v}(3v - 1) + c$.
6 (a) $x \tan x + \ln \cos x - \frac{1}{2}x^2 + c$.
7 (a) $\frac{1}{2}\pi - 1 \approx 0.571$, (b) $e - 2 \approx 0.718$, (c) $e^2 + 1 \approx 8.39$, (d) $\frac{1}{2}\pi - 1 \approx 0.571$,
   (e) $\pi^2/4 \approx 2.47$, (f) $50 - 99/(4 \ln 10) \approx 39.2$.

Exercise 13b, page 238

1 (a) $2(1 - 4x^2)^{-1/2}$, (b) $3(2 + 6x + 9x^2)^{-1}$, (c) $-\frac{2}{3}(1 - 4x^2)^{-1/2}$,
   (d) $2(8 + 2x - x^2)^{-1/2}$, (e) $(x^2 + 4)^{-1}$, (f) $2(4 - 9x^2)^{-1/2}$, (g) $-(1 + x^2)^{-1}$,
   (h) $\dfrac{1}{x\sqrt{(x^2 - 1)}}$, (i) $2x^3(1 + x^4)^{-1} + 2x \tan^{-1} x^2$, (j) 0.

2 (a) $\frac{1}{3}\tan^{-1}\dfrac{x}{3} + c$, (b) $3 \sin^{-1}\dfrac{y}{2} + c$, (c) $\frac{2}{3}\tan^{-1} 3u + c$, (d) $\frac{1}{2}\sin^{-1} 4x + c$,

   (e) $\dfrac{1}{\sqrt{3}}\tan^{-1}\dfrac{2t}{\sqrt{3}} + c$, (f) $\frac{1}{2}\sin^{-1}\dfrac{2x}{\sqrt{5}} + c$, (g) $\dfrac{1}{\sqrt{6}}\tan^{-1}\dfrac{\sqrt{3}y}{\sqrt{2}} + c$,

   (h) $\dfrac{1}{3\sqrt{6}}\sin^{-1}\sqrt{2}\,x + c$, (i) $\dfrac{1}{3\sqrt{2}}\tan^{-1}\dfrac{(y - 2)\sqrt{2}}{3} + c$,

   (j) $\dfrac{2}{\sqrt{3}}\sin^{-1}\dfrac{(x - 1)\sqrt{3}}{2} + c$.

3 (a) $2 \ln \tan \dfrac{x}{4} + c$, (b) $\frac{1}{2}\ln (\sec 2\theta + \tan 2\theta) + c$, (c) $\frac{1}{3}\ln \tan \frac{3}{2}x + c$,

   (d) $\frac{1}{4}\ln (\sec 4\phi + \tan 4\phi) + c$, (e) $\ln \tan x + c$, (f) $\tan \frac{1}{2}y + c$,
   (g) $-(1 + \tan x)^{-1} + c$, (h) $\ln (1 - \cos \theta) + c$,
   (i) $\frac{1}{3}\ln (3 + \tan \frac{1}{2}x) - \frac{1}{3}\ln (3 - \tan \frac{1}{2}x) + c$, (j) $\tan^{-1}(\frac{1}{2}\tan \frac{1}{4}\theta) + c$.

4 (a) $\dfrac{1}{\sqrt{3}}\tan^{-1}(\sqrt{3} \tan x) + c$, (b) $\frac{1}{4}\ln (1 + 2 \tan x) - \frac{1}{4}\ln (1 - 2 \tan x) + c$,

   (c) $\sqrt{2}\tan^{-1}\left(\dfrac{1}{\sqrt{2}}\tan x\right) - x + c$, (d) $\frac{1}{6}\ln (1 + 3 \tan x) - \frac{1}{6}\ln (1 - 3 \tan x) + c$.

5 (a) $\frac{1}{2}\ln (x^2 + 3) + \dfrac{5}{\sqrt{3}}\tan^{-1}\dfrac{x}{\sqrt{3}} + c$, (b) $\ln (y + 3) - (y + 3)^{-1} + c$,

   (c) $\frac{3}{2}\ln (u^2 + 2u + 5) + \frac{5}{2}\tan^{-1}\dfrac{u + 1}{2} + c$, (d) $7\sqrt{(4x - x^2)} - 11 \sin^{-1}\dfrac{x - 2}{2} + c$,

   (e) $\frac{1}{2}\theta + \frac{1}{2}\ln (\sin \theta + \cos \theta) + c$, (f) $\frac{1}{2}x + \frac{5}{2}\ln (\sin x + \cos x) + c$.
6 (a) 1, (b) $\pi/6$.
7 (b) 2, (e) $\frac{1}{2}$, (f) 1, (g) $-\frac{1}{2}(\ln 2 + 1) \approx -0.847$, (h) -1,
   (i) $\frac{1}{4}\pi - \frac{1}{2}\sin^{-1}\frac{2}{3} \approx 0.4205$, (j) $\frac{1}{20}\pi$.

Page 239

8 (a) tends to π, (b) tends to infinity.

9 15π.

10 256/3.

Exercise 13c, page 241

1 (a) $\frac{1}{13}e^{3x}(3\cos 2x + 2\sin 2x) + c$, (b) $\frac{1}{25}e^{4x}(4\sin 3x - 3\cos 3x) + c$,

(c) $\frac{2}{5}e^{-t}(\sin \frac{1}{2}t - 2\cos \frac{1}{2}t) + c$, (d) $\frac{1}{5}e^{x}\{\sin(2x + 1) - 2\cos(2x + 1)\} + c$,

(e) $\frac{1}{8}e^{2\theta}(2 + \cos 2\theta + \sin 2\theta) + c$.

2 $\frac{1}{2}\tan x \sec x + \frac{1}{2}\ln(\sec x + \tan x) + c$.

3 (a) $\frac{1}{16}x^4(4\ln x - 1) + c$, (b) $y\tan^{-1} 2y - \frac{1}{4}\ln(1 + 4y^2) + c$, (c) $-\frac{1}{2}e^{-x^2} + c$,

(d) $\frac{1}{9}(\sin 3x - 3x\cos 3x) + c$, (e) $\frac{1}{4}\cos 2x(1 - 2x^2) + \frac{1}{2}x\sin 2x + c$,

(f) $\frac{1}{13}e^{3x}(3\sin 2x - 2\cos 2x) + c$, (g) $\frac{1}{4}e^{u^2}(u^2 - 1) + c$,

(h) $\frac{1}{168}(2x - 1)^6(12x + 1) + c$, (i) $\frac{1}{4}(x^2 - 1)\ln(x - 1) - \frac{1}{8}x^2 - \frac{1}{4}x + c$,

(j) $x(\ln 3x - 1) + c$, (k) $\frac{1}{4}e^{2x}(2x^2 - 2x + 1) + c$,

(l) $\frac{2}{5}e^{-y}(\sin \frac{1}{2}y - 2\cos \frac{1}{2}y) + c$, (m) $-\frac{1}{4}x^{-2}(1 + 2\ln x) + c$,

(n) $t\sin^{-1}\dfrac{t}{3} + \sqrt{(9 - t^2)} + c$, (o) $3x(\ln x - 1) + c$,

(p) $\frac{1}{6}y^3 + \frac{1}{8}(2y^2 - 1)\sin 2y + \frac{1}{4}y\cos 2y + c$, (q) $\frac{1}{2}\sin x^2 + c$,

(r) $\frac{1}{2}x^2(\ln x^2 - 1) + c$, (s) $\frac{1}{2}\sin \theta^2 - \frac{1}{2}\theta^2\cos \theta^2 + c$,

(t) $\frac{1}{4}(2x^3 - 3x)\sin 2x + \frac{3}{8}(2x^2 - 1)\cos 2x + c$.

4 $C = \dfrac{e^{ax}}{a^2 + b^2}(a\cos bx + b\sin bx)$, $S = \dfrac{e^{ax}}{a^2 + b^2}(a\sin bx - b\cos bx)$.

6 $\frac{64}{21}$.

7 $(\frac{1}{2}\pi - 1, \frac{1}{8}\pi)$.

8 $(\pi/4 - 1/\pi, 0)$.

10 $\frac{1}{10}(1 + e^{3\pi})$.

Exercise 13d, page 245

1 $-\frac{1}{4}\cos x \sin^3 x - \frac{3}{8}\cos x \sin x + \frac{3}{8}x + c$, $3\pi/16$.

2 $-\frac{1}{5}\cos x \sin^4 x - \frac{4}{15}\cos x \sin^2 x - \frac{8}{15}\cos x + c$, $\frac{8}{15}$.

3 (a) $\frac{2}{3}$, (b) $5\pi/32$, (c) $\frac{128}{315}$, (d) $3\pi/16$, (e) $63\pi/512$, (f) $\frac{16}{35}$.

4 (a) $\frac{8}{35}$, (b) $\frac{8}{5}$, (c) $5\pi/96$.

5 (a) $\frac{32}{35}$, (b) $3\pi/8$, (c) $5\pi/16$, (d) 0, (e) 0, (f) $\frac{256}{315}$, (g) $63\pi/256$, (h) $35\pi/128$.

6 (a) $I_{m-2,n} = \dfrac{m - 3}{m + n - 2}I_{m-4,n}$, $I_{m,n} = \dfrac{(m - 1)(m - 3)}{(m + n)(m + n - 2)}I_{m-4,n}$,

(b) $I_{m-4,n} = \dfrac{(n - 1)(n - 3)(n - 5)}{(m + n - 4)(m + n - 6)(m + n - 8)}I_{m-4,n-6}$, (c) $\frac{8}{693}$,

(d) (i) $\frac{8}{1287}$, (ii) $5\pi/4096$, (iii) $\frac{16}{3003}$, (iv) $\frac{1}{120}$.

7 (a) $\frac{1}{504}$, (b) $7\pi/2048$.

8 (a) $I_n = (n/a)I_{n-1}$, (b) $\frac{2835}{8}$.

9 $I_n = \dfrac{1}{n - 1} - I_{n-2}$, (a) $\frac{5}{12} - \frac{1}{2}\ln 2 \approx 0.070$, (b) $\frac{1}{4}\pi - \frac{76}{105} \approx 0.062$.

10 (b) $1328\sqrt{3}/2835$.

Exercise 13e, page 248

1 (a) $\frac{1}{3}\sqrt{(x^2+1)^3}+c$, (b) $\frac{2}{3}\sqrt{(x^3+3x-4)}+c$, (c) $\sin u-\frac{2}{3}\sin^3 u+\frac{1}{5}\sin^5 u+c$,
   (d) $\tan\theta+\frac{2}{3}\tan^3\theta+\frac{1}{5}\tan^5\theta+c$, (e) $\frac{1}{5}\sec^5 x-\frac{2}{3}\sec^3 x+\sec x+c$,
   (f) $-\frac{1}{2}\cos x^2+c$, (g) $2\tan\sqrt{x}+c$, (h) $\frac{1}{4}\ln(2x^2+3)+c$, (i) $-\frac{1}{2}e^{-x^2}+c$,
   (j) $\ln\sec^2\dfrac{\theta}{2}+c$.

2 (a) $\frac{1}{5}(x+1)\sqrt{(2x-3)^3}+c$, (b) $\frac{1}{324}(24x+1)(3x-1)^8+c$,
   (c) $y+16(y-4)^{-1}+c$, (d) $\dfrac{1}{\sqrt{5}}\sin^{-1}\dfrac{\sqrt{5}\,y}{2}+c$, (e) $\dfrac{1}{3\sqrt{3}}\tan^{-1}\sqrt{3u}+c$,
   (f) $\dfrac{1}{2\sqrt{2}}\tan^{-1}\dfrac{u-3}{2\sqrt{2}}+c$, (g) $\dfrac{1}{\sqrt{2}}\sin^{-1}\dfrac{(x-1)\sqrt{2}}{3}+c$,
   (h) $\frac{1}{2}y\sqrt{(4-y^2)}+2\sin^{-1}\dfrac{y}{2}+c$, (i) $\sec^{-1}3x+c$, (j) $\frac{2}{3}\tan^{-1}(\frac{1}{3}\tan\frac{1}{2}\theta)+c$.

3 (a) $\frac{1}{3}e^{3x}+c$, (b) $(\ln 10)^{-1}10^y+c$, (c) $-\frac{1}{3}e^{-x^3}+c$, (d) $\frac{1}{3}\ln x+c$,
   (e) $\frac{1}{3}\ln(3x+4)+c$, (f) $-\frac{1}{2}\ln(2x-3)+c$, (g) $\frac{1}{3}\ln(x+3)+c$,
   (h) $\frac{1}{2}\ln\dfrac{1+x}{1-x}+c$, (i) $x(\ln x-1)+c$, (j) $2e^{\sqrt{x}}(\sqrt{x}-1)+c$.

4 (a) $\frac{1}{3}\ln\dfrac{3+x}{3-x}+c$, (b) $\frac{1}{3}\ln\dfrac{y-3}{y}+c$, (c) $\dfrac{1}{x}+\ln\dfrac{x-1}{x}+c$,
   (d) $4(4-x)^{-1}+\ln(4-x)+c$, (e) $-\dfrac{4x+5}{2(x+1)^2}-\ln(x+1)+c$,
   (f) $\ln\dfrac{(x+1)^3}{x^2-x+1}+c$.

5 (a) $2x\sin\frac{1}{2}x+4\cos\frac{1}{2}x+c$, (b) $\frac{1}{2}e^x(x-1)+c$, (c) $\ln\sin y-y\cot y+c$,
   (d) $-\frac{1}{252}(21y+1)(1-3y)^7+c$, (e) $(\ln 3)^{-2}\,3^x(x\ln 3-1)+c$,
   (f) $\frac{1}{4}x^2(2\ln 2x-1)+c$, (g) $t(\ln t-1)+c$, (h) $x\tan^{-1}3x-\frac{1}{6}\ln(1+9x^2)+c$,
   (i) $4^x(\ln 4)^{-1}+c$, (j) $x(6-x^2)\cos x+3(x^2-2)\sin x+c$.

6 (a) $\frac{1}{4}\ln(4x^2+3)-\dfrac{1}{2\sqrt{3}}\tan^{-1}\dfrac{2x}{\sqrt{3}}+c$,
   (b) $4\sqrt{(1+2y-y^2)}-3\sin^{-1}\dfrac{y-1}{\sqrt{2}}+c$, (c) $\frac{2}{5}\theta-\frac{1}{5}\ln(2\cos\theta-\sin\theta)+c$,
   (d) $\frac{2}{5}\ln(4\sin x+3\cos x)-\frac{1}{5}x+c$.

7 (a) $\pi/9$, (b) $\dfrac{1}{2\sqrt{2}}\tan^{-1}\dfrac{18-10\sqrt{2}}{31}$, (c) $\frac{1}{2}$, (d) $\frac{256}{693}$, (e) $231\pi/2048$, (f) $5\pi/128$,
   (g) $35\pi/128$, (h) 0, (i) 128/230 945, (j) $\frac{1}{2}\ln\frac{1}{5}$.

8 (a) $-\frac{1}{5}\cos 5x+c$, (b) $3\sin\frac{1}{3}x+c$, (c) $\frac{1}{5}\ln\sec 5x+c$, (d) $2\ln\sin\frac{1}{2}x+c$,
   (e) $\ln\tan\frac{1}{2}x+c$, (f) $\ln(\sec x+\tan x)+c$ or $\ln\tan(\frac{1}{2}x+\frac{1}{4}\pi)+c$.

9 (a) $3\tan\frac{1}{3}x+c$, (b) $-\frac{1}{4}\cot 4x+c$, (c) $\frac{1}{2}x-\frac{1}{4}\sin 2x+c$,
   (d) $\frac{1}{2}x+\frac{1}{4}\sin 2x+c$, (e) $\tan x-x+c$, (f) $-\cot x-x+c$.

10 (a) $\frac{1}{3}\cos^3 x-\cos x+c$, (b) $\sin x-\frac{1}{3}\sin^3 x+c$, (c) $\frac{1}{2}\tan^2 x+\ln\cos x+c$,
    (d) $-\frac{1}{2}\cot^2 x-\ln\sin x+c$, (e) $\frac{1}{2}\tan x\sec x+\ln\sqrt{(\sec x+\tan x)}+c$,
    (f) $\frac{1}{2}\ln\tan\frac{1}{2}x-\frac{1}{2}\cot x\cosec x+c$.

11 (a) $\frac{1}{32}(12x-8\sin 2x+\sin 4x)+c$, (b) $\frac{1}{32}(12x+8\sin 2x+\sin 4x)+c$,
    (c) $x-\tan x+\frac{1}{3}\tan^3 x+c$, (d) $-\frac{1}{3}\cot^3 x-\cot x+c$,

Page 250

(e) $\frac{1}{3}\tan^3 x + \tan x + c$, (f) $x + \cot x - \frac{1}{3}\cot^3 x + c$.

12 (a) $x\sin^{-1} x + \sqrt{(1-x^2)} + c$, (b) $x\cos^{-1} x - \sqrt{(1-x^2)} + c$,

(c) $x\tan^{-1} x - \frac{1}{2}\ln(1+x^2) + c$, (d) $x\cot^{-1} x + \frac{1}{2}\ln(1+x^2) + c$,

(e) $x\sec^{-1} x - \ln\{x + \sqrt{(x^2-1)}\} + c$, (f) $x\operatorname{cosec}^{-1} x + \ln\{x + \sqrt{(x^2-1)}\} + c$.

13 (a) $\dfrac{1}{2\sqrt{3}}\tan^{-1}\dfrac{2x}{\sqrt{3}} + c$, (b) $\frac{1}{8}\sqrt{(5+8x^2)} + c$, †(c) $\ln\{x + \sqrt{(1+x^2)}\} + c$,

(d) $\frac{1}{6}\ln(2+3x^2) + c$, (e) $\frac{1}{3}\sqrt{(3+x^2)^3} + c$,

(f) $\frac{1}{4}\ln(3+2x^2) + \dfrac{1}{\sqrt{6}}\tan^{-1}\dfrac{x\sqrt{2}}{\sqrt{3}} + c$, (g) $\frac{1}{2}\ln(x^2-4x+7) + c$,

(h) $\frac{1}{2}x\sqrt{(x^2+2)} + \ln\{x + \sqrt{(x^2+2)}\} + c$,

(i) $\frac{3}{2}\ln(x^2-4x+5) - 5\tan^{-1}(x-2) + c$, (j) $\frac{2}{135}(9x-4)\sqrt{(2+3x)^3} + c$.

14 (a) $\dfrac{1}{\sqrt{5}}\sin^{-1}\dfrac{\sqrt{5}\,x}{2} + c$, (b) $-\frac{2}{27}(3x+2)\sqrt{(1-3x)} + c$, (c) $\frac{1}{3}\ln\dfrac{3+x}{3-x} + c$,

(d) $3(16-x)^{-1} + c$, (e) $-\frac{1}{3}\sqrt{(6-x^2)^3} + c$, (f) $-\frac{3}{2}\ln(4-x^2) + c$,

(g) $\frac{1}{2}x\sqrt{(4-x^2)} + 2\sin^{-1}\dfrac{x}{2} + c$, (h) $-\frac{1}{2}\sqrt{(7-2x^2)} + c$,

(i) $-\frac{1}{4}\sqrt{(3-4x^2)} - \sin^{-1}\dfrac{2x}{\sqrt{3}} + c$, †(j) $\ln\{x + \sqrt{(x^2-9)}\} + c$.

15 (a) $\dfrac{180}{\pi}\sin x° + c$, (b) $\frac{1}{32}\sin 4x - \frac{1}{8}x\cos 4x + c$, (c) $2\ln\tan\frac{1}{2}\theta + c$,

(d) $-\frac{1}{7}\cos^7 x + \frac{2}{9}\cos^9 x - \frac{1}{11}\cos^{11} x + c$, (e) $y\tan y + \ln\cos y + c$,

(f) $\sin x - x\cos x + c$, (g) $-\frac{1}{2}\cos x^2 + c$, (h) $(u^2-2)\sin u + 2u\cos u + c$,

(i) $\frac{1}{8}y - \frac{1}{32}\sin 4y + c$, (j) $-\frac{1}{42}(3\cos 7x + 7\cos 3x) + c$.

16 (a) $\tan\frac{1}{2}\theta + c$, (b) $\frac{1}{4}\ln\dfrac{1+2\tan\theta}{1-2\tan\theta} + c$, (c) $\tan x - \sec x + c$,

(d) $\ln(\cos\theta + 3\sin\theta) - \theta + c$, (e) $-2\cot\frac{1}{4}x + c$, (f) $\frac{4}{3}\tan^{-1}(3\tan x) + c$,

(g) $\frac{1}{4}\ln(\sec 4y + \tan 4y) + c$, (h) $\tan x + \sec x + c$, (i) $\frac{1}{2}\ln\dfrac{1+\tan\frac{1}{2}\theta}{3-\tan\frac{1}{2}\theta} + c$,

(j) $\frac{1}{2}x + \frac{1}{2}\ln(\cos x + \sin x) + c$.

17 (a) $-e^{-x}(x^3 + 3x^2 + 6x + 6) + c$, (b) $(x+2)\ln(x+2) - x + c$, (c) $2e^{\sqrt{y}} + c$,

(d) $2\sqrt{\ln t} + c$, (e) $\frac{1}{18}(1+9x^2)\tan^{-1} 3x - \frac{1}{6}x + c$, (f) $\frac{1}{2}(\sin^{-1} x)^2 + c$,

(g) $(\ln 4)^{-1} 4^x + c$, (h) $x(\ln 10)^{-1} 10^x - (\ln 10)^{-2} 10^x + c$,

(i) $\frac{1}{4}x^4 \ln 2x - \frac{1}{16}x^4 + c$, (j) $\frac{1}{2}e^x(x^2-1) + c$.

Exercise 13f, page 252

1 (a) $\frac{1}{5}x^5 - 2x^2 - 4x^{-1} + c$, (b) $\frac{1}{16}(4\cos 2x - \cos 8x) + c$,

(c) $2\ln x - \ln(1+x^2) + c; \frac{1}{3}$.

2 (a) $\frac{1}{5}(1+x^2)^{5/2} + c$, (b) $\frac{1}{2}\ln(1+2x) - \ln(1-2x) + c$,

(c) $\frac{1}{3}x^3 \ln x - \frac{1}{9}x^3 + c$, $\frac{1}{16}\pi^2 - \frac{1}{36}$.

3 $\dfrac{1}{4(1+x)} + \dfrac{1}{4(1-x)} + \dfrac{1}{2(1+x^2)}$.

†See also pp. 346, 348.

Page 252

4 (a) $\ln \dfrac{x-2}{x-1} + c$, (b) $\tfrac{1}{2}x^2 \tan^{-1} x - \tfrac{1}{2}x + \tfrac{1}{2} \tan^{-1} x + c$,

   (c) $x - \ln(x^2 + 2x + 2) + c$.

5 (a) $2 \ln(x+2) - \ln(x+4) + c$, (b) $\tfrac{1}{2}x^2 \ln x - \tfrac{1}{4}x^2 + c$,

   (c) $x \sin^{-1} x + \sqrt{(1-x^2)} + c$; $2 + \sqrt{2}$.

6 $\ln \tfrac{4}{3} - \tfrac{14}{65}$.

8 (a) $-\tfrac{1}{2} \operatorname{cosec}^2 x + c$, (b) $-\sqrt{(7 - 6x - x^2)} + c$.

9 (a) $\sin^{-1} \dfrac{x+2}{3} + c$, (b) $-\tfrac{1}{2}e^{-x^2}(x^2 + 1) + c$.

10 $\tfrac{1}{4}\pi + \tfrac{1}{2}$.

11 $\tfrac{1}{2}\sqrt{19} - 2$.

12 (a) $x + \ln(x-2) - \ln(x+2) + c$, (b) $-e^{-x^2/2}(x^2 + 2) + c$,

   (c) $\sin^{-1} \dfrac{x-2}{3} + c$; $\dfrac{\pi\sqrt{6}}{12}$.

13 $\tfrac{1}{13}(3 - 2e^{\pi})$.

14 (a) $-(2x - 3)^{-1/2} + c$, (b) $\sin x - x \cos x + c$, (c) $\tfrac{2}{15}(3x + 8)(x - 4)^{3/2} + c$; π.

15 (a) Positive, (b) zero, (c) negative.

16 (a) Positive, (b) zero, (c) positive.

18 (a) $\tfrac{1}{4}\pi - \tfrac{1}{2} \ln 2$, (b) $\tfrac{1}{4} + \tfrac{3}{32}\pi$.

Chapter 14

Qu. 1 $y = Ax + B$, $y = 3x - 5$.

Qu. 2 $y = x^3 + A$.

Qu. 3 $x^2 + y^2 = A$.

Qu. 4 $s = \tfrac{1}{2}at^2 + At + B$; $s = ut + \tfrac{1}{2}at^2$.

Qu. 5 (a) $\dfrac{d^2 y}{dx^2} = 0$, (b) $\dfrac{dy}{dx} = \dfrac{y}{x}$, (c) $\dfrac{dr}{d\theta} + r \tan \theta = 0$, (d) $\dfrac{dy}{dx} = -\dfrac{y}{x}$, (e) $\dfrac{dy}{dx} = y$,

   (f) $x\dfrac{dy}{dx} = y \ln y$, (g) $y\dfrac{d^2 y}{dx^2} = \left(\dfrac{dy}{dx}\right)^2$, (h) $x\dfrac{dy}{dx} \ln x = y$,

   (i) $(1 + x^2)\dfrac{dy}{dx} \tan^{-1} x = y$.

Qu. 7 (a) $x^2 - y^2 + A = 0$, (b) $y = Ax$, (c) $x = Ae^{y^2/2}$, (d) $x = A \sin y$,

   (e) $\ln \sqrt{\left|\dfrac{y-1}{y+1}\right|} = e^x + A$, (f) $y^2 = 2\sqrt{(x^2 + 1)} + A$.

Qu. 8 $v^2 = u^2 + 2as$.

Qu. 9 (a) $x^2 y = x + A$, (b) $t^2 \ln x = 3 \sin t + A$, (c) $x^2 \sin u = \ln(kx)$,

   (d) $xe^y = 2x + A$.

Qu. 10 (a) $x, x^2 y = \tfrac{1}{2}e^{x^2} + A$, (b) $x, x^2 e^y = \tfrac{1}{3}x^3 + A$, (c) $\dfrac{1}{x}, xy^2 = \ln(kx)$,

   (d) $r, r^2 \tan \theta = 2\theta + A$.

Qu. 11 $y = \tfrac{1}{2} + Ae^{-x^2}$; $y = \tfrac{1}{2} - e^{-x^2}$.

Qu. 12 x^2.

Qu. 13 (a) $y = 1 + x \tan x + A \sec x$, (b) $y = x - 4 + Ae^{-x}$.

Qu. 14 (a) $\left(\dfrac{y}{x}\right)^2$, (b) $\dfrac{d^2y}{dx^2} + xy$, (c) $x\sqrt{(x^2 + y^2)}$.

Qu. 15 (a), (b), (d), (e).

Qu. 16 (a) $xe^{x/y} = A$, (b) $x^2 - 2xy = A$, (c) $x - 2y + Axy = 0$.

Qu. 17 $y = x \ln (Ax^2)$.

Qu. 19 (a) $y = 2x \ln x + Ax + B$, (b) $y = 2 \sin x - x \cos x + Ax + B$,
   (c) $y = x^3 + A \ln x + B$, (d) $y^2 = Ax + B - 2 \cos x$.

Qu. 20 (a) $y = Ae^{Bx}$ or $y = C$, (b) $y = -3x^2 + 3x + 2$.

Qu. 21 $y = \ln x^2 + Ax^{-1} + B$.

Qu. 22 (a) $x = a \cos (2t + \varepsilon)$, (b) $y = a \cos (3x + \varepsilon)$, (c) $y = -\tfrac{8}{3}x^3 + Ax + B$.

Qu. 23 $2\dfrac{dx}{dt}$.

Qu. 24 $x = 2 \cos (\tfrac{3}{2}t)$, (a) $\dfrac{dx}{dt} = \tfrac{3}{2}\sqrt{(4 - x^2)}$, (b) $\dfrac{dx}{dt} = -3 \sin (\tfrac{3}{2}t)$.

Qu. 25 (a) $x = a \cos nt$, (b) $x = a \sin nt$.

Qu. 26 (a) $x = a \cos nt$, (b) $x = a \sin nt$.

Qu. 27 $\pm \pi/3$.

Qu. 28 (a) $y = a \cos (2x + \varepsilon) - 1$, (b) $\theta = a \cos (\sqrt{2}t + \varepsilon) + 3$,
   (c) $x = A + Bt - \tfrac{1}{2}t^2 - \tfrac{3}{8}t^3$.

Exercise 14a, page 261

1 (a) $\dfrac{dy}{dx} = \tfrac{3}{2}$, (b) $\dfrac{dy}{dx} = \dfrac{y + \tfrac{1}{2}}{x}$, (c) $\dfrac{dy}{dx} = \dfrac{y}{x}$, (d) $\dfrac{dy}{dx} = -\dfrac{x}{y}$, (e) $\dfrac{dy}{dx} = -\dfrac{y}{x}$,
   (f) $\dfrac{dy}{dx} = \dfrac{y}{x - 4}$.

2 (a) $\dfrac{d^2y}{dt^2} = -9y$, (b) $\dfrac{d^2y}{dt^2} = 3\dfrac{dy}{dt}$, (c) $\dfrac{d^2y}{dx^2} = 9y$, (d) $\dfrac{d^2y}{dx^2} - \dfrac{dy}{dx} - 6y = 0$,
   (e) $\dfrac{d^2y}{dx^2} - 8\dfrac{dy}{dx} + 16y = 0$.

3 $3x - 10y - 35 = 0$.

4 $y = x^2 - 3x + 1$.

5 $s = A - 3t^2$, $s = 12 - 3t^2$.

6 (a) $y = e^x - 3x \cos x + 3 \sin x - 1$, (b) $y = e^x - 3x \cos x + 3 \sin x - e^{\pi/2}$.

7 (a) $y = Ae^x$, (b) $y = \tfrac{2}{15}(x - 1)(3x + 2)\sqrt{(x - 1)} + A$, (c) $y = A(x + 2)$,
   (d) $x = \tfrac{1}{2}y + \tfrac{1}{4} \sin 2y + A$, (e) $v - 1 = Ave^u$, (f) $y = x \ln x - x + A$,
   (g) $\sin y = Ae^x$, (h) $x = y \tan^{-1} y - \ln \sqrt{(1 + y^2)} + A$, (i) $y^2 = x^2 - 2x + A$,
   (j) $y = A\sqrt{\dfrac{x - 1}{x + 1}}$, (k) $r = \ln \left(A \tan \dfrac{\theta}{2}\right)$, (l) $y + 3 = Ae^{-1/x}$, (m) $y = Axe^x$,
   (n) $\cos \theta \sin \phi = A$, (o) $r = \theta \tan \theta + \ln (A \cos \theta)$, (p) $2y^2 = x^2(\ln x^2 - 1) + A$,
   (q) $r = -\theta - \ln (\cos \theta - \sin \theta) + A$, (r) $2y + 3 = A(x - 2)^2$,
   (s) $x = A - \tfrac{1}{2}e^{-t}(\cos t + \sin t)$, (t) $y = 2 \tan (2e^{-x} + A)$.

8 (a) $y = \tan \theta$, (b) $(y - 2)^2 = 9e^{x^2}$, (c) $y = \dfrac{7x + 1}{7 - x}$, (d) $y = \sin (x - \tfrac{1}{6}\pi)$.

9 83.4 minutes.

10 9.05 kg, 34.7 minutes.

Exercise 14b, page 269

1 (a) $y^2 = \dfrac{A}{x} - \dfrac{1}{x^2}$, (b) $y^2 = \dfrac{1}{x^2}(\tan 2x + A)$, (c) $x \ln y = \sec x + A$,

    (d) $(1 - 2x)e^y = \tan x + A$, (e) $t^2 e^s = t \sin t + A$, (f) $r^2 e^u = \cot u + A$.

2 (a) $x^2 \sin y = 3x^2 + A$, (b) $xy = e^x + A$, (c) $x \tan y = e^{x^2} + A$,

    (d) $ye^x = \ln (Ay)$.

3 (a) $y = e^{-2x}(\sin x + A)$, (b) $s = \frac{1}{2} + Ae^{-t^2}$, (c) $y = e^{-x^2}(1 + Ae^{-x})$,

    (d) $r = (\theta + A)\operatorname{cosec}^2 \theta$, (e) $r = (\theta + A)\cos \theta$, (f) $y = x^{-2}(\sin x + A)$,

    (g) $y = x \ln \dfrac{A(x - 1)}{x}$, (h) $y = \dfrac{x}{3} + 3 + Ax^{-1/2}$, (i) $y = (x - \sin x + A)\cot \frac{1}{2}x$,

    (j) $y = (x - 2)^{-1} + A(x - 2)^{-3}$.

4 (a) $x^3 = Ae^{y/x}$, (b) $x^2(x^2 - 2y^2) = A$, (c) $\tan^{-1}\dfrac{y}{x} = \ln (Ax)$, (d) $3x - y + Axy = 0$,

    (e) $y = Ae^{x^2/(2y^2)}$, (f) $2x = (2x - y)\ln \{A(2x - y)\}$, (g) $\sin^{-1}\dfrac{y}{x} = \ln (Ax)$,

    (h) $y = x(Ax - 1)$, (i) $(x + y)(2x - y)^2 = A$, (j) $\dfrac{2}{\sqrt{3}} \tan^{-1}\left(\dfrac{2y - x}{x\sqrt{3}}\right) = \ln (Ax)$.

5 $x^2 - y^2 - 2xy + 4x = A$.

6 (a) $y - 2 = Ae^{(x-3)/(y-2)}$, (b) $(x - y - 3)^2(x + 2y - 3) = A$.

7 $x^2 + y^2 - 2xy - 4x - 8y + A = 0$.

8 (a) $x - y + A = \ln (2x + y)$, (b) $x + y - 1 = Ae^{x-y}$.

9 (a) $y = (x + A)(x + 3)^2$, (b) $x = (x - y)\ln (Ax)$, (c) $(y + 3)\sin x = A - \frac{1}{2}e^{-2x}$,

    (d) $\sin y = \dfrac{Ax - 1}{x(x + 3)}$, (e) $\tan^{-1}\dfrac{2(y - 1)}{x - 2} = \ln \{A(x^2 + 4y^2 - 4x - 8y + 8)\}$,

    (f) $y + 2 = xe^{2x} + Ae^x$, (g) $y^{x^2} = A \sin x$, (h) $(2r + 3)\tan \theta = 3\theta + A$, (i) $xy = Ae^{y/x}$,

    (j) $x^2 + y^2 - 2xy + 2x - 6y + A = 0$.

10 (a) $y = (x + 1)^4$, (b) $u = \sin \theta + 2 \operatorname{cosec} \theta$, (c) $x^2 - 2xy - y^2 = 17$,

    (d) $4y^2 - x^2 = 2y^2 \ln \dfrac{y}{2}$, (e) $e^t(x - 1) = 1 - 1/t$.

Exercise 14c, page 278

1 (a) $y = x \ln x + Ax + B$, (b) $y = \ln \sec \theta + A \ln (\sec \theta + \tan \theta) + B$,

    (c) $y = 2e^{-x} + Ax + B$, (d) $y = A \ln \{B(2x + 1)\}$, (e) $y = e^x(x - 1) + Ax^2 + B$.

2 (a) $y = Ax^4 + B$, (b) $A + Bx = e^{-2y}$, (c) $6x = y^3 + Ay + B$ or $y = C$,

    (d) $y = A \tan^{-1} x + B$, (e) $y = A \sin^{-1} x + B$.

3 $s = \ln \{A(t + B)^{10}\}$.

4 (a) $y = \ln \operatorname{cosec} x + A \ln \tan \dfrac{x}{2} + B$, (b) $\frac{1}{9}x^3 + \frac{1}{6}x^2 - \frac{2}{3}x + A \ln (x + 2) + B$,

    (c) $36y = 6x^2 \ln x - 5x^2 + Ax^{-1} + B$.

5 $y + \pi/4 = \frac{1}{2} \sin^{-1} x + \frac{1}{2}x\sqrt{(1 - x^2)}$.

6 (a) $s = a \cos (5t + \varepsilon)$, (b) $y = a \cos (\frac{3}{2}x + \varepsilon)$, (c) $s = A + B\theta - \frac{1}{27}\theta^3$,

    (d) $y = a \cos \left(\dfrac{\sqrt{3}}{2}t + \varepsilon\right)$.

7 $s = 4 \cos \dfrac{3t}{4}$.

Page 279

8 $\dfrac{d^2x}{dt^2} = -9x,\ x = 2\cos\left(3t \pm \dfrac{\pi}{6}\right).$

9 At O; 12 s.

10 0, 6 s.

11 $x = \sqrt{2}\cos(2t - \pi/4),\ \sqrt{2}$ m, $-\pi/4$ s, π s.

12 5 m, 0.927 s, 4π s.

13 (a) $\pi/2$, (b) $\pi/12$, (c) $\pi/2$, (d) π, (e) $7\pi/12$.

14 (a) $y = a\cos(2x + \varepsilon) - 3$, (b) $\theta = a\cos\left(\dfrac{3}{\sqrt{2}}t + \varepsilon\right) + \dfrac{1}{3}$,

(c) $s = A + Bt + \frac{1}{6}t^2 - \frac{2}{9}t^3$, (d) $x = \sqrt{2}\cos\left(2t + \dfrac{\pi}{4}\right) - 2$.

15 (a) $y = A + Be^{2x} - x$; (b) $y = A + Be^x$ or $y = C$, (c) $x = a\cos(t + \varepsilon)$,
(d) $y = A\ln(Bx)$ or $y = C$, (e) $(3y - 1)^{2/3} = Ax + B$ or $y = C$.

Exercise 14d, page 280

1 $x = A(\sec t + \tan t)$.   **2** $y = x^2\ln x - x^2 + Ax$.

3 $y = \frac{1}{2}e^x\sin x + Ax + B$.   **4** $r = 2\tan\dfrac{\theta}{2} - \theta + A$.

5 $(y + 1)^x = \dfrac{Ax}{x + 1}$.   **6** $y = a\cos\left(\dfrac{2}{\sqrt{3}}x + \varepsilon\right)$.

7 $y = (2x - 1)^{-2} + A(2x - 1)^{-4}$.   **8** $y + \dfrac{1}{A}\ln(Ay - 1) = 3x + B$ or $y = C$.

9 $x + 2y = Ax^6(x - 2y)^3$.   **10** $v = \frac{1}{2}(\ln u)^2 + A$.

11 $y\sqrt{(1 - x^2)} = \sin^{-1}x + A$.   **12** $x = 4\cos(\frac{2}{3}t + \pi/3)$.

13 $y = \frac{5}{3}\cos 3x$.   **14** $s = 2\cos(3t - \pi/3) + 1$.   **15** $x^2y = (x - 1)e^x + 1$.

16 $\ln y = \frac{1}{2}x^2\ln x - \frac{1}{4}x^2 + \frac{1}{4}$.   **17** $y^2 = 4\frac{1}{2} - xe^{-2x} - \frac{1}{2}e^{-2x}$.

18 (a) $y = Ae^{-2x} + \frac{1}{4}e^{2x}$, $y = \frac{1}{4}(e^{2x} + e^{-2x})$, (b) $\dfrac{d^2y}{dx^2} - m^2y = 0$.

19 $y = e^{-x^2}$.

20 $y\sin x = (-2\cos 4x + 4\cos 2x + 5)/16$.

21 $y = 36e^{-t/6} + 27$; 11min; $27°$.

22 $k = 20$, $x = 4\sin 2t + 3\cos 2t + 5$, max. speed $= 10$;
(a) $\frac{1}{2}(\pi/2 - \tan^{-1}0.75)$(b) $\pi - \tan^{-1}0.75 \approx 2.50$.

24 $x = ae^{-2t}$; $y = 4a(e^{-t/2} - e^{-2t})$.

Chapter 15

Qu. 1 (a) $y = Ae^x + Be^{-x}$, (b) $y = Ae^{2x} + Be^{10x}$, (c) $y = Ae^{-x/2} + Be^{3x}$,
(d) $y = Ae^{x/3} + Be^{x/5}$.

Qu. 2 (a) $z = Ae^{5t} + Be^{-5t}$, (b) $z = Ae^{t/2} + Be^{-t/3}$.

Qu. 3 (a) $z = e^{5t} - e^{-5t}$, (b) $z = 12(e^{t/2} - e^{-t/3})$.

Qu. 4 $f(x) = 2e^{5x} - e^x$.

Qu. 5 (a) $V = (At + B)e^{-3t}$, (b) $r = (At + B)e^{3t/10}$.

Qu. 6 $y = xe^x$.

Qu. 7   (a) $y = e^x(A \cos 7x + B \sin 7x)$, (b) $V = e^{-3t}(A \cos 5t + B \sin 5t)$,

(c) $r = A \cos \dfrac{t}{6} + B \sin \dfrac{t}{6}$.

Qu. 8   (a) $y = 5e^x \sin 7x$, (b) $V = e^{-3t}(\cos 5t + 2 \sin 5t)$, (c) $r = 2 \sin \dfrac{t}{6}$.

Qu. 11 $y = 2x^2 - 3x + 1.6$.

Exercise 15a, page 293

1   $y = Ae^{2x} + Be^{5x}$.   2 $y = Ae^{3x} + Be^{-2x}$.

3   $y = e^{2x}(A \cos 5x + B \sin 5x)$.   4 $y = Ae^{x/5} + Be^{-x/5}$.

5   $x = (At + B)e^{5t}$.   6 $x = (At + B)e^{t/4}$.   7 $u = Ae^{t/2} + Be^{t/3}$.

8   $y = A + Be^{-5x}$.   9 $x = Ae^{3t/2} + Be^{-t}$.   10 $r = A + Be^{\theta}$.

11   $y = e^{-2x}(A \cos 4x + B \sin 4x)$.   12 $r = A \cos \frac{1}{10}t + B \sin \frac{1}{10}t$.

13   $y = Ae^{-x} + Be^{-2x} + C$.   14 $y = (Ax + B)e^{2x} + 2$.

15   $x = e^{-t}(A \cos t + B \sin t) + \frac{1}{2}$.   16 $y = 3e^{6x} + 2e^{-x}$.

17   $u = 5 \sin 3t + 4 \cos 3t$.   18 $r = (1 - t)e^{6t}$.   19 $z = 4 \cos \frac{1}{2}t$.

20   $u = e^{\theta}$.

Exercise 15b, page 301

1   $y = Ae^{3x} + Be^{4x} + \frac{1}{3}$.

2   $y = Ae^{5x} + Be^{-x} - 3x - 1$.

3   $y = A \cos 3x + B \sin 3x + 2e^x$.

4   $y = (Ax + B)e^{-3x} + \sin x$.

5   $y = e^{3x}(A \cos x + B \sin x) + x^2 + 3x + 1$.

6   $r = (At + B)e^{t/4} + 3e^t$.

7   $z = A \cos 5r + B \sin 5r + 0.4$.

8   $u = e^{-t/5}(A \cos \frac{2}{5}t + B \sin \frac{2}{5}t) + 4e^{-t}$.

9   $V = A + Be^{-3\theta} + 4\theta - (3 \cos \theta + \sin \theta)/10$.

10   $x = Ae^{t/5} + Be^{-3t/2} + \frac{1}{3}e^{2t} + 2t + 8$.

11   $y = \frac{2}{5}e^{-2x} - \frac{1}{2}e^{-x} + (3 \sin x + \cos x)/10$.

12   $y = 2x^2 - 4x + 3 - (2x + 3)e^{-2x}$.

13   $y = 2 \cos 5x + 3 \sin 5x + e^{5x}$.

15   $x = g(e^{-kt} - 1)/k^2 + gt/k$.

16   $x = (\cos \omega t - \cos nt)/(n^2 - \omega^2) + (V/n) \sin nt$.

19   (a) 3, 9, (b) 5.

20   $x = -5e^{-t} + 4e^{-2t} + \cos t + 3 \sin t; \sqrt{10}$.

21   $x = 12 \sin 3t + 5 \cos 3t + 2; 15, 39; 0.967$.

22   $y = A \sin 3x^2 + B \cos 3x^2 - \frac{2}{3}$.

23   $y = A \sin 2x + B \cos 2x + 2 - \cos x; y = 2 - \cos x + \frac{1}{2} \cos 2x$.

24   $\dfrac{d^2u}{dx^2} + 4u = 24; y = x(\sin 2x + 6)$.

Chapter 16

Qu. 1   1.0209

Qu. 2   0.809.

Qu. 8 $f(x) \approx f(a) + f'(a)(x - a) + \dfrac{f''(a)}{2}(x - a)^2.$

Qu. 9 $f(a + h) \approx f(a) + f'(a)h + \dfrac{f''(a)}{2}h^2.$

Qu. 11 $1 + 2h + 2h^2 + \frac{8}{3}h^3.$

Qu. 12 $\frac{3}{5} - \frac{4}{5}(x - \alpha) - \frac{3}{10}(x - \alpha)^2 + \frac{2}{15}(x - \alpha)^3.$

Qu. 13 (a) $1 + x + \dfrac{x^2}{2!} + \dfrac{x^3}{3!} + \dfrac{x^4}{4!} + \dfrac{x^5}{5!}$, (c) $1 - \dfrac{x^2}{2!} + \dfrac{x^4}{4!}.$

Qu. 15 (a) $0, 1, 0, -\frac{1}{3}, 0$, (b) $x - \frac{1}{3}x^3 + \frac{1}{5}x^5 - \frac{1}{7}x^7.$

Qu. 16 $1 + x + \frac{1}{2}x^2 - \frac{1}{8}x^4.$

Qu. 17 $1 + \frac{1}{2}x - \frac{1}{8}x^2 + \frac{1}{16}x^3 + \frac{49}{384}x^4.$

Qu. 18 (a) $-2 < x < 2$, (b) $-1 \leqslant x < 1$, (c) $-2 \leqslant x \leqslant 0$, (d) all values of x,
(e) $-\frac{1}{2} < x \leqslant \frac{1}{2}$, (f) $-1 \leqslant x \leqslant 1$, (g) $-\frac{1}{3} < x < \frac{1}{3}$, (h) $-1 < x < 1$,
(i) $-1 < x < 1$.

Qu. 19 (a) $x < -2$ or $x > 2$, (b) $x < -1$ or $x \geqslant 1$.

Qu. 20 (a) $0 < x \leqslant 4$, (b) $-1 < x < 1 + \pi$.

Exercise 16a, page 312

1 $\ln 2 - \frac{3}{2} + x - \frac{1}{8}x^2.$

2 $\sin \alpha + (\cos \alpha)(x - \alpha) - \dfrac{\sin \alpha}{2}(x - \alpha)^2.$

3 (a) $1 + \dfrac{1}{e}(x - e) - \dfrac{1}{2e^2}(x - e)^2 + \dfrac{1}{3e^3}(x - e)^3 - \dfrac{1}{4e^4}(x - e)^4.$
(b) $1 + \frac{1}{2}(x - \pi/2)^2 + \frac{5}{24}(x - \pi/2)^4.$

4 0.581.

5 $f(0) + f'(0)x + \dfrac{f''(0)}{2!}x^2 + \dfrac{f'''(0)}{3!}x^3 + \dfrac{f''''(0)}{4!}x^4 + \dots.$

6 (a) $1 + 2x + 2x^2 + \frac{4}{3}x^3 + \dots$, (b) $-x - \dfrac{x^2}{2} - \dfrac{x^3}{3} - \dfrac{x^4}{4} - \dots,$
(c) $1 - \dfrac{x^4}{2!} + \dfrac{x^8}{4!} - \dfrac{x^{12}}{6!} + \dots$, (d) $\dfrac{x}{2} - \dfrac{x^3}{48} + \dfrac{x^5}{3840} - \dfrac{x^7}{645\,120} + \dots.$

8 (a) 1.4918, (b) 0.1823, (c) 0.955, (d) 0.199.

9 (a) $x^2 - \frac{1}{3}x^4$, (b) $1 + nx + \dfrac{n(n - 1)}{2}x^2 + \dfrac{n(n - 1)(n - 2)}{6}x^3,$
(c) $1 + x \ln 2 + \dfrac{(x \ln 2)^2}{2!} + \dfrac{(x \ln 2)^3}{3!}$, (d) $\pi/2 - x - x^3/6,$
(e) $x + x^2 + \frac{1}{3}x^3 - \frac{1}{30}x^5$, (f) $x - \frac{1}{6}x^3.$

10 $x + x^2 + \frac{1}{3}x^3 - \frac{1}{30}x^5 - \frac{1}{90}x^6.$

11 $\ln 4 + \frac{1}{4}(x - 4) - \frac{1}{32}(x - 4)^2 + \frac{1}{192}(x - 4)^3$; 1.3913.

Exercise 16b, page 318

1 (a) $x + \frac{1}{6}x^3 + \frac{3}{40}x^5 + \frac{5}{112}x^7$, (b) $x + \frac{1}{6}x^3 + \frac{1}{24}x^5,$
(c) $\frac{1}{4}\pi + \frac{1}{2}x - \frac{1}{4}x^2 + \frac{1}{12}x^3 - \frac{1}{40}x^5$, (d) $\frac{1}{2}\pi - x - \frac{1}{6}x^3 - \frac{3}{40}x^5.$

Page 318

2 (a) $1 + \dfrac{x^2}{2!} + \dfrac{x^4}{4!} + \dfrac{x^6}{6!}$, (b) $1 + \dfrac{x^2}{6} + \dfrac{7x^4}{360}$, (c) $1 - \frac{3}{2}x^2 + \frac{7}{8}x^4$,

  (d) $x + \frac{1}{3}x^3 + \frac{2}{15}x^5$, (e) $-\frac{1}{6}x^2 - \frac{1}{180}x^4$, (f) $x - \frac{1}{2}x^2 + \frac{1}{6}x^3 - \frac{1}{12}x^4$,

  (g) $\ln 2 + \frac{1}{2}x + \frac{1}{8}x^2 - \frac{1}{192}x^4$.

3 (a) None, (b) $-1 < x < 1$, (c) $-1 < x < 1$, (d) $-2 < x < 2$, (e) $-1 < x \leqslant 1$.

5 3.142.

6 3.142.

7 0.7494.

8 $\dfrac{\pi}{6} - \dfrac{2}{\sqrt{3}}x + \dfrac{2}{3\sqrt{3}}x^2 - \dfrac{8}{9\sqrt{3}}x^3$; $-\frac{1}{2} < x < \frac{3}{2}$.

Exercise 16c, page 319

1 $\dfrac{-2}{x+3} + \dfrac{2x+4}{x^2+1}$, $\frac{10}{3} + \frac{20}{9}x - \frac{110}{27}x^2 - \frac{160}{81}x^3$, $-1 < x < 1$.

2 $x + \frac{1}{2}x^2 - \frac{2}{3}x^3 + \frac{1}{4}x^4 + \frac{1}{5}x^5 - \frac{1}{3}x^6$.

6 $\dfrac{1}{x+2} - \dfrac{3}{(x+2)^2} + \dfrac{1}{x-3}$.

8 $\ln 2 + \frac{1}{2}x + \frac{3}{8}x^2 + \frac{7}{24}x^3$, $-1 \leqslant x < 1$; $2 \ln \frac{3}{2}$, 1.098.

9 $x - 2x^2 + 2x^3 - \frac{8}{3}x^4$.

11 $(-1)^{r-1}\dfrac{(r-1)}{r!}3^r$.

12 $\frac{1}{4}x^2 + \frac{1}{4}x^3 + \frac{7}{32}x^4$; $-1 \leqslant x < 1$; $\dfrac{2^{n-1}-1}{n2^{n-1}}$.

13 0.080 04.

14 $p = q = 2$.

15 $1 + x^2$.

16 $1 - 3x + 5x^2$; $y = 1 - 3x$.

17 $x + x^3/3$; 0.0014%, 14.4%.

18 (a) $\frac{1}{2}(e^x + e^{-x})$, $\frac{1}{2}(e^x - e^{-x})$; $\frac{1}{2}(e^{2x} + e^{-2x})$, $\dfrac{2^{2n}}{(2n)!}$.

  (b) $3x - \frac{3}{2}x^2 + 3x^3 - \frac{15}{4}x^4$, $-\frac{1}{2} < x \leqslant \frac{1}{2}$.

19 $y_1 = ax - a^2x^2/2 + (a^3/3 - a/6)x^3 + (a^2/6 - a^4/4)x^4$,

  $y_2 = ax - a^2x^2/2 + a^3x^3/6$; $a = \pm 1$.

20 1, 0; $1 + x - x^3/3 - x^4/6 - x^5/30$; 1.099650.

Chapter 17

Qu. 1   7.7 m.

Qu. 2   25.1 m.

Qu. 3   0.816, $304/375 \approx 0.811$.

Qu. 4   (a) $\frac{1}{2}d(y_1 + 2y_2 + 2y_3 + 2y_4 + 2y_5 + 2y_6 + 2y_7 + y_8)$,

  (b) $\frac{1}{2}d(y_1 + 2y_2 + 2y_3 + 2y_4 + 2y_5 + 2y_6 + 2y_7 + 2y_8 + y_9)$.

Qu. 5   240, to nearest 10. (First two ordinates are further apart than the others.)

Qu. 6 (a) $a(y_1 + 2y_2 + 2y_3 + 2y_4 + 2y_5 + 2y_6 + 2y_7 + y_8)/14$,

(b) $a(y_1 + 2y_2 + \ldots + 2y_{n-1} + y_n)/(2n - 2)$.

Qu. 7 0.694.

Qu. 8 37.6 m.

Qu. 9 (a) $\frac{1}{3}d(y_1 + 4y_2 + 2y_3 + 4y_4 + y_5)$,

(b) $\frac{1}{3}d(y_1 + 4y_2 + 2y_3 + 4y_4 + 2y_5 + 4y_6 + 2y_7 + 4y_8 + y_9)$.

Qu. 10 0.6931.

Qu. 11 304/375.

Exercise 17, page 335

1 (a) 0.347, (b) 0.350, (c) 0.347.

2 (b) 1.49, (c) 1.46.

3 $166\frac{2}{3}$.

4 0.7468.

5 (a) 49.4 cm², (b) 49.9 cm²; 3.12.

6 310 cm³.

7 1.86 litres.

8 (a) 0.2983, (b) 0.2983.

9 3.988, 0.997.

10 3.142.

11 $1\frac{1}{3}$.

13 $1 + 10x^3 + 45x^6 + 120x^9$; 0.204; 0.204.

14 (a) 0.879, (b) 0.879.

15 0.867.

16 0.467, 0.475; $\pi \approx 3.12$.

17

(a) x	y	(b) x	y
0	1	0	1
0.1	1	0.01	1
0.2	1.01	0.02	1.0001
0.3	1.0302	0.03	1.0003
0.4	1.0611	0.04	1.0006
0.5	1.1036	0.05	1.0010

18

(a) x	y	(b) x	y
0	1	0	1
0.1	1.2	0.01	1.02
0.2	1.45	0.02	1.0405
0.3	1.76	0.03	1.0615
0.4	2.142	0.04	1.0830
0.5	2.6104	0.05	1.1051

19

(a) x	y	(b) x	y
0.5	0	0.5	0
0.6	0.025	0.51	0.0025
0.7	0.0611	0.52	0.0051
0.8	0.1104	0.53	0.0078
0.9	0.1757	0.54	0.0106
1.0	0.2597	0.55	0.0135

Page 338

20 (a)

x	y
0	1
0.1	1.1
0.2	1.1990
0.3	1.2951
0.4	1.3865
0.5	1.4717

(b)

x	y
0	1
0.01	1.01
0.02	1.0200
0.03	1.0300
0.04	1.0400
0.05	1.0500

21

x	y
0	1
0.1	1
0.2	1.02
0.3	1.0408
0.4	1.0824
0.5	1.1274

22

x	y
0	1
0.1	1.2
0.2	1.5
0.3	1.84
0.4	2.296
0.5	2.8384

23 $y = 1 + x^2/2$ (the coefficients of x and x^3 are zero); 1.005, 1.02.

24 $y = 1 + 2x + \frac{5}{2}x^2 + \frac{5}{3}x^3$; 1.227, 1.513.

25 $y = 1 + x + x^2 + \frac{4}{3}x^3$; 1.111, 1.251.

Chapter 18

Qu. 1   (a) $\sinh A + \sinh B = 2 \sinh \frac{1}{2}(A + B) \cosh \frac{1}{2}(A - B)$,
   (b) $\cosh A + \cosh B = 2 \cosh \frac{1}{2}(A + B) \cosh \frac{1}{2}(A - B)$,
   (c) $\cosh A - \cosh B = 2 \sinh \frac{1}{2}(A + B) \sinh \frac{1}{2}(A - B)$,
   (d) $\operatorname{sech}^2 \theta = 1 - \tanh^2 \theta$, (e) $\operatorname{cosech}^2 \theta = \coth^2 \theta - 1$,
   (f) $\cosh 3\theta = 4 \cosh^3 \theta - 3 \cosh \theta$,
   (g) $\tanh 3\theta = (3 \tanh \theta + \tanh^3 \theta)/(1 + 3 \tanh^2 \theta)$.

Qu. 2   (a) $2 \sinh 2x$, (b) $\frac{1}{2} \cosh \frac{1}{2}x$, (c) $\sinh \frac{1}{3}x$, (d) $2 \cosh 4x$,
   (e) $2 \sinh x \cosh x = \sinh 2x$, (f) $6 \cosh^2 2x \sinh 2x$.

Qu. 3   (a) $x = a \cosh \theta$, (b) $x = a \sinh \theta$.

Qu. 5   (a) $x = a \tan \theta$ (or $x = a \sinh \theta$),
   (b) $x = a \tanh \theta$ (or $x = a \sin \theta$ or $x = a \cos \theta$).

Qu. 7   (a) L^0, (b) L^{-1}.

Qu. 8   The two expressions differ by a constant (possibly zero).

Qu. 10   (a) 0.8813, (b) 1.3169, (c) 0.5515.

Qu. 11 $\frac{1}{12} \cosh 3\theta - \frac{3}{4} \cosh \theta + c$.

Exercise 18a, page 340

3 $-1 < \tanh x < 1$.

14 $\dfrac{\tanh A + \tanh B}{1 + \tanh A \tanh B}, \dfrac{\tanh A - \tanh B}{1 - \tanh A \tanh B}$.

15   ln 1.8.

16 $a^2 = b^2 + c^2$.

Exercise 18b, page 345

2 $\sinh x + c, \cosh x + c$.

Page 345

3 (a) 3 sinh 3x, (b) 2 cosh 2x, (c) sinh 2x, (d) 6 sinh^2 x cosh x, (e) 6 sech^2 2x,
(f) $-$sech^2 x tanh x, (g) 3 sinh 6x, (h) $-\frac{1}{2}$ cosech^2 $x\sqrt{}$(tanh x),
(i) 2 tanh $\frac{1}{2}x$ sech^2 $\frac{1}{2}x$.

4 Domain $= \{x : x \in \mathbb{R}, -1 < x < +1\}$, range $= \mathbb{R}$; tanh x is odd.

5 (a) 2 cosech 2x, (b) e^{2x}, (c) $\frac{1}{2}$ sech^2 $\frac{1}{2}x$.

6 (a) $\frac{1}{2}$ tanh 2x + c, (b) $-$sech x + c.

7 4.

8 $1/\sqrt{(1 + x^2)}$.

10 0.

11 (a) $\frac{1}{10}$ cosh 5x + $\frac{1}{2}$ cosh x + c, (b) $\frac{1}{8}$ sinh 4x + $\frac{1}{4}$ sinh 2x + c.

12 $2/(e - 1)^2$.

13 bx cosh $\theta - ay$ sinh $\theta - ab = 0$,
ax sinh $\theta + by$ cosh $\theta - (a^2 + b^2)$ sinh θ cosh $\theta = 0$,
$$\frac{4a^2x^2}{(a^2 + b^2)^2} - \frac{b^2}{4y^2} = 1.$$

14 $\left(\dfrac{d^2y}{dx^2} - 4y\right)$ (3 coth 3x coth 2x $-$ 2) $= 5\left(\text{coth } 2x \dfrac{dy}{dx} - 2y\right)$.

15 $x - \dfrac{x^3}{3} + \dfrac{2x^5}{15}$.

16 $x + \dfrac{x^3}{3} + \dfrac{x^5}{5} + ... + \dfrac{x^{2n+1}}{2n + 1} + ...,$ ln $\sqrt{\dfrac{1 + x}{1 - x}}$.

17 ln $\sqrt{\dfrac{1 + x}{1 - x}}$.

18 $(\pm\sinh^{-1} \frac{1}{2}, -5\sqrt{5})$ min., $(0, -11)$ max.

Exercise 18c, page 350

1 $\sinh^{-1} \frac{1}{3}x$ + c.

2 3 $\sin^{-1} (x - 2)$ + c.

3 $\cosh^{-1} (2x)$ + c.

4 $\cosh^{-1} \dfrac{x + 2}{2}$ + c.

5 $\frac{2}{3}\sqrt{3}$ tan$^{-1} \dfrac{2x + 1}{\sqrt{3}}$ + c.

6 $\sinh^{-1} (x - 3)$ + c.

7 $\frac{1}{2}\cosh^{-1} (8x + 1)$ + c.

8 $\frac{1}{2}\sin^{-1} \dfrac{8x - 3}{3}$ + c.

9 $\sinh^{-1} \frac{1}{2} \approx 0.481$.

10 $\cosh^{-1} \frac{4}{3} \approx 0.795$.

11 $\sinh^{-1} 4 - \sinh^{-1} 3 \approx 0.2763$.

12 $\frac{1}{2}\sqrt{2}$ $\sinh^{-1} \sqrt{2} \approx 1.146$.

13 $\frac{1}{2}x + \frac{1}{4}$ sinh 2x + c.

14 $\frac{1}{3}$ sinh^3 x + sinh x + c = $\frac{1}{12}$ sinh 3x + $\frac{3}{4}$ sinh x + c.

Page 350

15 $(12x - 8 \sinh 2x + \sinh 4x)/32 + c.$

16 $x - \tanh x + c.$

17 $\ln \cosh x + c.$

18 $\ln \sinh x - \frac{1}{2} \operatorname{cosech}^2 x + c.$

19 $x - \tanh x - \frac{1}{3} \tanh^3 x + c.$

20 $\ln \{(e^x - 1)/(e^x + 1)\} + c = \ln \tanh \frac{1}{2}x + c.$

Exercise 18d, page 351

2 (a) $0, -\ln 3$, (b) $\ln (2 + \sqrt{3}), \ln (3/2 + \sqrt{5}/2).$

3 $b.$

4 $\dfrac{d^4 y}{dx^4} - (p^2 + q^2)\dfrac{d^2 y}{dx^2} + p^2 q^2 y = 0.$

5 (a) (i) $\dfrac{2}{1 - 4x^2}$, (ii) $\dfrac{1}{\sqrt{(1 + x^2)}}$; (b) $\dfrac{-b(3 \sinh^2 \theta + 1)}{a^2 \sinh^3 \theta \cosh^3 \theta}.$

6 $a^2\{\sqrt{2} - \ln (1 + \sqrt{2})\}.$

7 $ab(\frac{1}{2} \sinh 2\theta - \theta), \left(\dfrac{4a \sinh^3 \theta}{3(\sinh 2\theta - 2\theta)}, 0\right).$

8 $(1.20, 1.81).$

9 $4\sqrt{3}\pi a^2 b.$

10 (a) (i) $2x + \dfrac{8x^3}{3!} + \dots + \dfrac{(2x)^{2n-1}}{(2n-1)!} + \dots,$ (ii) $\dfrac{2x^2}{2!} + \dfrac{8x^4}{4!} + \dots + \dfrac{2^{2n-1}x^{2n}}{(2n)!} + \dots,$

(b) $1 + \dfrac{x^2}{2!} + \dfrac{x^4}{4!} + \dots + \dfrac{x^{2n}}{(2n)!} + \dots, 1 - \dfrac{x^2}{2} + \dfrac{5x^4}{24}.$

11 (a) $(\sinh x - x)/x^2$, (b) $\cosh 2 - 1.$

12 $y = 5 \cosh 2x + 4 \sinh 2x.$

13 $\pm \dfrac{1}{\sqrt{2}}.$

14 (b) $k^2 < 1; \sqrt{5}.$

15 (a) $\sinh u$, (b) $\operatorname{sech} u; \frac{1}{2}\pi.$

16 $\ln (7 + 5\sqrt{2}).$

18 (a) $\cosh^{-1}\left(\dfrac{x + 2}{3}\right) + c.$

19 $\ln 2; l^2 a^2 = n^2 + m^2 b^2.$

Chapter 19

Qu. 2 $\frac{3}{2}\pi a^2.$

Qu. 3 $\dfrac{a^2}{2k} \sinh 2k\pi.$

Qu. 4 $\frac{1}{12}\pi a^2.$

Qu. 5 $2\pi + \frac{3}{2}\sqrt{3}.$

Qu. 8 $\pi ab.$

Qu. 9 $17\frac{1}{15}.$

Qu. 10 $\frac{4}{3}$.

Qu. 11 $3\pi a^2$.

Qu. 12 (a) $\dfrac{dy}{dx}$, (b) $\dfrac{ds}{dx}$, (c) $\dfrac{ds}{dx} = \sqrt{\left\{1 + \left(\dfrac{dy}{dx}\right)^2\right\}}$.

Qu. 13 $\frac{14}{27}$.

Qu. 14 ln 3.

Qu. 15 $\frac{1}{2} \sinh^{-1}(2\sqrt{2}) + 3\sqrt{2}$.

Qu. 16 $a\alpha$.

Qu. 17 $\dfrac{a}{27}\{(9t^2 + 4)^{3/2} - 8\}$.

Qu. 18 $6a$.

Qu. 19 $8a$.

Qu. 20 (a) δr, $r\delta\theta$.

Qu. 21 $a\sqrt{(1 + k^2)}(e^{2k\pi} - 1)/k$.

Qu. 22 $\frac{1}{2}a\{\sinh^{-1}\pi + \pi\sqrt{(1 + \pi^2)}\}$.

Qu. 23 $8a$.

Qu. 24 $28\sqrt{5}\pi$.

Qu. 25 (a) $4\pi a^2$, (b) $2\pi ah$.

Qu. 26 $\dfrac{8\pi a^2}{3}(2\sqrt{2} - 1)$.

Qu. 27 $\frac{12}{5}\pi a^2$.

Qu. 28 $2\pi b\sqrt{(2b^2 + a^2)} + \sqrt{2}\pi a^2 \sinh^{-1}(\sqrt{2b/a})$.

Qu. 29 $\frac{64}{5}\pi a^2$.

Qu. 30 15π cm^3.

Qu. 31 $\frac{1}{3}\pi r^2 h$, πrl.

Qu. 32 $r(r + h)/(2r + h)$.

Qu. 33 $(2/\pi) \times$ (radius) from centre along axis of symmetry.

Qu. 34 (a) $2\pi^2 a^2 b$, (b) $4\pi^2 ab$.

Qu. 38 $1/(4a)$.

Qu. 39 (a) $2\sqrt{5}/25$, (b) 2.

Qu. 41 $-1/\{2a(t^2 + 1)^{3/2}\}$.

Qu. 42 $-1/(4a)$.

Qu. 43 $\delta\psi$, $\dfrac{ds}{d\psi}$.

Qu. 44 $x^2 + y^2 - 4cx - 4cy + 6c^2 = 0$.

Qu. 45 The centre of curvature.

Qu. 46 kx^2.

Qu. 47 (a) $\frac{1}{2}$, (b) $\frac{1}{2}$, (c) $\frac{1}{4}$, (d) $\frac{1}{2}a^2/b$.

Exercise 19, page 367

1 $9\pi/2$.

2 a^2.

3 $2c^2 \ln a$.

4 $2\pi a^2(\cos\alpha + 2\sin\alpha)$, $\frac{1}{2}a(\cos\alpha + 2\sin\alpha)/(1 + \alpha)$.

6 $\frac{3}{8}\pi a^2$.

Page 367

7 $\frac{4}{3}\pi(a^2 - r^2)^{3/2}$.

8 $\left(\dfrac{a^2 - b^2}{a}, 0\right)$.

9 $c \sinh \dfrac{x}{c}$, $2\pi c\left(x \sinh \dfrac{x}{c} - c \cosh \dfrac{x}{c}\right) + 2\pi c^2$.

10 $\frac{1}{2} \sinh 2\pi$.

11 $\frac{176}{105}\pi$.

12 (a) $\frac{3}{2}\pi a^2$, (b) $16a/(9\pi)$, (c) $\frac{8}{3}\pi a^3$.

13 $24a$, $12\pi a^2$.

14 (a) $1/(8a)$, (b) $16a$, (c) $216\sqrt{3}\pi a^2/35$.

15 $1\frac{1}{2}$, $x^3 + y^3 - 3xy = 0$.

16 2.58.

18 $ct^4 - t^3 x + ty - c = 0$, $\left(\dfrac{c(3t^4 + 1)}{2t^3}, \dfrac{c(t^4 + 3)}{2t}\right)$.

19 $\sqrt{(x^2 + 1)} + \dfrac{1}{2} \ln \dfrac{\sqrt{(x^2 + 1)} - 1}{\sqrt{(x^2 + 1)} + 1} - \sqrt{2} - \dfrac{1}{2} \ln(3 - 2\sqrt{2})$,

$\pi\{\sinh^{-1} x + x\sqrt{(1 + x^2)} - \ln(1 + \sqrt{2}) - \sqrt{2}\}$.

20 $\frac{1}{2}\sqrt{2}\pi a^2\{\sqrt{2} \cosh \theta\sqrt{(2 \cosh^2 \theta - 1)} - \cosh^{-1}(\sqrt{2} \cosh \theta) - \sqrt{2} + \cosh^{-1}\sqrt{2}\}$.

Chapter 20

Qu. 2 $u^2 + (v - \frac{1}{2})^2 = \frac{1}{4}$.

Qu. 3   (a) $\cos 5\theta + i \sin 5\theta$, (b) $\cos 2\theta - i \sin 2\theta$, (c) $\cos 3\theta + i \sin 3\theta$, (d) $\cos 5\theta + i \sin 5\theta$, (e) $\cos(\theta - \phi) + i \sin(\theta - \phi)$, (f) $\cos(\theta + \phi) + i \sin(\theta + \phi)$.

Qu. 4   (a) $\cos \theta + i \sin \theta$, (b) $\cos \frac{2}{3}\pi + i \sin \frac{2}{3}\pi = -\frac{1}{2} + \frac{1}{2}\sqrt{3} i$, (c) $\cos \frac{3}{2}\theta + i \sin \frac{3}{2}\theta$.

Qu. 5 1, $-\frac{1}{2} \pm \frac{1}{2}\sqrt{3} i$.

Qu. 7 -1, $\frac{1}{2} \pm \frac{1}{2}\sqrt{3} i$.

Qu. 10 $\displaystyle\sum_{r=0}^{n} \dfrac{w^r z^{n-r}}{r!(n-r)!}$.

Qu. 13 $\cosh z = \cos iz$, $\sinh z = -i \sin iz$.

Exercise 20a, page 372

1 (a) $4 + 5i$, (b) $2 + 3i$, (c) $-1 + 7i$, (d) $3\frac{1}{2} + \frac{1}{2}i$, (e) $-7 + 24i$, (f) $-117 + 44i$, (g) $(3 - 4i)/25$, (h) $(1 - i)/2$.

2 5, 0.927; $\sqrt{2}$, 0.785. (a) $\sqrt{41}$, 0.896; (b) $\sqrt{13}$, 0.983; (c) $\sqrt{50}$, 1.713; (d) $\sqrt{12.5}$, 0.142; (e) 25, 1.855; (f) 125, 2.782; (g) 0.2, -0.927; (h) $1/\sqrt{2}$, -0.785.

3 $2 \pm 7i$.

5 (a) $(x - 10)^2 + y^2 = 25$, (b) $x = y$.

Exercise 20b, page 376

1 (a) enlargement $\times a$, (b) translation $\begin{pmatrix} a \\ 0 \end{pmatrix}$, (c) translation $\begin{pmatrix} a \\ b \end{pmatrix}$,

Page 376

(d) rotation through arctan (b/a) and enlargement $\times |c|$,

(e) reflection in real axis.

5 $1 \leqslant |w| \leqslant 4$.

6 $0 \leqslant \arg(w) \leqslant \pi/2$.

7 $\{w:|w| = 25,\ 0 \leqslant \arg(w) \leqslant \pi/3\}$.

8 (a) $v = 20$, (b) the positive v-axis, (c) the negative v-axis.

9 The region between the circles $(u - \frac{1}{2})^2 + v^2 = \frac{1}{4}$ and $(u - \frac{1}{4})^2 + v^2 = \frac{1}{16}$.

10 $x = k \mapsto$ the parabola $v^2 = 4k^2(k^2 - u)$.

$y = k \mapsto$ the parabola $v^2 = 4k^2(k^2 + u)$.

Exercise 20c, page 380

1 $\cos 5\theta + i \sin 5\theta$.

2 $\cos 4\theta + i \sin 4\theta$.

3 $\cos \theta + i \sin \theta$.

4 $\cos \theta + i \sin \theta$.

5 $\cos 8\theta + i \sin 8\theta$.

6 $\cos 3\theta + i \sin 3\theta$.

7 $\cos (2\phi + 3\theta) + i \sin (2\phi + 3\theta)$.

8 $\cos (6\theta - 6\phi) + i \sin (6\theta - 6\phi)$.

9 $1,\ -1,\ i,\ -i;\ \frac{1}{2}\sqrt{2}(1 \pm i),\ \frac{1}{2}\sqrt{2}(-1 \pm i)$.

10 $1,\ \cos \frac{2}{5}\pi \pm i \sin \frac{2}{5}\pi,\ \cos \frac{4}{5}\pi \pm i \sin \frac{4}{5}\pi$;

$(x^5 - 1) = (x - 1)(x^2 - 2x \cos \frac{2}{5}\pi + 1)(x^2 - 2x \cos \frac{4}{5}\pi + 1)$.

11 $1,\ -1,\ \frac{1}{2}(1 \pm \sqrt{3}\,i),\ \frac{1}{2}(-1 \pm \sqrt{3}\,i);\ (x^2 + x + 1)(x^2 - x + 1)$.

12 Rotate the radius vectors through an angle of $\pi/2n$.

13 When n is a prime number. If n is odd but not prime, the first property will hold for some roots but not for others. The second holds for all n.

14 (a) $(x - 1)\ (x^2 - 2x \cos \frac{2}{7}\pi + 1)(x^2 - 2x \cos \frac{4}{7}\pi + 1)(x^2 - 2x \cos \frac{6}{7}\pi + 1)$,

(b) $(x + 1)(x^2 - 2x \cos \frac{1}{5}\pi + 1)\ (x^2 - 2x \cos \frac{3}{5}\pi + 1)$,

(c) $(x - 1)(x + 1)\left(x^2 - 2x \cos \dfrac{\pi}{n} + 1 \right) \dots \left(x^2 - 2x \cos \dfrac{(n-1)\pi}{n} - 1 \right)$.

15 $\cos 3\theta = 4 \cos^3 \theta - 3 \cos \theta$.

16 $\sin 3\theta = 3 \sin \theta - 4 \sin^3 \theta$.

17 $\cos 4\theta = 8 \cos^4 \theta - 8 \cos^2 \theta + 1$.

21 $\tan 6\theta = \dfrac{6t - 20t^3 + 6t^5}{1 - 15t^2 + 15t^4 - t^6}$,

$$\tan 2n\theta = \dfrac{\binom{2n}{1}t - \binom{2n}{3}t^3 + \dots + (-1)^{n-1}\binom{2n}{2n-1}t^{2n-1}}{1 - \binom{2n}{2}t^2 + \dots + (-1)^n t^{2n}},$$

$$\tan (2n + 1)\theta = \dfrac{\binom{2n+1}{1}t - \binom{2n+1}{3}t^3 + \dots + (-1)^n t^{2n+1}}{1 - \binom{2n+1}{2}t^2 + \dots + (-1)^n \binom{2n+1}{2n}t^{2n}},$$

where $t = \tan \theta$.

Page 382

22 $-5.03, -0.668, 0.199, 1.50.$

23 $-1.69, -0.431, 0.225, 1.14, 10.8.$

24 $u = x^3 - 3xy^2, v = 3x^2y - y^3.$

25 $\dfrac{ac + bd + i(bc - ad)}{c^2 + d^2}.$

26 (a) $\pm(2 + 3i)$, (b) $\pm \dfrac{1}{\sqrt{2}}(1 + i).$

27 (b) $Z = \dfrac{(1 - z^n)(1 - z^*)}{4 \sin^2 \frac{1}{2}\theta}$, (c) $C = \dfrac{\sin \frac{1}{2}n\theta \cos \frac{1}{2}(n - 1)\theta}{\sin \frac{1}{2}\theta}$,

$S = \dfrac{\sin \frac{1}{2}n\theta \sin \frac{1}{2}(n - 1)\theta}{\sin \frac{1}{2}\theta}.$

28 $C = \dfrac{1 - a \cos \theta - a^{n+1} \cos (n + 1)\theta + a^{n+2} \cos n\theta}{1 - 2a \cos \theta + a^2}$,

$S = \dfrac{a \sin \theta - a^{n+1} \sin (n + 1)\theta + a^{n+2} \sin n\theta}{1 - 2a \cos \theta + a^2}.$

29 Multiply by $\sin \frac{1}{2}\theta$ and $1 - 2a \cos \theta + a^2$ respectively.

30 $0.$

Exercise 20d, page 385

1 (a) $x^6 - 15x^4y^2 + 15x^2y^4 - y^6 + i(6x^5y - 20x^3y^3 + 6xy^5)$,

(b) $\dfrac{3 \cos \theta - 1}{5 \cos^2 \theta - 6 \cos \theta + 5} + i \dfrac{-2 \sin \theta}{5 \cos^2 \theta - 6 \cos \theta + 5}$,

(c) $\cos \frac{3}{2}\theta + i \sin \frac{3}{2}\theta$, (d) $\dfrac{x^2 + y^2 - 1}{x^2 - 2x + 1 + y^2} + i \dfrac{-2y}{x^2 - 2x + 1 + y^2}.$

2 (a) $\cos \theta - i \sin \theta$, (b) $\cos 8\theta + i \sin 8\theta$,

(c) $\dfrac{\text{cis}\,(-\frac{1}{2}\theta) + (-1)^{n-1} \text{cis}\,(n - \frac{1}{2})\theta}{2 \cos \frac{1}{2}\theta}.$

4 The circle $x^2 + y^2 + 4x = 0$; $zz^* - 2(z + z^*) - i(z - z^*) - 4 = 0.$

5 (a) $\sin^4 \theta = \frac{1}{8}(\cos 4\theta - 4 \cos 2\theta + 3)$,

(b) $\tan (\theta_1 + \theta_2 + \theta_3 + \theta_4)$

$= \dfrac{\sum \tan \theta_1 - \sum \tan \theta_1 \tan \theta_2 \tan \theta_3}{1 - \sum (\tan \theta_1 \tan \theta_2) + \tan \theta_1 \tan \theta_2 \tan \theta_3 \tan \theta_4}.$

6 (a) $\cos \pi + i \sin \pi$; $(z + 1)(z - \frac{1}{2} - \frac{1}{2}\sqrt{3}\,i)(z - \frac{1}{2} + \frac{1}{2}\sqrt{3}\,i)$,

(b) $(x - a)(x^2 - 2ax \cos \frac{2}{5}\pi + a^2)(x^2 - 2ax \cos \frac{4}{5}\pi + a^2).$

7 $z^2 - 4z + 13 = 0$; $1, 1, \frac{1}{2}(1 \pm \sqrt{3}\,i).$

8 (a) $-2 \pm i, 2 \pm i$, (b) $1 - i, 2 \pm 3i.$

9 $-1.$

10 $\dfrac{4 + 2 \cos \theta}{5 + 4 \cos \theta}.$

11 $\dfrac{1 - 2 \cos 2\theta - 2^{n+1} \cos 2(n + 1)\theta + 2^{n+2} \cos 2n\theta}{5 - 4 \cos 2\theta}.$

12 $\dfrac{\sin^2 \theta}{1 + \sin^2 \theta - \sin 2\theta}.$

Page 386

13 (a) $u = 0$, (b) $u + v - 2 = 0$, (c) $|w| = 1$, (d) $|w - 4| = 2$.

14 (a) $u - 2v = 0$, (b) $v - 3 = 0$, (c) $|w| = 3$, (d) $|w - 6| = 6$.

15 (a) (u, v) moves round $|w| = 1$ twice in a counter-clockwise sense,
(b) (u, v) moves round $|w| = 4$ twice in a counter-clockwise sense.

16 Circle, centre $(5/3, 0)$, radius $4/3$.

17 Circle, centre $(0, -5/4)$, radius $3/4$.

18 (a) $(5 - 14i)/13$, (b) $\sqrt{34}$, -0.54, (i) circle, centre O, radius $\sqrt{34}$;
(ii) $3v = 5u - 34$ (half-line, from $(5, -3)$ with gradient $5/3$).

19 $1 + \sqrt{3}i$; $2, \pi/3$; -512.

20 (a) Circle, centre $(-\frac{1}{2}, 1)$, radius $3/2$; (b) circle, centre $(1, 0)$, radius $\sqrt{2}$;
(c) $4x^2 + 3y^2 = 12$; $y = 0$.

21 (a) (i) $\sqrt{2}$, $-\pi/4$, (ii) 4, π, (iii) 1, $\pi/3$, (b) $2^{1/6} \cos (2n\pi/3 - \pi/12)$, $n = 0, 1, 2$.

22 $v^2 = 4(4 - u)$.

23 (a) $(-2, 1)$, (b) (i) circle, centre $(0, 2/3)$, radius $4/3$, (ii) line $4v - 2u + 3 = 0$.

24 Centre $(0, 0)$, radius 2; $v = 0$.

25 (a) $u = -\frac{1}{2}$, v increasing; (b) $u = \frac{1}{2}$, v decreasing.

26 $\cos z = \cosh iz$, $\sin z = -i \sinh iz$.

27 $\cosh z = (e^z + e^{-z})/2$, $\sinh z = (e^z - e^{-z})/2$,
$\cosh z = \cos iz$, $\sinh z = -i \sin iz$,
$\cos^2 z + \sin^2 z = 1$, $\sin (w + z) = \sin w \cos z + \cos w \sin z$.

28 $\sqrt{(x^2 + y^2)}$, $\tan^{-1} (y/x)$ if z lies in 1st or 4th quadrants, $\tan^{-1} (y/x) + \pi$ if z lies in 2nd quadrant, $\tan^{-1} (y/x) - \pi$ if z lies in 3rd quadrant;
(a, b), $\sqrt{(a^2 + b^2 - c)}$.

30 (a) $\pm \sqrt{\left(\dfrac{1}{CL} \times \dfrac{L - CR_1{}^2}{L - CR_2{}^2} \right)}$, (b) $\frac{1}{5}(3 \pm 4i)$, $\frac{1}{2}(1 \pm i\sqrt{3})$.

Chapter 21

Qu. 3   (a) O, A, B, C coplanar, (b) when **a**, **b**, **c** form a *left*-handed set.

Qu. 5 $7i - 4j - 3k$.

Qu. 7   5.

Exercise 21a, page 392

1 (a) 19, 2, 21, (b) 33, -34, -1.

2 45.9°.   **3** $(1, -3, 1)$.   **4** $(-3, -5, -9)$.

5 The diagonals of a rhombus are perpendicular.

9 $\mathbf{r} = \frac{1}{4}\mathbf{p} + \frac{3}{4}\mathbf{q}$.

10 (a) $\cos \alpha \, \mathbf{i} + \sin \alpha \, \mathbf{j}$, (b) $\cos \beta \, \mathbf{i} + \sin \beta \, \mathbf{j}$, (c) $\alpha - \beta$.

Exercise 21b, page 397

1 (a) $\mathbf{i} - \mathbf{j}$, (b) $-2\mathbf{k}$, (c) $5\mathbf{j} - 15\mathbf{k}$, (d) $17\mathbf{i} + 5\mathbf{j} - 9\mathbf{k}$.

2 (a) 1, (b) -16, (c) 0, (d) -2.

3 (a) 1, (b) -16, (c) 0, (d) -2.

Page 397

5 (a) $-7\mathbf{i} + 7\mathbf{j} - 7\mathbf{k}$, (b) $3\mathbf{i} - 9\mathbf{j} - 15\mathbf{k}$.

6 (a) $-\mathbf{k}$, (b) $5\mathbf{i} - 5\mathbf{j} + 18\mathbf{k}$, (c) $2\mathbf{i} + 8\mathbf{j} - 10\mathbf{k}$, (d) 0.

8 $\dfrac{1}{\sqrt{6}}(\mathbf{i} - 2\mathbf{j} + \mathbf{k})$.

10 $4, 2, -6$.

Exercise 21c, page 401

1 (a) -4, (b) $-28\mathbf{i} - 16\mathbf{j} + 8\mathbf{k}$, (c) $3\mathbf{i} - 4\mathbf{j} + \mathbf{k}$.

2 $11\mathbf{i} + 18\mathbf{j} - 13\mathbf{k}$; $11x + 18y - 13z = 0$.

4 $x + 2y + 3z = 12$.

7 $\mathbf{r} = \begin{pmatrix} 2 \\ 1 \\ 1 \end{pmatrix} + t \begin{pmatrix} -1 \\ 2 \\ 1 \end{pmatrix}$; D$(4, -3, -1)$; $x - y + 3z = 4$.

8 (a) $16/\sqrt{310}$, (b) $\mathbf{r} = \begin{pmatrix} 0 \\ 5 \\ 2 \end{pmatrix} + \lambda \begin{pmatrix} 1 \\ -2 \\ 1 \end{pmatrix}$.

9 (a) $\mathbf{r} = \begin{pmatrix} 1 \\ 0 \\ -3 \end{pmatrix} + t \begin{pmatrix} 2 \\ 1 \\ -3 \end{pmatrix}$, (b) no solution.

10 (a) $(2, 4, 6)$, (b) $(4/7, 1/7, -2/7)$, (c) $x^2 + y^2 + z^2 = \tfrac{1}{2}x$, $x^2 + y^2 + z^2 = -x$.

12 $25t^2 - 30t + 46$.

Contents of Book 1

Index